22nd Chemnitz Seminar on Materials Engineering

22. Werkstofftechnisches Kolloquium

Held online due to COVID-19

IOP Conference Series: Materials Science and Engineering
Volume 1147

Chemnitz, Germany
24 March 2021

Editors:

Thomas Lampke
Andreas Undisz
Guntram Wagner Martin
F.-X. Wagner

ISBN: 978-1-7138-3417-5
ISSN: 1757-8981

Printed from e-media with permission by:

Curran Associates, Inc.
57 Morehouse Lane
Red Hook, NY 12571

Some format issues inherent in the e-media version may also appear in this print version.

Institute of Physics
Dirac House, Temple Back
Bristol BS1 6BE UK

Phone: 44 1 17 929 7481
Fax: 44 1 17 920 0979

techtracking@iop.org

Additional copies of this publication are available from:

Curran Associates, Inc.
57 Morehouse Lane
Red Hook, NY 12571 USA
Phone: 845-758-0400
Fax: 845-758-2633
Email: curran@proceedings.com
Web: www.proceedings.com

TABLE OF CONTENTS

JOINING

MATERIAL AND PROCESSING SIMULATION

MECHANICAL TESTING / FATIGUE / DAMAGE

LIGHTWEIGHT MATERIALS

SHAPE-MEMORY ALLOYS

THERMAL AND THERMOCHEMICAL TREATMENTS

IOP Conf. Series: Materials Science and Engineering 1147 (2021) 011001 doi:10.1088/1757-899X/1147/1/011001

Preface

The Institute of Materials Science and Engineering (Institut für Werkstoffwissenschaft und Werkstofftechnik, IWW) of Chemnitz University of Technology is proud to have hosted the Chemnitz Seminar on Materials Engineering – Werkstofftechnisches Kolloquium – for the 22nd time. Due to the long-lasting COVID-19 pandemic, this year's conference was staged as a one-day online-live event in late March.

Scientists and engineers, members of academia and industry, presented their most recent research results in 28 scientific talks, organized in two parallel webrooms.
The topics of our – from now on biennial – meeting ranged from general challenges in materials and surface engineering to specific applications of transmission electron microscopy for analysis of materials' surfaces. Further key sessions of this year's event were focused on the engineering of high entropy alloys, shape memory alloys, composite materials, additive manufacturing as well as material and process simulation.

In addition to the scientific talks, interdisciplinary lectures were given on the key topic "Industry 4.0" ("4th industrial revolution"). Prof. Detlev Zühlke, who was the main initiator of the technology initiative SmartFactory KL e. V. and is now retired, spoke about the necessary framework to bring the industrial revolution from vision to reality. Dr. Franziska Bocklisch, commissionary head of the professorship General Psychology and Human Factors of Chemnitz University of Technology, highlighted the importance of transdisciplinary collaboration for successful implementation of human-technology cooperation.

Following the encouraging trend of the last years, most of our contributors again decided to publish their work in this international, peer-reviewed journal. Consequently, this issue of *IOP Conference Series: Materials Science and Engineering* provides an excellent overview of the main topics of the Seminar.

Chemnitz, March 2021

Editors:

Prof. Thomas Lampke, Prof. Andreas Undisz, Prof. Guntram Wagner, Prof. Martin F.-X. Wagner
Institute of Materials Science and Engineering (IWW)
Chemnitz University of Technology (TU Chemnitz)

IOP Conf. Series: Materials Science and Engineering 1147 (2021) 011001 doi:10.1088/1757-899X/1147/1/011001

Digital meeting of the 22nd Chemnitz Seminar on Materials Engineering using ZOOM. Overview of participants. The heads of the institute, Prof. Guntram Wagner (1st row, middle), Prof. Thomas Lampke (1st row, 2nd from right), Prof. Andreas Undisz (2nd row, left) und Prof. Martin F.-X. Wagner (2nd row, 2nd from left) welcomed, amongst others, Prof. Detlev Zühlke (1st row, left), Prof. Ulrich Prahl (3rd row, left), Prof. Bernhard Wielage (4th row, left), Prof. Raether (4th row, right) and Prof. Siegfried Steinhäuser (5th row, left).

IOP Conf. Series: Materials Science and Engineering 1147 (2021) 011002 doi:10.1088/1757-899X/1147/1/011002

Peer review declaration

All papers published in this volume of *IOP Conference Series: Materials Science and Engineering* have been peer reviewed through processes administered by the Editors. Reviews were conducted by expert referees to the professional and scientific standards expected of a proceedings journal published by IOP Publishing.

- **Type of peer review: Single-blind**

- **Describe criteria used by Reviewers when accepting/declining papers. Was there the opportunity to resubmit articles after revisions?**
  - Does the paper contain relevant scientific findings suitable for a conference proceeding?
  - Is there an adequate number of references to support the statements given in the paper?
  - Is the presentation of the results logical and comprehensible?
  - Do figures and tables contain relevant information and is their quality appropriate?
  - Do the wording and grammar meet our standard?
  - There was the opportunity to resubmit articles after revisions.

- **Conference submission management system:**
  - Upload via conference website

- **Number of submissions received:**
  - 37

- **Number of submissions sent for review:**
  - 35

- **Number of submissions accepted:**
  - 34

- **Acceptance Rate (Number of Submissions Accepted / Number of Submissions Received X 100):**
  - 97%

- **Average number of reviews per paper:**
  - 2

- **Total number of reviewers involved:**
  - 30

- **Any additional info on review process (ie plagiarism check system):**

- **Contact person for queries:**
  - Nadja Berndt: nadja.berndt@mb.tu-chemnitz.de

Improvement of the properties of additively manufactured steel parts by combination of heat treatment and hard coatings

H Frank*, M Ambos, S Lutze and M Scholz

GFE – Gesellschaft für Fertigungstechnik und Entwicklung Schmalkalden e.V.

*e-mail: h.frank@gfe-net.de

Abstract. In additive manufacturing using SLM (selective laser melting), the maraging steel 1.2709 is often used to produce metallic components for various applications. The work piece properties depend on the process control in the additive process, but also on the heat treatment and the final processing. Here, the heat treatment and the used finishing processes lead to changes in the work piece and material properties (e.g. residual stresses, strength). As part of the international "Ad-Proc-Add" project, the presented work took into account individual process steps of the process chain with consideration of: additive manufacturing - pre-treatment - functionalization - final processing when using different materials. In investigations of additive manufacturing of the material 1.2709, it could be proven that the thermal treatment determines the mechanical properties of the components. A PVD coating process was used to replace the usual heat treatment. Similar temperature-time regimes during the coating resulted in mechanical properties comparable to conventionally heat-treated additively manufactured components. In addition to modifying the microstructure and the component properties, the PVD coatings improve the surface wear resistance up to more than 60%. Further investigations to improve the adhesion strength of the coatings are planned.

1. Introduction

In Additive Manufacturing (AM) three-dimensional objects are built directly by layer-wise material deposition. Complex shaped parts with internal structures can be generated, which cannot be realized with conventional manufacturing methods or only at considerable expense [1, 2]. One way of additive manufacturing is the laser powder bed fusion (LPBF) method. Here a laser is used to selectively heat and melt powder in order to build up the layers sequentially. Depending on the source power and scanning velocity, a significant energy impact on the work piece takes place. This energy influences structure and properties, e.g. strength, hardness and residual stresses. In powder bed based AM, properties can be modified by heat treatment of the generated part [3].

Main challenges of AM technologies are the achievable surface quality and dimensional accuracy. Functional surfaces often have to be reworked in order to reach the required tolerances [4]. In addition, any existing support structures must be removed from the part. Post-processing steps are usually necessary for functional surfaces, due to the inherently insufficient surface quality after AM. Control of the so called-additive subtractive process chain is dependent on several printing and post-processing strategies. In particular, heat treatment for hardening of the material and removal of residual stresses are often used as well as machining processes such as milling, drilling and grinding for finishing [4, 5].

Based on these challenges, the European research project "Ad-Proc-Add" investigates additive-subtractive manufacturing chains in order to obtain a detailed understanding of the dependencies and interactions of material and component properties of additively manufactured and reworked work pieces with regard to process parameters, manufacturing strategies and boundary conditions. The aim is to be able to consciously adapt geometry, surface and boundary zone properties via additive-subtractive production chains in order to be able to meet predefined requirements. This enables the targeted design and implementation of additive-subtractive production chains in various industrial applications.

One of the tasks in the “Ad-Proc-Add” project is to determine the influence of the LPBF process and the process chain on the material properties of additively manufactured components of the maraging steel 1.2709 (X3NiCoMoTi18-9-5). This material is often used for the production of metallic components for various applications [6]. In addition to the printing direction, the heat treatment is an essential aspect. Of special interest are the mechanical and structural properties of the additively manufactured parts, depending on the used process chain. The objective is to investigate a correlation between the AM parameters and the material properties. From this, statements about the design of the additive-subtractive process chains can be derived.

2. Experimental setup

2.1. Process chain

In order to get a detailed understanding of the correlations in additive manufacturing, the additive-subtractive process chain shown in Figure 1 was used. The LPBF process is used to produce components from the material 1.2709. After the AM process, the components were separated from the build plate. To modify the mechanical properties, the work piece was heat treated. The final step was the finishing by machining processes (e.g. milling, drilling), as well as surface functionalization, e.g. by hard coatings.

The following investigations will focus on different heat treatment procedures and their influence on the properties of the additively manufactured parts. A PVD coating deposition process was used to combine heat treatment and surface functionalization in a single step.

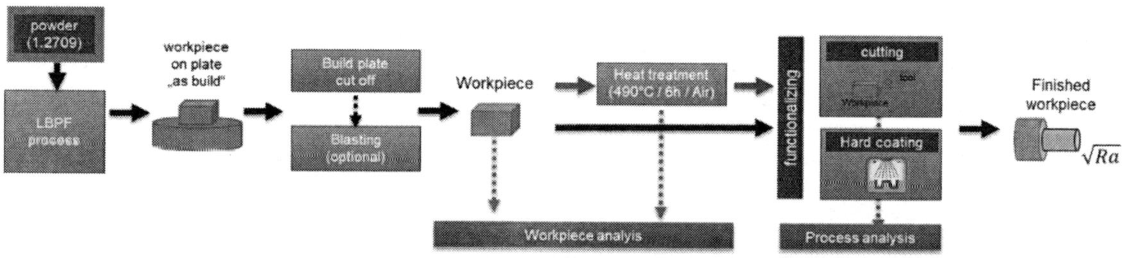

Figure 1. Additive-subtractive process chain.

2.2. Additive manufacturing

The LPBF machine Lasertec 30 SLM 2nd from the company DMG MORI Germany was used to produce the AM samples from the material 1.2709. For quality assurance, the powder used is examined for its composition and particle size distribution and it is determined that it is in the range of a typical LPBF powder with the diameters: D10 = 24.45 µm, D50 = 35.47 µm and D90 = 47.79 µm. Simple geometries (discs, hollow cylinders, rods) were produced for the basic analyzes. Used process parameters, based on the recommendations of the manufacturer of the AM system, are shown in Table 1. Regarding the AM process parameters, only the influence of the buildup-direction was investigated (see Table 1).

Table 1. LPBF process conditions and work piece geometry.

Process conditions		work piece geometry
Material	1.2709	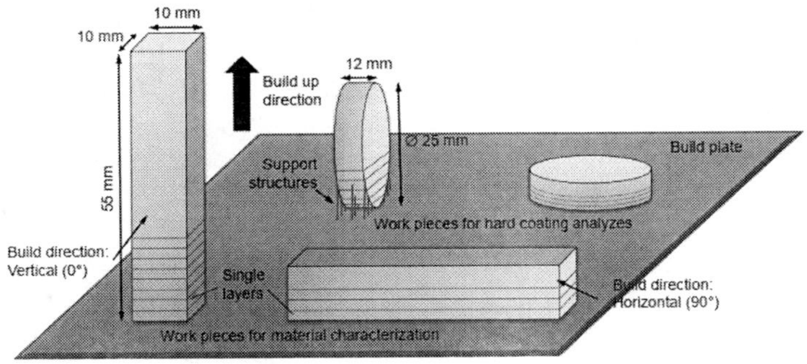
Layer thickness	65 µm	
Atmosphere	Argon	
Build plate temperature	200 °C	
Laser power (hatch)	275 W	
Scan speed (hatch)	750 mm/s	

2.3. Heat treatment and hard coating

After the AM process, the components were separated from the construction platform and the support structures were removed by blasting and grinding operations. For mechanical investigations and the coating deposition, 26 samples were manufactured, each in horizontal and vertical build up direction. Then half of the samples were age hardened at 490°C for 6 hours which is a common heat treatment for the steel 1.2709 [7]. Currently it is uncertain if a solution annealing at temperatures of 820-850°C before hardening is really necessary to achieve sufficient mechanical properties. Some researchers found that the solution treatment before hardening plays only a minor role in the fracture process [7]. So according to the aim of the current research, no solution treatment has been carried out. This is planned for further investigations, but only for comparison purposes.

In order to examine the influence of different heat treatment procedures on the work piece properties, a PVD hard material coating was used instead of the heat treatment. This combination makes it possible to modify the material properties as well as to change the surface with regard to defined properties such as wear resistance or friction coefficient. For the investigations, the following PVD coatings were chosen: Diamond-like carbon (ta-C), a nano-structured (nACo2), oxygen-containing (AlCrON) and a wear-protective coating (TiCN-LT). The coatings differ in their special properties and the coating temperature. Table 2 shows the selected coating systems, the deposition temperature and time for the specific coating thickness as well as special coating features. The coating deposition was done using an arc-PVD process on commercial coating systems π^{211} and π^{411} from the company PLATIT AG from Switzerland.

Table 2. Properties and features of used PVD-hard coatings [8].

Coating	Deposition temperature	Deposition time	Thickness	Hardness	Features
ta-C	110 °C	370 min	0,8 µm	65 GPa	Tetraedrical amorphous DLC Extremely high hardness
TiCN-LT	300 °C	360 min	2,5 µm	34 GPa	Wear protective coating Low friction coefficient
nACo2	480 °C	365 min	2,5 µm	41 GPa	Nanostructured wear protection coating High wear resistance
AlCrON	550 °C	370 min	4,0 µm	39 GPa	Oxygen containing coating High thermal resistance

The various temperature-time profiles for coating are compared with conventional heat treatment in Figure 2. The depositing temperature and the temperature profile of the nACo2 coating as well as the AlCrON- coating are nearly comparable with the conventional heat treatment. The holding times are almost the same for all variants. Differences can be seen in the heating and cooling rates - the heating rates of the hard coating systems are higher (14-16 K/min) compared to the conventional heat treatment (6 K/min), the cooling rates are higher as well (hard coatings: 5-6 K/min, conventional HT 2 K/min). For the ta-C coating, the heating (4 K/min) and cooling rate (1 K/min) are lower compared to the conventional HT.

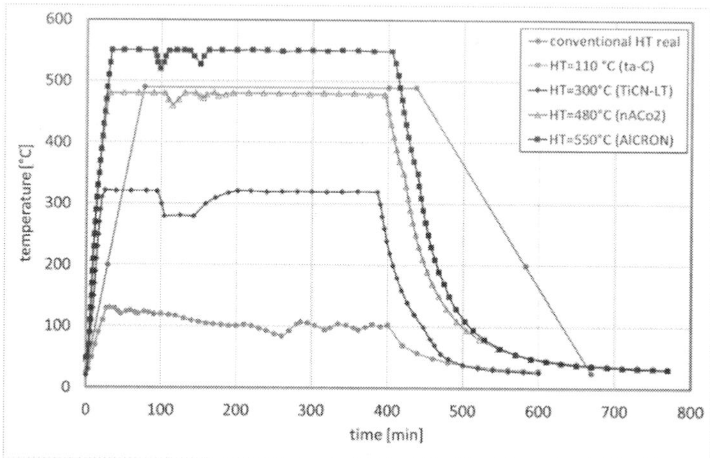

Figure 2. Temperature-time-profile of different hard coating deposition processes compared to conventional heat treatment for the material 1.2709.

2.4. Material and surface characterization

After additive manufacturing and the heat treatment as well as the coating deposition, the modified work pieces were characterized. Due to the large number of influencing parameters, the analysis is limited to essential parameters (hardness, strength). With regard to the mechanical properties of the AM work pieces, the notched impact strength at room temperature (pendulum impact tester PSd 300, company WPM Werkstoffprüfsysteme) and the hardness (DiaTestor 7521, company Wolpert) were determined according to the standards DIN EN ISO 6508-1, and ISO 148-1. An analysis of the failure behavior was also carried out on the fracture surface using a SEM (Phenom XL, company Thermo Fischer Scientific).

Furthermore the modified and coated surfaces were analyzed with regard to roughness (Taylor Hobson PGI 1500E, DIN EN ISO 4287:2009-11), coating hardness (Picodentor HM500, DIN EN ISO 14577-1) and coating adhesion (Rockwell indentation, DiaTestor 7521, DIN EN ISO 26443) to determine the influence of the different heat treatments on the mechanical properties of the AM work pieces. The abrasive wear resistance was also determined by spherical grinding with defined boundary conditions (ball on disc, grinding velocity 6m/min, grinding length 18 m, DIN EN ISO 26423). The influence of the heat treatment on the microstructure was determined on using SEM and EDX. The internal stresses of the components under the different thermal boundary conditions were also determined using X-ray diffraction (XRD) methods based on the $\sin^2\psi$ method.

3. Results

3.1. Mechanical properties of the work piece

In the first step, the mechanical properties notched impact strength and hardness of horizontally (90°) and vertically (0°) printed samples were determined. The characterization was carried out on samples without and with conventional heat treatment (6h / 490 °C / air) after the AM process.

Figure 3 shows the hardness and impact energy of the analyzed samples for the different buildup directions with and without heat treatment (age hardening). The measured values correlate with information from the literature. For the work pieces, no significant difference in hardness between the different buildup directions can be determined. The heat treatment leads to a significant increase in hardness, which can be explained by the intermetallic segregations formed during age hardening [9].

With regard to the impact energy, there are small differences between the horizontally and vertically built work pieces without heat treatment. The specimens with a vertical direction (0° to the buildup direction) with a notch in the direction of the layer have slightly higher notched impact energies compared with a buildup direction of 90° with a notch vertical to the layer. As result of the heat treatment, the impact energy is reduced, and there is no longer any significant difference between the work pieces produced in different orientations. The reduced impact energy after the heat treatment correlates with the increase in hardness and is due to the formed intermetallic segregations, which limit the deformability of the material and lead to a more brittle material behavior.

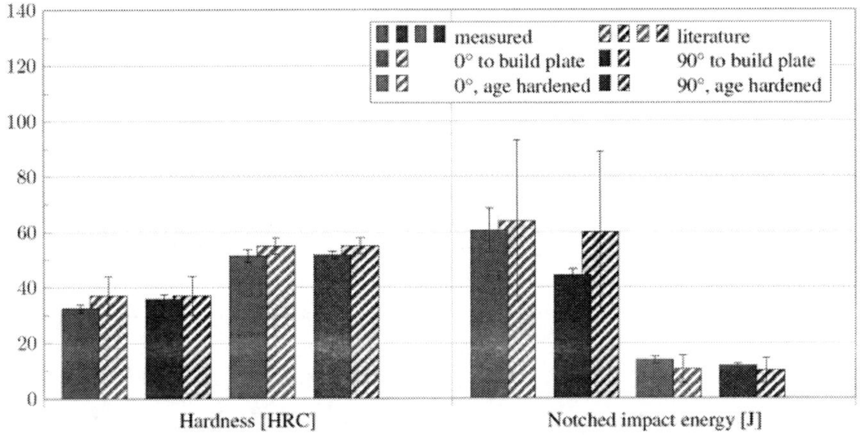

Figure 3. Hardness and notched impact strength of AM work pieces of 1.2709 with differently heat-treatment, comparison with information from literature [7, 10, 11].

The fracture surfaces generated during the notched impact tests (Figure 4) can be used to give information about ductility. With a buildup direction of 0° without heat treatment, more dimples occur and the work pieces are more ductile. Heat-treated samples have fewer or smaller dimples, regardless of the buildup direction, and therefore have a significantly lower ductility [9]. These results are consistent to the measured impact energy.

Figure 4. SEM images of the fracture surfaces of the notched impact work pieces with and without heat treatment.

The results of the material characterization show that after the AM process, work pieces of the material 1.2709 with a buildup direction of 0° to the build plate have a higher notched impact energy (notch parallel to the layer) than with a buildup direction of 90°. In the hardened state, the hardness and impact energy are independent on the buildup direction.

3.2. Combination of heat treatment and hard coating
In order to examine the influence of different heat treatment procedures on the work piece properties, the heat treatment was replaced by a PVD coating deposition process. The resulting microstructure of the work piece as well as the work piece and coating properties were analyzed after each PVD coating process.

3.2.1. Microstructure. Figure 5 shows a cross-section through the structure of the differently heat treated and coated work pieces in a horizontal buildup direction. For a better visibility of grain boundaries and structural components, the structure was etched using using the etchant Nital (solution of 3% HNO_3 in ethanol). The grain boundaries of the individual layers are particularly visible in the variants without heat treatment and after the ta-C coating (deposition temperature 110 °C). The closer the deposition temperature meets the temperature of the conventional heat treatment the more comparable is the resulting microstructure to the conventional heat treatment. In particular, at a coating temperature of 480 °C, martensitic structures result that are comparable to those of conventional heat treatment.

The same results apply to a horizontal buildup direction. It can be concluded, that a PVD hard coating is suitable for replacing the heat treatment. For best results in terms of hardening of the coating substrate the coating temperature should be as close to 490 °C as possible. In the present case, the nano-structured nACo2 coating with a deposition temperature of 480 °C meet this requirement. The buildup direction has no influence on the resulting structure.

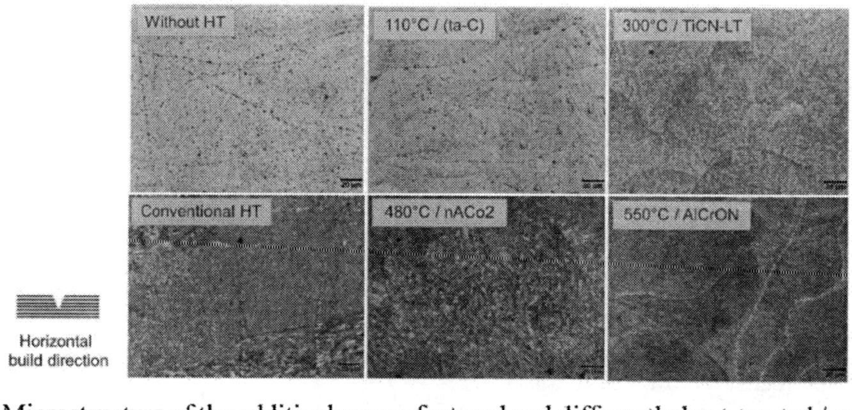

Figure 5. Microstructure of the additively manufactured and differently heat-treated / coated work pieces (horizontal build direction, 90° to build plate).

3.2.2. Mechanical properties. Figure 6 shows the hardness of the additively manufactured work pieces after coating deposition. It can be seen that work pieces have a low hardness of approx. 370 HV10 directly after the AM process. With a following martensitic age hardening (6h / 490 °C / air), the hardness increases to 550 HV10. Using a PVD coating, similar results can be achieved due to the analogous temperature profile. At temperatures of 110 °C and 300 °C, the hardness of 370 HV10 (110 °C) and 490-510 HV10 (300 °C) does not reach the level of conventional heat treatment; the usual hardness values can be achieved at temperatures of 480 °C and 550 °C. For 480 °C (coating nACo2), slightly higher hardness can be achieved based on the slightly higher cooling rate at the PVD process. Only small differences in hardness can be seen between the horizontal and vertical buildup direction.

The residual stresses in the surface near areas of the work pieces is shown in Figure 7. Due to the effort of the XRD measurement, only horizontal samples have been analyzed. On the surface compressive stresses are measured, these have a favorable effect for improved wear resistance. There are only small differences in the residual stress between the AM work pieces with and without conventional heat treatment, which are within the tolerance for XRD methods. For the PVD coated work pieces, higher internal compressive stresses result at lower deposition temperatures. At higher coating deposition temperatures, the internal stresses are reduced compared to conventional heat treatment. This indicates that the higher heating and cooling rates under vacuum conditions of the PVD coatings have a significant influence on stresses close to the surface. Further analyses are necessary to verify the results.

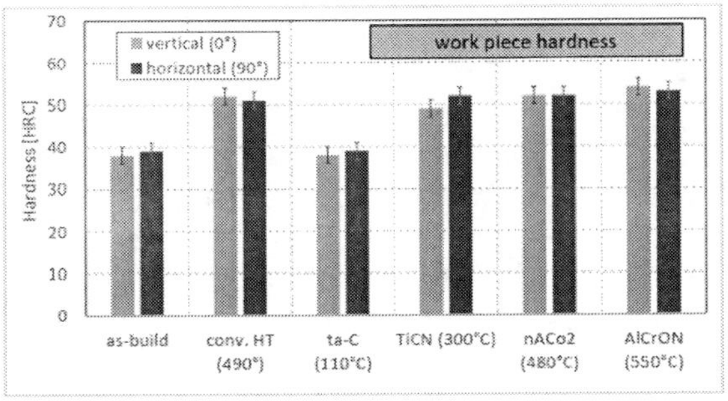

Figure 6. Hardness of the AM work pieces with different heat-treatments / coatings in different buildup directions.

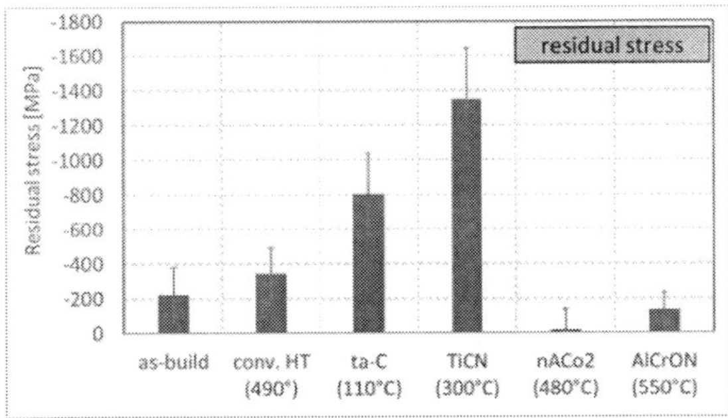

Figure 7. Residual stresses (XRD) of the AM work pieces with different heat-treatments / coatings.

3.2.3. Coating and surface properties. PVD coating can be used to increase the wear resistance of the work pieces, therefore the surface and coating properties are important. Figure 8 shows the coating adhesion determined by Rockwell indentation test as well as an example of a Rockwell indentation for the different deposition temperatures. In comparison with coatings on conventionally produced high speed steel (HSS) material 1.3433, which is a common substrate material for PVD coatings, on the AM material 1.2709 the coatings show a lower coating adhesion (higher adhesion class). For the ta-C coating with the lowest deposition temperature, large-scale flakes and insufficient coating adhesion are visible. The reason for this are the high internal stresses of the coating due to the extremely high coating hardness (see also figure 9). All other heat treatments / coatings show adequate adhesion, the coating adhesion is enhanced with increasing deposition temperature. For the further coating development on AM produced work pieces, an improvement is necessary to ensure coating adhesion during the use of coated AM parts.

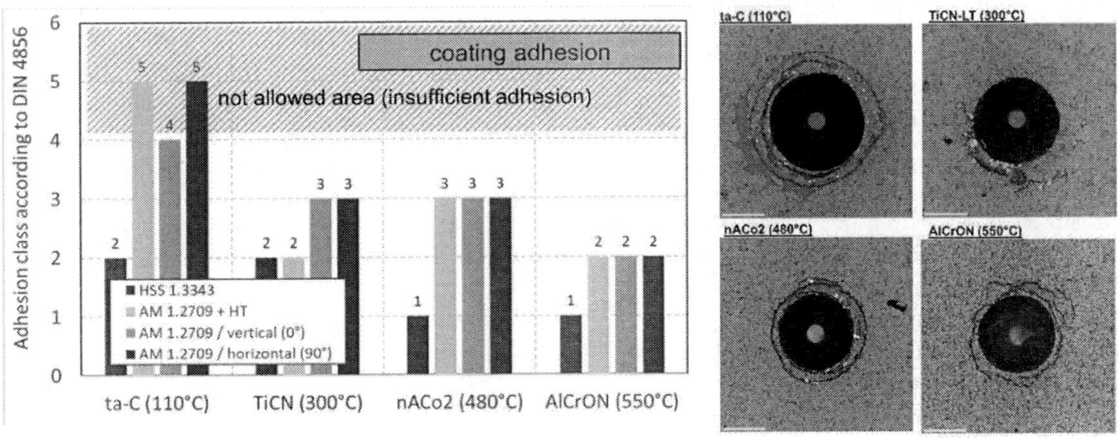

Figure 8. Coating adhesion (Rockwell indentation method) and sample Rockwell indentations for different coatings.

The hardness of the coatings with different deposition temperatures is shown in Figure 9. For the same coating system, there are only small variations with respect to the buildup direction and a previous heat treatment. The highest hardness of about 22-23 GPa shows the ta-C coating. The coating hardness for the ta-C coating can be increased to about 25 GPa by conventional heat treatment before coating deposition, because of the better support due to the higher substrate hardness. All other coatings show significantly lower coating hardness between 11 and 15 GPa. The hardness of coatings on AM work pieces is lower, compared with hard coatings on conventional base material due to different substrate properties (e.g. hardness, microstructure).

The wear resistance to abrasive wear is shown in Figure 10. Compared to an uncoated, additively manufactured component, all coatings reduce the wear rate (increase the wear resistance). The results show, that the ta-C coating can strongly reduce the abrasive wear. As well as it can be seen, that the buildup direction or a previous heat treatment has only a small influence on the wear rate.

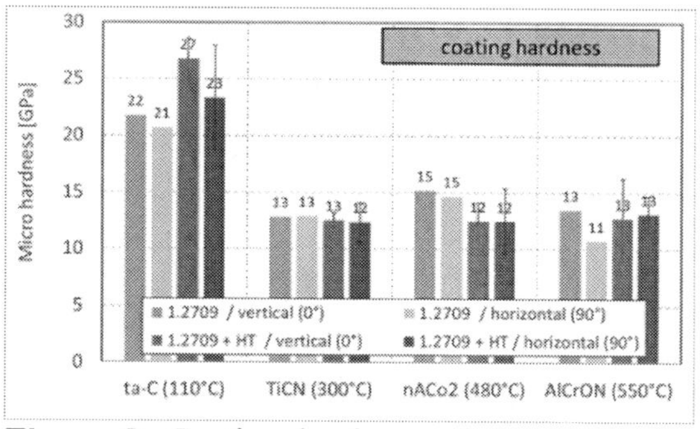

Figure 9. Coating hardness of the AM work pieces with different heat-treatments / coatings in different buildup directions.

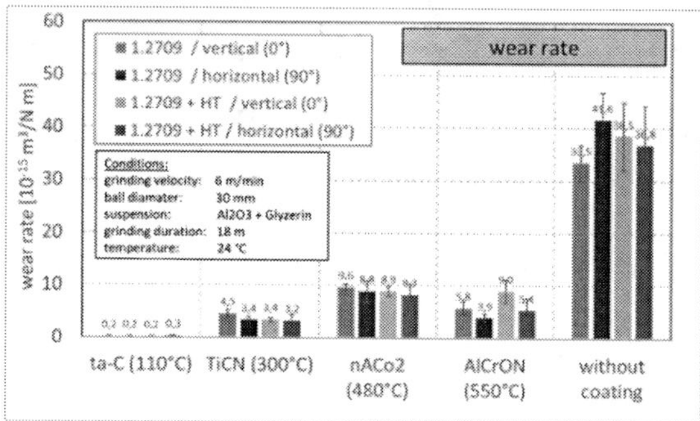

Figure 10. Wear rate of different hard coatings on AM work pieces with different heat-treatments in different buildup directions.

4. Discussion and Conclusion

The analyses of different heat treatments of AM work pieces of the martensitic steel 1.2709 show that heat treatment is essential for optimal work piece properties. It could be shown that the heat treatment not only influences hardness and ductility, but also the properties of a subsequent coating. Hence, the complete additive-subtractive process chain with the specific treatment steps must be taken into account. Investigations have shown that both the buildup direction and the heat treatment have an influence on the mechanical properties of additively manufactured components. Work pieces with a buildup direction of 0° to the buildup plate have a higher notched impact energy (notch parallel to the layer) compared to a buildup direction of 90°. After the heat treatment, the hardness is increased from 32-36 HRC up to 51-52 HRC and the notched impact energy is reduced from 44-60 J down to 12-14 J while the buildup direction has no significant influence on these properties.

If the usual heat treatment for the material 1.2709 is replaced by a PVD coating, comparable hardness can be achieved. Here a deposition temperature of about 480 °C is necessary. The combination of heat treatment and coating deposition allows to set similar microstructures, but also to improve the wear resistance of the coated work pieces. With this properties application areas comparable to conventionally manufactured components can be used.

Acknowledgments

This work is part of the research project "Ad-Proc-Add", which is funded by the Federal Ministry for Economic Affairs and Energy with the support of the VDMA via the project management organization FKM under the number 255 EBG as part of the Cornet / IGF research program. The work in this article was carried out in close coordination with the project committee of the "Ad-Proc-Add" project.

References

[1] Vaneker T H J 2017 The role of Design for Additive Manufacturing in the successful economical introduction of AM *Procedia CIRP* **60** 181-186
[2] Schmidt M, Merklein M, Bourell D, Dimitrov D, Hausotte T, Wegener K, Overmeyer L, Vollertsen F and Levy G N 2017 Laser based additive manufacturing in industry and academia *CIRP Annals* **66** 561-583
[3] Riemer A, Leuders S, Richard A R and Kullmer G 2017 *Optimierung der Werkstoffperformance lasergeschmolzener metallischer Werkstoffe.* In: Additive Fertigung von Bauteilen und Strukturen ed H A Richard and B Schramm (Wiesbaden: Springer Vieweg) p 173-188
[4] Klocke F, Arntz K, Teli M, Winands K, Wegener M and Oliari S 2018 State-of-the-art Laser Additive Manufacturing for Hot-work Tool Steels *Procedia CIRP* **63** 58-63
[5] Levy G N, Schindel R and Kruth J P 2003 Rapid manufacturing and rapid tooling with layer manufacturing (LM) technologies, state of the art and future perspectives *CIRP Annals* **52** 589-609
[6] Moshiri M, Candeo S, Carmignato S, Mohanty S and Tosello G 2019 Benchmarking of Laser Powder Bed Fusion Machines *Journal of Manufacturing and Materials Processing* **3** 85
[7] Bajaj P, Hariharan A, Kini A, Kürnsteiner P, Raabe D and Jägle E A 2020 Steels in additive manufacturing: A review of their microstructure and properties *Materials Science & Engineering A* **772** 138633
[8] Compedium, Platit AG 2021, 62nd Edition
[9] Kučerová L, Zetková I, Jandová A amd Bystrianský M 2019 Microstructural characterisation and in-situ straining of additive-manufactured X3NiCoMoTi 18-9-5 maraging steel *Materials Science & Engineering A* **750** 70-80
[10] Simson, T, Koch J, Rosenthal J, Kepka M, Zetek M, Zetková I, Wolf G, Tomčík P and Kulhánek J 2019 Mechanical properties of 18Ni-300 maraging steel manufactured by LPBF *Procedia Structural Integrity* **17** 843-849
[11] Tan C, Zhou K, Kuang M, Ma W and Kuang T 2018 Microstructural characterization and properties of selective laser melted maraging steel with different build directions *Science and technology of advanced materials* **19** 746-758

IOP Conf. Series: Materials Science and Engineering 1147 (2021) 012002 doi:10.1088/1757-899X/1147/1/012002

Process-related influences and correlations in wire arc additive manufacturing of high-strength steels

D Schroepfer[1,*], R Scharf-Wildenhain[2], A Haelsig[2], K Wandtke[1], A Kromm[1] and T Kannengiesser[1]

[1] Bundesanstalt für Materialforschung und -prüfung (BAM)
[2] Technische Universität Chemnitz, Professur Schweißtechnik

[*]e-mail: dirk.schroepfer@bam.de

Abstract. High-strength fine-grained structural steels have great potential for weight-optimized, efficient structures in many modern steel applications. Further advances in efficiency can be achieved through additive manufacturing and bionic design. Commercial high-strength filler materials for wire arc additive manufacturing (WAAM) are already provided by the consumable producers. Today, application would be strictly limited due to absence of quantitative findings or any guidelines for the industry regarding welding-related stresses and component safety during manufacturing and service. Hence, process- and material-related influences and design-related restraint conditions associated with formation of residual stresses and cold cracking risk are investigated. The aim is the accessibility of special WAAM self-restraining cold cracking tests and easy applicable processing recommendations, enabling an economical, fit-for-purpose and crack-safe WAAM of high-strength steels. This first study focuses on determination of interactions between WAAM process parameters, resulting layer geometry, microstructure and residual stresses, analyzed via X-ray diffraction. Defined reference specimens are automated welded using a special WAAM solid wire (yield strength >820 MPa). Geometric properties can be specifically adjusted by wire feed and welding speed, but cannot be varied arbitrarily, since a high heat input causes local overheating, inadmissible changes of microstructure and mechanical properties, defects and comparable high tensile residual stresses.

1. Introduction

Due to economic and design aspects, high-strength fine-grained structural steels are nowadays increasingly applied in modern steel structures, e.g. in building construction, plant engineering and crane construction. By utilizing fine-grain structural steels with higher yield strengths (≥690 MPa), considerable weight reductions and lower processing costs can be achieved by reducing wall thickness. For example, by substituting S960QL for S235J2, the weight can be reduced by up to 78% [1, 2]. Current efforts towards lightweight construction result from the goal of reducing greenhouse gas emissions by up to 62% by 2030 compared to 1990. These requirements mean that process emissions in industry must be reduced by increasing energy, raw material and material efficiency [3]. This often leads to more complex components and structures. Today, additive manufacturing for the modification or production of high-strength components offers many economic advantages over conventional processes and a significant gain in design freedom, for example, to create bionic designs [4]. In addition, near-net-shape components can often save costly machining steps [5]. In particular, the use of high-strength fine-grain structural steels has great potential for weight-optimized, efficient structures with high mechanical strength for many areas of steel construction. In conventional production, the most important welding process is gas metal arc welding (GMAW). Based on GMAW, the shape-giving process wire arc additive manufacturing (WAAM) has been developed in recent years. Furthermore, special welding wires for the production of high-strength components using WAAM, developed on the base of high-strength filler materials for joint welding, are already available on the market [6].

WAAM is particularly suitable for generating complex structures due to its high deposition rates, process reliability and good automatability [4, 5]. The basic applicability for large-format components using welding robots has already been proven in several studies [4, 7]. Depending on the dimensions of the component, commercially available welding consumables as well as finished WAAM welding systems in various sizes are now available, with which components can be generated by multi-axis portals or robots [8]. Modified GMAW short arc processes are particularly suitable for large-volume components (e.g. Cold Metal Transfer, CMT) due to their high deposition rate at low heat input [9]. In welding experiments and accompanying numerical simulations, the high dependence of the temperature distribution and the resulting component geometry on the production parameters could be demonstrated [10, 11]. These studies confirmed that adequate heat conduction in the WAAM process represents a major challenge for the special microstructure of high-strength fine-grained structural steels. The necessary cooling rates to achieve the required microstructure and mechanical properties (e.g. $t_{8/5}$ time for S690: approx. 5 s to 20 s) can only be achieved by limiting the component temperatures and heat input. For this purpose, build-up strategy and heat control (energy per unit length and preheating or interpass temperature) must be adapted in such a way that as little heat as possible is introduced during production [12]. With regard to the heat control of fine-grained structural steels, there are processing guidelines (e.g. EN1011-2, SEW088 [13, 14]) based on carbon equivalents, such as the CET. However, these are mainly based on tests on steels with lower strength. In addition, in guidelines for the design of structures, residual stresses are generally assumed to be equal to the yield strength in the weld [15].

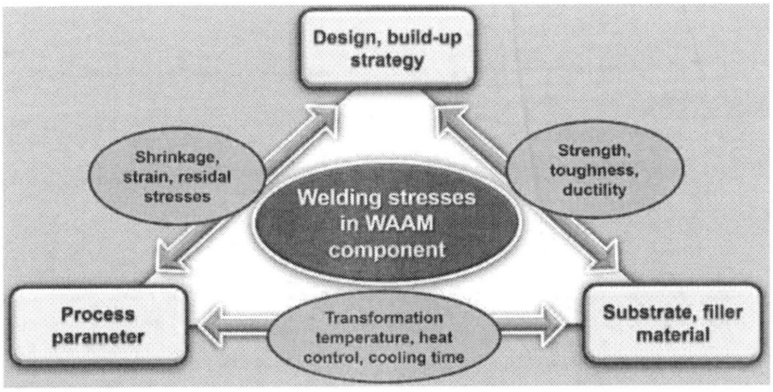

Figure 1. Complex interaction of influences on residual stresses in WAAM components.

Hence, there is still a lack of quantitative information on the production-related stresses and component safety during manufacture and operation that would prevent extensive industrial application [12, 16]. Moreover, many applications require high dimensional accuracy and component integrity due to safety, cost and technological aspects. For the production of high-strength components, generative and machining production steps can be used in a complementary manner. For this purpose, however, as shown in figure 1 comprehensive knowledge of the complex interactions between the welding process and heat conduction during production, the metallurgical processes and the existing design influences on the residual stresses arise is necessary [17, 18]. Thus, in order to reliably avoid high welding-related residual stress levels and possible resulting crack formations during production or a pre-mature component failure, recommendations regarding welding process, heat control, build-up strategy, design and material selection are necessary [12, 19]. Furthermore, the effects of separating processes by removing the substrate plate or by machining the preforms should be clarified, since these affect the residual stress state and may cause significant distortion of the components [20].

In this regard, systematic investigations are currently being carried out in a research project of TU Chemnitz and BAM (FOSTA P1380/IGF 21162 BG). The present work deals with the basic relationships between the welding parameters, the resulting cooling time and the generated layer geometry. Furthermore, the resulting residual stresses are quantified and compared regarding the application of different welding or heat input values, respectively.

2. Experimental and materials

2.1. *WAAM processing system*

The configuration of the robot-supported WAAM system at TU Chemnitz consists of a 6 axis-robot system (KR 6 by KUKA) and a welding machine (TransPuls Synergic, TPS 5000, by Fronius). The robot controller, the central processing unit of the robot system, interprets the robot program codes and moves the manipulator along the programmed paths. A programmable logic controller (PLC) communicates with the external system components. Figure 2 shows the block diagram of the complete welding system. A robot interface serves as a link between the robot controller and the welding system. The commands for the welding system and motion of the robot are located in the corresponding program line of the robot program, which are interpreted and send by the controller to the robot (movement) and the welding system (welding commands). Besides the welding speed due to traveling speed of the robot, the modified short arc welding process CMT, used for these investigations, is adjusted due to the parameters wire feed speed, arc correction and dynamics. The used programming language for the robot system interface enables the integration of several functions allowing special commands for different movement modes or for the control of external systems components. The program is basically extracted and compiled from a CAD model via a slicer, which generates a tool path for the motion planning. This process is also parametric and real-time capable to react on differences or changes to the CAD geometry. These procedures were developed at TU Chemnitz.

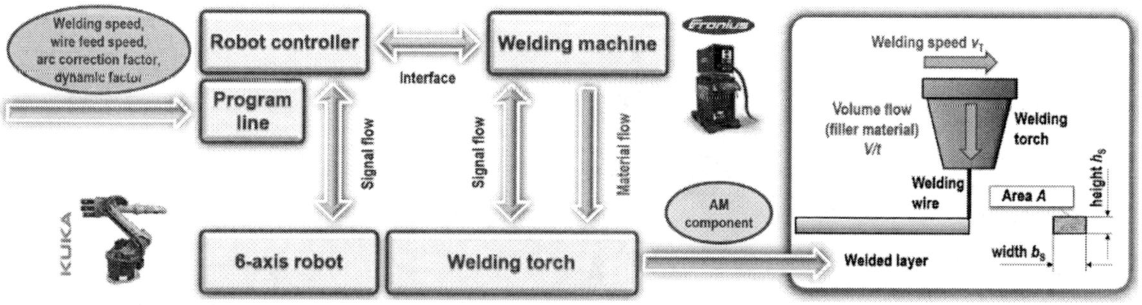

Figure 2. Signal and material flow of the robot-supported WAAM welding system; welding parameters (welding speed v_T, volume flow per time V/t) determining layer width b_S and height h_S.

2.2. *Materials*

In the study, a special for WAAM developed high-strength filler material 10NiMnMoCr8-7-6 (3Dprint AM 80 HD, abbreviation: AM 80, by voestalpine Böhler Welding Germany) having a diameter of 1.2 mm is used. Compared to conventional high-strength solid wire consumables of similar yield strength with a minimum of 790 MPa, e.g. G 79 5 M21 Mn4Ni1.5CrMo (EN ISO 16834-A), the modification of the composition for WAAM welding allows optimised deposition conditions for layer-by-layer build-up. Furthermore, an increase in the manufacturing degrees of freedom is achievable and, with regard to welding heat control, extended processing windows, which still ensure the required

IOP Conf. Series: Materials Science and Engineering 1147 (2021) 012002 doi:10.1088/1757-899X/1147/1/012002

mechanical properties. Table 1 shows the chemical composition and mechanical properties of the investigated welding filler material. The substrate was a 4 mm thick plate of high-strength steel S690QL with a minimum yield strength of 690 MPa. According to the steel producers' recommendations this allows welding in a similar working range regarding heat control for both substrate and filler material to ensure a proper fusion of the component to the substrate. In addition, due to the lower strength a rather higher ductility and thus moderate rigidity conditions are obtained when welding the components.

Table 1. Chemical composition (in wt.-%, Fe balanced) and mechanical properties of filler material 3Dprint AM 80 HD according to the producer (Böhler) in the recommended $t_{8/5}$ range of 5 s to 20 s [6].

C	Si	Mn	Cr	Mo	Ni	$R_{p0.2}$	R_m	A_5
0.09	0.40	1.70	0.35	0.60	2.00	820 MPa	920 MPa	20%

2.3. Welding experiments

During additive manufacturing, the robot moves the welding torch at a constant speed along a programmed path. The welding system presented in figure 2 feeds the filler wire with a defined wire feeding speed or volume flow, respectively. Assuming a rectangular layer cross-section, equation (1) describes the relationship between the layer width b_S, height h_S and the cross-section A:

$$A = b_S \cdot h_S = \frac{V}{t} \cdot \frac{1}{v_T} \quad (1)$$

Consequently, both wire feed and welding speed have an influence on the cross-section of the applied weld bead [4]. Furthermore, these parameters are the main influencing factors of the energy per unit length and thus also determine the flow and solidification behavior of the metallic melt by the amount of heat applied. In order to meet the requirements on geometric specifications, a systematic knowledge of these correlations in the desired parameter range is required. Hence, the focus of this study is on the systematic investigation of these welding parameters and their influence on the layer geometry. For this purpose, component geometries were manufactured as open hollow cuboids by applying 11 layer-wise seams, as shown in figure 3a. The $t_{8/5}$ cooling time was determined during the process by means of a thermocouple (type K) in the 5th layer. The process parameters for these experimental test series are shown in table 2.

Table 2. Constant welding parameters for the experiments.

Welding machine (TPS 5000) characteristic	Arc correction factor	Dynamic factor	Preheating temperature	Interpass temperature	Shielding gas
C1622	–15%	4	20 °C	200 °C	98% Ar, 2% CO_2

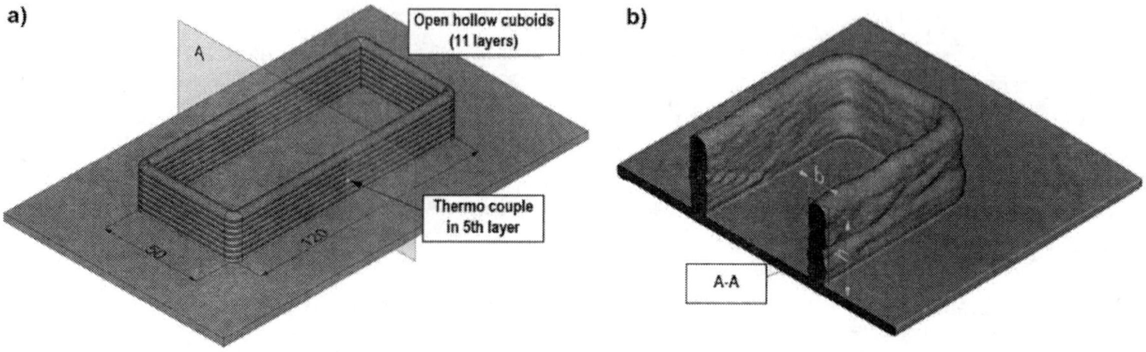

Figure 3. a) Open hollow cuboid with 11 layers, thermo couple type K (determination of $t_{8/5}$ cooling time); b) non-destructive evaluation of test specimens using 3D scan: determination of web height and width in cross section A-A using bounding box.

By systematically varying the parameters of the welding and wire feed speed, a total of 16 open hollow cuboids were produced (V1 to V16). The test matrix is presented in table 3. To evaluate these tests, a 3D scan of the components was performed (accuracy: ±0.05 mm). The side walls of the open hollow cuboids were analyzed with regard to web height and width in the symmetry planes of the test samples. Figure 3b shows this exemplary for one side by means of a digitized sample geometry in cross section A-A. For each hollow cuboid, web heights and widths were measured at four defined positions. The averaged layer dimensions height h_S and width b_S can be calculated using equations (2) and (3) with web height h_i, web width b_i and amount of layer n:

$$h_S = \frac{1}{4 \cdot n} \sum_{i=1}^{4} h_i \quad (2)$$

$$b_S = \frac{1}{4} \sum_{i=1}^{4} b_i \quad (3)$$

Table 3. Test matrix for determining the layer geometry dependent on welding and wire feed speed (bold test numbers: residual stress analysis performed).

		Welding speed [cm/min]			
		20	30	40	50
Wire feeding speed [m/min]	2	V1	V5	**V9**	V13
	3	V2	V6	V10	V14
	4	V3	**V7**	V11	V15
	5	V4	V8	V12	V16

2.4. Residual stress analysis

X-ray diffraction (XRD) was used to analyze the local residual stresses transverse and longitudinal to the weld via the $\sin^2\Psi$-method. This was performed at several positions on the top layer of the hollow cuboid specimens to determine the effect of the heat input due to varied wire feeding and welding speed. The measuring lines are shown in figure 4. Two test specimens were analyzed from the test matrix: V7 (heat input E=0.43 kJ/mm, $t_{8/5}$=20.2 s) and V9 (E=0.18 kJ/mm, $t_{8/5}$=9.2 s) to observe the effect in lower and upper scope regarding the recommended $t_{8/5}$ cooling time range of $t_{8/5}$. Table 4 shows the used parameters for the residual stress analysis.

Table 4. Parameters for residual stress analysis by means of XRD ($\sin^2\Psi$-method).

Radiation:	Tupe power:	Collimator:	Detector:
CrKα	30 kV/6.7 mA	2 mm	Linear solid state detector
Diffraction line:	Ψ-tilting:	Ψ-steps:	Measuring time:
{211}α	0° to ±45°	9	5 s

3. Results and discussion

3.1. Microstructure

Figure 5 shows a microsection of the weld metal of a generated wall after production with a constant interlayer temperature of 200 °C and a $t_{8/5}$ cooling time of approx. 10 s. The cross section shows a suitable, fine-grained microstructure. Compared to conventional filler materials, e.g. G 79, the especially for WAAM developed welding wire (AM 80) tends to solidify more fine-grained at similar welding parameters in the recommended $t_{8/5}$ cooling time range of 5 s to 20 s.

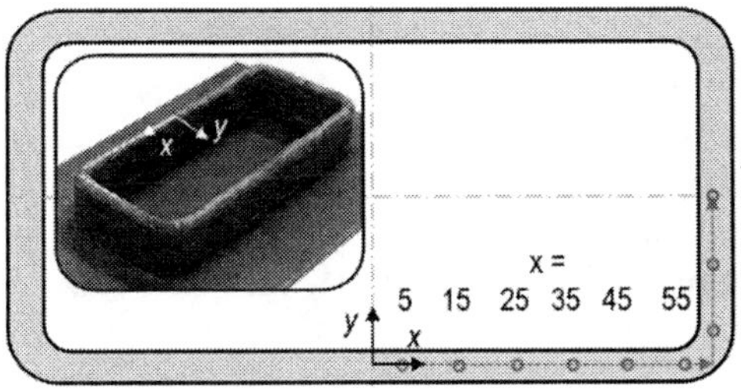

Figure 4. Measuring lines for residual stress analyses at the test specimen.

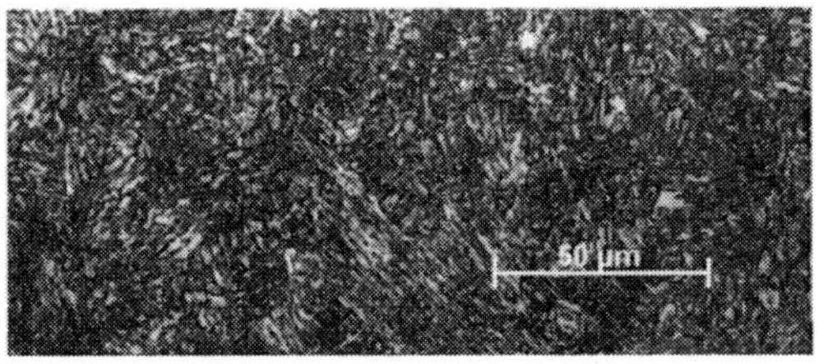

Figure 5. Cross section of the weld metal (AM 80) for $t_{8/5}$ cooling time of approx. 10 s.

3.2. Layer and component dimension

Table 5 summarizes the averaged layer widths b_s and heights h_s as a function of the welding and wire feed speed. Based on these measurement data for the component or layer geometry, the characteristics of the WAAM production in the investigated parameter range can be modelled. Figure 6 shows the contour plot of the average layer dimensions (width (a); height (b)) as a function of wire feeding v_W and welding speed v_T for the derived regression model. In the regression model v_W and v_T are highly significant. Also, the squared interaction of v_T is significant for both b_S and h_S with a p value of 0.04 and 0.01 respectively. The model qualities are high with R^2=0.988 and R^2=0.996, respectively.

Table 5. Results for averaged layer geometry (b_s, h_s), heat input E and resulting cooling time $t_{8/5}$ dependent on welding speed v_T and wire feed speed v_W.

			Welding speed v_T [cm/min]			
		Results	20	30	40	50
Wire feeding speed v_W [m/min]	2	b_s [mm] / h_s [mm]	5.4 / 2.2	4.6 / 2.0	**3.9 / 1.8**	3.4 / 1.6
		E [kJ/mm] / $t_{8/5}$ [s]	0.36 / 12.6	0.24 / 10.4	**0.18 / 9.2**	0.14 / 5.8
	3	b_s [mm] / h_s [mm]	6.9 / 2.5	5.6 / 2.2	4.9 / 2.0	4.5 / 1.8
		E [kJ/mm] / $t_{8/5}$ [s]	0.51 / 19.4	0.34 / 14.0	0.25 / 11.8	0.20 / 9.1
	4	b_s [mm] / h_s [mm]	7.8 / 2.7	**6.7 / 2.4**	5.8 /2.2	5.3 / 2.0
		E [kJ/mm] / $t_{8/5}$ [s]	0.65 / 25.6	**0.43 / 20.2**	0.32 / 14.2	0.26 / 13.0
	5	b_s [mm] / h_s [mm]	8.8 / 2.9	7.7 / 2.6	7.4 / 2.4	6.2 / 2.2
		E [kJ/mm] / $t_{8/5}$ [s]	0.79 / 40.0	0.53 / 26.4	0.40 / 18.2	0.32 / 13.8

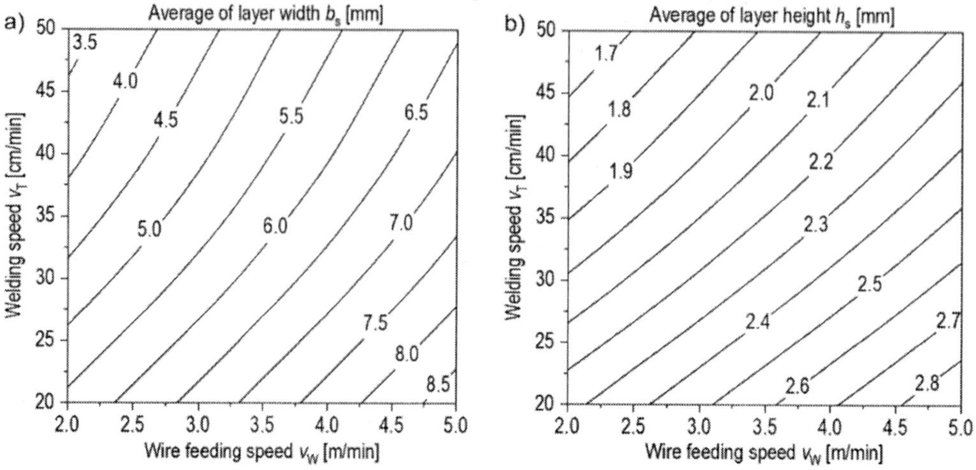

Figure 6. Average layer width (a) and height (b) as a function of wire feeding and welding speed in the investigated parameter range (contour plots of the regression models).

The layer width almost linearly increases with higher values for wire feeding speed and lower values for welding speed. For the layer height the same trend can be observed. Comparing both, it reveals that for higher values the influence of welding speed on the layer width decreases and then depends to a greater extent on the wire feeding speed, in contrast to the effect on the layer height. This finally also exhibits that the width and height of the generated layers depend largely on the energy per unit length used. In this context, figure 7a shows the contour diagram of energy per unit length as a function of wire feed rate and welding speed. The derived regression model for heat input shows highly significant effects for the wire feed rate, welding speed, their interaction, and a highly significant quadratic effect of welding speed with a model quality of R^2=0.995. Compared to figure 6, as expected, increased layer heights and widths correlate with a higher heat input, so that an almost linear relationship exists. In addition, the cooling time $t_{8/5}$ correlates with these results which is presented in figure 7b.

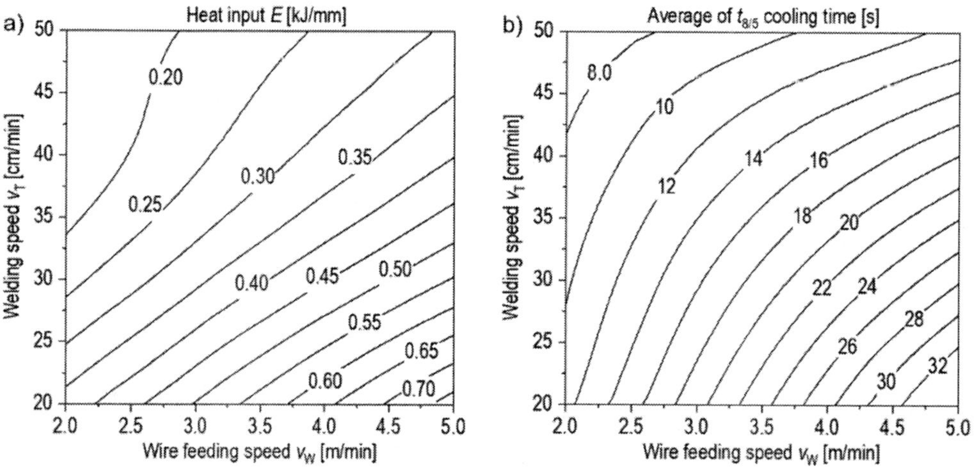

Figure 7. Heat input E (a) and average of $t_{8/5}$ cooling time (b) as a function of wire feeding and welding speed in the investigated parameter range (contour plots of the regression models).

Moreover, figure 7a reveals that at higher welding speeds and also at higher wire feed rates, the layer width and the energy per unit length decouple to a certain extent. The influence of the welding speed remains more or less the same even at higher values for the energy per unit length but is less pronounced on the layer width. This is due to the direct relationship between welding speed and heat input. A lower heat input ultimately results in reduced dilatation of the layers and an earlier solidification. A similar effect can be seen at higher wire feeding speeds. Here the influence of the welding speed on the heat input is even somewhat greater, but at the same time the influence on the layer width decreases somewhat in comparison, especially for high welding speeds. When welding high-strength fine-grained structural steels, the special microstructure requires a cooling rate to be kept within a tight interval ($t_{8/5}$ cooling time range). The cooling time and thus the heat control during welding significantly affects the resulting mechanical-technological properties of the generated component. During WAAM, similar to multi-layer welding, each layer undergoes a complex heat treatment consisting of multiple cyclic heating and cooling.

Figure 7b shows the contour plot of the last measurable $t_{8/5}$ cooling times (i.e. with peak temperature >800 °C) of the 5th layer as function of welding speed and wire feed rate. In the derived regression model of the resulting cooling times, the wire feed rate and the interaction of wire feed rate and welding speed are highly significant. The model quality is high with R^2=0.953. By comparison with the determined average $t_{8/5}$ cooling times, a correlation with the heat input reveals. It is apparent that the influence of the wire feed rate on cooling time decreases as the welding speed increases towards lower values for the cooling time. This is mainly due to the relatively large influence on the layer width in this area. As a result, at high welding speed and with increasing wire feed rate, the layer width increases at a higher rate than the heat input. The increased layer width results in a better adhesion to the layers below and an enhanced heat transfer, which results in a faster heat dissipation from the welded layer.

3.3. Residual stress analysis

The residual stress analysis by means of XRD at the top layer of the two WAAM test specimens is presented in figure 8. Remarkably, the residual stresses in the transverse direction of the welded layers or walls for both specimens are significantly higher than those in the direction longitudinal to the weld, despite the high shrinkage restraint in the longitudinal direction caused by the underlying layers. Yet, it reveals that an increasing restraint in transverse direction in near the corner of the hollow cuboids causes for both specimen bodies the highest tensile residual stresses of approx. 350 MPa at these locations. The maximal longitudinal residual stresses are present in a certain distance of the corners.

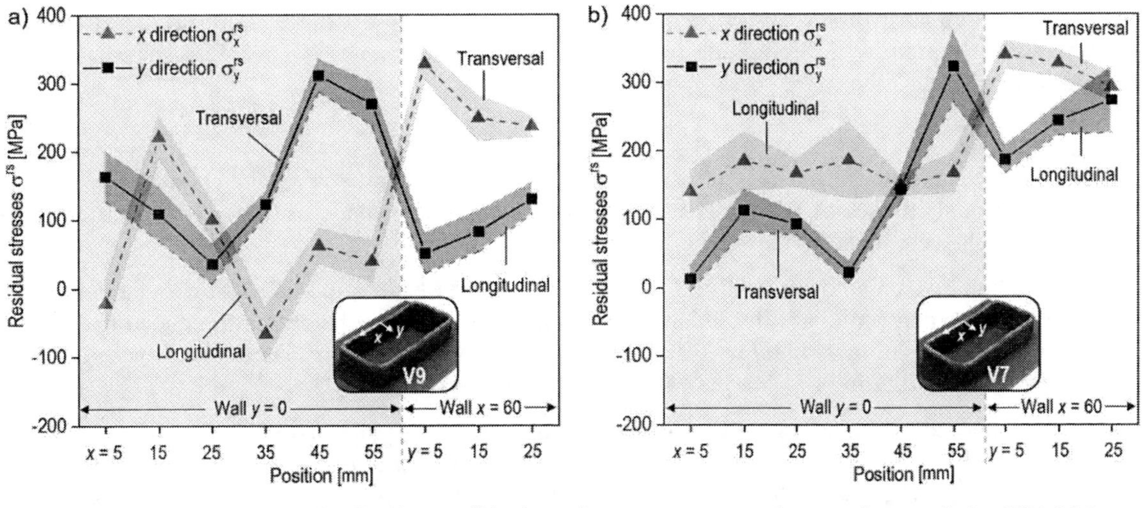

Figure 8. Residual stress distributions $\sigma^{rs}(x,y)$ and error range at the top layer of the WAAM test specimens: a) V9 (E=0.18 kJ/mm, $t_{8/5}$=9.2 s) and b) V7 (E=0.43 kJ/mm, $t_{8/5}$=20.2 s).

By comparison of both components, longitudinal tensile residual stresses reveal to be considerably higher in case of increased heat input for both side walls of the hollow cuboid, as shown for test V7 in figure 8b. Figure 9 reveals in particular for the short side walls a difference of approx. 150 MPa.

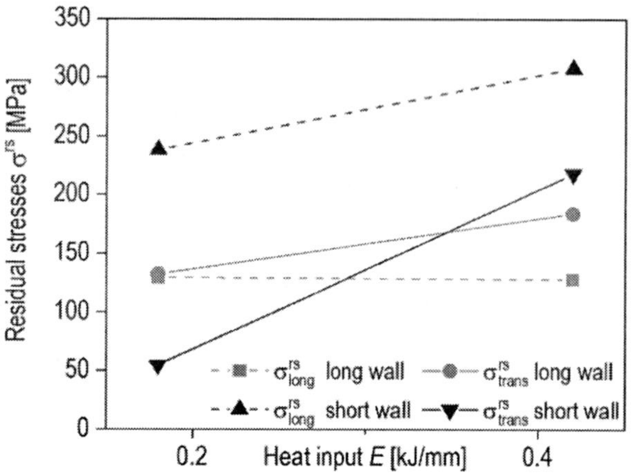

Figure 9. Effect diagram of averaged transversal and longitudinal residual stress levels at the top layer of the short and long side walls of the hollow cuboids depending on the applied heat input E during welding.

The higher heat input as well leads to higher transverse residual stresses, especially at the short side wall. According to the common concepts of residual stress evolution in steels with undergoing phase transformation this is primarily due to the higher volume of heated material combined with an increased restrained shrinkage initiating higher tensile stresses while cooling and, furthermore, due to the different microstructure developing at higher cooling times [17, 18]. It is assumed that shorter cooling times lead to a microstructure with a higher share of bainite and martensite instead of a ferritic microstructure. Thereby a higher volume expansion occurs during the phase transformation, the hindrance of which leads to the formation of compressive stresses or to compensation of arising tensile stresses [19, 21]. However, as figure 8a reveals for test V9, the gradients of the longitudinal stresses are higher at the long side wall of the specimen welded with lower heat input, which is due to local effects in connection with the phase transformation. The averages for the residual stresses longitudinal and transverse to the weld are shown in figure 9 in an effect diagram for the short and the long side walls dependent on the applied heat input during welding. It is noticeable that the mean longitudinal residual stresses over the long side wall are hardly affected depending on the heat input, which is assumed to be due to the relatively thin substrate plate of 4 mm and the resulting deformation during welding. The greatest effect can be observed with the short side walls which are supposed to be effectively restrained by the long side walls. Longitudinal and transversal residual stresses increase due to the higher heat input.

4. Conclusions

In the present study, the influences welding speed and wire feed rate on the layer geometry, cooling time and residual stresses of additively manufactured components made of high-strength fine-grained structural steels were investigated.

(1) Welding consumables specially developed for WAAM have a modified flow behavior, and in contrast to conventional filler metals, lead to an increased layer height with reduced widths. The cross sections of the additively welded walls show a suitable microstructure and that the special WAAM welding filler material AM 80 solidifies fine-grained.

(2) An enlarged parameter range for the application of the filler material serves facilitating additive production steps, especially in the case of complex geometries. It can be assumed that this ultimately has a positive effect on the mechanical-technological properties, which is subject to further investigations in the research project.

(3) By deriving regression models, effects and interactions of the welding parameters on the target quantities were found, interpretations as operating characteristics of the WAAM system used are achievable. With these characteristic curves, suitable starting points for the calibration of the generation process can be identified and slicer programs can be optimized for the WAAM process.

(4) The relationships shown also clarify the fundamental influence of welding and wire feeding speed on the cooling time and the residual stress build-up, which is superimposed by cooling time dependent effects, e.g. phase transformation.

(5) The residual stresses are depending on the heat input or cooling time, respectively. For higher heat input, an increase in residual stress level were found, especially in transverse direction to the weld. The highest increase was found in the short side wall. The low effect in longitudinal direction of the long side wall is assumed to be due to low restraint of the applied substrate plate.

These findings and correlations form the basis for the next steps of the planned investigations, e.g. with focus on the influences of component design and build-up strategy on the residual stress evolution. Additionally, the restraint during welding of the component will be significantly increased due to application of heavy and rigid substrate plates for further experiments. Furthermore, the effect of the heat control on the microstructure formation has to be clarified in detail, especially considering a homogeneous distribution with respect to the resulting component properties.

Acknowledgments
The studies were funded by the AiF-project project IGF-Nr. 21162 BG/FOSTA P1380. Sincere thanks are given for this support and to the representing companies actively involved in the project board.

References
[1] Raoul J 2005 *Use and Application of High-Performance Steels for Steel Structures* ed Guenther H-P (Zürich: IABSE) p 152
[2] Hulka K, Kern A and Schriever U 2005 *Materials Science Forum* **500-501** 519-526
[3] German Federal Ministry for the Environment (BMU) 2019 *Climate Action in Figures*
[4] Plangger J, Schabhüttl P, Vuherer T and Enzinger N 2019 *Metals* **9**
[5] Frazier W E 2014 *Journal of Materials Engineering and Performance* **23** 1917-1928
[6] voestalpine Böhler Welding Group GmbH 2018 *The Future of Productivity: Wire Arc Additive Manufacturing (WAAM)* p 12
[7] Ding D, Pan Z, Cuiuri D and Li H 2015 *International Journal of Advanced Manufacturing Technology* **81** 465-481
[8] Rodrigues T A, Duarte V, Miranda R M, Santos T G and Oliveira J P 2019 *Materials (Basel)* **12**
[9] Xizhang C, Su C, Wang Y, Siddiquee A N, Sergey K, Jayalakshmi S and Singh R A 2019 *Journal of Surface Investigation: X-ray, Synchrotron and Neutron Techniques* **12** 1278-1284
[10] Graf M, Hälsig A, Höfer K, Awiszus B and Mayr P 2018 *Metals* **8** 1009
[11] Ogino Y, Asai S and Hirata Y 2018 *Welding in the World* **62** 393-401
[12] Schroepfer D, Kromm A, Schaupp T and Kannengiesser T 2018 *Welding in the World* **63** 647-661
[13] European Standard EN 1011-2 2001 *Welding - Recommendation for welding of metallic materials - Part 2: Arc welding of ferritic steels*
[14] Technical guideline SEW 088 2017 *Weldable non-alloy and low-alloy steels Recommendations for processing, in particular for fusion welding* (Düsseldorf: Stahleisen GmbH) p 52
[15] European Standard EN 1993 2010 *Eurocode 3: Design of steel structures*
[16] Hönnige J, Seow C E, Ganguly S, Xu X, Cabeza S, Coules H and Williams S 2020 *Materials Science and Engineering: A* [https://doi.org/10.1016/j.msea.2020.140368]

[17] Wohlfahrt H 1987 *Advances in Surface Treatments Vol. 4 - International Guidebook on Residual Stresses,* p 40-58
[18] Schroepfer D, Kromm A and Kannengiesser T 2018 *Welding in the World* **63** 43-51
[19] Schroepfer D, Kromm A and Kannengiesser T 2015 *Welding in the World* **59** 455-464
[20] Fuchs C, Baier D, Semm T and Zaeh M F 2020 *Production Engineering* **14** 629-637
[21] Bhadeshia H K D H 2004 *Materials Science and Engineering A* **378** 34-39

Electrodeposition of FeCrNi and FeCr alloys and influence of heat treatment on microstructure and composition

V Meinhold*, D Höhlich, A Dittes, T Mehner and T Lampke

Chemnitz University of Technology, Materials and Surface Engineering Group, Erfenschlager Straße 73, 09125 Chemnitz, Germany;

*e-mail corresponding author: vanessa.meinhold@mb.tu-chemnitz.de

Abstract. The electroplating of iron-chromium and iron-nickel-chromium layers is an economic alternative to mild steel and hard-chrome layers from chromium (VI) electrolytes. Iron-chromium and iron-nickel-chromium layers were electrodeposited using an environment friendly chromium (III) electrolyte. The layers were heat-treated at different temperatures (150 °C, 300 °C, 450 °C and 600 °C) in order to determine the temperature at which recrystallization takes place, which phases are formed and to study the influence on the element content. The phase analysis was conducted by X-ray diffraction, the chemical composition and the microstructure were characterized by the scanning electron microscopy. Both layer systems show an X-ray-amorphous structure that begins to recrystallize at a temperature of 450 °C. From a heat-treatment temperature of 600 °C, the organic additives decompose and the oxygen forms chromium oxide with the chromium.

1. Introduction

Since 2017, the use of hexavalent chromium is banned by the European Union regulation for Registration, Evaluation, Authorisation and Restriction of Chemicals (REACH) [1]. In the past, chromium layers were usually electroplated from these chromium compounds. Therefore, there is a carcinogenic and mutagenic risk during the deposition process or if chromium (VI) compounds end up in the environment and groundwater [2]. The advantage of hard chrome coatings, which are deposited from hexavalent chromium, lies in their good corrosion resistance, high hardness, wear resistance, and constant shine [3]. That is why electroplated hard-chrome layers have been used on a large scale in countless products for many years by the industry. Due to REACH, it is of great importance to find alternatives to hard chrome as quick as possible.

A potential substitute for hard-chrome coatings could be stainless steel. The advantage of these iron-nickel-chromium bulk alloys is their good corrosion resistance, especially against acids. Due to their biocompatibility, they are often used in the food industry, in systems for heat treatment and in steam generators. Because of this purpose of high-temperature applications, it is of interest to examine the iron-nickel-chromium and iron-chromium alloys with regard to their behaviour under the influence of heat [4]. However, bulk stainless steel is an expensive material. Electroplating is an economical process by which stainless-steel coatings can be applied to less expensive materials like mild steel. The excellent surface properties are retained while the material costs are reduced.

So far, only a few studies have been carried out on the galvanic deposition of stainless steel. The studies investigated the bath composition of the electrolyte and its influence on the layer composition and the microstructure of the deposited layers [5-8]. There has only been minor research activity in one study from 2019, where iron-nickel-chromium coatings are examined precisely concerning their structure and the influence of heat treatment at a temperature of 600 °C on the microstructure of the coatings. It was found that the deposited layers in the initial state have an amorphous α-Fe structure. The iron and nickel exist mainly in metallic form with small amounts of chromium oxide and the heat-treated samples have larger cracks in relation to their counterparts in the untreated state [5]. In comparison, stainless-steel bulk alloys have a grain size of around 50 µm [9]. Furthermore, in addition to large amounts of oxygen, carbon and hydrogen, the researchers also found small amounts of nitrogen in the layers. The heat treatment reduced the carbon and hydrogen content in the layer. In addition, a crystallite size of 20–25 nm was measured. From this, it was concluded that the high carbon and hydrogen concentration is mainly responsible for the amorphous character and that the annealing outgases the only weakly bound elements. The new structure, which arose after the heat treatment, is characterized by a nanocrystalline γ-Fe phases with a few amorphous chromium oxide phases [5].

Based on the work of Bertero et al. [5], the influence of the heat treatment on the structure and the element content will be investigated further in this paper. It is now necessary to investigate if these effects already occur at lower temperatures. For example, this is the case with electrodeposited copper layers, which recrystallize at significantly lower temperatures than it is the case with bulk materials due to their high internal energy and defect density [10]. A reduction of internal stresses and the formation of new grains through heat treatment can increase the strength of the layers while at the same time increase the ductility.

2. Experimental details

2.1. Preparation of the samples

In order to study the influence of heat treatment on element composition and microstructural evolution, two chromium (III) electrolytes were used for the production of iron-nickel-chromium and iron-chromium layers. The coatings were electroplated on nickel-brass tubes with a diameter of 1 cm. At first, the tubes were cleansed in an ultrasonic bath with ethanol. After this treatment, the substrate was etched by concentrated sulphuric acid for 2 minutes and rinsed thoroughly in deionized water. A platinum-plated titanium anode was used for electrodeposition. The electrolyte solutions were magnetically stirred at 100 rpm. The temperature of the electroplating baths was stabilized by a tempered water bath.

The iron-nickel-chromium coatings were deposited at a current density of 3.5 A/dm² under direct current. The composition of the glycine-based electrolyte for the iron-nickel-chromium coatings are chromium (III) chloride, nickel (II) chloride, iron (II) chloride, glycine, ammonium chloride, sodium chloride and boric acid. In order to give the thermodynamic equilibrium setting in the electrolyte sufficient time during the production of the electrolyte, a multi-stage process of component addition must be followed. At first, the chromium (III) chloride and the glycine has to be dissolved in deionized water. During the process of complex formation, the solution was heated to 80 °C for 30 minutes. After the solution cooled down to 25 °C, the remaining components were added. The pH value of the electrolyte was adjusted to 1 with hydrochloric acid. The deposition took place at a temperature of 35 °C.

The iron-chromium coatings were performed by using pulsed-current electroplating. A pulse-current source POWER-PULSE pe86CB-40-10-50-S/GD (plating electronic GmbH, Sexau, Germany) was used. The electrolyte consists of Cr(III) and Fe(II) chlorides, boric acid, a carboxylic acid, amide complexing agents, alcohol, a reducing agent, brighteners and wetting agents. The process of the electrolyte production was performed in two steps. In the first step, only the metal salts were completely dissolved in deionized water with boric acid and complexing agent, and in the second step the other organics were added and the pH was adjusted. The pH value of the electrolyte was set to a range between 1.5 and 1.6 with hydrochloric acid. Layers were deposited at a temperature of 30 °C. They were produced at a pulse-current density of 97 A/dm². The pulsed-current plating took place at a frequency of 109 Hz and an on–off ratio of 3:6.

From each electrolyte, a layer was deposited and cut into different samples. Each sample has a dimension of 0.5 cm x 0.5 cm. The samples were heat-treated for one hour at 150 °C, 300 °C, 450 °C or 600 °C, or 24 hours at a temperature of 450 °C. A Torvac vacuum furnace (Torvac, St. Paul, MN, USA) with a vacuum between 10^{-4} mbar and 10^{-5} mbar was used during the heat treatment. Two investigation methods (listed in chapter 2.2) were carried out to understand the microstructural changes during heat treatment of the iron-nickel-chromium and iron-chromium electroplated films and the element content, the content of impurities in particular, which is caused by the use of organic additives. For the scanning electron microscope examination, the samples were embedded conductively. The embedded samples were ground, polished and thoroughly cleaned under the influence of ultrasound.

2.2. Layer characterization
Firstly, the layer composition with regard to its average over the entire surface and its course over the layer thickness was examined. The elements of the layers were determined with an EDAX GENESIS energy-dispersive X-ray spectroscopy (EDXS) spectrometer at a LEO1455VP (Zeiss, Oberkochen, Germany) scanning electron microscope. The acceleration voltage was 25 kV. The integration time per element analysis was two minutes. The layer compositions were determined by area analysis and by point analysis as shown in figure 1 to determine the element content of the layer as a function of the depth under the surface. In addition, images were taken using scanning electron microscopy to identify microcracks.

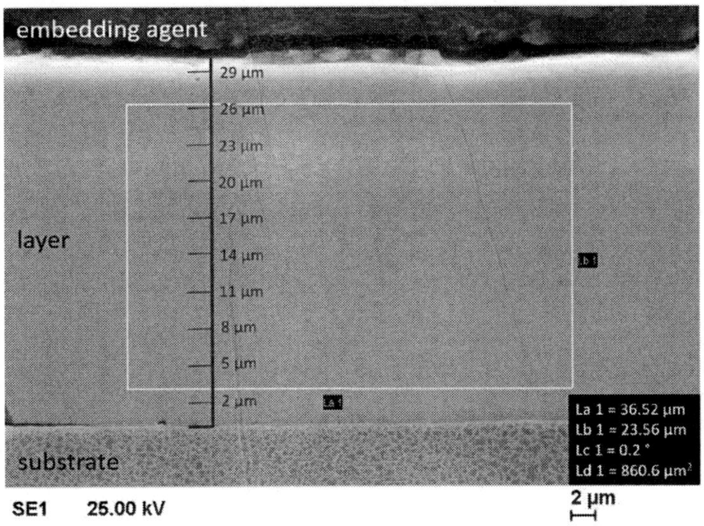

Figure 1. Method of point analysis (red) and method of area analysis *(white) for EDXS measurements.*

The phases and the existence of crystallites of the deposited coatings were analysed by the X-ray diffraction (XRD, point measurements) with a D8 Discover (Bruker AXS, Billerica, MA, USA) with Co Kα radiation (tube parameters: point focus, 40 kV, 40 mA), polycap optics, a pinhole aperture with a diameter of 0.5 mm, and a 1D detector LYNXEYE XE with a step size of 0.02° and 17.8 s/step (overall measurement time per sample: 8.9 hours).

3. Results and Discussion

3.1. Element composition
By using EDXS, the composition of the iron-nickel-chromium and iron-chromium layers was determined (mass fractions w). The elemental compositions for all samples were measured by area analysis, including the metallic species iron, chromium and nickel and the contamination by oxygen, carbon and phosphorus, and are listed in table 1. The representative individual measured mass values represent the element content over the entire layer thickness. It is noticeable that the carbon as well as the oxygen content are in a similar range for both layer systems.

Table 1. Composition and layer thickness of the as-deposited layers.

Sample	w_{Cr} (%)	w_{Fe} (%)	w_{Ni} (%)	w_{C} (%)	w_{O} (%)	w_{P} (%)	s (µm)
FeCrNi	15	50	19	6	10	<0,1	19
FeCr	21	58	< 0,1	4	9	8	29

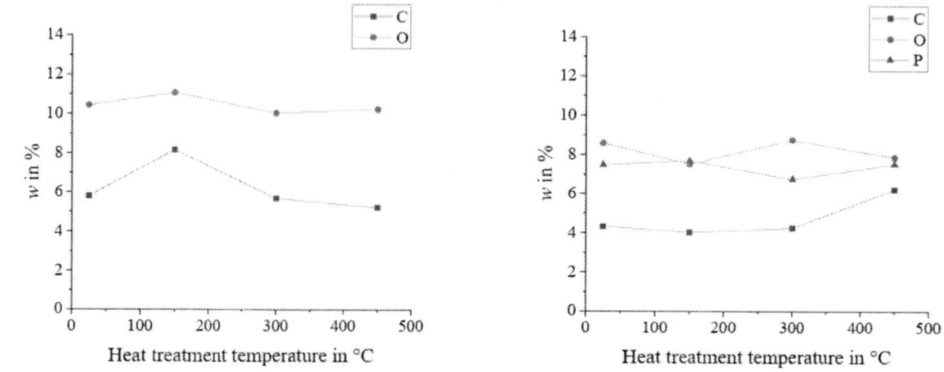

Figure 2. EDXS measurements of the impurities over the entire layer surface of the iron-nickel-chromium (left) and iron-chromium coatings (right).

As seen in the profiles of the impurity contents over the entire layer cross-section in figure 2, there is no proof for diffusion of the organic elements during the heat treatment, as it was observed by Bertero et al. as well [5]. The reason for this may be small differences in the compositions of the initial states of the individual samples. Therefore, a diffusion of carbon, oxygen and phosphorus through heat treatment cannot be excluded. It is interesting to note that based on the course of the curves of the iron-nickel-chromium samples for all heat treatments, the oxygen and carbon contents possess an almost constant ratio. This indicates that the two elements are largely present in chemical compounds in the layers. Furthermore, one can assume that this compound originates from the used additives, which are incorporated into the layer. Compared to the studies by Bertero et al., the oxygen content of the as-deposited iron-nickel-chromium layers measured in this study is around $w_O = 4$ % higher. The carbon content of around $w_C = 6.4$ % is in the same range. After the heat treatment of 600 °C of the iron-nickel-chromium layers a oxygen content of $w_O = 8{,}2$ % and a carbon content of $w_C = 2{,}4$ % were measured [5]. In this investigation such a decrease in the carbon content as a result of heat treatment could not be proven. The curves of the iron-chromium layers show a different ratio. The oxygen and phosphorus contents possess an almost constant ratio. Up to a heat-treatment of 300 °C the carbon content is nearly constant. However, the carbon content of the sample heat-treated at 450 °C ist $w_C = 2$ % higher. The value of the heat-treatment at 450°C, 5 µm can be seen as a measurement deviation.

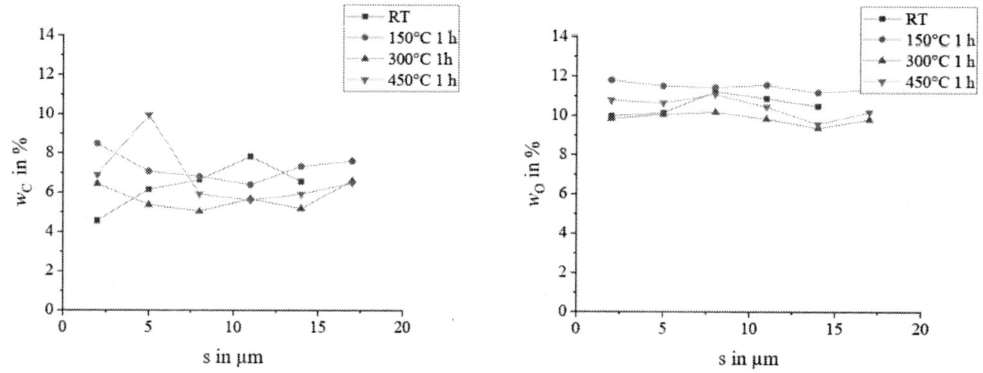

Figure 3. EDXS measurements of the carbon content (left) and the oxygen content (right) of the iron-nickel-chromium layers. The measuring points show the distance to the substrate.

In the figure 3, s means the layer thickness starting from the substrate. The illustrated data are representative individual measured values. This also applies to figures 4 and 5. Figure 3 shows that the oxygen contents over the layer thickness are approximately the same for all heat-treatment stages. It is noticeable that the carbon content near the substrate and the surface of the heat-treated layers is higher than in the middle of the layer (area between 5 µm and 14 µm). The samples at room temperature are an exception. A possible cause is that the carbon diffuses during the heat-treatment to the rim of the layers. As with the area analysis, there is no significant influence of the heat treatment on the total carbon and oxygen contents in the layers.

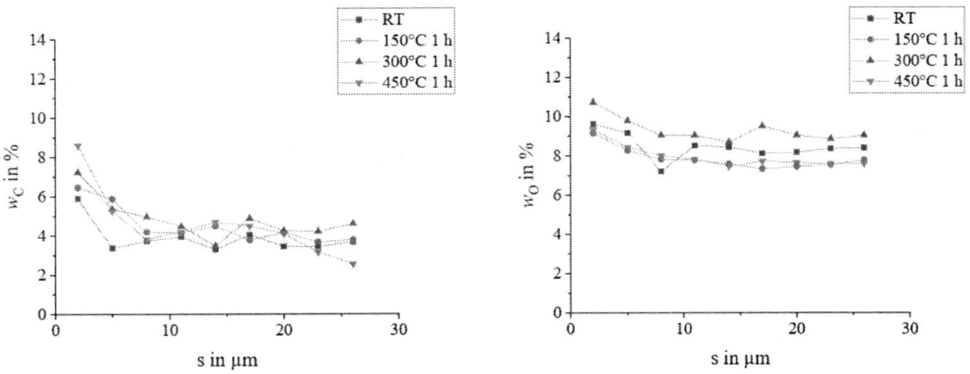

Figure 4. EDXS measurements of the carbon content (left) and the oxygen content (right) of the iron-chromium layers. The measuring points show the distance to the substrate.

The curves in figure 4 show that the carbon content of the iron-chromium layers seems to increase due to the heat treatment. The reason for this trend could be the inhomogeneous composition along the entire sample because of a poor electrolyte throwing power and uneven incorporation of electrolyte additives. A comparison of the two layers highlights that the iron-chromium layers show a trend towards fewer impurities with increasing distance from the substrate, while the iron-nickel-chromium-layers show an indifferent incorporation of carbon and oxygen. Consequential one can deduce a strong influence of the used individual electrolyte additives on the degree and distribution of the impurities. However, all of the layers have in common that the carbon is significantly higher near the substrate than near the surface. From this, one can conclude that in the beginning, more carbon is co-deposited. Above a distance from the substrate of approximately 8 µm, it can be observe that the amount of carbon is significantly less but remains constant. Similarly, the amount of oxygen in the layers, which is around w_O = 10 % near the substrate and nearly constant above a layer thickness of 8 µm (between w_O = 7.5 % and w_O = 9 %). This indicates also a diffusion of oxygen during the heat-treatment.

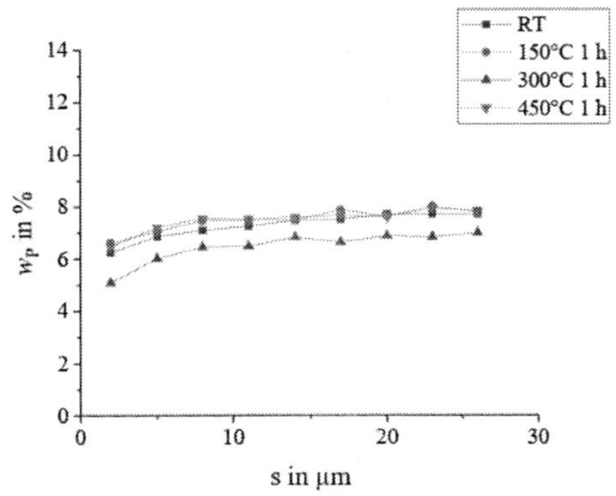

Figure 5. EDXS measurements of the phosphorus content of the iron-chromium layers. The measuring points show the distance to the substrate.

Considering figure 5, the phosphorus content shows the opposite trend to the carbon and oxygen contents. Looking at the figures, the phosphorus content is lowest at the beginning of the deposition and that more phosphorus is incorporated during the further the layer grows. The largest amount of phosphorus is found near the surface.

3.2. Phase analysis
Figure 6 shows that an X-ray amorphous or nanocrystalline structure for RT, 150 °C and 300 °C exists for the iron-chromium layers. An iron-chromium austenite phase only forms after a heat treatment at 450 °C for one hour. Iron phosphide and cubic chromium-iron phases (austenitic) are forming by heat-treating samples for 24 hours at a temperature of 450 °C. At a temperature of 600 °C, chromium oxide phases are added. The large peak at $2\theta = 30°$ can most likely be assigned to carbon that is incorporated into the lattice. This hypothesis is supported by an increase of the peak intensity due to the heat treatment and the longer holding time. Based on the previously discussed course of the oxygen content over the layer thickness, chromium oxides form at 600 °C. A chromium-oxide formation due to oxygen from the atmosphere can be ruled out based on these curves. The small peaks marked with the number 1 derive from the substrate.

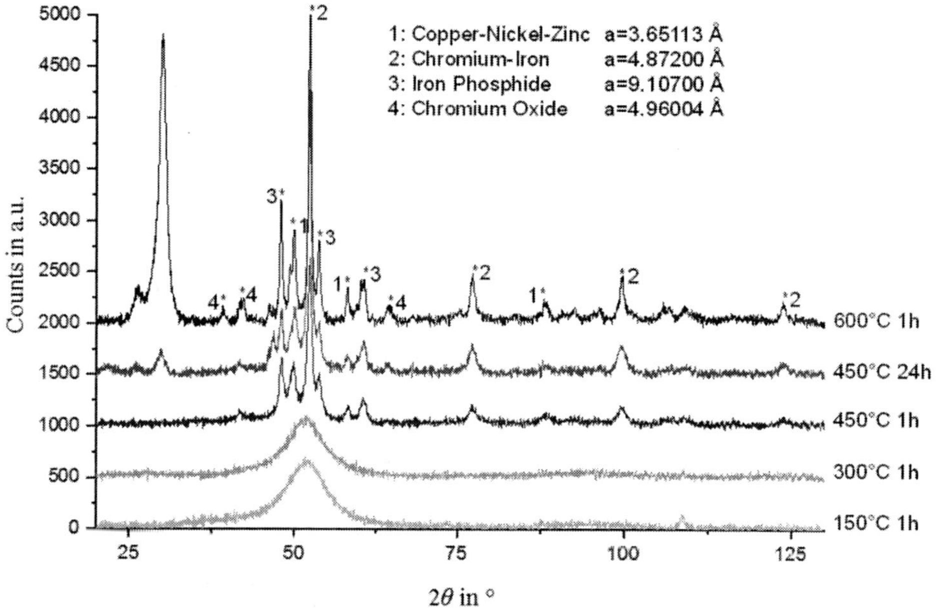

Figure 6. XRD measurements of electrodeposited iron-chromium layers in dependence of the heat-treated condition.

For the iron-nickel-chromium layers shown in figure 7, an X-ray amorphous or nanocrystalline structure for RT, 150 °C and 300 °C can be observed. The individual visible small peaks originate from the substrate. A monoclinic iron-nickel-chromium phase and an austenitic lattice can only be detected above a heat-treatment temperature of 450 °C. It is noticeable that the heat-treatment processes at 450 °C for one hour and for 24 hours are difficult to distinguish from each other. It can be concluded that the duration of the treatment has no influence on the type of phases, but the slightly reduced peak width indicates a further-progressed recrystallization of these phases. At a heat-treatment temperature of 600 °C, pronounced peaks appear which probably indicate grain growth. Bertero et al. also found that the annealing caused recrystallization. The mean crystallite size is projected at 20 - 25 nm [5]. The heat treatment probably leads to recrystallization of the iron-nickel-chromium austenite and the chromium oxide. Furthermore, chromium-oxide peaks are present at a heat-treatment temperature of 600 °C. Presumably, these compounds of carbon and oxygen only decompose above a temperature of 600 °C

IOP Conf. Series: Materials Science and Engineering 1147 (2021) 012003 doi:10.1088/1757-899X/1147/1/012003

and the oxygen forms chromium oxide with the chromium The large peak at $2\theta = 30°$ is also seen for the iron-nickel-chromium layers. Most likely, this can be assigned to carbon that is incorporated into the lattice. The peaks with the number one correspond to the substrate.

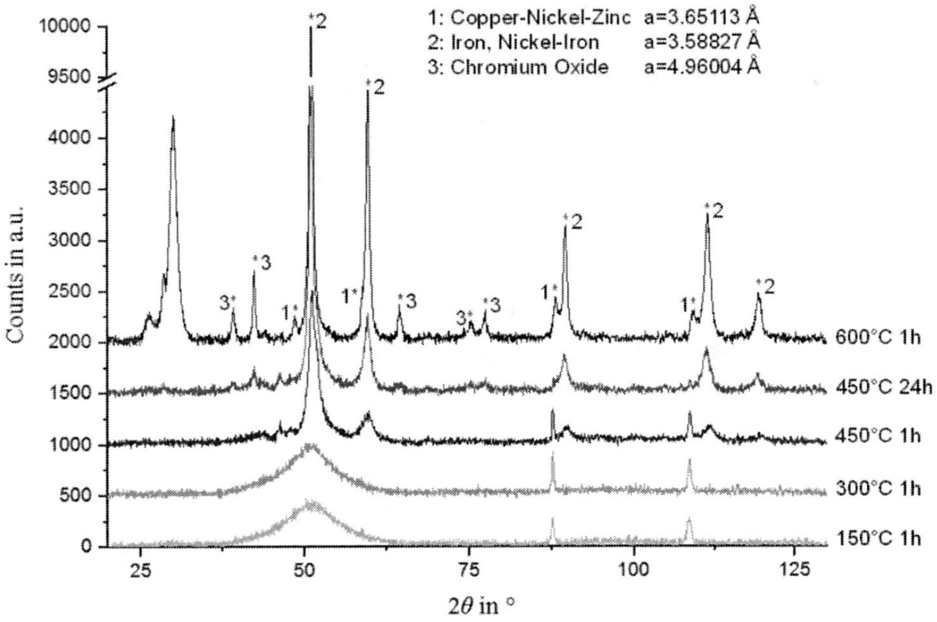

Figure 7. XRD measurements of electrodeposited iron-nickel-chromium layers in dependence of the heat-treated condition.

3.3. *Microstructure*

The deposited iron-chromium-nickel layers and iron-chromium layers show a uniform and bright surface. To investigate the tightness of the layers, the results of the investigations of the detection of microcracks are presented in figure 8. Clearly, it shows that heat treatment increases the number and the size of the microcracks in the iron-chromium layer as well as in the iron-nickel-chromium layers. The cracks run through from the surface to the substrate. This indicates an increase in the internal stresses of the layers due to heat treatment. When comparing the as-deposited iron-nickel-chromium samples with the ones that were heat-treated at 300 °C or 450 °C, it can be seen that the cracks are getting more pronounced with rising heat-treatment temperature. This observation can not be made for the iron-chromium samples. Previous similar work reports that cracks are only observed in layers thicker than 5 mm. However, the crack formations occurred a short time later or in the next few days after exposure to the atmosphere after the deposition [8]. It is assumed that the galvanic deposition of chromium (III) alloys creates high overpotentials, which lead to a rise in the pH value at the cathode and the incorporation of chromium hybrids. These chromium hybrids are hexagonal close packed (hcp) and unstable. Over time, they decompose into metallic chromium, which is body-centered cubic (bcc). During this phase transition, the volume of the chromium shrinks by approximately 15 %, which results in a cracking of the layer [5]. This process was accelerated and reinforced by a heat treatment, which resulted in increased crack growth and widening of the cracks. The results of this investigation support this theory because recrystallization only starts at a temperature of 450 °C. Therefore, the recrystallization does not contribute to crack growth at lower heat-treatment temperatures.

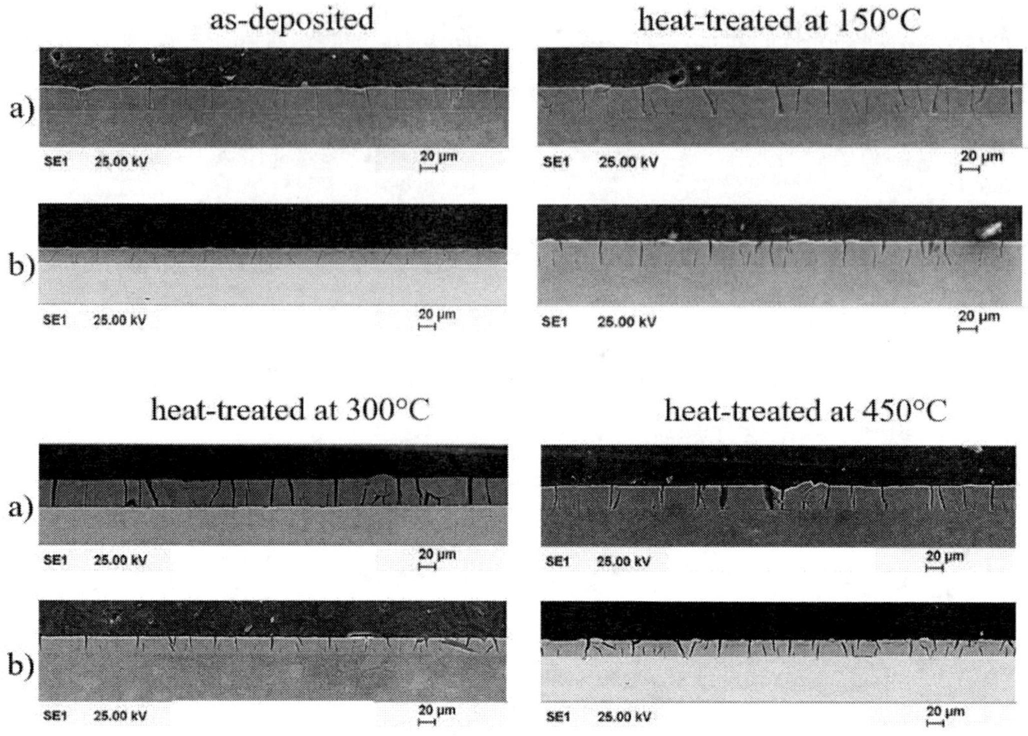

Figure 8. Scanning electron microscopy micrographs of cross-sections of electrodeposited a) Iron-chromium coatings and b) Iron-nickel-chromium coatings.

4. Conclusions

Galvanic depositions of iron-nickel-chromium and iron-chromium were examined. Heat treatments were carried out on these deposits in order to investigate their influence on the microstructure and the layer composition as well as the crystallinity of the layers. The appearance of the electrodeposited layers shows a uniform and bright surface with a silver-gray color which is suitable for decorative applications. The heat treatments had a negative effect on crack growth. Therefore, one can assume that this does not relieve any internal stresses in the layers, but rather intensify them. However, these internal stresses could have positive effects on wear behaviour and result in an increase in hardness. When considering the element analyses in combination with the results of the crystallinity, we assume that the additives used in the iron-nickel-chromium and the iron-chromium layers are incorporated into the layer and decompose at a temperature of about 600 °C. The released oxygen along with the chromium form chromium oxide. Moreover, there is no evidence that the heat treatment leads to a reduction of the co-deposited organic compounds. High proportions of carbon and oxygen were measured in both layers, the origin of which can be traced back to the additives used and the formation of passive layers due to the influence of the atmosphere can be largely excluded. X-ray amorphous layers were obtained for the layers at room temperature and the heat treatment at 150 °C and 300 °C. It has been observed that recrystallization already starts at a temperature of 450 °C. An increase in the holding time probably leads to a progressive recrystallization and thus improves ductility despite massive internal stresses. On the basis of future wearing and hardness tests it is possible to examine the influence of heat treatment on the tribological properties of the layers. Further investigations can prove if an increase in the holding time also enables a reduction in the amount of the co-deposited organic compounds.

Acknowledgements
The authors gratefully acknowledge the funding of the project sub-Music–Reko-HCE by Federal Ministry for Economic Affairs and Energy (BMWE) and the Central Innovation Programme for small and medium-sized enterprises (SMEs) under grant number 16KN074320. We also would like to

acknowledge the support of the VDI technology centre and all project partners (Chemopur H. Brand GmbH, LKS Kronenberger GmbH, Jürgen Voigt Meisterwerkstatt für Metallblasinstrumente and the Musicon Valley e.V.) and thank Marc Pügner for the XRD measurements and Steffen Clauß for the EDXS and scanning electron microscopy measurements.

References

[1] Alfort H J 2016 Wann kommt das Verbot für Chrom(VI)? *Journal für Oberflächentechnik* ***56*** 68-71

[2] Pellerin C and Booker S M 2000 Reflections on hexavalent chromium: health hazards of an industrial heavyweight *Environmental Health Perspectives* **108** A402-A407

[3] Schade C and Käszmann H 2013 Galvanisch abgeschiedenes Chrom - Ein Blick auf die mechanischen Eigenschaften *WOMag* 1-7

[4] Kalweit A, Paul C, Peters S and Wallbaum R *2006 Handbuch für Technisches Produktdesign* (Berlin Heidelberg: Springer Verlag)

[5] Bertero E, Manzano C V, Pellicer E, Sort J, Ulfig R M, Mischler S, Michler J and Philippe L 2019 Green' Cr(III)-glycine electrolyte for the production of FeCrNi coatings: electrodeposition mechanisms and role of by-products in terms of coating composition and microstructure *RSC Advances* **9** 25762-25775

[6] Bertero E, Manzanoa C V, Bürki G and Philippe L 2020 Stainless steel- like FeCrNi nanostructures via electrodeposition into AAO templates using a mixed-solvent Cr(III)-based electrolyte *Materials and Design* **190** 1-7

[7] Hasegawa M, Yoon S, Guillonneau G and Zhang Y 2014 The electrodeposition of FeCrNi stainless steel: microstructural changes induced by anode reactions *Physical Chemistry Chemical Physics* **16** 26375-26384

[8] Philippe L, Heiss C and Michler J 2008 Electroplating of Stainless Steel *Chem. Mater.* **20** 3377-3384

[9] Baldenebro-Lopez R J, Gomez-Esparza C D, Corral-Higuera R, Arredondo-Rea S P, Pellegrini-Cervantes M J, Ledezma-Sillas J E, Martinez-Sanchez R and Herrera-Ramirez J M 2015 Influence of Size on the Microstructure and Mechanical Properties of an AISI 304L Stainless Steel - A Comparison between Bulk and Fibers *Materials* **8** 451- 461

[10] Brongersma S H, Richard E, Vervoort I, Bender H, Vandervorst W, Lagrange S, Beyer G and Maex K 1999 Two-step room temperature grain growth in electroplated copper," *Journal of Applied Physics* **86** 3642-3645

IOP Conf. Series: Materials Science and Engineering 1147 (2021) 012004 doi:10.1088/1757-899X/1147/1/012004

Electrolyte design and characterization of REACh-compliant Zn-W and Zn-W-Cu electrodeposits

R Morgenstern[1*], M Müller[2], D Höhlich[1], T Mehner[1] and T Lampke[1]

[1] Materials and Surface Engineering group, Chemnitz University of Technology, Erfenschlager Straße 73, 09125 Chemnitz
[2] 3D-Micromac AG, Technologie-Campus 8, 09126 Chemnitz

*e-mail: roy.morgenstern@mb.tu-chemnitz.de

Abstract. Within this work, an entirely new electrolyte system for the electrochemical deposition of REACh-compliant Zn-W and Zn-W-Cu alloy layers was developed based on thermodynamic calculations. The required constants of complex formation were determined using potentiometric titrations. The layers were analyzed with regard to the layer microstructure (including coating thickness, chemical and phase composition), the optical properties (including gloss and color) and the passivation and corrosion behaviors. The Zn, Zn-W, and Zn-W-Cu layers are characterized by a decorative appearance, high uniformity, a nanocrystalline structure and natural passivation during the exposure in the climate chamber. Due to their particularly favorable optical properties and their pronounced natural passivation, the Zn-W alloy layers exhibit a great potential for industrial applications.

1. Introduction

Regarding the risk of a possible restriction in the use of Zn-Ni alloy layers as a result of the REACh regulation, there is a need to establish REACh-compliant alternatives. At the present time, Zn-Fe alloy layers are primarily used as a substitute for Zn-Ni, although these are inferior to Zn-Ni layers in terms of corrosion resistance (according to DIN 50962) and layer hardness [1]. In the scientific literature, there is also some information available on Zn-Mn [2, 3], Zn-Cr [4] and Zn-Fe-Cr [5]. However, those alloys have not been marketable so far. Alloy deposits of Cu and Zn have a long tradition. In recent years, research focused on the development of cyanide-free Cu-Zn electrolytes. Alternative complexing agents were investigated in order to align the strongly different potentials for the reduction of Cu and Zn ions and to enable the deposition of alloys with a high Zn content from aqueous solutions. Citrate is used as a complexing agent both in strong alkaline [6] and in neutral to weak acidic electrolytes [7]. In similar systems, ternary Cu-Zn-Sn [8] and Cu-Zn-Mo [9] alloy deposits were obtained. Pyrophosphate was used as a complexing agent at approximately neutral pH values [10, 11]. Furthermore, the deposition of Cu-Zn alloys was investigated in gluconate [12] and EDTA [13] electrolytes at neutral to moderately acidic pH values. All of these studies on the Cu-Zn deposition from cyanide-free baths have in common that maximum Zn contents of 30 to 40 % were targeted and Zn contents of more than 50 % were rarely achieved. The deposition of the so called "white brass" with a Zn content of up to 90 % was only performed in cyanide electrolytes [14]. To our knowledge, Zn alloys with a Cu content of less than 5 % were not reported in the literature so far.

Theoretically, in the alloy systems mentioned above, all elemental compositions could be achieved under the appropriate conditions, ranging from 100 % of Zn to 100 % of the alloying elements. However, in case of the so called "induced co-deposition" [15], ignoble elements such as W can be co-deposited as alloying elements, although they cannot be deposited as a monometallic layer. In particular, Ni-W alloy deposits with up to about 25 % W have gained interest as potential alternatives to hard-chromium layers due to their high hardness [16]. In case of Fe-W-Zn, less than 10 % Zn were incorporated in order to improve the passivation behavior and the corrosion resistance of the Fe-W alloy deposit [17]. It was stated that the co-deposition of tungsten was related to the iron deposition and that the Zn incorporation took place independently [17]. Also, in case of an alkaline Zn-Ni-W electrolyte, the co-deposition of 4 % W could have been related to the deposition of 18 % Ni [18]. Nakano et al. observed no W co-deposition in case of an acidic electrolyte that contained only Zn and W ions [19]. Thus, no evidence was found that W ions can be co-deposited with Zn ions.

It can be concluded from the literature research that both Zn-Cu alloys from cyanide-free electrolytes with a low Cu content and binary Zn-W are potential alternatives to Zn-Ni coatings, but their properties have not been investigated yet. This article demonstrates the development of an electrochemical process for the deposition of Zn-rich Zn-Cu alloys from a cyanide-free bath using thermodynamic calculations. Furthermore, tungsten ions are added to the electrolyte in order to generate Zn-W and Zn-W-Cu alloys. As these alloys are entirely new, the microstructural and optical properties as well as the passivation behavior are described for the first time.

2. Experimental and materials

The basic thermodynamic electrolyte design was conducted for Zn and Cu ions and different complexing agents. The equilibrium state of the electrolytes was described by the law of mass action. Furthermore, the optimal pH working range of the electrolyte was calculated by considering the protonation behavior. The simplified chemical equations are summed up in table 1. The complex equilibrium equations were mathematically solved using a non-linear equation system. A MatLab routine was created for this purpose.

Table 1. Simplified thermodynamic equations and the resulting equilibrium constants *K*.

law of mass action	solubility behavior	protonation equilibrium
$[\mathrm{Me}] + [\mathrm{KB}] \rightleftarrows [\mathrm{KBMe}]$	$[\mathrm{Me}\,]^{2+} + [\mathrm{OH}]^{-} \rightleftarrows [\mathrm{Me(OH)_n}] \downarrow$	$[\mathrm{H}]^{+} + [\mathrm{KB}] \rightleftarrows [\mathrm{HKB}]$
$K = \frac{[\mathrm{KBMe}]}{[\mathrm{Me}^{2+}] + [\mathrm{KB}]}$	$K = [\mathrm{Me}^{2+}] \cdot [\mathrm{OH}^{-}]^{n}$	$K = \frac{[\mathrm{HKB}]}{[\mathrm{H}^{+}] + [\mathrm{KB}]}$

The required thermodynamic equilibrium constants *K* were determined using the potentiometric titration, which was first described by Martell [20]. A modified iterative algorithm was used to determine the equilibrium constants of unknown systems:

1) Definition of default equilibrium constants and proton concentrations,
2) definition of the law of mass action,
3) solving the equation using the Newton's method,
4) error optimization,
5) starting the next iteration step with the updated parameters or ending the process.

The experimental setup is shown in figure 1. The potentiometric acid-base-titrations were carried out in the pH range from 2 (adjusted by the addition of HCl or H_2SO_4) and 12 (incremental addition of NaOH). The experiments were carried out at a constant concentration of [KB] = 0.01 mol/l and different [KB]:[Me] ratios ranging from 1:0 to 1:5.

As a result of the electrolyte design, the basic Zn electrolyte consists of 27.26 g/l (0.2 mol/l) $ZnCl_2$ and 67 g/l (0.5 mol/l) $C_4H_6O_5$ (malic acid). The additives $C_9H_{10}O_3$, $C_{19}H_{42}BrN$ and H_3BO_3 were used in order to improve the coating quality and the process stability. The elements W and Cu were added as 2.94 g/l (0.01 mol/l) Na_2WO_4 and 0.5 g/l (0.002 mol/l) $CuSO_4 \cdot 7H_2O$. For further characterization, cold rolled sheets of the unalloyed steel CR1 (according to ISO 3574) were electroplated at a current density of 1 A/dm² at 30 °C and pH 6 for 60 minutes. A soluble Zn anode was used as the counter electrode.

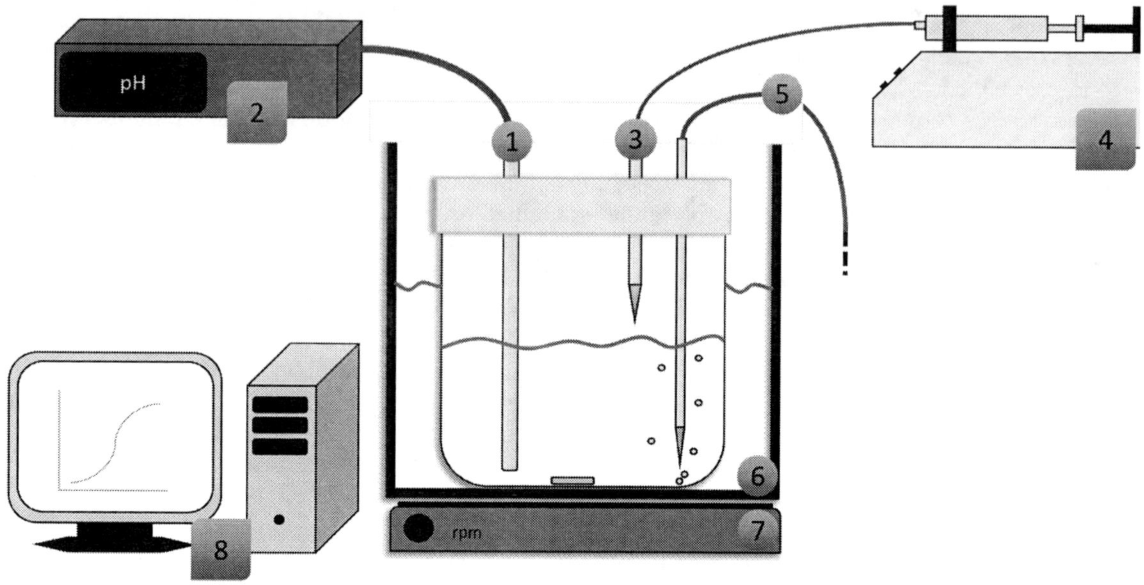

Figure 1. Experimental setup of the potentiometric titration measurements: 1 – pH electrode, 2 – pH recorder, 3 – dosing unit, 4 – syringe pump, 5 – argon inflow, 6 – heating unit, 7 – magnetic stirrer, 8 – PC.

As a standard procedure, the thickness and the chemical composition of the coatings were determined using the X-ray fluorescence analysis (XRF, FISCHERSCOPE - X-RAY XAN, Fischer, Sindelfingen, Germany). In favor of a higher precision, the chemical composition was also determined after chemical digestion using the inductively coupled plasma optical emission spectroscopy (ICP-OES, Optima 8300, Perkin Elmer, Waltham, USA). The microstructural analysis was performed at the metallographically prepared coating cross-section using optical microscopy (OM, GX51, OLYMPUS, Shinjuku, Japan) and scanning electron microscopy (SEM, LEO-1455VP, Zeiss, Oberkochen, Germany). Phase analyses and crystallite-size measurements were conducted utilizing X-ray diffraction (XRD, D8 DISCOVER, Bruker AXS, Karlsruhe, Germany). In order to quantitatively describe the optical appearance of the coatings, reflectometric gloss measurement (micro-TRI-gloss µ, BYK Gardner, Geretsried, Germany) was performed under an incidence angle of 20° as defined in DIN 67530 for high reflecting surfaces. The color of the coatings was measured using a spectrometer (spectro-guide, BYK Gardener, Geretsried, Germany) with the illuminant D65 and the 10° standard observation angle and visualized using the CIE Lab color space. Different commercial coatings were used as references in order to illustrate the gloss and color.

Prior to corrosion testing, the samples were exposed to artificial weathering in a climate chamber (KB 300, Liebisch, Bielefeld, Germany) for up to 21 days, which enabled the formation of a natural passive layer. Each climate cycle consisted of a humid phase (95 % relative humidity for 2 h at 28 °C) and a dry phase (60 % relative humidity for 4 h at 21 °C). The electrochemical corrosion measurements were performed by the Federal Institute for Materials Testing and Research (BAM) using a gel electrolyte based on agar, which does, in contrast to conventional aqueous electrolytes, not chemically dissolve the passive layer during the measurement. The testing conditions are described in more detail in [21]. For each coating, the open circuit potential (OCP) was recorded for 5 min before performing potentiodynamic polarization tests. As an indicator of the corrosion resistance of the passive layer, the polarization resistance was determined using electrochemical impedance spectroscopy (EIS).

3. Results and discussion

3.1. Process design

The equilibrium state of free metal ions and metal ions, which are bound in complexes, can be calculated using the thermodynamic law of mass action. It is generally assumed, that the metal deposition can only take place by the reduction of free metal ions. With increasing complex stability (high equilibrium constants *K*), the concentration of free metal ions decreases. Consequently, the Nernst potential for the metal deposition is shifted to more negative values. By the use of selective complexing agents, the Nernst potentials of the generally more noble metal Cu and the ignoble metal Zn can be aligned, which enables the simultaneous deposition of both metals. By considering the protonation equilibrium and the solubility of the complexes, optimal pH working ranges can be predicted, which are characterized by a high electrolyte stability and a low probability of precipitation and cementation.

As shown in figures 2 and 3, free metal ions are dominating at pH values below 2. Within a range of pH 4 to pH 9, malic acid forms stable complexes with both Zn and Cu ions and the precipitation of insoluble hydroxides can be suppressed. The free metal concentrations of both Zn and Cu are reduced due to complexation, which is generally related with an increasing throwing power of the electrolyte. The difference of the Nernst potentials of Cu and Zn is still higher than 1 V. The Cu concentration in the electrolyte is significantly lower than the Zn concentration. Hence, the deposition of Cu is limited due to the ion diffusion and Zn-rich alloy coatings can be achieved. Above pH 9, insoluble Zn and Cu hydroxides are formed. In order to ensure a high electrolyte stability and a sufficient complexation, the further experiments were conducted within the stability range of the malic acid complex at pH 6.

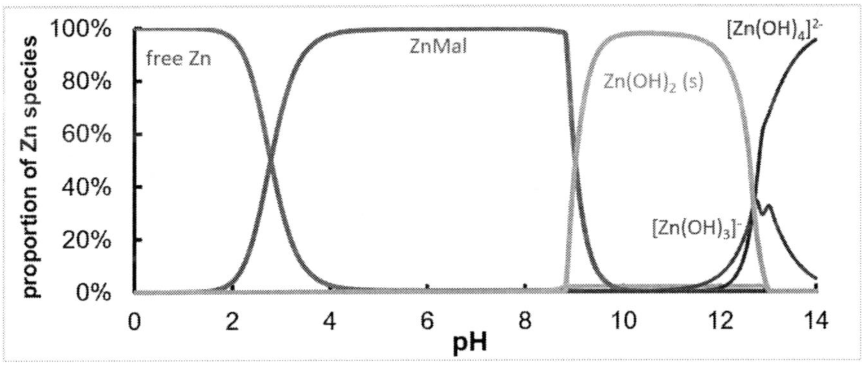

Figure 2. Zn species as a function of the pH value in 0.5 mol/l malic acid + 0.2 mol/l Zn.

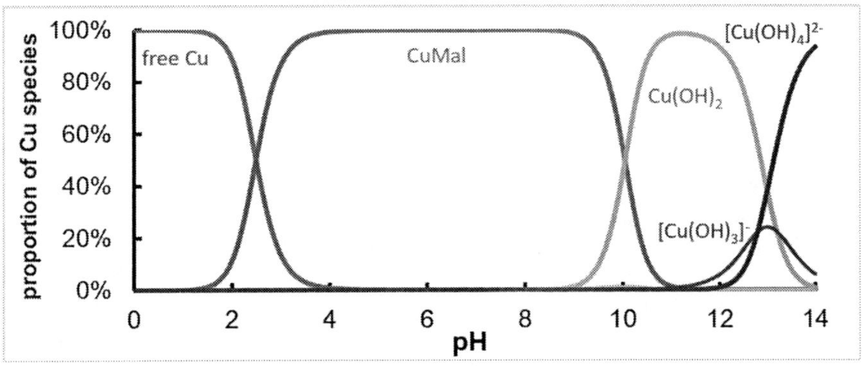

Figure 3. Cu species as a function of the pH value in 0.5 mol/l malic acid + 0.002 mol/l Cu.

3.2. Microstructure

Both the Zn-W and the Zn-W-Cu coatings are characterized by a smooth surface and a high uniformity of the coating thickness, as can be seen from figure 4. At increased magnification, some pores can be recognized within the Zn-W-Cu coatings (figure 5) with the aid of the BSD detector (material contrast). This indicates that the side-reaction of hydrogen evolution took place. However, the current efficiency of the metal deposition was still about 85 % at a current density of 1 A/dm^2, which is comparable to a commercial alkaline Zn electrolyte.

The ICP-OES measurements revealed a W content of about 0.7 % for the Zn-W alloy coating. The Zn-W-Cu alloy coating contains about 0.1 % W and 0.5 % Cu. All Zn and Zn alloy coatings from malic acid electrolytes exhibited a crystalline structure. No other phases could be detected by XRD besides the hexagonal Zn phase, due to the low contents of the alloying elements below 1 %. Because of the strong texture of the Zn alloy coatings, the crystalline size was only quantifiable at the preferred crystal plane. Each of the new Zn and Zn alloy layers exhibits a crystallite size of about 50 nm, which is similar to the crystallite size of a commercial bright Zn layer. The nano-crystalline microstructure is achieved by the addition of brighteners, which inhibit the crystal growth in thickness direction and it is beneficial with respect to the mechanical properties such as hardness and ductility.

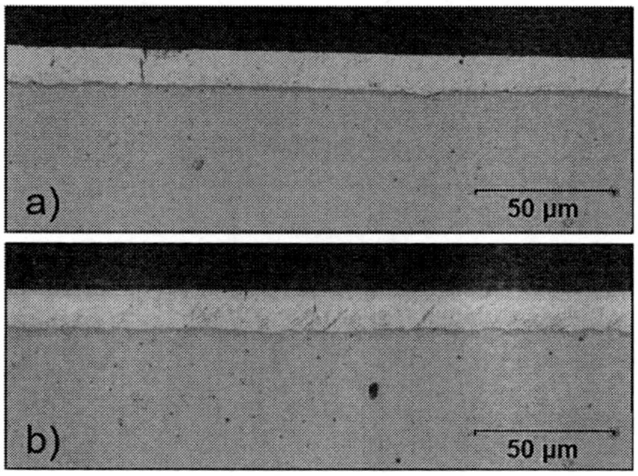

Figure 4. LM images of the cross-section of a) Zn-W and b) Zn-W-Cu layers.

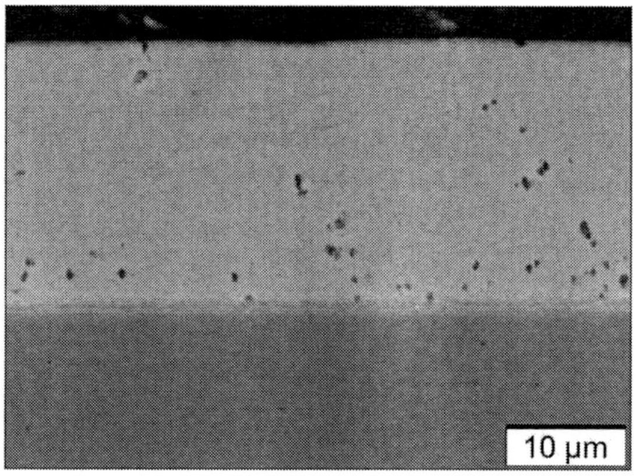

Figure 5. SEM image (BSD) of the coating cross-section of Zn-W-Cu.

3.3. Optical appearance

The optical appearance is one of the most important properties of Zn layers as they are being used for the corrosion protection of steel parts. Although all Zn and Zn-alloy layers from the malic acid electrolyte and the commercial Zn layer are characterized by the same crystallite size, they exhibit strongly different gloss values. This can be ascribed to the different levelling ability. In the case of the commercial Zn layer, there is only some geometric leveling of the substrate roughness, which results in a slight increase of the surface reflectance (figure 6). In contrast to this, the Zn and Zn-alloy layers exhibit a significantly higher gloss. This indicates that metal deposition preferably took place in deepenings which leads to a strong leveling of the surface roughness. The highest gloss of more than 900 gloss units (GU) was observed for Zn-W. This gloss level is almost comparable to a mirror finish, which is typically characterized by a gloss level of more than 1000 GU.

The CIE Lab color space is described by a cartesian coordinate system with the axis “a” ranging from -150 (green) to 100 (red), the axis “b” ranging from -100 (blue) to 150 (yellow) and the axis “L” ranging from 0 (black) to 100 (white). The standard illuminant D65 corresponds with a color temperature of 6504 K, which is comparable to the average daylight according to DIN 5033-7. As can be seen from figure 7, the brightness (L value) is similar for all Zn and Zn-alloy coatings. In comparison with the monometallic Zn reference coating from an alkaline electrolyte, the color of the Zn coating from the malic-acid electrolyte is shifted towards yellow. When considering both color and gloss, the appearance of the new Zn coating is comparable to conventional bright nickel deposits. The color of the Zn-W-alloy coating is comparable with the color of a Zn-Fe reference layer, which was deposited from a commercial alkaline electrolyte. In this case, the b-value is similar to the color of bright nickel and the a-value indicates a slight green color portion, which is comparable to Cr coatings from Cr(VI) electrolytes. Compared with the color of Zn-W coatings, Zn-W-Cu coatings are characterized by a color shift towards yellow and red due to the addition of Cu. All Zn and Zn-alloy coatings are characterized by a decorative appearance, which might be further optimized by a subsequent passivation treatment.

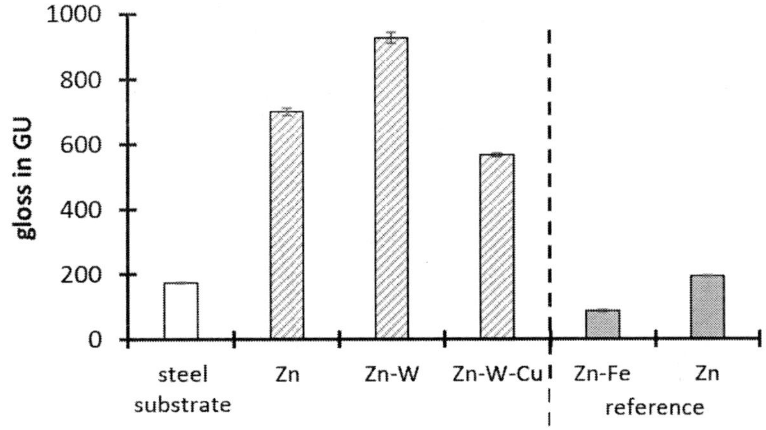

Figure 6. Gloss values of different Zn and Zn alloy coatings under an incidence angle of 20°.

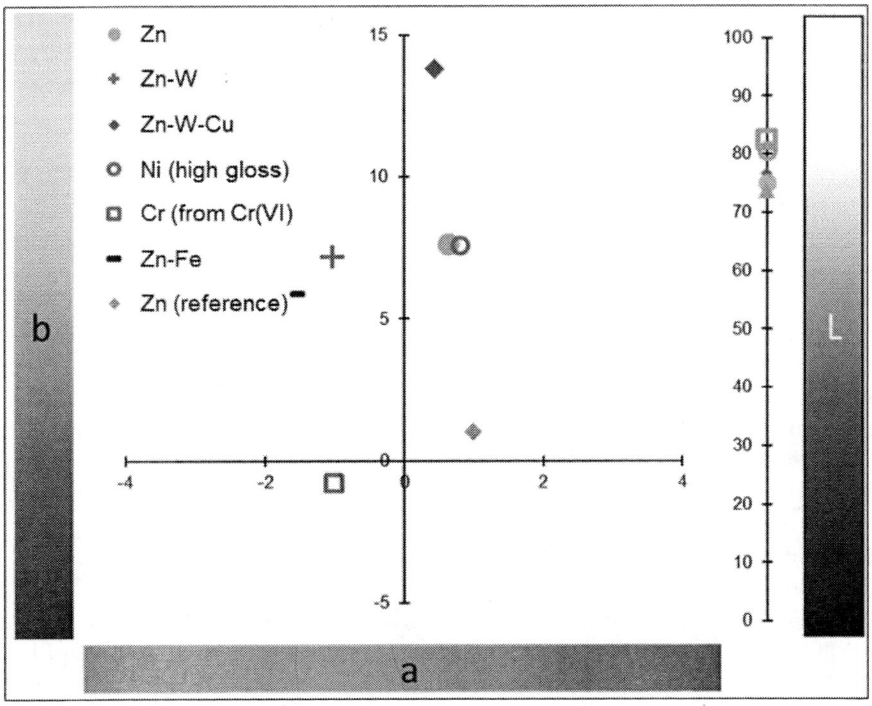

Figure 7. CIE Lab coordinate system showing the color of Zn, Zn alloy and reference coatings; illuminant D65, 10° standard observation angle.

3.4. Passivation behavior

The potentiodynamic polarization measurements, which were performed with the gel electrolyte, generally show a decrease of the corrosion current density with increasing exposition time in the climate chamber, as shown for the Zn-W alloy coating in figure 8. This already indicates that the corrosion rate, which is directly related to the corrosion current density by the Faraday's law, decreases due to the growth of the natural passive layer. In theory, the corrosion rate of a system also decreases inversely proportional to the increasing polarization resistance, which is obtained from EIS measurements [22]. This correlation can be verified (figure 9). The polarization resistance increases over the exposition time for each of the Zn and Zn-alloy layers. The highest polarization resistance was measured for Zn-W coatings, while the polarization resistance of Zn-W-Cu is almost at the same level compared to a monometallic Zn layer which was obtained from the malic acid electrolyte. This observation indicates that the polarization resistance, and consequently the corrosion resistance, strongly depends on the chemical composition of the Zn alloy coating. Even a W-content of 0.7 % (compared to 0.1 % W for Zn-W-Cu) obviously enhances the polarization resistance of the natural passive layer significantly.

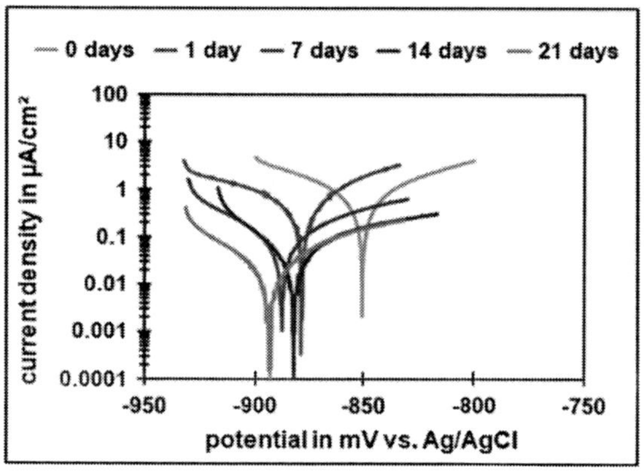

Figure 8. Potentiodynamic polarization curves of Zn-W layers after exposition in the climate chamber for up to 21 days.

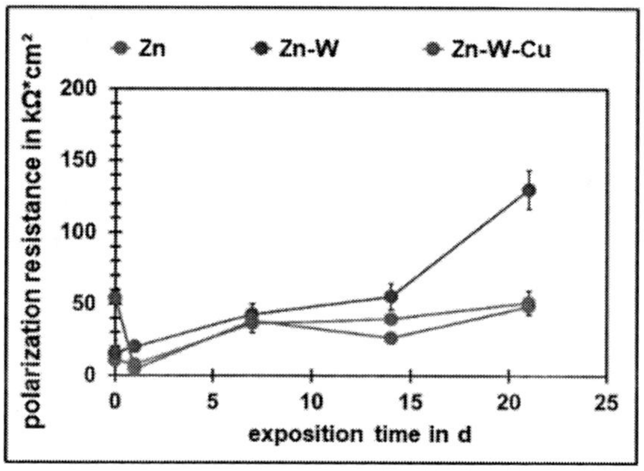

Figure 9. Development of the polarization resistance of Zn and Zn alloy coatings after exposition in the climate chamber for up to 21 days.

4. Conclusions

As potential alternatives to Zn-Ni coatings, REACh-compliant Zn-W and Zn-W-Cu alloy coatings were successfully deposited and characterized for the first time. In addition to the existing scientific literature, it was shown that Zn-rich Zn-Cu alloy coatings can be deposited from a cyanide-free bath and that the induced co-deposition of tungsten is not only related to the deposition of iron-group elements or Cu, but can also take place in combination with the Zn deposition. Within the stage of the development of a new electrolyte, which is based on the complexing agent malic acid, the calculation of the thermodynamically stable Zn and Cu species as a function of the pH value proved to be a valuable instrument. In this way, extensive experimental depositions at different pH values were not required. A sufficient complexation of the metal ions was achieved within the stability range of the malic acid complexes at pH 6 without the formation of insoluble hydroxides. Due to the addition of further electrolyte components, such as a brightener, a wetting agent and a pH buffer, nanocrystalline coatings with a high homogeneity and compactness were achieved. All Zn and Zn alloy coatings from the new malic acid electrolyte are characterized by a high gloss and a slightly yellowish color. The Zn-W coating exhibits the highest gloss and a color, which is comparable to already established alkaline Zn-Fe coatings. By applying electrochemical characterization techniques using gel electrolytes, it was observed that for all Zn and Zn alloy layers, natural passivation takes place during exposition in a climate chamber. However, the corrosion resistance strongly depends on the chemical composition of the Zn alloy coating. It turned out

IOP Conf. Series: Materials Science and Engineering 1147 (2021) 012004 doi:10.1088/1757-899X/1147/1/012004

that even a small W content of 0.7 % is suitable in order to significantly raise the polarization resistance of the natural passive layer. From the perspective of both the decorative appearance and the pronounced natural passivation, the Zn-W layers have a high potential for the substitution of Zn-Ni-alloy coatings in industrial applications.

Acknowledgments

The authors gratefully acknowledge funding by the German Federal Ministry of Education and Research (BMBF) and the project carrier VDI Technologiezentrum GmbH within the joint project 13XP5031E. The support of S. Kaufmann, P. Seidel and M. Pügner is gratefully acknowledged.

References

[1] Hedge A C, Venkatakrishna K and Eliaz N 2010 Electrodeposition of Zn–Ni, Zn–Fe and Zn–Ni–Fe alloys *Surf. Coat. Tech.* **205** 2031–2041
[2] Sylla D, Savall C, Gadouleau M, Rebere C, Creus J and Refait P 2005 Electrodeposition of Zn–Mn alloys on steel using an alkaline pyrophosphate-based electrolytic bath *Surf. Coat. Tech.* **200** 2137–2145
[3] Wykpis K, Bierska-Piech B and Kubisztal J 2014 Electrodeposition of Zn–Mn coatings from a sulphate bath in the presence of complexing additives *Surf. Interface Anal.* **46** 740–745
[4] Boiadjieva T, Kovacheva D, Petrov K, Hardcastle S, Sklyarov A and Monev M 2004 Electrodeposition, composition and structure of Zn–Cr alloys *J. Appl. Electrochem.* **34** 315–321
[5] Karahan I H 2008 Preparation and characterization of the electrodeposited Zn-Cr-Fe ternary alloys *Optoelectron. Adv. Mat.* **2** 788–791
[6] Assaf F H, Abd El Rehim S S, Mohamed A S and Zaky A M 1995 Electroplating of brass from citrate-based baths *Indian J. Chem. Techn.* **2** 147–152
[7] Silva F L G, do Lago D C B, D'Elia E and Senna L F 2010 Electrodeposition of Cu–Zn alloy coatings from citrate baths containing benzotriazole and cysteine as additives J. *Appl. Electrochem.* **40** 2013–2022
[8] Salhi Y, Cherrouf S, Tellal J and Cherkaoui M 2020 Electrodeposition of Cu-Zn-Sn coating in citrate medium *Mediterr. J. Chem.* **9** 456–467
[9] Santos H S, Correa A A, Gromboni M F and Mascaro L H 2019 Electrodeposition study of the Cu-Zn-Mo system in citrate/sulfate medium *Ecletica Quim J.* **44** 26–35
[10] Senna L F, Díaz S L and Sathler L 2003 Electrodeposition of copper–zinc alloys in pyrophosphate-based electrolytes *J. Appl. Electrochem.* **33** 1155–1161
[11] Senna L F, Diaz S L and Sathler L 2005 Hardness Analysis and Morphological Characterization of Copper-Zinc Alloys Produced in Pyrophosphate-Based Electrolytes *Mater. Res.* **8** 275–279
[12] Survila A, Mockus Z, Kanapeckaitė S and Stalnionis G 2013 Kinetics of zinc and copper reduction in gluconate–sulfate solutions *Electrochim. Acta* **94** 307–313
[13] Redjechta A, Loucif K, Mentar L, Khelladi M R and Beniaiche A 2014 Electrodeposition and Characterization of Cu-Zn Alloy Films Obtained From a Sulfate Bath *Mater. Technol.* **48** 221–226
[14] Weiner R 1943 Verfahren zur Erzeugung galvanischer Niederschläge auf Zinklegierungen *D.R.P.* 741 131
[15] Brenner A 1963 *Electrodeposition of Alloys* vol 2 (Cambridge: Academic Press) 347–412
[16] Allahyarzadeh M H, Aliofkhazraei, M, Rezvanian, A R, Torabinejad, V, and Sabour Rouhaghdam, A R 2016 Ni-W electrodeposited coatings: Characterization, properties and applications *Surf. Coat. Tech.* **307** 978–1010
[17] Park J H, Hagio T, Kamimoto Y and Ichino R 2020 Electrodeposition of a Novel Ternary Fe–W–Zn Alloy: Tuning Corrosion Properties of Fe–W Based Alloys by Zn Addition *J. Electrochem. Soc.* **167** 132508

[18] Popczyk M and Budniok A 2010 Structure and Corrosion Resistance of Zn-Ni and Zn-Ni-W Coatings *Mater. Sci. Forum* **636-637** 1042–1046
[19] Nakano H, Oue S, Taniguchi S, Kobayashi S and Fukushima H 2008 Effects of Small Amounts of Molybdenum, Tungsten, or Tin Additives on the Morphology and Orientation of Electrodeposited Zinc *ISIJ Int.* **48** 634–639
[20] Martell A E and Chaberek S 1954 Use of Chelating Agents as Reagents in Titrimetric Analysis *Anal. Chem.* **26** 1692–1696
[21] Babutzka M and Heyn A 2017 Dynamic tafel factor adaption for the evaluation of instantaneous corrosion rates on zinc by using gel-type electrolytes, IOP conf. *ser., Mater. sci. eng.* **181** 012021
[22] Mehner T, Morgenstern R, Scharf I and Lampke T 2017 Zeitaufgelöste Korrosionsprüfung mittels elektrochemischer Impedanzspektroskopie *Womag* **5** 14–17

IOP Conf. Series: Materials Science and Engineering 1147 (2021) 012005 doi:10.1088/1757-899X/1147/1/012005

Conversion layers by plasma-electrolytic oxidation of aluminum in acrylate and benzoate electrolytes

R Morgenstern[1*], O Selyshchev[2], T Mehner[1], T Lampke[1], D R T Zahn[2], W A Goedel[3] and J Schreckenbach[3]

[1] Materials and Surface Engineering, Chemnitz University of Technology, Erfenschlager Straße 73, 09125 Chemnitz
[2] Semiconductor Physics, Chemnitz University of Technology, Reichenhainer Straße 70, 09126 Chemnitz
3 Physical Chemistry, Chemnitz University of Technology, Straße der Nationen 62, 09111 Chemnitz

*e-mail: roy.morgenstern@mb.tu-chemnitz.de

Abstract. Within this work, aluminum is oxidized via plasma-electrolytic oxidation (PEO) in the presence of organic substances, including reactive monomers. The aim of this approach is to generate polymers and simultaneously bind them to the surface of the freshly generated oxide layers. For this purpose, sheets of aluminum were immersed into electrolytes that comprised either 4 % of sodium acrylate or 10 % of sodium benzoate. The aluminum sheets were oxidized by anodic pulse current at 0.25 A/cm² for 30–90 s. By this process, predominantly oxidic conversion layers were produced. The presence of the monomers in the electrolyte influenced the passivation and discharge behavior and finally the microstructure of the layers. It further gave rise to organic material in the layer. In particular, infrared attenuated total reflection spectroscopy (ATR-IR) and X-ray photoemission spectroscopy (XPS) show that layers which were generated in the presence of acrylate or benzoate comprised -C-C-, - C=C-, -C-O, and C=O bonds.

1. Introduction

In general, adhesion-promoting surface structures or interlayers have to be used in order to provide a strong adhesion between the metallic and polymer counterparts of hybrid material compounds [1]. In the case that the polymer surface is newly formed during the compound production (*e.g.* by injection molding) or at least reshaped (*e.g.* by hot pressing), the adhesion-promoting structure or layer can only be applied to the metallic partner. A wide variety of mechanical or chemical surface treatments and coatings has been discussed in the literature, whereby the highest adhesion strength of thermally joined aluminum-polyamide hybrid laminates was found for surface structures with pronounced undercuts in general [1] and for plasma-electrolytic oxidation (PEO) in particular [2]. This process enables the generation of highly adherent oxide-ceramic conversion layers due to the reaction of the locally molten substrate material and the anodically generated oxygen in the direct vicinity of microarc discharges [3]. PEO layers exhibit an open-pored surface, which is suitable for the mechanical interlocking of polymers and therefore provides an excellent adhesion of polymer counterparts on PEO-treated aluminum and magnesium alloys [4-6]. Moreover, the PEO layers are generally characterized by a compact electrically insulating barrier layer at the substrate-oxide interface, which is suitable for the protection of susceptible substrates in corrosive environments [7]. Due to the electrical insulation, oxidic conversion layers also prevent contact corrosion of base metals in the presence of more noble carbon fibers [8].

To the best of our knowledge, there was no attempt so far to create thin polymer films on structured or coated metals in order to promote the chemical adhesion of the polymer counterpart and to further increase the adhesion strength. The innovative approach of this investigation comprises the production of an adhesion promoting porous oxide surface, which is covered with a thin polymer film. Besides the formation of an oxide-ceramic conversion layer at high local temperature, light emission and radicals are associated to the microarc discharges. Similar to conventional plasma polymerization, the radicals shall be used to initiate the polymerization of monomers, which are dissolved in the electrolyte. The polymerization of the reactive species might be initiated at the PEO surface (grafting from) or in the electrolyte (grafting to). The aim of this research work is to illustrate the PEO-layer topography and to prove the existence of adherent organic compounds.

2. Materials and experimental methods

Pure aluminum (Al99.9) sheets with a sample area of 4 cm^2 were used as substrates. As a pretreatment, the samples were pickled into a solution of 10 % by weight of NaOH in water for 2 minutes, rinsed 2 times with distilled water for 30 seconds, one time with isopropyl alcohol, and dried in air. The PEO electrolytes were aqueous solutions of 4 % by weight of sodium acrylate ($H_2CCHCOONa$) or 10 % by weight of sodium benzoate (C_6H_5COONa). An aqueous solution of 5 % NaOH was used to produce reference PEO layers. The PEO experiments were conducted in a beaker (electrolyte volume: 250 ml) using rectangular pulsed current with a peak current density of 0.25 A/cm^2 and a pulse frequency of 50 Hz at room temperature. The process duration was limited to 60 s as the early low-energy micro-arcs tend to be more suitable for the plasma-induced polymerization of monomers. The sample acted as the anode; a platinum electrode was used as the cathode. The bath was agitated by a magnetic stirrer. After the PEO treatment, the samples were rinsed by distilled water for 100 s.

Prior to the scanning electron microscopy (SEM) investigations, the samples were rinsed in pure water, a mixture of water and isopropyl alcohol (50:50), and pure isopropyl alcohol in order to dissolve soluble compounds off the surface and to ensure that only insoluble and adherent compounds remained at the sample surface. Subsequently, the samples were dried at 55 °C in air for at least 4 h and again dried at 55 °C in rough vacuum (> 1 mbar). Overview images were captured and energy-dispersive X-ray spectroscopy (EDX) was performed at a working distance of 14.5 mm and an amplification voltage of 25 kV using the SEM LEO1455VP (Zeiss). More detailed views of selected PEO surfaces were taken at a working distance of 3 mm and an amplification voltage of 1 kV using a field emission (FE) SEM NEON40EsB (Zeiss).

Infrared attenuated total reflection spectroscopy (ATR-IR) was conducted using a Nicolet 6700 FT-IR (Thermo Scientific). The sample chamber is additionally equipped with an ATR unit (type MIRacle-ATR, PIKE Technologies), which is permanently flushed with dry, CO_2-free air and enables the recording of spectra in ATR mode. A single reflection using a diamond ATR crystal was used. The ATR-IR spectra were recorded using 20 scans at a resolution of 4 cm^{-1}. The ATR-IR data were analyzed by means of the software Omnic7.4.

X-ray photoemission spectroscopy (XPS) measurements were performed with an ESCALAB 250Xi (Thermo Scientific) photoelectron spectrometer equipped with a monochromated Al K_α X-ray source (photon energy 1486.7 eV). The system was operated at a base pressure of $5 \cdot 10^{-10}$ mbar. The spectra were collected in a constant analyzer pass energy mode, at pass energies of 200 and 20 eV for survey and high-resolution spectra (with a spectral resolution of 0.5 eV), respectively. A stripe of a half PEO-treated a half pristine Al sample was mounted on a metallic sample holder to achieve an electrical contact with the unprotected Al part. To minimize charging effects, a built-in charge compensation setup was used. Fitting of the spectra was performed with Voigt profiles using the Avantage software.

IOP Conf. Series: Materials Science and Engineering 1147 (2021) 012005 doi:10.1088/1757-899X/1147/1/012005

3. Results and discussion

3.1. Scanning electron microscopy investigations

In the early stage of a PEO process, a passive layer is formed under anodic polarization in a suitable substrate-electrolyte system and an oxygen-rich gas layer is formed around the substrate. After exceeding a critical potential gradient, microarc discharges are initiated that are ranging from the outer surface of the ionized gas layer (quasi-cathode) to the substrate. Regarding short-time PEO processes, the oxide surface typically exhibits an irregular porosity which represents the discharge channels of the fine dispersed early microarcs. Furthermore, the early PEO layers are typically characterized by a low thickness and a high content of amorphous phases. A topography characteristic for this stage is visible in the secondary electron (SE) image of aluminum treated with plasma-electrolytic oxidation in presence of sodium acrylate (figure 1). If oxidation progresses further, the oxide thickness increases and the surface topography becomes more complex, including an increasing number of undercuts. These surface characteristics can already be observed for samples, which were PEO-treated with benzoate addition to the electrolyte (figure 2), although the process time was the same. This indicates that different monomers influence the initial stage of the PEO process and the corresponding oxide microstructure in different ways. This can be explained by adsorbed benzoate molecules, which act as an electrical insulator, resulting in a faster increase of the potential gradient over the dielectric layer and an earlier ignition of the microarc discharges. Therefore, the plasma-induced layer formation took place for a longer time and an advanced oxide formation state was reached.

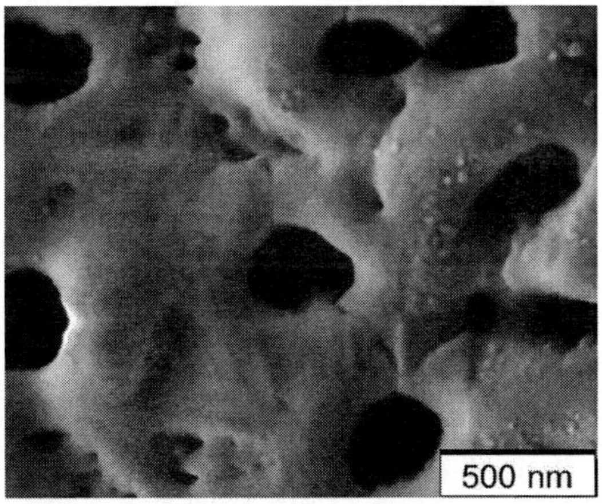

Figure 1. SE image of the sample surface after PEO on the presence of 4 % sodium acrylate showing the discharge channels of the initial microarcs

Figure 2. SE image of the sample surface after PEO in presence of 10 % sodium benzoate showing a more complex topography, being characteristic for an advanced PEO stage

Despite using a low amplification voltage of 1 kV, the existence of organic compounds or a thin polymer film could not be verified by SEM imaging. However, the presence of carbon-rich compounds was evidently shown by EDX spectroscopy. In the case of the PEO in aqueous NaOH solution, only the usual environmental trace contamination of carbon (approximately one percent of C) was detected at the oxide surface. After PEO in a solution of 4 % sodium acrylate (topography shown in figure 1), a molar fraction of 7 % C was measured, which could be explained by the presence of a thin adherent organic layer. An exceptionally high C-content of 66 % was detected after PEO in a solution of 10 % sodium benzoate (topography shown in figure 2). This could also indicate the thermal decomposition of organic molecules and the incorporation of carbon into the PEO layer. Clear evidence for the presence of organic compounds on the PEO surface can only be provided by the analysis of the atomic bonds using ATR-IR and XPS.

IOP Conf. Series: Materials Science and Engineering 1147 (2021) 012005 doi:10.1088/1757-899X/1147/1/012005

3.2. Attenuated total reflection infrared spectroscopy

The generation of organic species on the sample surface during the PEO process is demonstrated for the addition of sodium benzoate to the electrolyte in figure 3. To check for organic compounds, infrared light was guided within the diamond crystal by total reflection at its surfaces. The light excites vibrations of organic functional groups on the surface of the samples and in turn is attenuated via absorption at wavelengths that are characteristic for these groups. Thus, the resulting spectra show characteristic extinction peaks that allow for the identification of organic groups on the sample surface. The black graph shows the ATR-IR spectrum of a reference sample after PEO in pure NaOH solution without any additive (referred to as "NaOH"). In this case, no characteristic bindings of organic compounds can be recognized. Another reference sample was dipped in a PEO solution containing 10 % sodium benzoate after PEO in the pure NaOH solution (referred to as "dip benzoate"). As can be seen from the blue-colored graph in figure 3, this sample contains additional bands of the adherent, unaffected benzoate structures. Finally, the red-colored graph is related to a sample after PEO in an electrolyte containing 10 % sodium benzoate (referred to as "benzoate"), hence, reaction products of the sodium benzoate as well as unaffected but bound or strongly adsorbed benzoate are present at the sample surface. By comparing the spectra of "benzoate" and "dip benzoate", significant deviations of the molecular structure can be observed. In the case of the "benzoate" spectrum, a new carbonyl band (C=O) can be observed in the range of 1700 cm^{-1} (marked by "1" in figure 3), which also exhibits a small shoulder at higher wavenumbers (about 1740 cm^{-1}). The strong signals between 1600 and 1300 cm^{-1} can be attributed to different C-H, C=C, C-C, and C-O bonds. It is noteworthy that these bonds were shifted towards higher wavenumbers after the PEO process, which can be explained by currently unknown structural changes or polarization effects. A sharp new signal appears at about 1500 cm^{-1} (marked by "2" in figure 3) in the "benzoate" spectrum, which means that new bonds formed during the PEO process, which were not present in the original benzoate molecule. Additionally, new signals can be observed at 1300–1100 cm^{-1} (marked by "3" in figure 3) and further spectral changes are present at 1100–800 cm^{-1}.

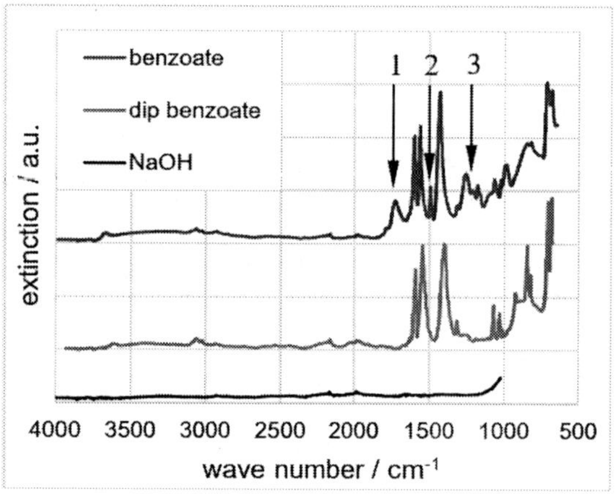

Figure 3. ATR-IR spectra corresponding to: PEO in NaOH solution without any additions ("NaOH"), dipping in 10 % sodium benzoate solution after PEO in NaOH solution ("dip benzoate"), and PEO in 10 % sodium benzoate solution ("benzoate") showing characteristic changes due to the PEO process

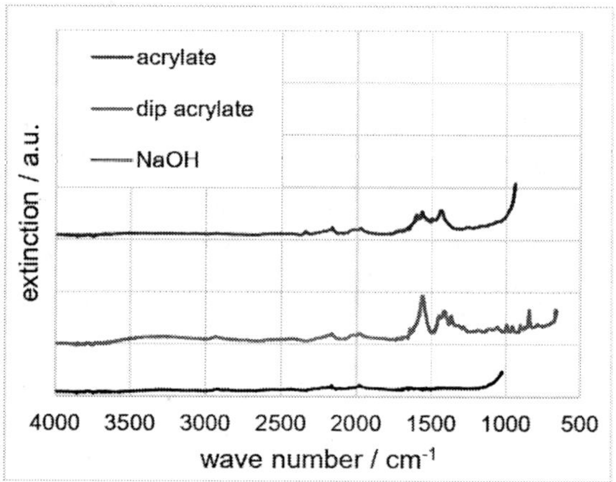

Figure 4. ATR-IR spectra corresponding to: PEO in NaOH solution without any additions ("NaOH"), dipping in 10 % sodium acrylate solution after PEO in NaOH solution ("dip acrylate"), and PEO in 4 % sodium acrylate solution ("acrylate") showing minor changes due to the PEO process

It is evident from the ATR-IR measurements of the "benzoate" and "dip benzoate" samples that organic compounds are present at the sample surface, which can be clearly differentiated from the original compounds in the PEO electrolyte according to their characteristic ATR-IR fingerprint.

Apart from the C=O bond, a clear allocation of the C-H, C=C, C-C, COH, and C-O bonds to distinct organic compounds is subject of further research. The interpretation of the ATR-IR spectra is further complicated as the possible bonds might be present in unaffected, adherent monomers as well as in newly developed polymers.

The ATR-IR spectra after PEO in NaOH solution ("NaOH"), dipping in 10 % sodium acrylate solution after PEO in NaOH solution ("dip acrylate"), and PEO in 4 % sodium acrylate solution ("acrylate") are shown in figure 4. In comparison to "dip acrylate", no additional ATR-IR bands can be observed in the "acrylate" spectrum, indicating that no new chemical bonds were formed during the PEO process. However, the two peaks around 1500 cm^{-1} are wider and less specific for the "acrylate" sample. This can be explained by the formation of polymers from the activated monomers. The "acrylate" sample was further investigated by XPS measurements in order to gain more detailed information on the molecular bonds.

3.3. X-ray photoemission spectroscopy

Composition and chemical states of the elements on the surface were additionally investigated by X-ray photoemission spectroscopy (XPS). Since the inelastic mean free path (IMFP) of electrons which are excited by Al K_α radiation is in the range of several nanometers, the information provided corresponds to a surface layer thinner than approximately 10 nm [9]. The survey spectrum of the "acrylate" sample (figure 5(a)) reveals carbon, oxygen, and aluminum with molar contents of 32.4, 44.0, and 22.0 %, respectively. This fact indicates the presence of an organic layer in addition to an aluminum compound. The surface additionally contains 0.7 % of nitrogen, 0.7 % of sodium, and 0.2 % of calcium, which are residues adsorbed on the surface during the electrochemical reaction.

In order to clarify the chemical states, the high-resolution XPS spectra were analyzed. The Al2p spectrum (figure 5(d)) consists of two spin-orbit doublets with $Al2p_{3/2}$ peaks at (74.3 ± 0.2) eV and (75.8 ± 0.2) eV, respectively. A reference spectrum taken on a clean metallic Al surface reveals an $Al2p_{3/2}$ peak at (71.5 ± 0.2) eV. Therefore, the chemical states at the PEO-treated Al surface (4 % sodium acrylate solution) are unambiguously attributed to cationic aluminum compounds. According to the chemical shifts, the higher binding-energy peak can be attributed to bulk-like Al_2O_3, while the stronger peak at lower binding energy corresponds to surface non-stoichiometric oxides. Complementary, in the O1s spectra (figure 5(e)), the peak at (530.8 ± 0.2) eV stems from aluminum oxides [9]. The O1s peak at (532.1 ± 0.2) eV corresponds to binding-energy values typical for C-O and O-C=O groups [10]. However, according to the literature, it may be attributed to surface Al hydroxides or carbonates formed due to exposure to the atmosphere as well [9]. In the C1s spectrum (figure 5(c)), three major peaks are revealed by fitting. The peak at (288.5 ± 0.2) eV stems from organic O-C=O functional groups or inorganic $Al(CO_3)_n$ compound. The peak at (285.5 ± 0.2) eV is typical for C-O or C-N groups. The peak at (284.6 ± 0.2) eV corresponds to aliphatic hydrocarbons. The high-resolution N1s spectrum (figure 5(b)) is fitted with two components at (399.4 ± 0.2) eV and (400.7 ± 0.2) eV, which usually correspond to free and protonated amine or amide groups, respectively [10]. Thus, the XPS study shows that the surface is composed of Al-containing cationic compounds, *e.g.* oxides, while metallic Al is not detected. The presence of carbon and oxygen indicates the adsorption of organic molecules. The sample contains sodium as a side product of the electrochemical process which was performed in the acrylate containing electrolyte.

As the detected O-C=O and C-O bonds are present both in the sodium acrylate monomer and possible polyacrylates, there is no clear evidence that polymerization took place during the PEO in 4 % sodium acrylate solution.

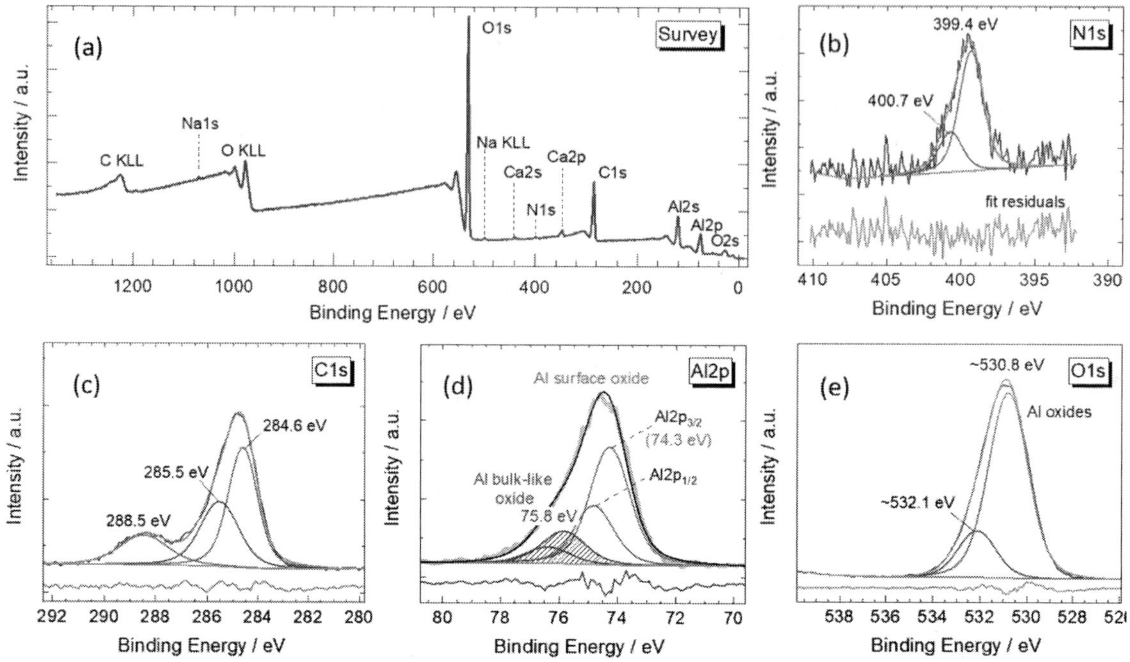

Figure 5. (a) Survey XPS spectrum of the "acrylate" sample showing the composition and the chemical states of the elements and (b-e) high resolution N1s, C1s, Al2p, and O1s spectra proving the existence of Al_2O_3 and organic compounds, respectively

4. Conclusions

Within this research work, orienting investigations were conducted, regarding the PEO in electrolytes that contain sodium acrylate or sodium benzoate. After a short process time of 60 s, the PEO layers exhibited a typical surface topography. There is some evidence that the presence of the organic compound in the PEO electrolyte influences the process initiation. More detailed studies of the passivation behavior in presence of absorbed monomers are planned. A significant increase of the carbon concentration on the top of the oxide was detected by EDX measurements. This might also be explained by the absorption of unchanged organic molecules and the existence of carbon-rich decomposition products or reaction products. However, by comparing the ATR-IR spectra of samples after PEO in the benzoate electrolyte and suitable reference samples, it was clearly shown that new organic bonds were formed during the PEO process. Thus, the feasibility of the plasma-induced activation of monomers and the generation of adherent, new organic compounds during the PEO process was successfully proved. The presence of organic compounds at the sample surface was also verified by XPS measurements after PEO in an acrylate-containing electrolyte, although they might not have formed during the PEO process. Additional research work will be performed in the future in order to identify the newly formed organic compounds and to evaluate their suitability for the adhesion promotion of polymers.

Acknowledgments

The authors gratefully acknowledge co-financing from the budget, which was decided by the members of the Saxon state parliament. The technical assistance of S. Clauß is also gratefully acknowledged.

References

[1] Saborowski E, Steinert P, Dittes A, Lindner T, Schubert A and Lampke T 2020 Introducing fractal dimension for interlaminar shear and tensile strength assessment of mechanically interlocked polymer-metal interfaces *Materials* **13** 2171

[2] Nestler D, Arnold S, Jung H, Wielage B and Kroll L 2014 Untersuchung geeigneter Oberflächenbehandlungsverfahren der Metallkomponente thermoplastbasierter hybrider Laminate *Wissenschaftliche Schriftenreihe Werkstoffe und Werkstofftechnische Anwendungen* **53** 208–216

[3] Simchen F, Sieber M, Kopp A and Lampke T Introduction to Plasma Electrolytic Oxidation—An Overview of the Process and Applications *Coatings* **10** 628

[4] Tang Y, Zhao X, Jiang K, Chen J and Zuo Y 2010 The influences of duty cycle on the bonding strength of AZ31B magnesium alloy by microarc oxidation treatment *Surf. Coat. Tech.* **205** 1789–1792

[5] Pan Y, Wu X, Huang G, Wu S, Sun S, Ye H and Zhang Z 2018 A new approach to enhancing interlaminar strength and galvanic corrosion resistance of CFRP/Mg laminates Compos. *Part A-App.* **105** 78–86

[6] Aliasghari S, Ghorbani M, Skeldon P, Karami H and Movahedi M 2017 Effect of plasma electrolytic oxidation on joining of AA 5052 aluminium alloy to polypropylene using friction stir spot welding *Surf. Coat. Tech.* **313** 274–281

[7] Sieber M, Simchen F, Scharf I, and Lampke T 2016 Formation of a Spinel Coating on AZ31 Magnesium Alloy by Plasma Electrolytic Oxidation *J. Mater. Eng. Perform. 25* 1157–1162

[8] Shan M, Guo K, Gou G, Fu Z, Yang B and Lu W 2019 Effect of anodizing on galvanic corrosion behavior of T300 CFRP/5083P-O Al bolted joints *Mater. Corros.* **71** 409–418

[9] Strohmeier B R 1990 An ESCA Method for Determining the Oxide Thickness on Aluminum Alloys *Surf. Interface Anal.* **15** 51–56

[10] Naumkin A V, Kraut-Vass A, Gaarenstroom S W and Powell C J 2012 *NIST X-ray Photoelectron Spectroscopy Database* (NIST)

IOP Conf. Series: Materials Science and Engineering 1147 (2021) 012006 doi:10.1088/1757-899X/1147/1/012006

Mathematical modeling and simulation of the dissolution of zinc anodes in industrial electroplating

S D Schwöbel*, V Meinhold, T Mehner and T Lampke

Materials and Surface Engineering Group, Institute of Materials Science and Engineering, Chemnitz University of Technology, D-09107 Chemnitz, Germany

E-mail: stephan-daniel.schwoebel@mb.tu-chemnitz.de

Abstract. The simulation of large-scale industrial electrodeposition of zinc is of major importance, since through the simulation, important data about the positioning of electrodes, thickness of layers, etc. can be generated. But the models used in practical applications indirectly assume small anodes and cathodes for which the electrical potential or the charge-transfer current are constant over the cathode, resp. anode. In this paper, a new type of model for the dissolution of zinc anodes is described. Furthermore, a numerical scheme to treat the model will be described and verified. The new model will be compared to a commonly known model. The paper gives an alternative model for the calculation of the current density on anodes and formulates a FEM for the approximation of Poisson equations on multiple domains with Robin-interfaces. Besides the application to the dissolution of zinc-anodes during electro-plating, the basic model and FEM can be applied to the dissolution of anodes made of a different material and to the modelling of corrosion processes.

1. Introduction

1.1. Motivation

Functional corrosion-protection coatings are one of the major industrial applications for electroplated Zn coatings. Besides the suitability of the produced coating for corrosion protection, it is also of major interest under which configurations these Zn coatings can be produced efficiently. In addition to the question of the evaluation and scaling of the actual plating process at the cathode, the electrochemical dissolution processes at the anodes have a major effect to the development of the whole process.

As industrial-scale experiments are very complicated and not affordable in practical applications, there is a major interest for simulations, for which measurements at a laboratory scale are sufficient. In this article, a mathematical model and a numerical scheme for the simulation of the dissolution of zinc anodes are presented, which only use impedance spectra as an input.

IOP Conf. Series: Materials Science and Engineering 1147 (2021) 012006 doi:10.1088/1757-899X/1147/1/012006

For the evaluation of the dissolved mass of zinc at a point in time m(t), one can use Faraday's law, cf. [1, 2], given through $m(t) = (M \cdot j \cdot t)/(z \cdot F)$, where M is the molar mass of zinc, z the corresponding valence of Zn and F is Faraday's constant, and j is the current density, which is the only part in the formula which is not known and thus needs to be simulated.

As described in section 1.2., the simplest and most common ways to simulate the current density is based on a model with the Poisson equation of electrostatic, cf. [3], in its core. As to be seen, the corresponding models assume constant electric potentials or constant charge transfer currents on the anode, resp. cathode. This assumption will be omitted at the cathode where neither the current nor the electrical potential are fixed but the relation of exchange current and exchange voltage, which is connected via the classical Ohm's law and the charge-transfer resistance. Hence in this paper, polarization data, as in [4] are used, for which good experimental results were obtained already.

1.2. State of the art

Most models of the current density in electroplating rely on the simulation of the electric field in the bath, which itself relies on the description of a static electric fields E_B , cf. [1, 3, 5]. The assumption of a static electric field simplifies the setting as static electric fields are conservative, i.e. they can be interpreted as gradient field of an electric potential ϕ_B , i.e. one assumes $E = \nabla \phi_B$.

Assuming as in [1, 6], that the bath is homogenously electrical neutral, i.e. that for the charge density ρ_B, the equation $\rho_B = 0$ holds true, the classical Ohm's law yields a description for the electric potential ϕ_B through the following Poisson equation:

$$\sigma_E \Delta \phi_B = 0. \tag{1}$$

Where σ_E is the permittivity of the electrolyte in the bath, i.e. the reciprocal of the resistance of the electrolyte. The two most common models for the current density are the primary and the secondary current distribution, cf. [7], [3] which define the boundary data of the Poisson equation (1) but differ in their boundary data on the anodes and cathodes. In both models, one splits the boundary $\partial\Omega_B$ of Ω_B into three sub-surfaces: an isolation part Γ_I, an anodic part Γ_A and a cathodic part Γ_C. In both models it is assumed that over the isolation Γ_I no current flows, i.e. one assumes that the Neumann boundary condition $\nabla \phi_B|_{\Gamma_I} \cdot n = 0$ is fulfilled, where n is the outer unit normal on Γ_I.

The *primary current density* interprets the conditions on the electrodes as Dirichlet boundary conditions for the electric potential ϕ_B. On the cathode the potential $\phi|_{\Gamma_C} = -E_{0,C}$ and on the anode, the potential $\phi|_{\Gamma_A} = E_{Cell} - E_{0,A}$, is assumed, where $E_{0,C}$ is the equilibrium potential of the cathode, $E_{0,A}$ is the equilibrium potential on the anode and E_{Cell} is the cell potential. In this section, it is assumed that the overpotential on cathode and on anode η vanishes.

The *secondary current density* interprets the conditions on the electrodes as Neumann boundary conditions for the electric potential ϕ_B. Additionally, it is assumed that the overpotentials η do not vanish. Furthermore, it is assumed that the charge transfer current j_{CT} is given through the Butler-Volmer equation, cf. [1]

$$i_{ct} = i_0 \left(\exp\left[\frac{(1-\beta)F\eta}{RT}\right] - \exp\left[\frac{\beta F\eta}{RT}\right] \right) \tag{2}$$

In the formula (2) i_0 is the exchange current density, β is the charge exchange coefficient, R the universal gas constant and T is the absolute temperature. Thus on the electrodes, the following Neumann boundary condition is given by: $\nabla\phi \cdot n|_{\Gamma_E}$, with $E \in \{A, C\}$.

Both models, *primary* and *secondary current density*, then evaluate the current density j on the anode and on the cathode via $j = |\nabla\phi_B|_2$, where $|\bullet|_2 : \mathbb{R}^3 \to \mathbb{R}_{\geq 0}$ denotes the Euclidean norm.

Note that the boundary data are assumed to be homogenous along the electrodes and furthermore note that the corresponding mathematical formulations of the boundary value problems are uniquely solvable, cf. [8].

2. On a mathematical model and a numerical scheme

2.1. Mathematical Background

Before a model for the dissolution of the zinc anodes is discussed in section 2.2 and the numerical strategy in section 2.3, some notation and mathematical notions have to be introduced.

First let $\Omega, D \subset \mathbb{R}^3$ be open, connected, bounded Lipschitz domains and let a decomposition $\mathscr{X} := \{\Omega_1, \ldots, \Omega_N\}$ into N open bounded and domains, with $\overline{\Omega} = \bigcup_{j=1}^{N} \Omega_j$ and $\Omega_j \cap \Omega_k$, for $j \neq k$ be given. For an open set D, the set $\overline{D}$ denotes the topological completion of D. First denote with $L^p(D)$ the Banach space of p-integrable functions, with $1 \leq p \leq \infty$, cf. [9]. In the following, one denotes, as in [9] and in [10], the spaces $W^{k,p}(\Omega)$ the space of functions $v \in L^p(D)$ for which the weak derivative $\frac{\partial^n}{\partial x_j^n}$ exists for all $j = 1,2,3$ and each $1 \leq n \leq k$. For the notion of weak differentiability that is used in this section, a definition can be found in [9, 10, 11], for example. Weak differentiability basically states that for a weak differentiable function $v \in L^p(\Omega)$ partial integration is well defined. The Banach spaces $L^p(\Omega)$ and $W^{k,p}(\Omega)$ are considered with their usual norms, cf. [9, 10]. Furthermore, note that $L^2(\Omega)$ and $H^1(\Omega) := W^{1,2}(\Omega)$ are even Hilbert spaces.

Furthermore, let the broken space of once weakly differentiable functions w.r.t. the decomposition $\mathscr{X}$ be given by cf. [12]:

$$H^1(\Omega; \mathscr{X}) := \left\{ v \in L^2(\Omega) \mid \forall D \in \mathscr{X} : v|_D \in H^1(D) \right\} = \prod_{D \in \mathscr{X}} H^1(D). \tag{3}$$

Furthermore, let $u^{(j)} = u|_{\Omega_j}$ additionally define the jump over an interface be define through $[u]_{j,k} := \left(\gamma_{\partial\Omega_j} u^{(j)} - \gamma_{\partial\Omega_k} u^{(k)}\right)$, where for all $1 \leq j \leq n$, the operator is the trace operator $\gamma_{\partial\Omega_j}$, as defined in [10].

2.2. *A model for the current density*

In this section, a new mathematical model for the dissolution of soluble anodes will be discussed. For that, consider an industrial galvanic bath as in figure 1a and figure 1b and the test volume Ω, filled with an electrolyte in the subdomain $\Omega_B \subset \Omega$, where the box inside the box does not belong to the domain. In this context the surface of the inner box represents the cathode surface, cf. figure 1c. Furthermore, let zinc anodes, be placed in the path, cf. figures 1a and 1d, the corresponding subdomain is denoted by $\Omega_A \subset \Omega$. In industrial applications, the zinc anodes are contacted over a titanium conductor, cf. figure 1e, the corresponding subdomain is denoted by $\Omega_Z \subset \Omega$.

Firstly, it is assumed, as in section 1.2 that during the electroplating, the electric field is static in each domain. It directly follows that for all $S \in \{A, C, Z\}$, the electric field $E^{(S)}$ in Ω_S can be represented as the gradient field of the electric potential $\phi^{(S)}$ in Ω_S, i.e. one has $E^{(S)} = \nabla\phi^{(S)}$. Then, the corresponding Ohm's laws in the separate domains can be formulated as in the following:

$$\sigma_B \Delta\phi^{(B)} = 0, \qquad in\ \Omega_B \tag{4}$$

$$\sigma_A \Delta\phi^{(A)} = 0, \qquad in\ \Omega_A \tag{5}$$

$$\sigma_Z \Delta\phi^{(Z)} = 0, \qquad in\ \Omega_Z \tag{6}$$

As described in section 1.2, the boundary and interface conditions must be fixed. Firstly, note, cf. [13], that the electric field is continuous over a common interface of two conductors. On $\partial\Omega^{(A)} \cap \partial\Omega^{(Z)}$, inside the anode in figure 1e, the following identity is assumed:

$$\nabla\phi^{(Z)} - \nabla\phi^{(A)}, \qquad on\ \partial\Omega^{(A)} \cap \partial\Omega^{(Z)} \tag{7}$$

At the interface of bath and anode, cf. figure 1d, i.e. on $\partial\Omega^{(A)} \cap \partial\Omega^{(B)}$, one assumes that Ohm's law $U = R_{CT} I$ is assumed, where R_{CT} is the charge-transfer resistance, U the potential jump on the interface and I is the current on the interface. In this article, the following is assumed:

$$R_{CT}\left(\nabla\phi^{(B)} - \nabla\phi^{(A)}\right) \cdot n - \left(\phi^{(B)} - \phi^{(A)}\right) = 0, \qquad on\ \partial\Omega^{(A)} \cap \partial\Omega^{(B)} \tag{8}$$

On the cathode surface Γ_K, cf. figure 1c, one assumes the following identity as in section 1.2.:

$$\phi^{(B)} = -E_{0,C}, \qquad on\ \Gamma_K. \tag{9}$$

Furthermore on the power supply of the Ti parts of the anodes Γ_{SZ}, cf. figure 1d, is given by:

$$\phi^{(Z)} = E_{Cell}, \qquad on\ \Gamma_{SZ} \tag{10}$$

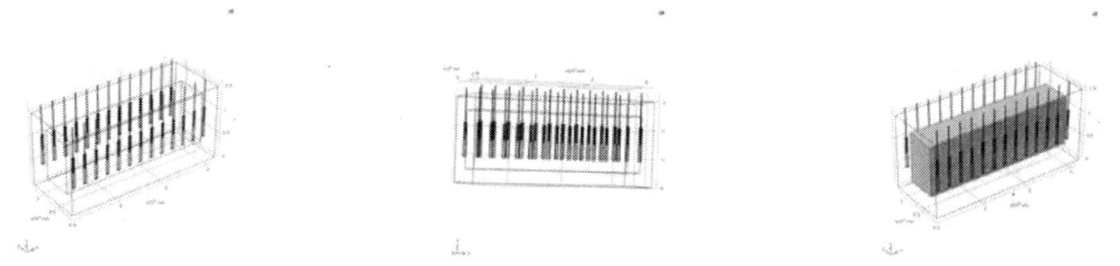

(a) Construction of the galvanic bath

(b) Side view on the construction

(c) Surface of the cathode

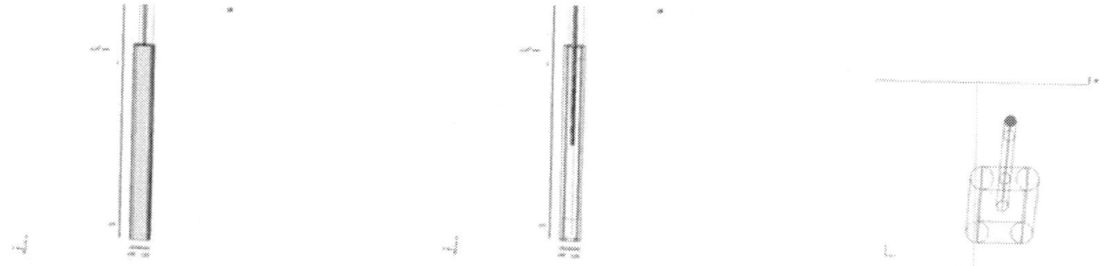

(d) Surface of the anode

(e) Surface of the Ti-kernel of the zinc anode

(f) Area of the power supply (blue area) on top of the Ti-kernel

Figure 1: Geometric configuration of a Zinc bath

Additionally, it is assumed that the other boundaries are insulating. As in section 1.2., one assumes that there is no current that flows over the insulation, i.e. the following boundaries are assumed:

$$\nabla\phi^{(B)}\cdot n = 0, \qquad on\ \partial\Omega_B \cap \partial\Omega, \tag{11}$$

$$\nabla\phi^{(A)}\cdot n = 0, \qquad on\ \partial\Omega_A \cap \partial\Omega, \tag{12}$$

$$\nabla\phi^{(Z)}\cdot n = 0, \qquad on\ \partial\Omega_Z \cap \partial\Omega \backslash \Gamma_{SZ}, \tag{13}$$

Furthermore, note that the permittivity of the Ti kernels of the zinc anode is given by $\sigma_Z = 2.5\cdot 10^6\ Sm^{-1}$ and the permittivity of the zinc anodes is given by $\sigma_A = 16.7\cdot 10^6\ Sm^{-1}$.

2.3. *Numerical strategy*

For the numerical strategy, a generalized mathematical formulation of (4)-(13) was considered. For the description of the solver, let $\Omega \subset \mathbb{R}^3$ be an open, bounded, polyhedral Lipschitz domain and let a decomposition $\mathscr{Z}$, as described in section 2.1., of Ω be given. Furthermore, let $N \in \mathbb{N}$ be the number of elements in $\mathscr{X}$. Furthermore, let $\Gamma_j := \partial\Omega_j \partial\Omega$ and $\Gamma_{j,k} = \partial\Omega_j \partial\Omega_k$. Additionally, let an set of indices J with $\emptyset \neq J \subset \{(j,k) \mid \Gamma_{j,k} \neq \emptyset\}$ be fixed and let $I := \{(j,k) \mid \Gamma_{j,k} \neq \emptyset\}\{(j,k) \mid \Gamma_{j,k} \neq \emptyset\} \backslash J$ be defined.

Furthermore, for all $1 \leq j,k \leq N$ let the values $0 < \sigma_j$ be arbitrary fixed, let $f_j \in L^2(\Omega_j)$, $a_j, b_j, g_j \in L^\infty(\Gamma_j)$ and $a_{j,k}, b_{j,k}, g_{j,k} \in L^\infty(\Gamma_{j,k})$ be arbitrary fixed. Then, the in this subsection considered, generalized problem will be given as:

Find $\phi \in H^1(\Omega; \mathscr{X})$ such that

$$-\sigma_j \Delta\phi^{(j)} = \quad f_j \quad in\, \Omega_j, \quad \forall 1 \leq j \leq N, \tag{14}$$

$$a_j \phi^{(j)} + b_j \nabla\phi^{(j)} = \quad g_j \quad on\, \Gamma_j, \quad \forall 1 \leq j \leq N, \tag{15}$$

$$a_{j,k}[\phi]_{j,k} + b_{j,k}[\nabla\phi]_{j,k} \cdot n = g_{j,k} \quad on\, \Gamma_{j,k}, \quad \forall (j,k) \in J, \tag{16}$$

$$\nabla\phi^{(j)} - \nabla\phi^{(k)} = \quad 0 \quad on\, \Gamma_{j,k}, \quad \forall (j,k) \in I. \tag{17}$$

As it can be easily seen, the model problem (14)-(17) generalizes the model (4)-(13) and even the models for the *primary* and the *secondary current density* described in section 1.2. Based on the weak formulation for the system of **P**artial **D**ifferential **E**quations, the numerical approximation searches for a weak solution of (14)-(17), which is given as the solution of the variation problem below, cf. [8, 11]:

Find $\phi \in H^1(\Omega; \mathscr{X})$ such that for all $v \in H^1(\Omega; \mathscr{X})$, the following identity holds true

$$a(\phi, v) = b(v), \qquad v \in H^1(\Omega; \mathscr{X}). \tag{18}$$

In the notation above, the bilinear form $a : H^1(\Omega; \mathscr{X}) \times H^1(\Omega; \mathscr{X}) \to \mathbb{R}$ is defined by

$$\begin{aligned} a(\phi, v) := \sum_{j=1}^{N} \Big[\sigma_j \big(\nabla\phi^{(j)}, \nabla v^{(j)}\big)_{L^2(\Omega_j)} + \big(a_j \phi^{(j)} + b_j \nabla\phi^{(j)} \cdot n, v^j\big)_{L^2(\Gamma_j)} \Big] \\ + \sum_{(j,l) \in J} \big(a_{j,k}[\phi]_{j,k} + b_{j,k}[\nabla\phi]_{j,k} \cdot \, n, [v]_{j,k}\big)_{L^2(\Gamma_{j,k})} \\ + \sum_{(j,k) \in I} \big(\nabla\phi^{(j)} - \nabla\phi^{(k)}, \nabla v^{(j)} - \nabla v^{(k)}\big)_{L^2(\Gamma_{j,k})} \end{aligned}$$

and the linear functional $b : H^1(\Omega; \mathscr{X}) \to \mathbb{R}$ is given by

$$b(v) := \sum_{j=1}^{N} \Big[\big(f_j, v^{(j)}\big)_{L^2(\Omega_j)} + \big(g_j, v^{(j)}\big)_{L^2(\Gamma_j)} \Big]. \tag{19}$$

For the numerical treatment of (18), a discretization X_h of $H^1(\Omega; \mathscr{X})$ has to be fixed. In this paper, a conforming discretization was chosen, i.e. $X_h \subset H^1(\Omega; \mathscr{X})$. Due to the definition of $H^1(\Omega; \mathscr{X})$, a discretization of $H^1(\Omega_j)$ suffices to be defined for each $1 \leq j \leq N$. For the definition of a discretization, let $\mathscr{T}$ be a triangulation, for a definition c.f. [14], of Ω given such that for each $1 \leq j \leq N$, there exists a subtriangulation $\mathscr{T}_j \subset \mathscr{T}$ such that $\mathscr{T}_j$ is a triangulation of Ω_j. For the discretization of $H^1(\Omega_j)$, the linear space of spines $S^1(\mathscr{T}_j)$ is used. Where $S^1(\mathscr{T}_j)$ is, as in [15], defined as $S^1(\mathscr{T}_j) := P_1(\mathscr{T}_j) \cap C_0(\Omega)$. In the notation above the space of piecewise affine functions $P_1(\mathscr{T}_j)$is defined as

$$P^1(\mathscr{T}_j) := \left\{ v \in L^2(\Omega_j) \mid \forall T \in \mathscr{T}_j \exists a \in \mathbb{R}, b \in \mathbb{R}^3 : \, v|_T \equiv b \cdot x + a \right\} \tag{20}$$

In this paper X_h is defined as $X_h := \prod_{j=1}^{N} S^1(\mathscr{T}_j)$.

An important step of the numerical scheme is the application of an adaptive algorithm, which allows to work very efficiently since it generates a mesh $\mathscr{T}$, which is fine where it needs to be to minimize the error of the numerical scheme. The adaptive FEM loop that is used in this paper, is given through the following steps, given as in [16, 17]:

(1) **Input:** Initial triangulation $\mathscr{T}^0$ as well as problem data w.r.t. (14)-(17) and define $k := 0$.

(2) **Solve:** Solve the discrete problem:
Find $\phi_h \in X_h$ such that for all $v_h \in X_h$ the following equality holds true:

$$a(\phi_h, v_h) = b(v_h) \tag{21}$$

(3) **Estimate:** Estimate the error $\|\phi - \phi_h\|_X^2$ with $\eta^2 := \sum_{T \in \mathscr{T}} \eta_T^2$, where η_T^2 is analogous to the discussion in [18], defined through

$$\begin{aligned}
\eta_T^2 := |T|^{1/3} \sum_{j=1}^{N} \Bigg[\sum_{F \in \mathscr{F}} \|\sigma_j [\nabla \phi_h^{(j)}] \cdot n\|_{L^2(F)}^2 \\
+ \sum_{F \in \mathscr{F}(\Gamma_j \cap \partial T)} \|a_j \phi_h^{(j)} + b_j \nabla \phi_h^{(j)} \cdot n - g_j\|_{L^2(F)}^2 \Bigg] \\
+ |T|^{1/3} \sum_{(j,k) \in J} \sum_{F \in \mathscr{F}(\partial T \cap \Gamma_{j,k})} \|a_{j,k} [\phi_h]_{j,k} + b_{j,k} [\nabla \phi_h]_{j,k} \cdot n - g_{j,k}\|_{L^2(F)}^2 \\
+ |T|^{1/3} \sum_{(j,k) \in I} \sum_{F \in \mathscr{F}(\partial T \cap \Gamma_{j,k})} \|\nabla \phi^{(j)} - \nabla \phi^{(k)}\|_{L^2(F)}^2
\end{aligned}$$

(4) **Mark:** Mark the elements with Dörfler marking, cf. [19], with associated bulk parameter $\theta = 0.5$.

(5) **Refine:** Refine the triangulation $\mathscr{T}^{k+1}$ =`Refine`($\mathscr{T}^k$), with the newest vertex bisection, cf. [20].

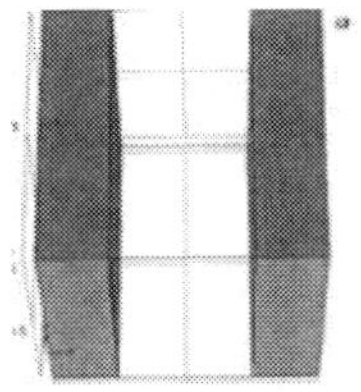

(a) Splitted domain for the numerical verification of the solver.

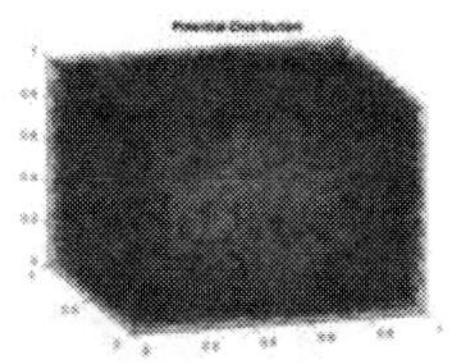

(b) Approximation to the solution to the generalized problem (14)-(17).

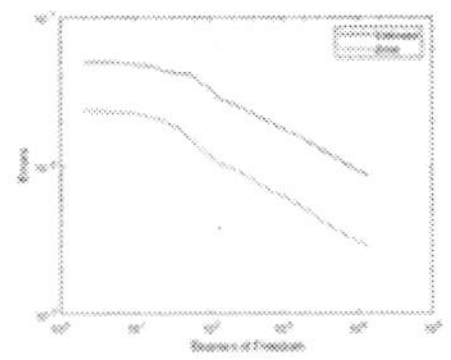

(c) Convergence plot due to adaptive refinement, with exact error (red) and defined estimator η (blue).

Figure 2: Splitted domain, approximate solution and error estimate for the verification of the solver.

(6) **Break:** If η^2 is lower than a given tolerance ε or if the number of degrees of freedom is higher than a given lower bound then break the condition. Else go to (2) and set $k = k+1$.

Note that in step (2), the problem (21) corresponds to a linear system of equations, which can be solved with standard software.

2.4. *Numerical Validation of the software*

Since the numerical scheme for the approximation of the solution of (14)-(17) was programmed in Matlab and no commercial software was used, hence, a numerical validation of the software is necessary. For the numerical validation let $\Omega :=]0,1[^3$ be fixed, with a decomposition $\mathscr{X} = \{\Omega_1, \Omega_2, \Omega_3\}$, where $\Omega_1 = \{x \in \Omega \mid x_1 < 1/4\}$, $\Omega_1 = \{x \in \Omega \mid 1/4 < x_1 < 3/4\}$ and $\Omega_1 = \{x \in \Omega \mid x_1 > 3/4\}$, cf. figure 2a. Furthermore, let an exact solution of (14)-(17) be defined through $u(x,y,z) = x(1-x)y(1-y)z(1-z)$. Furthermore, let $\sigma_1 = 1/2$, $\sigma_1 = 1/3$ and $\sigma_1 = 1/4$ are fixed for this problem. Furthermore, it is assumed that the identities $u|_{\partial\Omega} = 0$, $[\nabla u]_{1,2} \cdot n = [\nabla u]_{2,3} \cdot n = 0$. Additionally $f \in L^2(\Omega)$ is chosen such that u is the exact solution to the corresponding problem (14)-(17). An approximation of the solution of (14)-(17) yields the approximation given in figure 2b and one obtains the convergence graph given in figure 2c. As it can be seen from the convergence graph, the numerical scheme chosen seems to converge and reaches an asymptotic phase. The formal proof of convergence and calculation of the order of convergence will be omitted in this paper.

IOP Conf. Series: Materials Science and Engineering 1147 (2021) 012006 doi:10.1088/1757-899X/1147/1/012006

Table 1: Parameters for the EIS measurements with a zinc anode and a steel cathode.

Parameter	Setup
Process time	4h, continuous measurement of impedance spectra in ongoing time
Mode	Galvanostatic, constant anodic current density of 2 A/dm^2
Working electrode	Zinc, 99.995 %
Reference electrode	Ag/AgCl (saturated KCl)
Cathode	Steel
Frequency	100 kHz – 10 mHz
Amplitude	10 mV

3. Experimental, materials and simulations

To apply the model given in section 2.2, it suffices to measure the conductivity of the electrolyte, or equivalently the electrolyte resistance, as well as the charge-transfer resistance. A common and intelligent way to determine the charge-transfer resistance and the electrolyte resistance is to measure impedance spectra, cf. [21], and fit the corresponding measurement results to an equivalent circuit. The corresponding measurements are given in subsection 3.1.

3.1. Electrochemical impedance spectroscopy (EIS) measurements

For the impedance measurement, the experiment parameters are given in table 1 for a weakly sour electrolyte.

As mentioned in table 1, the measurements of the impedance spectra were done over 4 h. In figure 3a and 3b, the results for the first measured impedance spectrum are shown. As it can be seen in the Nyquist-plot, the results do not represent a half cycle, as one could expect. When evaluating the other spectra, one obtains a very similar behavior, which is known in the literature, cf. [22]. As discussed in [22], the spectra can be interpreted as the overlap of two semi-cycles which can be identified with the reactions $Zn \rightarrow Zn^{+} + e^{-}$ (I) and $Zn^{+} \rightarrow Zn^{2+} + e^{-}$ (II). Furthermore as described in [22], the equivalent cycle for the situation discussed above is given by two constant-phase elements, parallel to a resistor, and in series with another resistor R_{DW} representing the electrolyte resistance. The resistor associated to the reaction (I) is denoted by R_{CP_1} and the resistance associated to the reaction (II) is denoted by R_{CP_2}. Then, the total charge-transfer resistance is given by $R_{ct} = R_{tot} = R_{CP_1} + R_{CP_2}$.

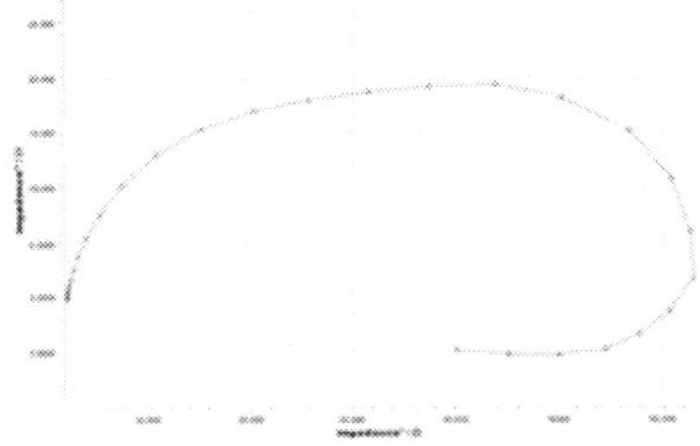

(a) Nyquist plot of the first EIS-Spectrum associated to the measurements gained with the parameters defined in table 1.

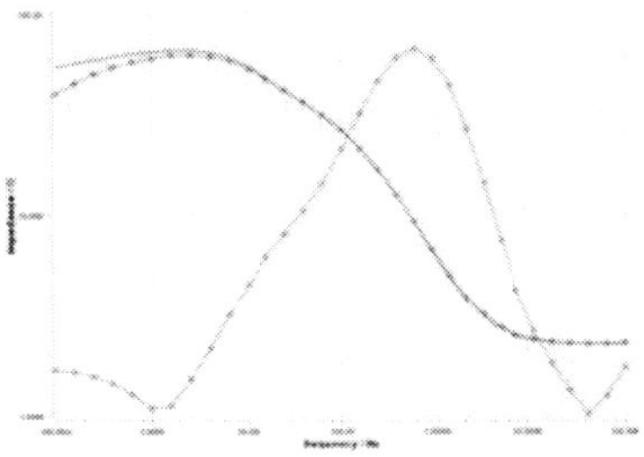

(b) ZHIT-approximation of the impedance, in the Bode plot associated to the first measurement in time associated to the parameters defined in table 1.

Figure 3: Nyquist plot and ZHIT approximation in the Bode-plot.

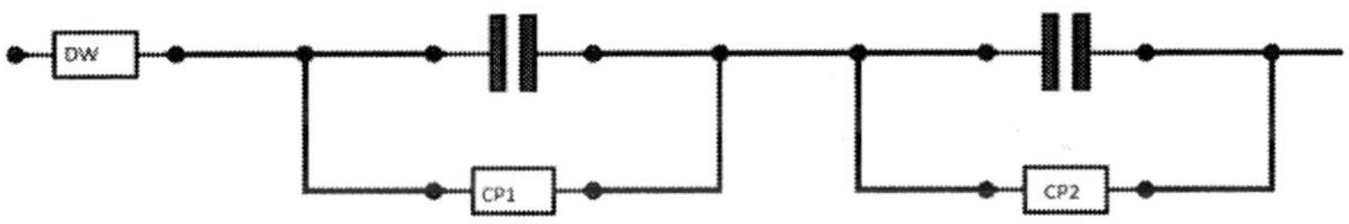

Figure 4: Equivalent circuit for the EIS spectrum given for the anode reactions $Zn \rightarrow Zn^{+} + e^{-}$ (I) and $Zn^{+} \rightarrow Zn^{2+} + e^{-}$ (II) for the measurements obtained for the parameters given in table 1. R_{DW} represents the electrolyte resistance, R_{CP_1} the resistance of the electrode reaction (I) and R_{CP_2} the resistance of the electrode reaction (II).

When evaluating the measurements over time with the equivalent circuit above, one obtains the resistances shown in figure 5a and figure 5b. As it can be seen in the figures, the resistance R_{DW} and R_{ct} can be considered almost constant during the process.

3.2. *Simulation results for the model*

For the comparison of models the primary current density was simulated with COMSOL as references for the anodic current density, cf. figure 6a and the cathodic current density, cf. figure 6c, In the figures 6b and 6d the results of the new model are presented.

As it can be seen in both figures for the primary current density, especially in figure 6a vertex and edge effects can be observed, which corresponds to experimental observations. Furthermore, it can be seen that the calculated anodic current density is highly homogeneous. No gradient

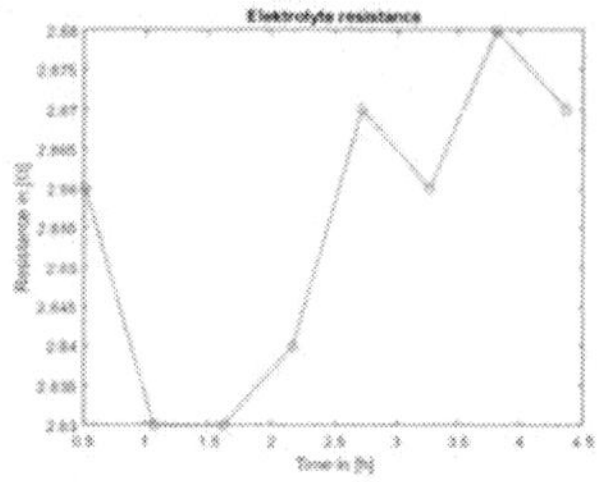

(a) Development of the Electrolyte resistance R_{DW} over the measurement-time from the EIS – spectra measured with the parameters defined in table 1.

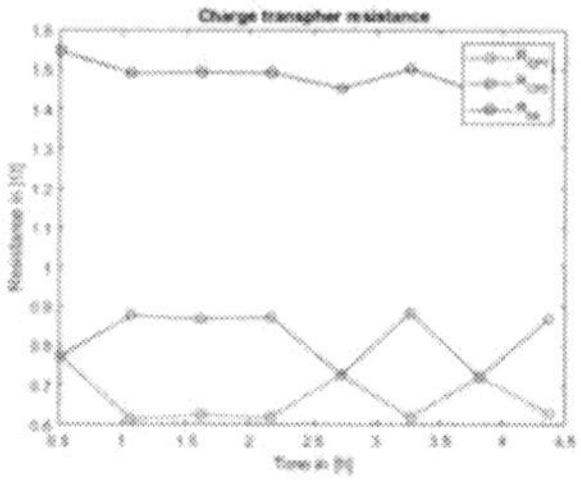

(b) Development of the resistances $R_{ct} = R_{tot}" = R_{CP_1} + R_{CP_3}$, R_{CP_1}, and R_{CP_2}, over the measurement time from the EIS – spectra measured with the parameters defined in table 1.

Figure 5: Fits of the electrolyte resistance R_{DW} and the charge transfer resistance R_{ct} with the equivalent circuit given in figure 4.

from the power supply to the end of the anode can be observed. This behavior can be expected from the model described in section 1.2, but it is not quite the behavior one could expect from observations that are commonly known in the real process: The anodes dissolve predominantly in a neighborhood of the power supply.

Furthermore, it can be seen that on the cathode surface, one obtains a highly homogeneous current density, with maxima directly opposite to the anodes. In addition, the cathodic current density is higher than the anodic current density.

Note, that the new model, simulated with the numerical scheme described in section 2.3 and implemented in Matlab, cf. figure 6b and 6d, predicts a more inhomogeneous current density than the model of the primary current density, cf. figure 6d. The inhomogenities on the cathode can be explained by the anodic current density, cf. figure 6b. As it can be seen from figure 6b, the anodic current density is highest in the proximity of the power supply. At the lower end of the anodes a further local maximum of the calculate current density can be observed, as result of the known edge-effects. While the high anodic current densities in the neighbourhood of the power supply can immediately be explained by the power supply, the higher current densities at the other end of the anode can be explained by edge and vertex effects. The high current densities at the power supply directly lead to the high current densities at the cathode surface.

(a) Anodic current density for the primary current density, with $\sigma_B = 1/1.6 \cdot 10^6$ $\mathrm{Sm}^{(} - 1)$, $E_{Cell} = 1.75$ V, $E_{0,C} = -0.04$ V and $E_{0,A} = -0.76$ V.

(b) Anodic current density for the new model, with $\sigma_B = 1/1.6 \cdot 10^6$ $\mathrm{Sm}^{(} - 1)$, $\sigma_Z = 2.5 \cdot 10^6$ $\mathrm{Sm}^{(} - 1)$ and the permittivity of the zinc anodes is given by $\sigma_A = 16.7 \cdot 10^6$ $\mathrm{Sm}^{(} - 1)$, $E_{Cell} = 1.75$V, $E_{0,C} = -0.04$V and $E_{0,C} = 0.76$V.

(c) Cathodic current density for the primary current density, with $\sigma_B = 1/1.6 \cdot 10^6$ $\mathrm{Sm}^{(} - 1)$, $E_{Cell} = 1.75$ V, $E_{0,C} = -0.04$ V and $E_{0,A} = -0.76$ V.

(d) Cathodic current density for the new model, with $\sigma_B = 1/1.6 \cdot 10^6$ $\mathrm{Sm}^{(} - 1)$, $\sigma_Z = 2.5 \cdot 10^6$ $\mathrm{Sm}^{(} - 1)$ and the permittivity of the zinc anodes is given by $\sigma_A = 16.7 \cdot 10^6$ $\mathrm{Sm}^{(} - 1)$, $E_{Cell} = 1.75$V, $E_{0,C} = -0.04$V and $E_{0,C} = 0.76$V.

Figure 6: Comparission of the anodic current densities and cathodic current densities calculated with the new model and the *primary current density*.

4. Conclusion

Concluding the results of the paper, one obtains an alternative model for the current density on the anode and on the cathode, which reflects the electric current in the anode and the cathode, as described in section 2.2. Furthermore in section 2.3, a solver for a generalized problem was discussed and verified in section 2.4.

In section 3, impedance measurements for a weakly sour zinc electrolyte and the evaluation of the spectra were discussed. In section 3.2, the simulations for the secondary current density and the new model were discussed. The differences between the simulations were significant, but could be explained by physical considerations. In contrast to the models described in section 1.2, the model presented in section 2.2 is devoted to the simulation of large-scale electrodepositions, using polarization data to enhance the effectiveness of the model.

In the core of this article a strategy was described to implement the charge transfer resistance R_{ct} at the interface regions of electrode and electrolyte in a model for galvanic cells basing on static electric fields and the Ohms law on multiple neighboring domains. Furthermore an adaptive conforming Courant FEM was described to approximate the new models.

The usage of charge transfer resistances gives the theoretical work on a huge number of physical processes in galvanic cells available or give a more holistic frame on the process, such as the dissolution of large scale zinc anodes discussed in this article, the actual electroplating process at the cathode, corrosion and processes in lithium-ion batteries or fuel cells etc.

Furthermore it can be seen in this article that the new model for the current density gives a more complete overview over the currents in the complete process, as the currents in the electrodes are contained in the model in contrast to the primary and secondary current density. This gives the chance to simulate the current densities on large scale structures more accurate than it could be done until now, since the resistances in the electrodes are respected. This can lead to answers to questions like where to contact large scale anodes and cathodes most effectively in galvanic cells.

Although the model in this article is described for the electrochemical dissolution of zinc, it can be immediately applied to the electroplating of zinc as well as adjusted to other settings, e.g. other electrolytes, corrosion. Additionally, the model allows for the prediction of suitable positions of electric contacts on more complex anodes like wire-shaped anodes.

Acknowledgments

The authors wish to acknowledge the financial support by the AIF in the context of the ZIM project ZF4131907 "Anokraftwire" for which the generalized model and the software were implemented.

Furthermore the authors want to thank Mr. Frank Simchen who took a great effort to the experiment design and the measurement of the EIS-spectra.

Literature

[1] J Bard A and R Faulkner L 1980 *Electrochemical methods : fundamentals and applications / Allen J. Bard, Larry R. Faulkner* vol 18
[2] Newman J S 1991 *Electrochemical systems* 2nd ed Prentice-Hall international series in the physical and chemical engineering sciences. (Prentice Hal)
[3] Mahapatro A and Suggu S 2018 *Advanced Material Science* **3**
[4] Bösch N C 2017 Numerische simulation von korrosionsprozessen für die industrielle anwendung
[5] Dickinson E, Ekström H and Fontes E 2014 *Electrochemistry Communications* **40** 71–74
[6] Hussein W 2018 *Journal of Engineering Science and Technology* **13** 3432–3444 cited By 0
[7] De Kubber D, Bortels L, Deconinck J and Daenen T 2001 *Electrochimica Acta* **47** 91–94 cited By 6
[8] Evans L 1997 *Partial Differential Equations* (*Graduate Studies in Mathematics* vol 166) (Providence: AMS)
[9] Alt H W 2012 *Lineare Funktionalanalysis* 6th ed (Heidelberg: Springer)
[10] Attouch H, Buttazzo G and Michaille G 2014 *Variational Analysis in Sobolev and BV Spaces* (Philadelphia, PA: Society for Industrial and Applied Mathematics)
[11] Glowinski R 2008 *Numerical Methods for Nonlinear Variational Problems* (Springer)
[12] Carstensen C, Demkowicz L and Gopalakrishnan J 2016 *Computers & Mathematics with Applications* **72** 494–522 ISSN 0898-1221

IOP Conf. Series: Materials Science and Engineering 1147 (2021) 012006 doi:10.1088/1757-899X/1147/1/012006

[13] Donnevert J 2015 *Die Maxwell'schen Gleichungen - Vom Strömungsfeld des Gleichstroms zum Strahlungsfeld des Hertz'schen Dipols* 1st ed (Wiesbaden: Springer Vieweg)
[14] Grossmann C, Roos H G and Stynes M 2007 *Numerical Treatment of Partial Differential Equations* (Springer)
[15] Braess D 2012 *Finite Elemente* 5th ed (Heidelberg: Springer) Theorie, schnelle Löser und Anwendungen in der Elastizitätstheorie
[16] Carstensen C, Feischl M, Page M and Praetorius D 2014 *ELSVIER, Computers and Mathematics with Applications* **67** 1195–1253
[17] Carstensen C and Rabus H 2016 *ArXiv e-prints* (*Preprint* `1606.02165`)
[18] Verfürth R 1994 *Journal of Computational and Applied Mathematics* **50** 67–83 cited By 251
[19] Dörfler W 1996 *SIAM Journal on Numerical Analysis* **33** 1106–1124 (*Preprint* `http://dx.doi.org/10.1137/0733054`) URL `http://dx.doi.org/10.1137/0733054`
[20] Traxler C 1997 *Computing (Vienna/New York)* **59** 115–137 cited By 62
[21] Ehm W, Göhr H, Kaus R, Röseler B and Schiller C 2000 *ACH - Models in Chemistry* **137** 145–157
[22] Bockelmann M, Becker M, Reining L, Kunz U and Turek T 2018 *Journal of The Electrochemical Society* **165** A3048–A3055

Corrigendum: Formation of corundum-rich alumina coatings on low-carbon steel by plasma electrolytic oxidation (2021 *IOP Conf. Ser.: Mater Sci Eng.* 1147 *012007*)

F Simchen, N Masoud-Nia, T Mehner, and T Lampke
Materials and Surface Engineering Group, Institute of Materials Science and Engineering, Chemnitz University of Technology, Gemany

The surface area of the test sample depicted in figure 2 was calculated wrong (without the contribution of the lateral area). That's why the current densities in Table 3, Figure 4 and Figure 5 were wrong also. We adjusted the mentioned article elements as well as the corresponding statement within the abstract:

Abstract: As a result of this, it was possible to obtain alumina layers of 80 micrometers in thickness, with a high corundum content of approximately 50 to 90%, after 37 minutes of treatment time, at a current density below 10 (instead of 25) A/dm² on C8C-steel.

Table 3. Parameters of the symmetric current-controlled alternating rectangular pulse for PEO.

I_1 / A	i_1/A·dm^{-2}	t_1/ms	U_1/V	I_2 / A	i_2/A·dm^{-2}	t_2/ms	U_2/V
1	10.85	5	500	-0.5	-5.42	5	-500

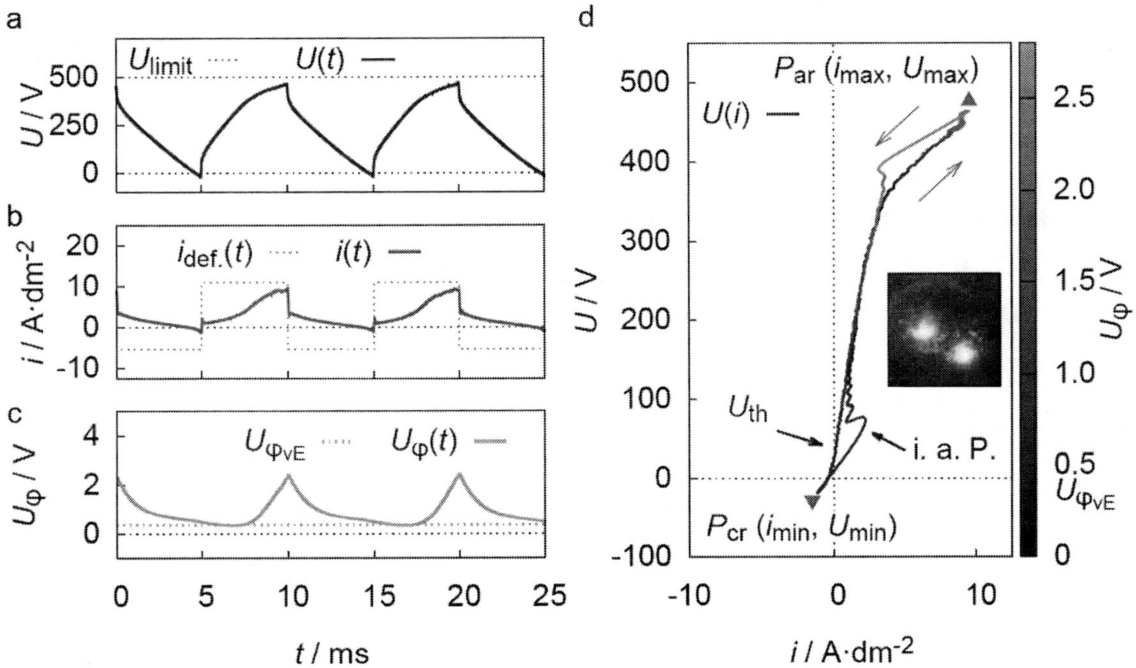

Figure 4. Graphical representation of the electrical pulse regime at 15 minutes of PEO treatment after initial voltage staircase: process-voltage (a), current-density (b), and photo-voltage (c) along with the associated limits, definitions, and pre-discharge values. These (a-c) are then summarised into a voltage current density-characteristic (d), with photo-voltage indicated by colour, combined with an image of the process.

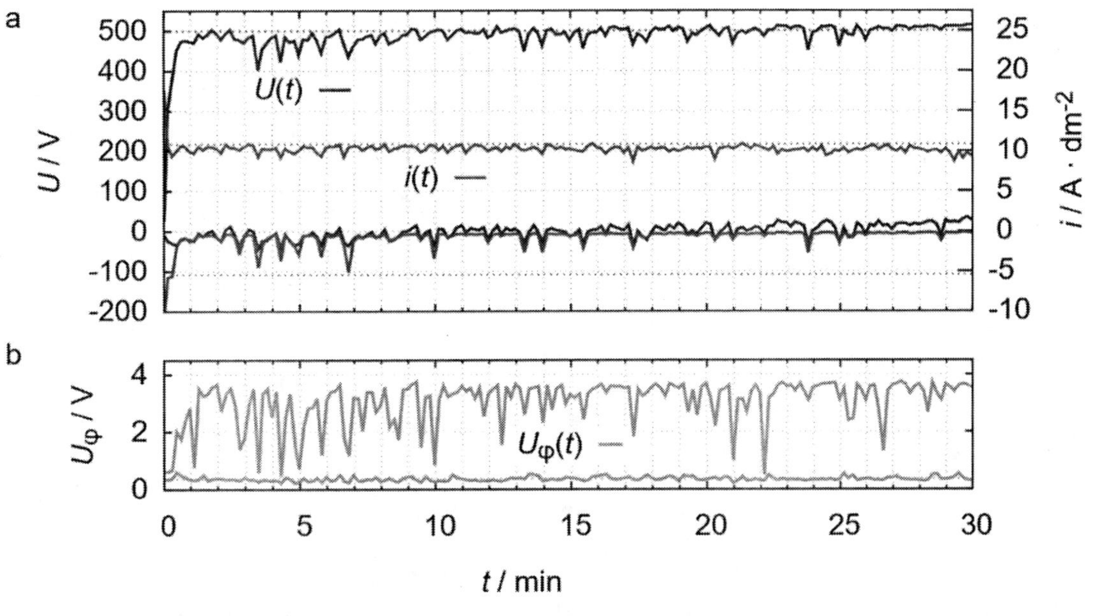

Figure 5. Course of the process voltage and current-density extremes (a) as well as the maximum and minimum values of the photo-voltage (b) as dependent on the process time.

Formation of corundum-rich alumina coatings on low-carbon steel by plasma electrolytic oxidation

F Simchen, N Masoud-Nia, T Mehner, and T Lampke
Materials and Surface Engineering Group, Institute of Materials Science and Engineering, Chemnitz University of Technology, 09107 Chemnitz, Germany

E-mail: Frank.Simchen@mb.tu-chemnitz.de

Abstract. The plasma electrolytic oxidation (PEO) is an innovative method for providing light metals and their alloys with protective ceramic surfaces. However, for iron-based materials, the process requires very high current densities and results in the formation of coatings which consist of less stable iron compounds. Therefore, it was the aim of this study to design a PEO procedure on low-carbon steel at moderate current densities, which allows for the formation of ceramic coatings whose chemical composition is dominated by the electrolyte constituents. The electrolyte used was based on aluminate and preselected by systematic electrochemical passivation experiments. The PEO treatment was monitored by electrical and optical process diagnostics. As a result of this, it was possible to obtain alumina layers of 80 micrometers in thickness, with a high corundum content of approximately 50 to 90%, after 37 minutes of treatment time, at a current density below 25 A/dm^2 on C8C-steel. However, the coating's microstructure was inhomogeneous and showed poor substrate bonding. Based on the results of the experimental work, explanatory approaches were provided and a course of action is suggested for counteracting these problems.

1. Introduction

Plasma electrolytic oxidation is an innovative method for the surface treatment of metallic materials. During the process, the workpiece is immersed within an aqueous electrolyte, opposite a counter electrode and receives a strong alternating polarisation. If suitable process parameters are chosen, this procedure leads to the formation of an electrically insulating passive layer on the sample surface. This results in a strong localised voltage drop between the substrate and an area of equal electrical potential, the so-called quasi cathode, close to the substrate/electrolyte interface. Generally, the initiation of the plasma electrolytic discharge occurs during the anodic partial pulse. They strike out of the electrolyte and into the substrate, leading to the formation of a ceramic layer Depending on the selected process parameters, the PEO coating can be composed of oxides of the substrate material or contain more complex reaction products of the substrate and electrolyte constituents. [1, 2]

While the method was initially developed for the surface treatment of aluminium alloys, its scope of applications has since then been extended to various metallic materials such as Mg, Ti, Zr, Hf, Nb, brass, and steels. The resultant coatings are in use for decorative, functional and protective applications. [3] However, especially in the case of steels, the PEO coatings provide low resistance against corrosive and tribological stresses. This is because the generated layers consist mainly of iron oxides with relatively little comprised of the more stable compounds formed

IOP Conf. Series: Materials Science and Engineering 1147 (2021) 012007 doi:10.1088/1757-899X/1147/1/012007

by the electrolyte constituent [4, 5, 6]. Furthermore, iron-based materials often require very high current densities between 40 [7, 8] and 300 [9, 10] A/dm^2, up to 3750 [11, 12] A/dm^2 to start and maintain the discharge process. This can be attributed to the poor passivation behaviour of the substrates used. In the context of PEO, this means that electrochemical processes dominate on the substrates under anodic polarisation in aqueous environments, which leads to substrate dissolution and oxygen development. The compounds formed from the released metal ions and electrolyte constituents, on the other hand, do not form a good electrically insulating reaction layer. This, however, would be a prerequisite for generating a strong localised potential drop between the bulk electrolyte and the substrate even at moderate current densities, which enables the formation of plasma electrolytic discharge phenomena. The few descriptions of PEO of low carbon steel under moderate current densities leading to the formation of coatings composed of mixed oxides, deal with the layer characterisation mainly [13]. However, the derivation of the electrolyte composition and investigations concerning the interplay of the substrate/electrolyte-interaction with rectifier control behaviour remain unaddressed. This impedes significantly to use them as a basis for further investigations.

Therefore, it is the aim of this study to develop an electrolyte for the PEO of low-carbon steel which allows for the formation of protective ceramic coatings composed of electrolyte constituents at moderate current densities. This could allow the transfer of excellent technological properties of oxidic protective coatings to steel substrates. According to the current state of the art, this is achieved primarily by thermal spraying of alumina or the plasma electrolytic oxidation of sprayed aluminium layers [14, 15, 16]. Direct PEO would be superior to these process routes, especially due to its good throwing power on complex component geometries. In addition, this would serve as an important intermediary step in the establishment of plasma electrolytic oxidation of steel-based material composites, which are becoming increasingly significant and applicable in lightweight construction [17, 18].

In this study, the preselection of the electrolyte composition is based on systematic polarisation measurements, which have already been proven to be suitable to evaluate the passivation behaviour of varied electrolyte/substrate combinations [3]. Afterwards, PEO experiments were carried out on low-carbon steel using optical and electrical process diagnostics. The resulting coating compositions and morphologies were characterised by X-ray diffraction (XRD), scanning electron microscopy (SEM), and energy-dispersive X-ray spectroscopy (EDX).

2. Experimental Procedure

The polarisation experiments were carried out in a three-electrode arrangement. A 3-mol/L-Ag/AgCl electrode was used as the reference and a platinum sheet as counter electrode. The test sample prepared by C8C low-carbon steel that acted as the working electrode was located at the bottom off the cell. This setup was used in order to minimise the influence of gas bubbles, which could be released during the experiments, on the recorded data. The electrolyte volume was 250 ml, and the diameter of the active sample area was 10 mm. The chemical composition of C8C steel (provided by Fr. u. H. LÜLING GmbH & Co. KG) is listed in table 1.

Table 1. Chemical composition of C8C steel, mass fraction in %.

Fe	C	Si	Mn	P	S	Al
balance	0.08	0.08	0.38	max. 0.020	max. 0.025	0.035

An electrochemical working station *Zennium* in combination with a voltage booster *CVB*120 (both ZAHNER-elektrik, Germany) were used to apply a voltage cycle according the vertices listed in table 2 and to record the resulting current. The current limit was set to 300 mA, which

IOP Conf. Series: Materials Science and Engineering 1147 (2021) 012007 doi:10.1088/1757-899X/1147/1/012007

corresponds to a current density of approximately 38.2 A/dm^2. The compliance voltage was 120 V.

Table 2. Vertices of the potential cycle used, depicted in figure 1.

n	1	2	3	4	5	6
U_n/V	0	100	100	0	100	0
t_n/s	0	100	220	320	420	≈ 420

Figure 1 shows a graphical representation of the two main types of measurement courses observed in this study. In case of poor passivation, the substrate forms no or just an unstable electrically resistive passive film in interaction with the electrolyte. The system under investigation reacts with a high flow of charge carriers and in excess of the current limit at a characteristic voltage (figure 1 a). This value is assigned as the breakthrough voltage U_b in this work. It should not be confused with the critical voltage for discharge initiation, sometimes termed breakdown or ignition voltage. An increase of the U_b voltage can be interpreted as a slight if still insufficient improvement in passivation. In case of better passivation, the experiments can be carried out completely (figure 1 b).

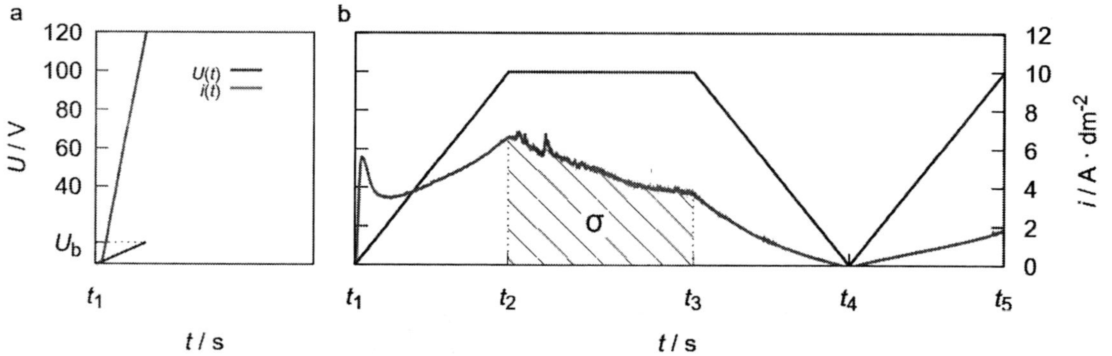

Figure 1. Graphical representation concerning the determination of the characteristic passivation values U_b (a) and σ (b).

The integration of the current over defined time ranges (see table 2) according equation (1) leads to a characteristic σ-value which can be used to quantify the observed passivation behaviour. Its lowering indicates an improved passivation.

$$\sigma = \int_{t_2}^{t_3} i(t)\ \mathrm{d}t \tag{1}$$

The test samples where pretreated by grinding on 4000 cm^{-2} paper, pickling in 1:1 hydrochloric acid, and then rinsing with deionised water. The electrolytes were solutions of disodium hydrogen phosphate and sodium aluminate with concentrations between 0.3 and 1.5 mol/l and varied pH values. The alkalinity was lowered through the addition of phosphoric or citric acid and elevated by potassium hydroxide solution. Table 4 and 5 list the investigated electrolyte compositions. All chemicals used were analytically pure (Merck, Germany).
The PEO experiments were carried out within an electrolyte container with a volume of 5 litres.

The solution was circulated by a mechanical stirrer and cooled by water of 5 °C that was circulating around the container. The cylindrical test samples had a diameter of 16.5 mm and a height of 10 mm. In order to avoid an interaction with the discharge phenomena, the specimen holder made of aluminium was covered with shrink tubing and pressed into a stepped bore in the lateral surface of the sample. Its geometry is depicted in figure 2 a. Figure 2 b shows the approximate position of the places where the SEM images depicted in figure 8 and 9 were taken.

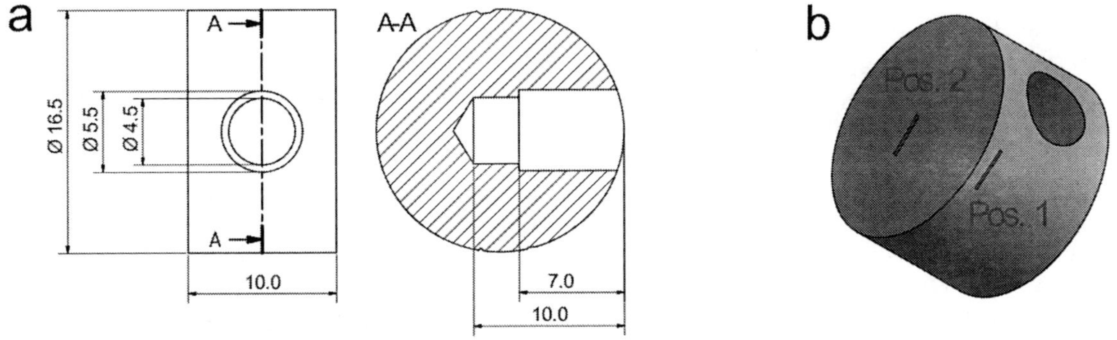

Figure 2. Technical drawing of the sample geometry (a) and approximate position (Pos. 1. and 2.) of the places where the SEM images shown in figures 8 and 9 were taken (b).

A *POWER PULSE* pe861UA-500-10-24-S rectifier (plating electronics, Germany) was used to apply a symmetric current-controlled alternating rectangular pulse, the parameters of which are summarised in table 3. The values of U_1 and U_2 represent the anodic and cathodic voltage limitation. Figure 4 a contains a corresponding graphical representation.

Table 3. Parameters of the symmetric current-controlled alternating rectangular pulse for PEO.

I_1/A	i_1/A·dm^{-2}	t_1/ms	U_1/V	I_2/A	i_2/A·dm^{-2}	t_2/ms	U_2/V
1	25	5	500	-0.5	-12.5	5	-500

The process showed a severely delayed discharge initiation of up to 25 min, even with the optimised electrolyte used. Therefore, the alternating regime was preceded by a voltage-controlled DC staircase in order to support the formation of an initial passive layer. Here, the voltage was elevated in steps of 25 V with a duration of 25 s per step up to 425 V over a time period of 7:05 min. The electrical process data were monitored by using a *DL*850 transient recorder (Yokogawa, Japan). The voltage was tapped directly at the electrodes with a 10:1 probe. The current signal was determined by recording the voltage drop over a shunt resistor (Isabellenhütte, Germany) of 0.1 Ω tied into the circuit. In order to obtain integral optical process informations in temporal synchronisation with the electrical process data, a photo diode was placed close to the steel sample. The device was mounted in a BNC-adapter and contained in a glass cylinder with darkened lateral surface. The photo voltage courses generated during the PEO were monitored with the transient recorder as well. Its measurements were triggered by the rise of the voltage edge over a level of 10 V every 10 seconds and were carried out with a sampling rate of 1 MS/s over 100 ms. Afterwards, the resulting data were processed by using a low-pass filter with a cut of frequency of 15 kHz, by using a MATLAB tool. Additionally the discharge phenomena during the experiments were filmed with a *Gopro Hero* 5 camera (Gopro, USA), which allows for a spatially resolved characterisation of the process.

The XRD investigations were carried out with a *D8 Discover* (Bruker AXS, USA). The samples were measured with Co-Kα radiation (40 kV, 40 mA) with point focus using a 0.5 mm collimator and a *LynxeyeXE−T* detector over an exposure time of 9.5 h. The phase allocation was based on the PDF-2 Release 2014 database of the International Centre for Diffraction Data (ICDD). The layer's morphology was made visible for selected examples with a scanning electron microscope 1455*VP* (LEO, Germany) using an excitation voltage of 25 kV based on secondary electron (SE) mode. The chemical composition was determined with an energy-dispersive X-ray spectroscopy system which was part of the experimental setup.

3. Results

3.1. Polarisation Experiments

The U_b values determined in the first series of polarisation experiments are listed in table 4. The pH was adjusted from its initial value by the addition of phosphoric acid or potassium hydroxide solution respectively. In the case that the electrolyte conductivity was so low that the compliance voltage of the booster was exceeded during the measurement the solution is marked with the ∘-sign. Electrolytes which precipitate during or shortly after the preparation are provided with a •-symbol. Electrolytes which allowed for the complete polarisation experiment and the determination of the σ-value are indicated by sigma. The statistical deviations of these measurements were less than 1 % and are not listed for the sake of clarity.

Table 4. Results of the first series of polarisation experiments.

substance		KOH	Na_2HPO_4			$NaAlO_2$		
c/mol·l^{-1}		varied	0.5	1.0	1.5	0.5	1.0	1.5
$pH_{initial}$			8.7	8.6	8.6	12.7	12.8	13.0
				U_b / V				
$pH_{adjusted}$	9		7	15	•			
	11		6	6	5	•		
	12	∘	6	4	4	σ	σ	σ
	13	11	4	3	•	29	31	31
	14	2	2	•	•	2	0.4	0.3

∘: conductivity too low; •: precipitation; σ: completed measurement (see table 5)

It can be seen that the phosphate electrolytes led to poor passivation behaviour even when directly compared to pure potassium hydroxide solution. The U_b values are between 2 V at pH = 14 and 15 V at pH = 9. In contrast to those within the aluminate electrolytes, breakthrough voltages of around 30 V can be achieved at pH = 13, while at a pH of 12 the polarisation experiment can be carried out completely. The corresponding $i(t)$ courses are depicted in figure 3 a. Inconveniently, it was unfeasible to prepare a volume of five litres for practical PEO experiments with the equivalent amount of phosphoric acid. The pH did not fall below 12.3. Further addition of phosphoric acid led to precipitation of the solution. This can be attribute to the buffering point of phosphates in this pH range.

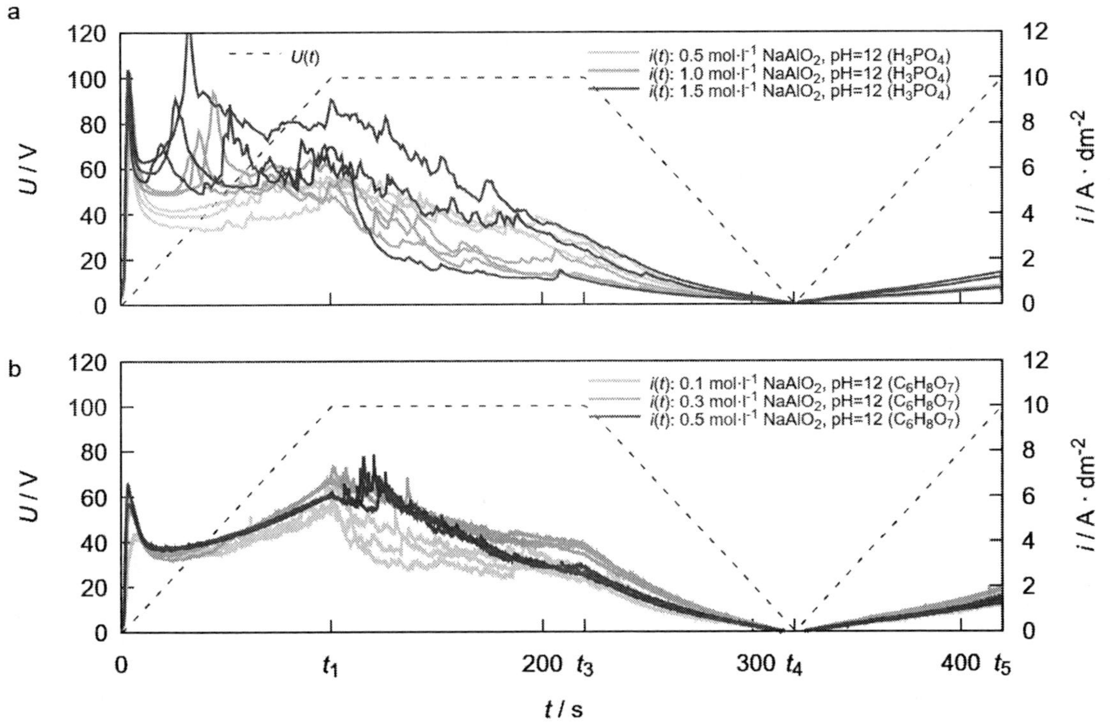

Figure 3. Results of the polarisation experiments obtained on C8C steel in aluminate solutions of pH = 12 adjusted through the addition of either phosphoric acid (a) or citric acid (b).

Therefore, a second series of polarisation experiments was carried out with aluminate solutions, the pH of which was adjusted by the addition of citric acid. Since electrolytes of 1 mol/l precipitated during this procedure, the experimental range was extended to 0.3 and 0.1 molare concentrations. The resulting $i(t)$ courses are depicted in figure 3 b. Table 5 summarises the experiments of the aluminate solutions with a pH of 12 and the determined σ-values.

Table 5. Results of the second series of polarisation experiments.

substance			Na_2AlO_2		
c/mol·l^{-1}	0.1	0.3	0.5	1.0	1.5
			σ/As·dm^{-2}		
pH = 12 by H_3PO_4			473 ± 14	318 ± 46	484 ± 213
pH = 12 by $C_6H_8O_7$	418 ± 52	495* ± 6	531±8	•	

•: precipitation, *electrolyte for the practical PEO experiments

The data only allow a statement regarding the influence of the substance used for pH adjustment on the resulting passivation behaviour for 0.5 molar solutions. The use of citric acid led to a slight deterioration. However, the σ-value decreased from 531 As/dm^2 to 418 As/dm^2 if the concentration was lowered down to 0.1 mol/l.

For the subsequent PEO experiments the solution of 0.3 mol/l aluminate with a pH of 12 adjusted by citric acid and a σ-value of 495 As/dm^2 was selected. This is a compromise between as-high-as-possible concentration for the better incorporation of electrolyte constituents into the layer, and the aim to initiate and maintain the plasma electrolytic discharges at moderate current densities by good passivation.

3.2. PEO process

Figure 4 shows typical recorded process informations at a PEO treatment time of 15 minutes after the initial voltage staircase. The data are represented with a temporal resolution in the range of milliseconds, which allows for the detailed evaluation of the generated pulse shapes.

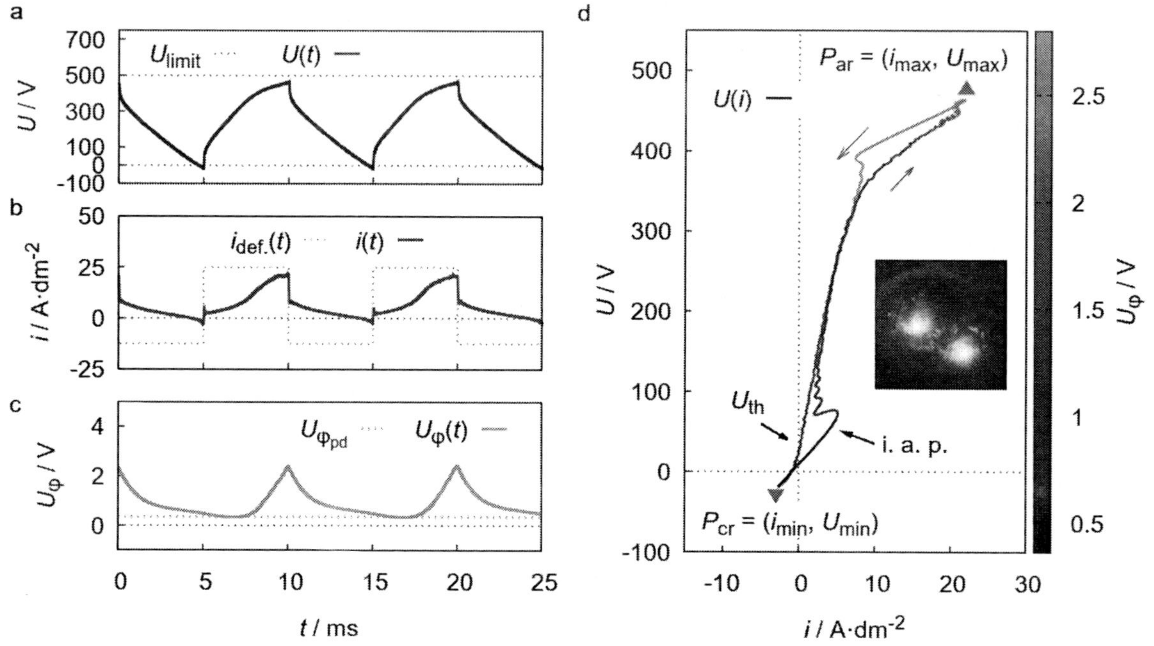

Figure 4. Graphical representation of the electrical pulse regime at 15 minutes of PEO treatment after initial voltage staircase: process-voltage (a), current-density (b), and photo-voltage (c) along with the associated limits, definitions, and pre-discharge values. These (a-c) are then summarised into a voltage current density-characteristic (d), with photo-voltage indicated by colour, combined with a image of the process.

In figures 4 a and b, the measured process-voltage and current-density patterns are depicted. Furthermore, the anodic voltage limitation and the shape of the predefined current-controlled pulse are marked by additional dashed lines. It can be seen that the set-point values are not achieved either during the anodic or the cathodic partial pulse. While the predefined current density is almost reached at the end of the positive half cycle, the following pulse part leads to polarity reverse only barely. The voltage curve shows a flattening at the end of the anodic partial pulse, which can be attributed to the approaching of the voltage limitation. In contrast, the amount of voltage change in the cathodic partial period remains largely constant. Figure 4 c shows the corresponding course of the photo-voltage which is proportional to the light emission. The horizontal dashed line at $U_{\phi\text{pd}} = 0.36$ V marks the photo-voltage value measured during the pre-discharge stage at the beginning of the experiment. The $U_{\phi\text{pd}}$ value can be attributed to background brightness of the experimental setup. It can be seen that rises of the photo voltage and thus discharge phenomena take place during the anodic partial pulse only. Despite the transition between the extremes being rather smooth, $U_{\phi\text{pd}}$ is achieved regularly during the cathodic partial pulse. This indicates a meantime interruption of the discharge phenomena.

Figure 4 d shows the combined data of 4 a–c without the time information. Furthermore, a screenshot of the process video is included in the diagram. The electrical process data are plotted according to a voltage–current-density characteristic (VCC) with the $U(i)$ curve. The grey arrows indicate the forward feed direction. The intensity of the measured photo voltage is depicted by the line colour and addressed by the corresponding colour bar. The achieved voltage and current density extremes are marked (by the anodic and cathodic reversal points P_{ar} and P_{cr}) as well as a characteristic voltage in the descending branch of the VCC where the current density is zero, which is referred to as the threshold voltage $U_{\text{th}} = 27$ V. It is thought that this value contains information regarding the PEO layers morphology and is rising with descending defect density [19]. Furthermore, an initial anodic peak (i. a. p.) can be observed in the rising branch of the $U(i)$ curve. Since the photo voltage is still in range of the $U_{\phi\text{pd}}$ value, this section can be attributed as belonging to the pre-discharge stage of the pulse. Therefore, the underlying processes are electrochemical and not plasma chemical in nature. While the ascending and descending branches are largely congruent in the later part of the $U(i)$-course, another loop appears around the anodic reversal point. Since the photo voltage assumes maximum values here, i.e. intense discharge phenomena take place, the system behaviour can be attributed to plasma chemical processes. The screenshot shows mainly two bright discharges, which illustrates a strong spatial localisation of the discharge phenomena.

Figure 5 represents the course of the characteristic values defined in 4 d as dependent on the process time. Additionally, the anodic voltage limitation, the predefined current-density extremes, and the $U_{\phi\text{pd}}$ value are depicted by dashed lines.

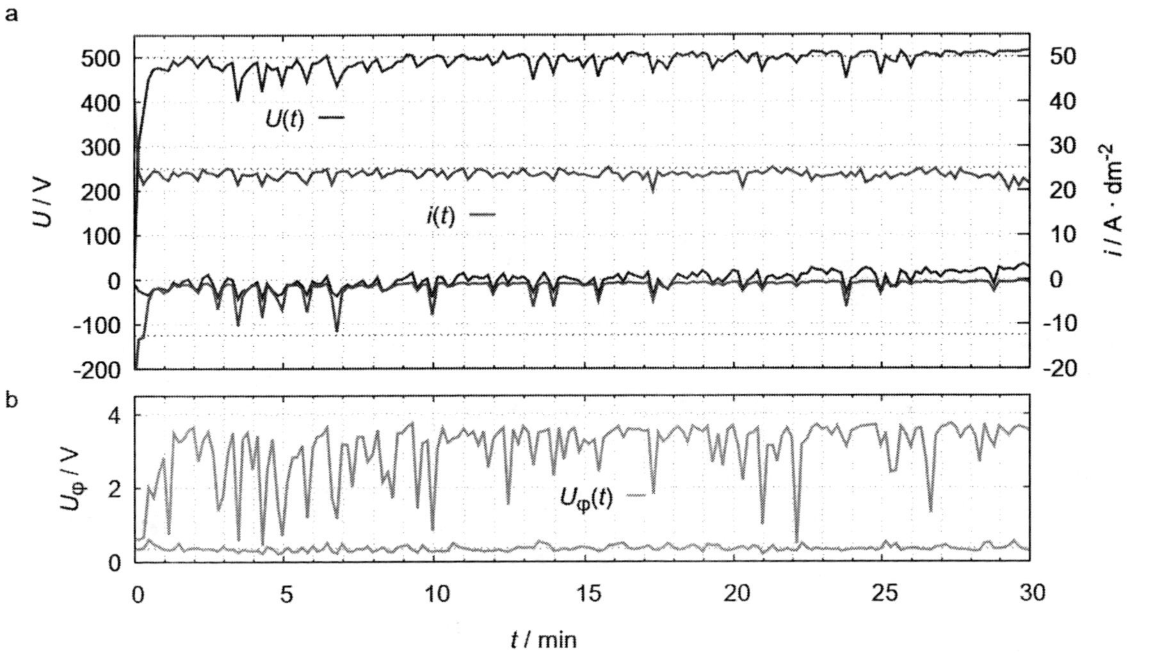

Figure 5. Course of the process voltage and current-density extremes (a) as well as the maximum and minimum values of the photo voltage (b) as dependent on the process time.

In figure 5 a, it can be seen that the anodic voltage limit is reached after about one minute, while the minimum voltage hardly reaches cathodic ranges over the entire process time. The current-density extremes show a similar behaviour in regards to reaching of the predefined vertices. The regularly occurring drops of the maximum values are noticeable, which coincide with more pronounced minima, which sometimes reach the negative range. Direct comparisons to the behaviour of the photo-voltage extremes, depicted in figure 5 b, show that these occur in temporal synchronisation with drops of the maximum photo-voltage, which can be attributed to "darker" process phases. The value of the minimum photo-voltage is in range of $U_{\phi \mathrm{pd}}$ during the entire process time. Hence it can be concluded that the discharges are interrupted regularly by the applied electrical regime.

The process-time dependent development of the discharge behaviour is shown for the initial DC voltage staircase in figure 6 and for the subsequent alternating pulse regime in figure 7. The photo series are based on the recorded video and allow for qualitative statements regarding the initiation and spatial distribution of the discharge phenomena. As can be seen in figure 6, plasma electrolytic discharge phenomena occurred for the first time at a DC voltage of 300 V. As the voltage was further increased up to a maximum potential/value/voltage of 425V, the light emission from the sample likewise increased.

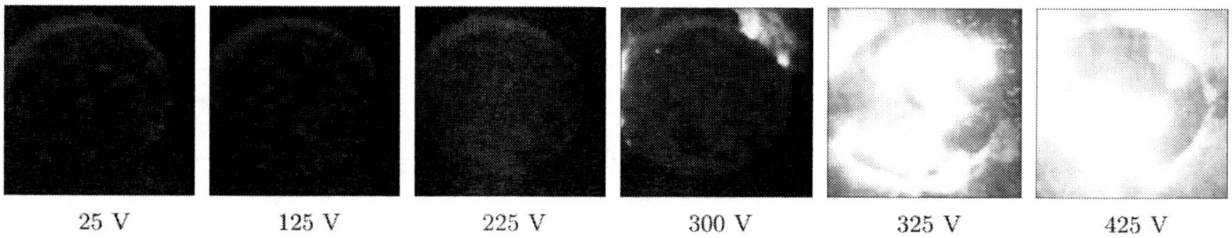

Figure 6. Figure 6. Process photo series of a PEO process during the initial DC voltage staircase with a 25V step every 25 s, with each individual photo labelled with the applied voltage.

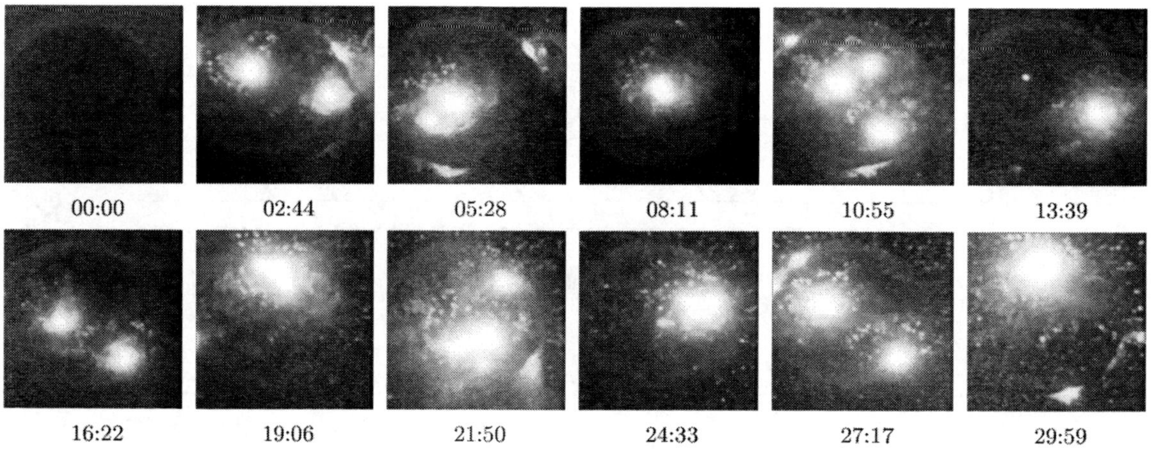

Figure 7. Process photo series of a PEO process during the alternating current-controlled pulse regime labeled by the process time in the mm:ss format.

During the subsequent PEO regime under alternating polarisation, the discharges initially came to a standstill, firstly. This can be seen in the first picture of the series in figure 7 and the low photo voltage in this process stage (5 b). After approximate 10 seconds, the discharges started again. Further experiments performed without the DC voltage staircase at the beginning or with lower maximum voltage during the DC stage (not presented in this study), showed much more pronounced delay in discharge initiation. Further on in the process, the discharge pattern remains largely qualitatively the same. There are rarely more than two or three discharges on the visible sample front. These move slowly over the substrate surface and burn for 5 to 30 seconds. In addition to the discharges, the video recordings seem to show significant gas evolution along the sample surface apart from the discharges. Unfortunately, this partial aspect cannot be clearly substantiated from the images due to their insufficient quality.

3.3. Coating characteristics

Micrographs of selected coating sections obtained by the SEM investigations are depicted in figure 8 and 9. Their respective extraction points are shown schematically in figure 2 b. Based on ten measurements of the sample extracted from position one, the layer thickness can be specified with an average of about 83 µm with a standard deviation of 4 µm. The coatings show a poor substrate bonding, and their microstructure seems to consist of two different constituents, one looking "glassy" with elongated defects, and one looking "foamy" with globular defects. While the characteristic areas are next to each other in Figure 8, a layered structure can be seen in Figure 9.

The results of the XRD measurements, shown in figure 10, indicate an absolute dominance of the α-Al_3O_3 corundum phase of approximate 90%. Hence, the particular microstructure cannot be explained by the presence of different phases. In order to trace the development of the layers phase composition during the process, additional PEO experiments were carried out with treatment times of 0, 10, 20, 30, and 40 minutes after the initial DC voltage staircase. It can be seen that α-Al_3O_3 is already present at the beginning of the process. The data clearly show that the layer consists almost exclusively of crystalline aluminium oxide, mainly the α- and η-phases. However, it was not possible to definitely assign a percentage of the layer composition to the two components.

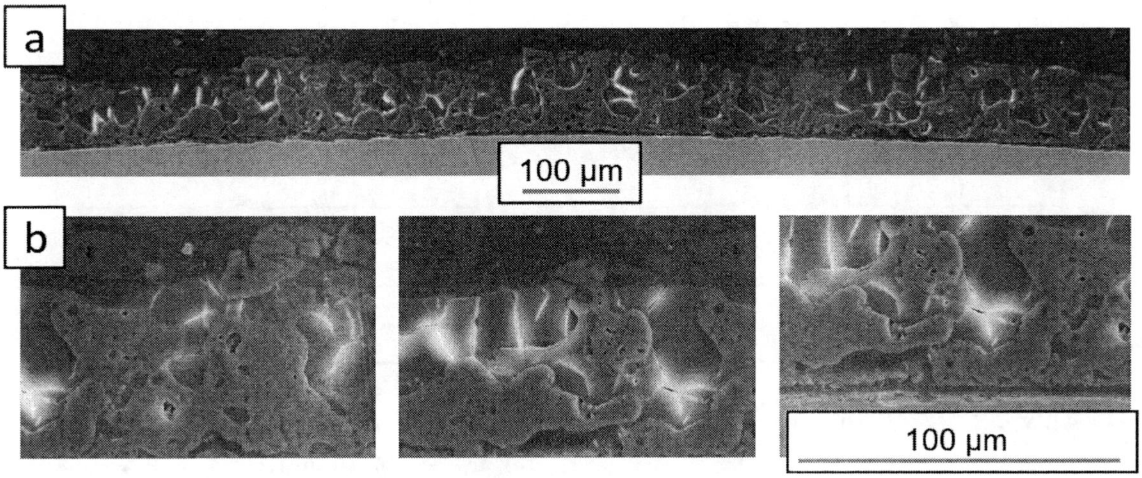

Figure 8. SEM images of the PEO coating at the sample's lateral surface (pos. 1 in figure 2 b), showing a microstructure consisting in two different areas with elongated and globular defects, arranged side by side.

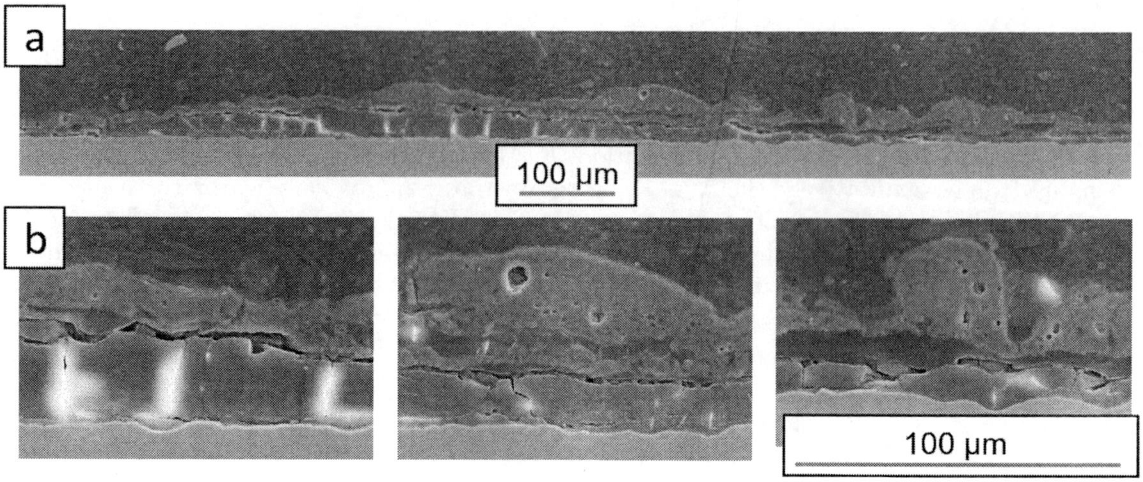

Figure 9. SEM images of the PEO coating at the samples top surface (pos. 2 in figure 2 b), showing a microstructure consisting in two different areas with elongated and globular defects, layered over each other.

While the visual appearance of the diffraction patterns indicate an absolute dominance of corundum, the software-supported evaluation lead to a ratio of 1:1 for the two phases. Iron oxides or other iron compounds formed with the electrolyte are not detectable. Therefore, the PEO, which in most of its variants is a conversion procedure, was successfully transferred into a deposition process. EDX measurements in both types of coating regions showed a composition of 56 % O and 43 % Al (atomic percent), which indicates Al_2O_3.

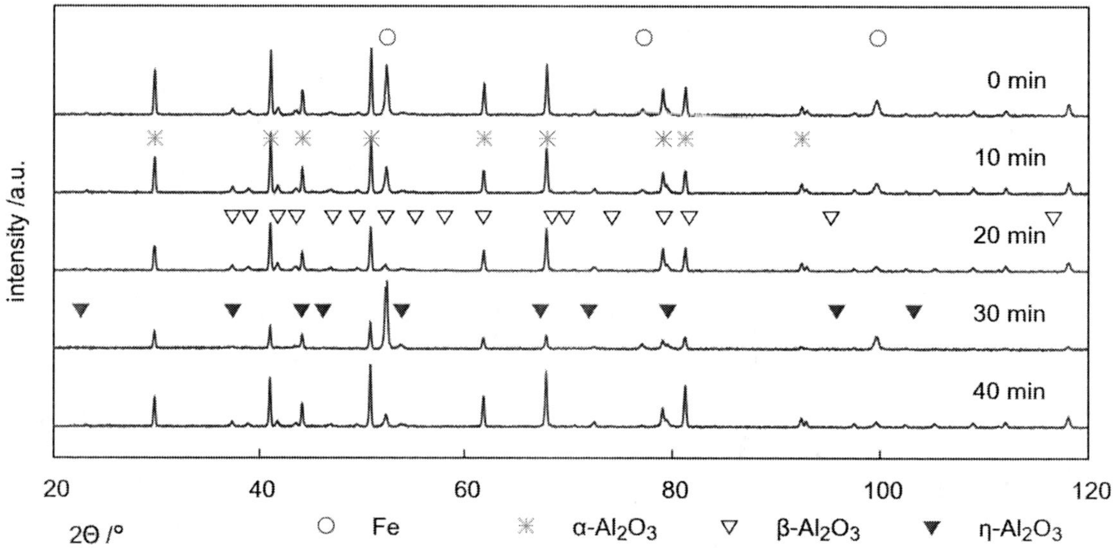

Figure 10. X-Ray diffractograms of PEO coatings obtained after different treatment times between 0 and 40 min after the initial DC voltage staircase, measured at the sample top area (position 2 in figure 2).

IOP Conf. Series: Materials Science and Engineering 1147 (2021) 012007 doi:10.1088/1757-899X/1147/1/012007

4. Discussion

The results of the polarisation experiments show that the systematic electrolyte variations led to an improved passivation on C8C steel. For aluminate electrolytes, it was possible to carry out complete measurements at moderate alkalinity of pH = 12. However, the σ-value of the electrolyte used for the practical PEO experiments in this study is, with 495 As/dm^2 still two times above the range which has been proven suitable for PEO experiments on Aluminium [20]. Therefore, the exhibited passivation behaviour may still be insufficient.

This presumption is supported by analysis of the process data shown in figure 4 d. The significant electrochemical anodic peak at the ascending branch of the $U(i)$ course stands for repassivation of bare metallic regions. However, since only very limited parts of the substrate surface are in direct interaction with the electrolyte even in the case of a porous layer with poor substrate bonding, the measured currents cannot be explained by passive film growth alone. This suggests the occurrence of parasitic reactions such as anodic metal dissolution and oxygen formation. After the initial anodic peak, the $i(U)$-curve rises again, first following a linear then exponential relationship to the applied voltage. Light emission can only be detected above 400 V. Thus, the aforementioned coating-deteriorating electrochemical processes become more pronounced within this pulse stage. The plasma-chemical occurrences around and after the anodic reversal point cannot be interpreted in more depth with the experimental setup used. The descending branch of the $U(i)$ course only crosses the zero line at a very low threshold voltage of U_{th}=27 V. This proves that the PEO coating consisting of crystalline alumina with a thickness of approximately 80 µm must have many defects that almost reach the substrate. Otherwise, its electrical resistance and the resulting threshold voltage would be significant higher.

This assumption is confirmed by the SEM images shown in figures 8 and 9. The layer structure shown here with its inhomogeneous phase composition and poor substrate bonding differs significantly from the alumina layers on aluminium materials produced by PEO [21]. The mentioned structural defects could have originated from strong localisation of the intense and long-lasting plasma electrolytic arc discharges during the process, visible in figures 6 and 7. The regularly appearing drops in the electrical process variables and the maximum photovoltage, as depicted in figure 5, occur in temporal synchronisation, which indicates that the PEO coating is damaged and subsequently healed several times during the process. A further hint is the uneven development of the substrate-related iron peak intensity over the process time in the diffraction diagrams in figure 10. The appearance and dominance of the arc discharges can be explained by the fact that the cathodic partial pulse was applied insufficiently on the experimental setup. As it can be seen in figures 4 and 5 the process-voltage and current density barely reach negative values at the end of the pulse. The course of the photo voltage shows that this is enough to interrupt the discharges, by reaching of $U_{\phi pd}$. However, it can be doubted that this is sufficient to repel the anions at the substrate/electrolyte interface and to pull them back evenly distributed in the subsequent anodic partial pulse. Such a homogenised distribution of the specimen along the quasi-cathode would be a prerequisite for uniformly distributed spark discharges.

According to figure 10, the particularly stable high-temperature modification α-Al_2O_3 is dominant within the layer composition. Furthermore, it appears not just in later process stages as is known from PEO of aluminium [22]. This can be interpreted as a positive result of the energy-intense arc discharges.

Based on the made observations and their interpretation, the following procedure is suggested for further investigations.The selection of the electrolyte used for PEO was a compromise between the best possible passivation and highest possible electrolyte concentration in order to incorporate electrolyte constituents into the resulting layer. Since it was possible to produce layers that consisted almost exclusively of aluminium oxide, the electrolyte concentration could be reduced further in favour of better passivation. This study does not provide final experimental evidence that the bubbles visible on the process recordings originated in any significant extent from electrochemical anodic gas formation. Since electrolyte evaporation in the vicinity of the discharges could also be the explanation, it is recommend that pulses with anodic voltage limitations below the process time dependent ignition voltage should be coupled with the electrical regime. If the gas formation persists, it can then be attributed solely to the electrochemical processes. In order to finally clarify the coatings phase structure, measurements using electron backscatter diffraction (EBSD) are recommended. A promising approach to homogenise the layer microstructure and to improve the substrate bonding is to shift the discharge behaviour in favour of more finely distributed spark discharges. Therefore, the cathodic part of the electrical regime should be more pronounced by adjusting the current value and/or the duration of the negative partial pulse.

5. Summary and conclusion

The results of this study can be summarised as follows.

- Systematic polarisation experiments have been used successfully to preselect an environment friendly electrolyte suitable for PEO of low-carbon steel.
- The resulting coatings have the following desirable characteristics:
  i) they grow fast (approximately 80 µm after 37 minutes),
  ii) consist exclusively of crystalline aluminium oxide,
  iii) and aim a high content of corundum.
- The coatings are limited in their suitability for practical applications by the following weaknesses:
  i) their inhomogeneous microstructure
  ii) and poor substrate bonding.
- The results of the electrochemical measurements and the analysis of the PEO process data indicate that this can be explained as follows:
  i) the passivation behaviour of the electrolyte is still insufficient,
  ii) to less pronounced cathodic partial pulses in the electrical regime lead to a strong discharge localisation during the PEO.
- An experimental procedure was proposed to check these assumptions and to solve the named issues.

The PEO, which in most of its variants is a conversion procedure, was successfully transferred into a deposition process in the course of this study. This allows thick, corundum-rich aluminium-oxide layers to be produced on low-carbon steel in a short process time. Further research is required to improve their technological properties. The knowledge gained can be used for the plasma electrolytic oxidation of steel and steel-based material composites.

Acknowledgments

The financial support of this work by the Deutsche Forschungsgemeinschaft (DFG) is gratefully acknowledged (project number La-1274/46-1). The authors thank the company Fr. u. H. LÜLING GmbH & Co. KG Stahldrahtwerk for providing the C8C steel material and Morgan Uland for the English proofreading.

References

[1] Yerokhin A L, Nie X, Leyland A, Matthews A and Dowey S J 1999 Plasma electrolysis for surface engineering *Surf. Coat. Technol.* **122**(2-3) 73–93
[2] Wielage B, Alisch G, Lampke T and Nickel D 2008 Anodizing – a key for surface treatment of aluminium *Key Eng. Mater.* **348** 263–281
[3] Simchen F, Sieber M, Kopp A and Lampke T 2020 Introduction to plasma electrolytic oxidation – an overview of the process and applications *Coatings* **10**(7)
[4] Wang Y, Jiang Z and Yao Z 2009 Microstructure, bonding strength and thermal shock resistance of ceramic coatings on steels prepared by plasma electrolytic oxidation *Appl. Surf. Sci.* **256**(3) 650–656
[5] Ma C, Liu J, Zhu X, Xue W, Yan Z, Cheng D, Fu J and Ma S 2020 Anticorrosive non-crystalline coating prepared by plasma electrolytic oxidation for ship low carbon steel pipes *Sci. Rep.* **10**(1)
[6] Yang W, Liu W, Peng Z, Liu B and Liang J. 2017 Characterization of plasma electrolytic oxidation coating on low carbon steel prepared from silicate electrolyte with al nanoparticles *Ceram. Int.* **43**(18) 16851–16858
[7] Pezzato L, Brunelli K, Magrini M and Dabalà M 2015 Corrosion resistance and surface characteristics of plasma electrolytic oxidation coating produced on steels *EUROCORR 2015* **3** 1498–1505
[8] Pezzato L, Brunelli K, Dolcet P and Dabalà M 2016 Plasma electrolytic oxidation coating produced on 39nicrmo3 steel *Surf. Coat. Technol.* **307** 73–80
[9] Karpushenkov S A, Shchukin G L, Belanovich A L, Savenko V P and Kulak A I 2010 Plasma electrolytic ceramic-like aluminum oxide coatings on iron *J. Appl. Electrochem.* **40**(2) 365–374
[10] Karpushenkov S A, Kulak A I, Shchukin G L, and Belanovich A L 2010 Microplasma electrochemical deposition of aluminum oxide-polyethylene composite coatings on iron surface *Prot. Met. Phys. Chem.* **46**(4) 463–468
[11] Wang Y, Zhu L, Wang M, Zhang M, Yang Y, Zhu Q and Lei Y 2016 Microstructure and photocatalytic activity of porous tio2–fe2o3 composite film by peo coupled with post heat treatment *Int. J. Hydrogen Energy* **41**(35) 15703–15709
[12] Wang J, Yao Z, Yang M, Wang Y, Xia Q and Jiang Z 2016 A fe3o4/feal2o4 composite coating via plasma electrolytic oxidation on q235 carbon steel for fenton-like degradation of phenol *Environ. Sci. Pollut. Res. Int.* **23**(15) 14927–14936
[13] Yang W, Li Q, Liu W, Liang J, Peng Z, and Liu B 2017 Characterization and properties of plasma electrolytic oxidation coating on low carbon steel fabricated from aluminate electrolyte *Vacuum* **144** 207–216
[14] Lampke T, Meyer D, Alisch G, Nickel D, Scharf I, Wagner L, and Raab U 2011 Alumina coatings obtained by thermal spraying and plasma anodising – a comparison *Surf. Coat. Technol.* **206**(7) 2012–2016
[15] Martin J, Akoda K, Ntomprougkidis V, Ferry O, Maizeray A, Bastien A, Brenot P, Ezo'o G and Henrion G 2020 Duplex surface treatment of metallic alloys combining cold-spray and plasma electrolytic oxidation technologies *Surf. Coat. Technol.* **392** 125756
[16] López-Ortega A, Arana J L, Rodríguez E and Bayón R 2018 Corrosion, wear and tribocorrosion performance of a thermally sprayed aluminum coating modified by plasma electrolytic oxidation technique for offshore submerged components protection *Corros. Sci.* **143** 258–280
[17] Thomä M, Wagner G, Straß B, Wolter B, Benfer S, and Fürbeth W 2018 Ultrasound enhanced friction stir welding of aluminum and steel: Process and properties of en aw 6061/dc04-joints *J. Mater. Sci. Technol.* **34**(1) 163–172
[18] Thomä M, Wagner G, Straß B, Wolter B, Benfer S, and Fürbeth W 2018 The influence

of ultrasound enhancement during friction stir welding of aluminum to steel *Key Eng. Mater.* **767** 351–359
[19] Suminow I W, Epelfeld A B, Ljudin W B, Krit B L and Borisow A M 2005 Micro arc oxidation - theory, applications, equipment. *Ekomet, Moscow*
[20] Simchen F, Sieber M, Mehner T and Lampke T 2020 Characterisation method of the passivation mechanisms during the pre-discharge stage of plasma electrolytic oxidation indicating the mode of action of fluorides in peo of magnesium *Coatings* **10**(10) 965
[21] Sieber M, Simchen F, Morgenstern R, Scharf I and Lampke T 2018 Plasma electrolytic oxidation of high-strength aluminium alloys—substrate effect on wear and corrosion performance *Coatings* **8**(5) 356
[22] Yerokhin A L, Snizhko L O, Gurevina N L, Leyland A, Pilkington A and Matthews A 2003 Discharge characterization in plasma electrolytic oxidation of aluminium *J. Phys. D: Appl. Phys.* **36**(17) 2110–2120

IOP Conf. Series: Materials Science and Engineering 1147 (2021) 012008 doi:10.1088/1757-899X/1147/1/012008

Influence of the cutting material on tool wear, surface roughness, and force components for different cutting speeds in face turning of CoCrFeNi high-entropy alloys

H Liborius[1*], T Uhlig[2], B Clauß[1], A Nestler[1], T Lindner[3], A Schubert[1], G Wagner[2] and T Lampke[3]

[1] Chemnitz University of Technology, Institute for Machine Tools and Production Processes, Professorship Micromanufacturing Technology, 09107 Chemnitz, Germany
[2] Chemnitz University of Technology, Institute of Materials Science and Engineering, Professorship of Composites and Material Compounds, 09107 Chemnitz, Germany
[3] Chemnitz University of Technology, Institute of Materials Science and Engineering, Materials and Surface Engineering Group, 09107 Chemnitz, Germany

[*]e-mail: hendrik.liborius@mb.tu-chemnitz.de

Abstract. High-entropy alloys (HEAs) provide advanced properties like a high resistance to corrosion and wear. Concerning this and the elevated material costs wear protection layers are a possible field of application. Finish machining of these layers is necessary to achieve adequate surface properties. In the experimental investigations face turning of high-entropy alloy CoCrFeNi layers generated by spark plasma sintering is regarded. In this context, the influence of the cutting material and the cutting speed is analysed. For this, CBN tipped (two types), PCD tipped, CVD diamond tipped, and solid cemented carbide indexable inserts are used. Additionally, the influence of four different cutting speeds in the range between 100 m/min and 400 m/min is analysed. The feed and the depth of cut are kept constant with 0.05 mm and 0.1 mm, respectively. The geometrical surface properties are determined by tactile measurements and 3D laser scanning microscopy. The tool wear is analysed microscopically. Regarding the surface roughness, CBN tipped tools with a high content of boron nitride lead to the best results with the lowest roughness values irrespective of the cutting speed. The tool wear is also significantly reduced compared to the other cutting materials tested. The investigations represent the first results analysing the influence of the cutting material in machining of high-entropy alloys. Hence, this contributes to the enhancement of the field of application of HEAs.

1. Introduction

High-entropy alloys (HEAs) represent a relatively new material group with properties like a high resistance to corrosion and wear. They are characterised by a composition of at least four different elements with nearly equimolar proportions. Regarding the comparatively high costs of these materials, the application of HEAs as protection layers is reasonable. Especially in tribological applications finish machining becomes necessary. Processes with geometrically defined cutting edges provide a high geometrical flexibility. Next to the geometrical surface properties, an adaption of the physical properties of the surface layer (e. g. hardness, residual stresses) by these processes is possible. Machining is generally associated with tool wear. Hence, the tool geometry and resulting surface properties change. Accordingly, a low tool wear rate and long tool life, respectively are aspired. Next to the workpiece material, the tool wear is influenced by the cutting material and the machining parameters. Knowledge of the relationships between machining parameters, tool wear, and resulting surface properties enables an adaption of the machining parameters depending on the state of tool wear. Additionally, the determination of an appropriate cutting material for machining of HEAs represents an important step for the application of HEA coatings in tribological systems.

Concerning the machining of high-entropy alloys only few results are published. In investigations of Clauß et al. [1] in turning of AlCoCrFeNiTi thermally sprayed HEA layers applied by atmospheric plasma spraying the influence of the cutting speed (100 m/min – 400 m/min) and the feed (0.05 mm, 0.1 mm) on the surface properties was analysed. The results showed a decrease of the surface roughness values with increasing cutting speed, as a consequence of the reduced proportion of pulled-out coating material. The minimum values for the surface roughness depth *Rz*, the reduced valley depth *Rvk,* and the valley void volume *Vvv* of the machined surfaces were reached when using a cutting speed of 300 m/min. The small increase of the mentioned values for higher cutting speeds of up to 400 m/min was attributed to increasing tool wear. Additionally, the absolute values of the compressive residual stresses in the axial / feed direction decreased with increasing cutting speed.

Huang et al. [2] analysed the influence of the cutting material (CBN and diamond), the rotational speed (1,000 min^{-1} – 2,500 min^{-1}), the feed (0.5 μm – 10 μm), and the depth of cut (0.5 μm – 20 μm) in face turning of Al80Li5Mg5Zn5Cu. Generally, the measured surface roughness depth was significantly higher in case of using CBN compared to using diamond tools. Increasing the rotational speed in the mentioned range led to slightly increased roughness values when using CBN tipped tools as a result of higher tool wear. For the use of diamond tools regardless of the cutting speed the measured roughness values correlated to the kinematic roughness. The use of CBN generally resulted in a higher tool wear and an increased thickness of the affected surface layer, as a consequence of the increasing friction between tool and workpiece compared to utilising diamond tools. Machining (rotational speed: 2,000 min^{-1}, feed: 5 μm, depth of cut: 8 μm) using CBN tipped tools led to a slight increase of the surface hardness (218 HV), whereas machining using diamond tools resulted in a decrease of the measured hardness (191 HV) compared to the raw material (211 HV).

Guo et al. [3] analysed the influence of different processes for finishing (EDM, electro and chemical polishing, grinding, milling, and process combinations) of CoCrFeMnNi high-entropy alloy specimens, produced by selective laser melting. The initial surface roughness (*Ra* about 30 μm) could be reduced significantly by finishing. Grinding led to an arithmetic mean surface roughness *Ra* of 4 μm. For EDM values of 3 μm were reached. However, milling allowed for the lowest surface roughness values represented by *Ra* of 1 μm. Mechanical polishing offered the possibility to significantly reduce the surface roughness *Ra* down to 0.003 μm. Additionally, grinding led to a surface hardness of about 350 HV0.2, EDM to about 400 HV0.2 and milling to about 450 HV0.2. As a result of the high forces and the plastic deformation, milling (about -700 MPa) and grinding (about -400 MPa) entailed strong compressive residual stresses after machining. It could be observed that after grinding and milling, respectively the microstructure was deformed in the cutting direction. Causing the higher cutting forces in milling compared to grinding, this disorientation was more severe in the case of machining with geometrically defined cutting edges.

The cutting material strongly influences the tool wear rate and the resulting surface properties after machining. This influence depends on the workpiece material. Additionally, the wear mechanisms of cutting tools are affected by the shear zone temperature. Depending on the cutting tool and the workpiece material composition softening of the cutting material as well as diffusion or adherence effects are possible. Concerning the few research studies and results in machining of high-entropy alloys, experimental investigations regarding the influence of the cutting tool material are necessary for upcoming tests analysing the effects of the machining parameters and the tool geometry. The material properties of the high-entropy alloy CoCrFeNi were properly analysed in numerous scientific material investigations. Accordingly, this alloy is appropriate for machinability investigations.

IOP Conf. Series: Materials Science and Engineering 1147 (2021) 012008 doi:10.1088/1757-899X/1147/1/012008

2. Experiments

2.1. Specimens

In the experimental investigations circular disc specimens of the high-entropy alloy CoCrFeNi were machined by face turning. To reduce the material consumption, substrates of the steel 1.4404 (AISI 316L) were utilised. These substrate discs exhibit a diameter of 40 mm and a thickness of 10 mm. The surface of these steel discs was roughened by grit blasting in order to achieve a good adhesion of the sintered coatings. On one face of the discs the high-entropy alloy layer was applied using spark plasma sintering (SPS). In this process, the HEA powder with a grain size of 20 µm to 50 µm was sintered using a constant pressure of 50 MPa. After a heating time of 10 min, a sintering temperature of 1050 °C was maintained for 5 min. The resulting thickness of the layers was about 3.5 mm. The substrates were used for clamping the specimens during the machining experiments. Hardness measurements at the unmachined specimens were conducted on the sintered surface. The values average out at 259 HV0.5.

2.2. Tools

For analysing the influence of the cutting material, five different grades were used applying indexable inserts of the type CCGW 09T304. The tools were characterised by a nominal rake angle of 0°. In connection with the tool holder used the cutting edge angle of the minor cutting edge was 5°.

Referring to the first published results for machining of thermally sprayed high-entropy alloys [1], a cutting tool with a CBN tipping characterised by about 90 % – 95 % boron nitride, a grain size of 1 µm, and a cobalt-based (metallic) binder was used (CBN 90). This cutting material is mostly used for machining grey cast iron. Additionally, a CBN tipped indexable insert with a proportion of about 50 % boron nitride, 1 µm grain size and a ceramic binder was utilised (CBN 50). The cutting material with the equal proportion of boron nitride and binder is generally used for machining of hardened steels. Furthermore, indexable inserts with tips consisting of Polycrystalline diamond (PCD) and CVD diamond (CVD) were tested in the experimental investigations. Additionally, a cemented carbide (Carbide) indexable insert, characterised by a grain size of the WC particles less than 1 µm and a mass proportion of cobalt (binder) of about 6 % was used.

To ensure comparable machining conditions, all indexable inserts provide a sharp cutting edge with a rounding of less than 10 µm. For a more detailed analysis, the cutting edges were recorded using a 3D laser scanning microscope Keyence type VK-9700. For the evaluation of the data the software MountainsMap® was used. In Table 1 the cutting edge radius for the different tools is shown. Although there were small differences regarding the cutting edge rounding, the change of the effective rake angle in relation to the parameters in the machining experiments was acceptable.

Table 1. Measured cutting edge radius of the tools before machining.

Tool	CBN 90	CBN 50	CVD	PCD	Carbide
Cutting edge rounding (µm)	2.8	7.6	3.6	3.7	5.7

2.3. Experimental investigations

The experimental investigations were carried out on a precision lathe SPINNER type PD 32. The specimens were clamped on the outer diameter of the substrate using a chuck. In the experimental investigations face turning was realised. To guarantee a constant cutting speed while machining, an increase of the rotational speed with decreasing radius of the tool engagement point was necessary. Concerning the limited rotational speed of the lathe and the acceleration of the rotational and the translational axes, machining the whole face with a constant cutting speed was not possible. Therefore the specimens were premachined by boring and internal turning. Afterwards, the specimens were characterised by an inner diameter of 25 mm and an outer diameter of 40 mm. The specimen geometry is shown in Figure 1.

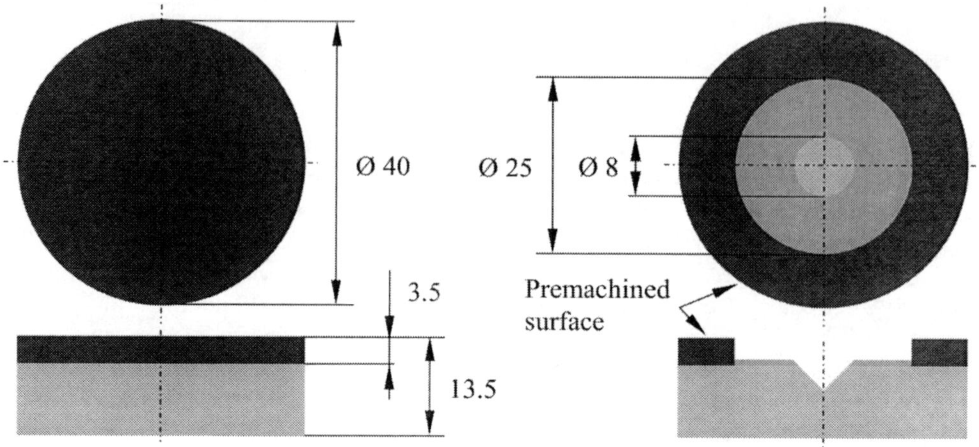

Figure 1. Geometry of the specimens used: before (left) and after premachining (right).

The machining experiments were carried out at the ring-shaped surface between the inner and the outer diameter which was characterised by a resulting width of 7.5 mm. This area was premachined in several steps to ensure similar starting conditions and to reduce the initial roughness using tools of the type CBN 90. Premachining and finish machining were done dry using a constant depth of cut of 0.1 mm and a feed of 0.05 mm. For premachining the cutting speed amounted to 200 m/min. The influence of the cutting material and the cutting speed was analysed according to the experimental design represented in Table 2.

Table 2. Experimental design.

Experiment number	Cutting material	Cutting speed (m/min)			
1 – 4	CBN 90	100	200	300	400
5 – 8	CBN 50				
9 – 12	CVD				
13 – 16	PCD				
17 – 20	Carbide				

For each cutting material used the influence of four different cutting speeds was analysed. Thereby it was possible to influence the temperature in the shear zone. Hence, it was possible to define appropriate, but also not suitable ranges of cutting speed and temperature for the respective cutting materials regarding the tool wear and the resulting geometrical surface properties. Generally, the shear zone temperature increases with increasing cutting speed (see [4]). For every combination of cutting material and cutting speed the specimens were machined three times (three cutting steps). The tool was not changed between these experimental investigations. Tool wear and geometrical surface properties were analysed after one and after three cutting steps. In the experimental investigations the components of the resultant force were measured by a three-axis dynamometer Kistler type 9257A on which the tool was mounted.

2.4. Analysis of the tools and the specimens

After machining the influence of the cutting material and the cutting speed on the tool wear was evaluated. The tool wear was analysed using a 3D microscope Keyence type VHX-100, an optical microscope Nikon type MM-400, and a 3D laser scanning microscope Keyence type VK-9700. The tool wear was evaluated qualitatively regarding the type of tool wear and also quantitatively by measuring the flank wear land width.

The geometrical surface properties of all specimens machined were analysed using a stylus instrument Mahr type LD 120. The stylus was characterised by a radius of 2 µm and an included angle of 90°. The measuring was done between the inner and outer diameter in the feed / radial direction. The measuring length was 4 mm. The filtering of the profile was done in accordance to ISO 11562. For a validation of the roughness values, each specimen was measured at five different positions. Additionally, the specimens were measured using a 3D laser scanning microscope Keyence type VK-9700. The size of the analysed area was 1 mm × 1 mm.

3. Results and discussion

3.1. Tool wear

After machining the tool wear was detected. As quantitative evaluation criterion the flank wear land width *VB* was analysed. Additionally, qualitative criterions like built-up edge formation or material adherence were regarded. The tool wear was detected after one and three machining steps. Generally, the relationship between cutting speed and tool wear depends on two simultaneously acting mechanisms. Increasing the cutting speed leads to an increased shear zone temperature resulting in a reduction of the strength of the cutting material or its binder. Additionally, increasing the cutting speed causes a quadratic increase of the kinetic energy of the workpiece material included hard phases like carbides.

The influence of the cutting material and the cutting speed on the flank wear land width is summarised in Table 3. Additionally, an indexable insert (CBN 90) before and after machining with the highest cutting speed is shown in Figure 2.

Table 3. Results of the tool wear analyses after three machining steps.

Cutting material	Tool wear *VB* (µm) depending on the cutting speed			
	100 m/min	200 m/min	300 m/min	400 m/min
CBN 90	30	40	50	65
CBN 50	45	50	45	55
CVD	110	65	110	Cutting edge breakage
PCD	60	40 – 60	50	60
Carbide	65 – 85	70	100	200

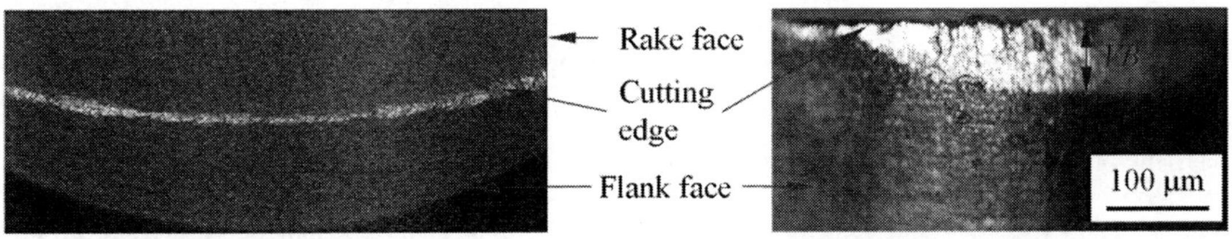

Figure 2. Indexable insert (CBN 90) detected by 3D laserscanning microscopy before machining (left) and by optical microscopy at the flank face after machining three specimens with a cutting speed of 400 m/min (right).

Generally, a continuous tool wear leads to an increase of the *VB*-values. The increase of the flank wear land width depends on the properties of the specimens, the tools, the process parameters, and the machined length. Regarding the predictability of this tool flank wear, the tool life time and the impact on the properties of the machined surfaces are also assessable.

The lowest tool wear was achieved by using the CBN tipped tools with the high proportion of boron nitride (CBN 90). The state of wear was characterised by flank wear. Nearly no chipping of the cutting edge occurred. The flank wear land width increased with increasing cutting speed. This was a result of the raising shear zone temperature and the subsequently decreasing strength of the cutting material binder. The decreasing strength and hardness with increasing temperature for cobalt and cobalt-based alloys was shown in [5] and [6]. Independent of the CBN properties (grain size, proportion of binder) there was a continuous decrease of hardness with increasing temperature (see [7]). Additionally, the increased kinetic energy of the workpiece material results in an easier separation of the BN grains from the binder. In [8] tools characterised by similar cutting material were used. In the experimental investigations a decrease of hardness and transverse rupture strength with increasing temperature was shown. In the analysed range of cutting speed the maximum flank wear land width was reached after three machining steps with the highest cutting speed ($VB = 65$ μm). Additionally, it should be mentioned that after the first cutting step the flank wear land width of the tools was naturally lower (up to 30 μm). For cutting speeds of 100 m/min and 200 m/min nearly no tool wear could be detected after the first cutting step. Additionally, in the case of using CBN 90 independent of the cutting speed no built-up edge occurred after machining.

When using CBN with the lower proportion of boron nitride (CBN 50), for all cutting speeds similar values concerning the flank wear land width could be detected. The largest value was reached after machining with the highest cutting speed ($VB = 55$ μm). All tools were characterised by severe cutting edge chipping after three machining steps. But the number and size of the microcracks increased with raising cutting speed and subsequently increasing shear zone temperature. This was caused by the reduced hardness and strength of the cutting material binder with increasing shear zone temperature. The effect of reducing the hardness with increasing temperature was shown for different ceramics in [9]. The ceramic binder of the CBN led to a less reduced transverse rupture strength with increasing temperatures compared to cutting materials with a cobalt-based binder. This was shown for similar tools in [8]. The transverse rupture strength for CBN with 85 % to 90 % boron nitride bounded by a metallic binder was about 140 kg/mm^2 at room temperature and decrease to 55 kg/mm^2 at 1000 °C. For CBN characterised by a proportion of 50 % - 55 % boron nitride and a ceramic binder, the transverse rapture strength (105 kg/mm^2 – 110 kg/mm^2) was constant in the same temperature range. This effect results in the similar values for the flank wear land width depending on the cutting speed in machining of CoCrFeNi. Additionally, there was a higher proportion of binder in the cutting material. The strength between the boron nitride and the binder was lower when using a ceramic binder compared to the also utilised cobalt binder. Hence, the increased occurrence of cutting edge chipping could be explained. Additionally, no diffusion effects between the cutting material binder and the workpiece material were expected in the case of using CBN characterised by a ceramic binder. The results conform to experimental investigations with similar cutting materials [10], in which a more strongly increasing tool wear rate with increasing cutting speed in the case of using CBN with a high content of boron nitride and metallic binder compared to CBN with a lower content of boron nitride and a ceramic binder was also visible. This was caused by a more intensive diffusion between the metallic binder and the specimen material with increasing cutting speed. Additionally, changes in the thermal conductivity, defect density within the grains of the tool, and tribo-chemical effects have to be taken into consideration.

Diffusion effects between the iron within the high-entropy alloy and the cutting material were expected to be an essential wear mechanism when using CVD diamond or PCD tipped tools. The HEA used consists of about 25 at.% iron. This value is significantly lower than in iron-based alloys. For machining of iron-based alloys the use of diamond tools is not common. Concerning other HEAs with a higher number of different elements, the proportion of iron is minor and it is expected, that the diffusion effect and subsequently the tool wear could be reduced. For a raising cutting speed, the increasing tool wear, for example the size and proportion of microcracks or the cutting edge chipping, could be explained by the higher temperatures in the shear zone. The increased temperatures accelerated the diffusion mechanisms. Additionally, when using PCD, there was again a softening of the cutting material binder. After machining, the CVD and PCD indexable inserts were characterised by material adherence (built-up edge). In case of machining with the highest cutting speed, the CVD tool broke at the cutting edge.

The cemented carbide tools showed the highest flank wear land width after machining. After three machining steps with a cutting speed of 400 m/min the highest flank wear land width occurred (*VB* = 200 µm). This was twice the flank wear land width compared to machining with a cutting speed of 300 m/min. The cutting material consisting of tungsten carbide particles bound by cobalt is typically not used in this range of cutting speeds. Cemented carbides do not belong to the super-hard cutting materials, like the others used in the experiments. Concerning this, the high tool wear and especially the differences between machining with cutting speeds of up to 300 m/min and 400 m/min were caused by the increased thermal softening of the cutting material. For all cutting speeds the formation of a built-up edge was visible after machining.

3.2. Components of the resultant force

The components of the resultant force were measured while machining. In Figure 3 the influence of the cutting material on the cutting force (F_c), the passive force (F_p), and the feed force (F_f) is shown. The mean values were calculated from the average values of the components of the resultant force disregarding the different cutting speeds. For the experiments, the respective mean values were calculated in the range of a theoretically constant cross-section of the undeformed chip. The first contact and subsequent raising of the cross-section of the undeformed chip as well as the decreasing cross-section of the undeformed chip at the end of the cutting process were omitted. The averaging for different cutting speeds was possible, because there was only a small difference between the mean values of the components of the resultant force for the different cutting speeds. This difference was smaller than the variation of the values during machining of the single specimens. The scattering bars shown in the diagram represent the highest and the lowest mean value out of the four single experiments.

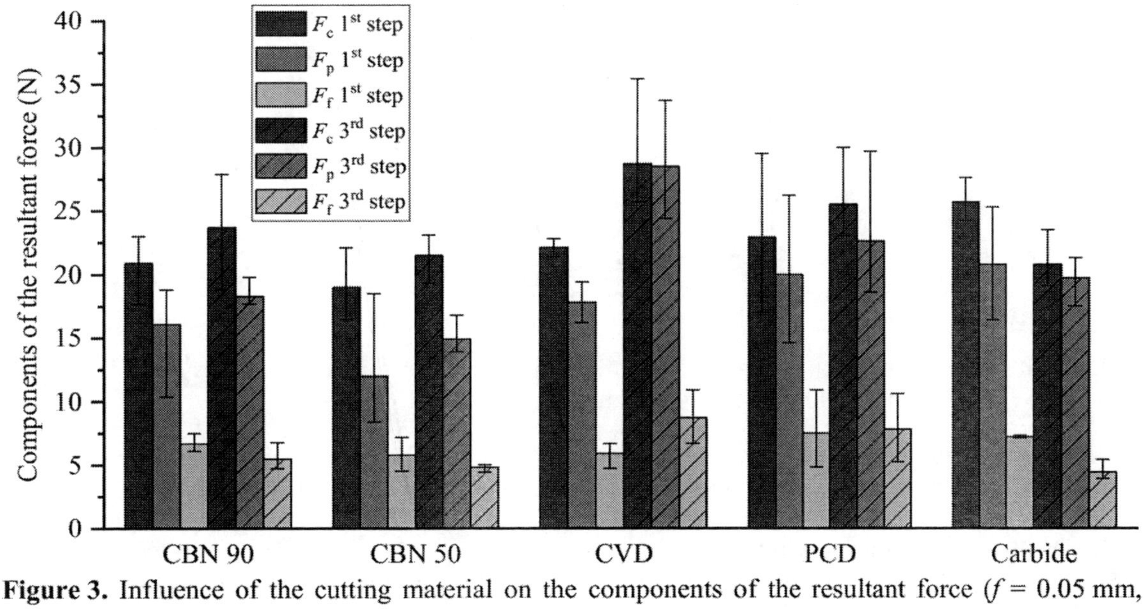

Figure 3. Influence of the cutting material on the components of the resultant force (f = 0.05 mm, a_p = 0.1 mm).

For every cutting material used the cutting force represented the largest component of the resultant force followed by the passive and the feed force. Regarding the super-hard cutting materials, a slight increase of the average force components for the third cut compared to the first cut was visible. This resulted as a consequence of the progressing tool wear. In case of using cemented carbide tools, the average components of the resultant force slightly decreased with a higher number of machining steps. This could be explained by the increasing tool wear as well. Regarding the comparatively high tool wear rates, the tool wear resulted in an offset of the cutting edge leading to a decreased cross-section of the undeformed chip. For all cutting materials the respective components of the resultant force were in a similar range, which can be explained by the constant cross-section of the undeformed chip. Although changes in friction between tool and specimen as well as the plastic deformation in the shear zone were expected (in particular comparing CBN and diamond) no tendencies were visible, especially regarding the spread of the values.

3.3. Geometrical surface properties

The geometrical surface properties of the machined specimens were determined by tactile measurements. Figure 4 shows the influence of the cutting material on the arithmetic mean surface roughness *Ra* and the surface roughness depth *Rz*. Regarding the relatively high number of different cutting conditions (cutting material, cutting speed, number of cutting steps), only the results of the cutting experiments performed with a cutting speed of 300 m/min are represented in the diagram. There were also small differences concerning the cutting speed detectable, but the influence of the cutting material was more dominant. The respective surface parameters were determined after machining one and three cuts to analyse the influence of the tool wear on the geometrical surface properties. The scattering bars shown represent the respective minimum and maximum of the five single measurements.

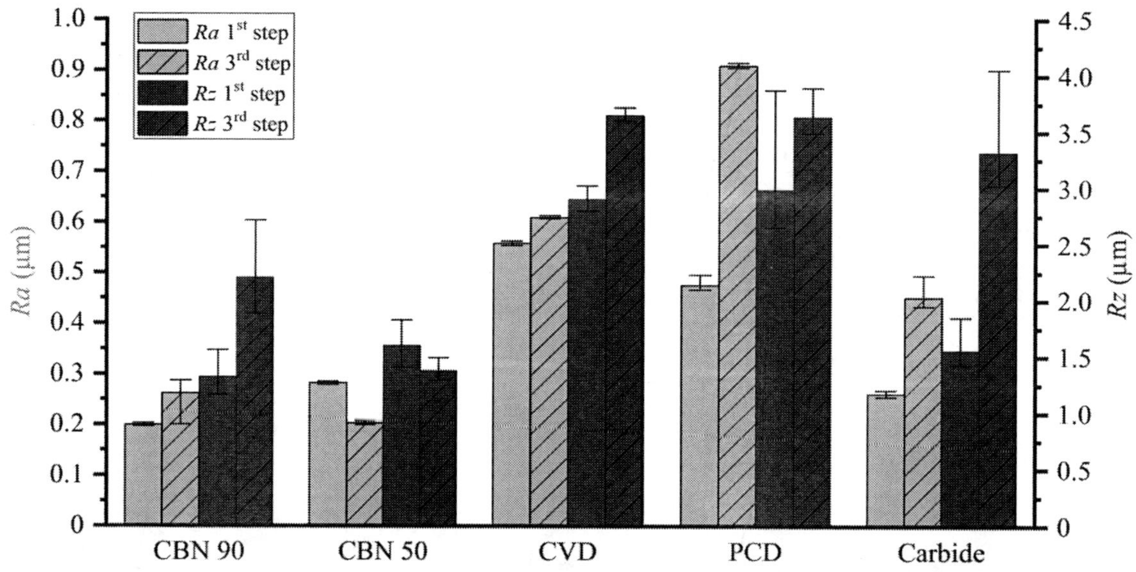

Figure 4. Influence of the cutting material on the arithmetic mean surface roughness *Ra* and the surface roughness depth *Rz* (v_c = 300 m/min, f = 0.05 mm, a_p = 0.1 mm).

For all cutting speeds and cutting materials tested a kinematic roughness surface profile was visible. This was characterised by valleys and peaks representing the corner radius of the tool. The distance between these valleys and peaks corresponded to the feed of 0.05 mm. Generally, the roughness values were higher than the calculated kinematic roughness (Rz_{kin} = 0.78 µm). The lowest values for one machining step were reached in the case of using the CBN with the high proportion of boron nitride (CBN 90) as cutting material. The mean values for *Ra* and *Rz* were only slightly higher in the case of utilising CBN with the higher proportion of binder (CBN 50) after the first machining step. Regarding that both cutting materials led to the best results concerning the geometrical surface properties and the tool wear resistance, the results depending on the cutting speed are discussed more in detail. Only concerning each firstly machined specimen the use of cemented carbide resulted in similar values like

CBN 50. Utilising indexable inserts with PCD or CVD diamond tips led to the highest values of the analysed surface roughness parameters.

Machining using CBN as cutting material led to slightly higher friction between tool and specimen resulting in a slight smoothing of the surface and a small compensation of the higher surface roughness due to the cutting edge chipping. Utilising PCD or CVD diamond resulted in a more defined cutting at the cutting edge, less friction between tool and specimen, and subsequently a higher surface roughness following to the cutting edge chipping. This chipping was higher in the case of using diamond tipped tools (PCD and CVD) compared to CBN tipped tools as a result of the increased tool wear. For all cutting material used, the surface roughness values were higher after three machining steps compared to the first step. The increasing tool wear and cutting edge chipping resulted in raising mean values for *Ra*, *Rz*, and *Sa*.

Concerning the geometrical surface properties and the tool wear CBN seemed to be the cutting material to prefer in the analysed range of the cutting speed. Hence, the influence of the cutting speed on the surface roughness depth *Rz*, the arithmetic mean surface roughness *Ra,* and the areal parameter *Sa* for using tools with tips from different types of CBN (CBN 50 and CBN 90) was analysed in detail (Figure 5). The diagram shows the highest and the lowest values of five tactile measurements and the results of the measurements using the 3D laser scanning microscope.

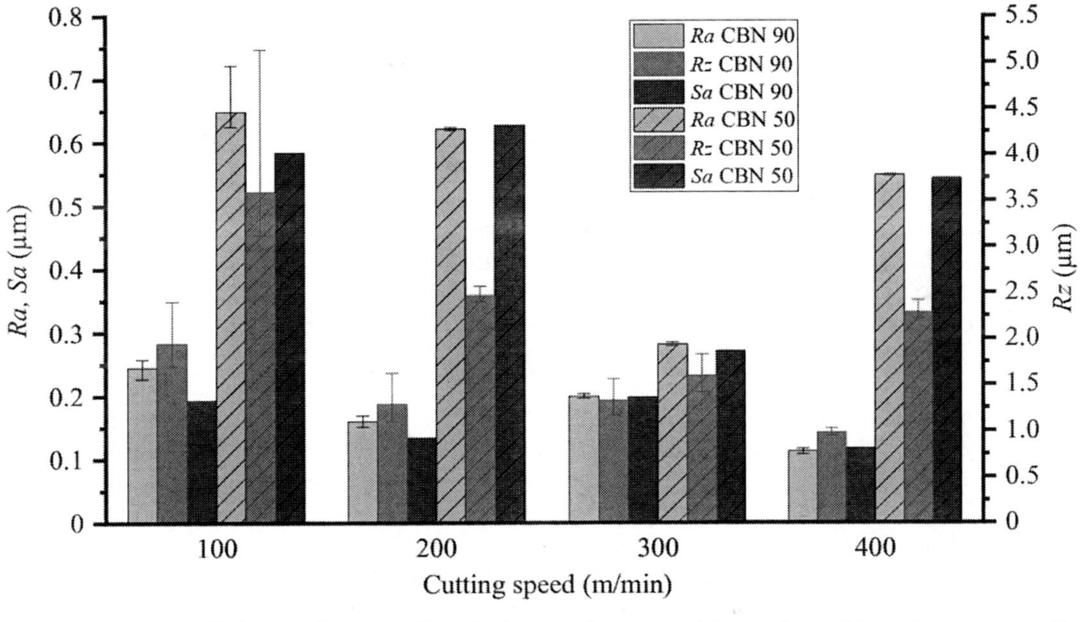

Figure 5. Influence of the cutting speed and the cutting material on the arithmetic mean surface roughness *Ra*, the average roughness *Sa,* and the surface roughness depth *Rz* ($f = 0.05$ mm, $a_p = 0.1$ mm, values after first machining step).

In the case of using CBN 90 similar mean values were detected irrespective of the cutting speed. However, higher surface roughness values were measured after machining with a cutting speed of 100 m/min, while a slight tendency of decreasing mean values with increasing cutting speed was apparent. This could be a result of the increased temperature in the shear zone and a subsequent softening of the specimen material. In the case of using indexable inserts with CBN 50 as cutting material, generally higher surface roughness values were measured. However, after machining with a cutting speed of 300 m/min similar surface roughness values compared to using CBN 90 were measured. For the three other cutting speeds regarded, significantly higher mean values for *Ra*, *Rz*, and *Sa* occurred when using CBN 50. These qualitative differences between machining with a cutting speed of 300 m/min and the other cutting speeds were also visible after three machining steps. The second highest cutting speed regarded represented a suitable range of the cutting speed for machining these high-entropy alloys with CBN 50. Additionally, the tool wear was the lowest after machining with this cutting speed when using CBN 50. The cutting edge chipping was also minor after machining with a cutting speed of 300 m/min resulting in decreasing roughness values.

4. Conclusions and outlook

The investigations and results represent the first experiments analysing the influence of the cutting material and the cutting speed when machining spark plasma sintered high-entropy alloys consisting of the elements cobalt, chrome, iron, and nickel. In the experimental investigations, comparable mean values of the resultant force could be detected while face turning the HEA specimens with different parameters. The lowest surface roughness values could be achieved when using CBN tipped tools characterised by a high proportion of boron nitride. This was possible in the full range of cutting speeds applied. Additionally, these tools showed the lowest wear. Summarising, for face turning of the HEA CoCrFeNi indexable inserts with CBN tips exhibiting a high content of boron nitride should be preferred. Additionally, the determination of the surface layer properties, like residual stresses, grain size, grain orientation, and hardness contributed to a deeper understanding in machining of HEAs with different cutting materials.

Acknowledgments

This study was funded via Sächsische Aufbaubank – Förderbank / SAB-100382175 by the European Social Fund ESF and the Free state of Saxony.

Europäische Union

Europa fördert Sachsen.
ESF
Europäischer Sozialfonds

Diese Maßnahme wird mitfinanziert durch Steuermittel auf der Grundlage des vom Sächsischen Landtag beschlossenen Haushaltes.

References

[1] Clauß B, Liborius H, Lindner T, Löbel M, Schubert A and Lampke T 2020 Influence of the cutting parameters on the surface properties in turning of a thermally sprayed AlCoCrFeNiTi coating *Procedia CIRP* **87** 19–24

[2] Huang Z, Dai Y, Li Z, Zhang G, Chang C and Ma J 2020 Investigations on surface morphology and crystalline phase deformation of $Al_{80}Li_5Mg_5Zn_5Cu_5$ high-entropy alloy by ultra-precision cutting 2020 *Materials and Design* **186** 1–9

[3] Guo J, Goh M, Zhu Z, Lee X, Nai M L and Wei J 2018 On the machining of selective laser melting CoCrFeMnNi high-entropy alloy *Materials and Design* **153** 211–220

[4] Junge T, Liborius H, Mehner T, Nestler A, Schubert A and Lampke T 2020 Measurement system based on the Seebeck effect for the determination of temperature and tool wear during turning of aluminum alloys *Procedia CIRP* **87** 1435–1441

[5] Dubos P A, Fajoui J, Iskounen N, Coret M, Kabra S, Kelleher J, Girault B and Gloaguen D 2020 Temperature effect on strain-induced phase transformation of cobalt *Materials Letters* **281** 128812

[6] Sato J, Omori T, Oikawa K, Ohnuma I, Kainuma R and Ishida K 2006 Cobalt-Base High-Temperature Alloys *Science* **312** 90–91

[7] Harris T K, Brookes E J and Taylor C J 2004 The effect of temperature on the hardness of polycrystalline cubic boron nitride cutting tool materials *International Journal of Refractory Metals & Hard Materials* **22** 105–110

[8] Sumiya H, Uesaka S and Satoh S 2000 Mechanical properties of high purity polycrystalline cBN synthesized by direct conversion sintered method *Journal of Materials Science* **35** 1181–1186

[9] Denkena B and Tönshoff H K 2011 Spanen Grundlagen vol 3 (Berlin: Springer) p 170

[10] Davies M A, Chou Y and Evans C J 1996 On Chip Morphology, Tool Wear and Cutting Mechanics in Finish Hard Turning *Annals of the CIRP* **45** 77–82

Influence of the production route on the phase formation, microstructure and wear behaviour of the high-entropy alloy $AlCoCrFeNiTi_{0.5}$

M Löbel*, T Lindner and T Lampke

Materials and Surface Engineering Group, Institute of Materials Science and Engineering, Chemnitz University of Technology, D-09107 Chemnitz, Germany

*e-mail: martin.loebel@mb.tu-chemnitz.de

Abstract. Different manufacturing approaches have been investigated regarding their suitability to process high-entropy alloys (HEAs). However, comprehensive investigations on the influence of the production route on the microstructure, phase formation and properties have not been conducted yet. For the current study the alloy $AlCoCrFeNiTi_{0.5}$ is considered. Previous investigations have proven the formation of phases with predominantly body centred cubic structure for this alloy. Castings are produced by arc-melting. Feedstock material for coating deposition and powder metallurgical processing is produced by inert gas atomisation. For the processing high-velocity-oxygen-fuel (HVOF) thermal spraying and spark plasma sintering (SPS) are applied. Due to the significantly differing process conditions and temperature-time profiles, differences of microstructure, phase formation and resulting properties can be observed. Wear investigations under various conditions have been conducted. Especially under sliding and reciprocating wear conditions the structural defects formed for the thermally sprayed coating cause a reduction of wear resistance. The formation of structural defects could be avoided by SPS. However, the additional tetragonal phase causes a reduction of the wear resistance. The current study contributes to a better understanding of the interaction between process, microstructure and properties.

1. Introduction

High-entropy alloys represent a new approach of alloy development. They are composed of at least five components mixed in approximately equimolar composition. A promising combination of properties could be proven e.g. high strength and hardness in combination with a pronounced ductility. Furthermore, a high corrosion and wear resistance was observed, making them an interesting candidate for coating applications [1].

One of the first HEAs identified to form a stable single-phase face centred cubic (fcc) structure was the equimolar alloy CoCrFeMnNi. Starting from this composition, the influence of many additional alloying elements has been investigated. One objective was to form a body centred cubic (bcc) structure to increase the hardness and strength. Aluminium showed a strong influence on the phase formation. For sufficiently high contents a bcc structure could be stabilised [2, 3].

A further improvement of mechanical and tribological properties could be achieved by the addition of titanium. Investigations by Zhou et al. revealed an increase of strength and hardness, caused by the high atomic radius of titanium and the resulting severe lattice distortion. However, high contents of titanium result in the formation of intermetallic phases and embrittlement [4]. The highest wear resistance was determined for the alloy $AlCoCrFeNiTi_{0.5}$ [5].

The suitability of different production routes has been investigated for HEAs. Early studies focused especially on the production from the liquid state by casting [6]. Also powder metallurgical processes have been applied. Spark plasma sintering (SPS) enables a fast processing and the suppression of grain coarsening [7]. Furthermore, the material usage can be reduced by limiting the application of HEAs to the surface. Thermal spray processes are widely used to produce coatings for corrosion and wear protection applications. Typical structural defects especially pores and oxides can be reduced by applying high kinetic processes e.g. high-velocity-oxygen-fuel (HVOF) thermal spraying [8]. Also HEAs have already been successfully processed by HVOF [9–11].

Despite various investigations on HEAs considering different production routes, the influence of the production route has not been investigated in detail. In the present comparative study the influence on the microstructure, phase formation and tribological properties of the alloy $AlCoCrFeNiTi_{0.5}$ is investigated. The wear behaviour is examined under sliding, reciprocating and abrasive conditions.

2. Experimental and materials

In the current investigations different production routes have been considered for the alloy $AlCoCrFeNiTi_{0.5}$. Castings were produced by arc-melting. The manufacturing conditions have been described in detail in a previous study [5]. Feedstock powder was prepared by inert gas atomisation. For further processing a particle size range of -45 +15 µm was used. Prior to the deposition of coatings, substrates (stainless steel EN 1.4404) were prepared by corundum blasting. The coatings were produced with the HVOF system K2 (GTV Verschleissschutz GmbH, Luckenbach, Germany), with the parameters summarised in Table 1.

Table 1. HVOF coating parameters of the $AlCoCrFeNiTi_{0.5}$ feedstock powder.

O_2	kerosene	Ar	nozzle	powder feed rate	spraying distance	relative traverse speed	spray path offset
850 l/min	22.5 l/h	2 x 11 l/min	100/12	2 x 40 g/min	360 mm	60 m/min	5 mm

Furthermore, the feedstock powder was compacted by spark plasma sintering (SPS). The production parameters have been described in a previous study [12].

A detailed characterisation of the microstructure and phase formation has been conducted for all production routes. Cross-sections were prepared by standard metallographic procedures. For the visualisation of material contrast the cross-sections were investigated in the scanning electron microscope (SEM) LEO 1455VP (Zeiss, Jena, Germany), using the backscattered electron detector (BSD). For the investigation of the powder morphology the secondary electron (SE) detector was used. The chemical composition was determined with the integrated energy-dispersive X-ray spectroscopy (EDS) system EDS GENESIS (EDAX, Mahwah, NJ, USA). For the analyses of the formed phases measurements by X-ray diffraction (XRD) have been conducted with a D8 Discover diffractometer (Bruker AXS, Billerica, MA, USA) using Co Kα radiation.

The influence of the microstructure and phase formation on the resulting properties was determined by microhardness measurements in a first step. The Wilson Tukon 1102 device (Buehler, Uzwil, Switzerland) was used to determine the Vickers hardness HV 0.1.

Furthermore, detailed studies to determine the influence on the tribological properties have been conducted. The sliding wear behaviour was investigated in ball-on-disk test, using the Tetra basalt tester (Tetra, Ilmenau, Germany). For the investigation of the wear behaviour under reciprocating conditions a Wazau SVT 40 device (Wazau, Berlin, Germany) has been applied. The abrasive wear behaviour has been investigated in scratch test with a CSM Revetest-RST instrument (CSM Instruments SA, Peseux, Switzerland). All applied wear test parameters are summarised in Table 2.

Table 2. Wear test parameters.

ball-on-disk test		reciprocating wear test		scratch test	
force	20 N	force	26 N	mode	progressive
radius	5 mm	frequency	40 Hz	force	1–200 N
speed	96 RPM	time	900 s	speed	2.5 mm/min
cycles	15,916	amplitude	0.5 mm	length	5 mm
counter-body	Al_2O_3 (ø 6 mm)	counter-body	Al_2O_3 (ø 10 mm)	tip	truncated diamond cone (ø 400 µm)

The wear depth of the ball-on-disk test was determined by tactile measurements with a Hommel-Etamic T8000 device (Jenoptik, Villingen-Schwenningen, Germany). For the evaluation of the reciprocating wear and scratch test the laser scanning microscope Keyence VK-X200 (Keyence, Osaka, Japan) was used.

3. Results and Discussion

3.1. Feedstock characterisation

The surface and cross-sections of the $AlCoCrFeNiTi_{0.5}$ feedstock powder were investigated in SEM, Figure 1.

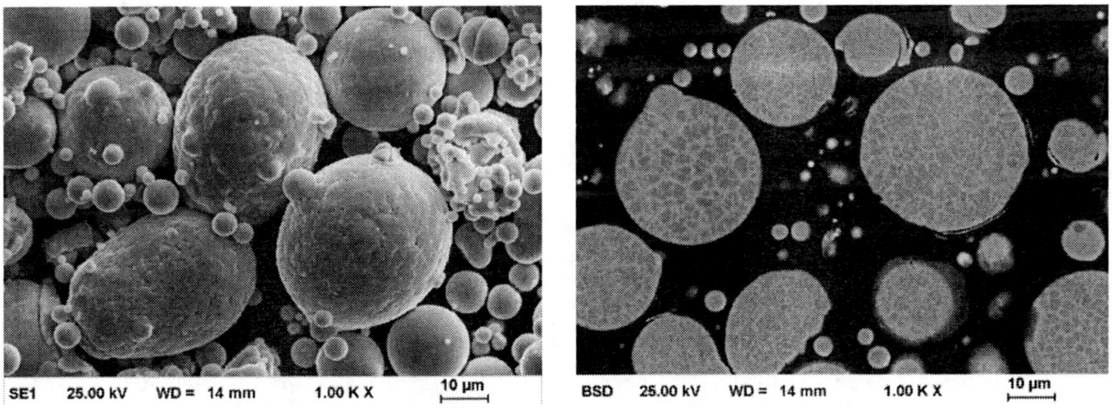

(a) (b)

Figure 1. $AlCoCrFeNiTi_{0.5}$ feedstock powder, investigated in SEM (SE/BSD): (a) surface and (b) cross-section.

Due to the feedstock production by gas atomisation a spherical powder morphology occurs, enabling a good flowability of the feedstock. In the cross-section a distinct material contrast can be observed, indicating the formation of a heterogeneous microstructure and a deviation of local chemical composition.

3.2. Chemical composition

The mean chemical composition was measured by EDS measurements for all production routes, Table 3.

Table 3. Chemical composition measured by EDS, in at.%.

sample	Al	Co	Cr	Fe	Ni	Ti
nominal	*18.2*	*18.2*	*18.2*	*18.2*	*18.2*	*9.1*
casting	20.0	17.8	17.4	17.6	18.2	9.0
feedstock powder	20.1	17.5	18.0	17.9	17.5	9.0
HVOF coating	19.7	17.7	17.6	18.2	17.8	9.1
SPS	21.0	17.3	17.7	17.5	17.5	8.9

The measured chemical composition is in accordance with the intended values. Only the aluminium contents show a deviation of ≥ 1 at.%. Processing of the feedstock powder by HVOF and SPS causes no distinct change of the chemical composition.

3.3. Microstructure and phase formation

The microstructure of the alloy $AlCoCrFeNiTi_{0.5}$ was investigated in dependence of the production route. Cross-sections were prepared and investigated in SEM, Figure 2.

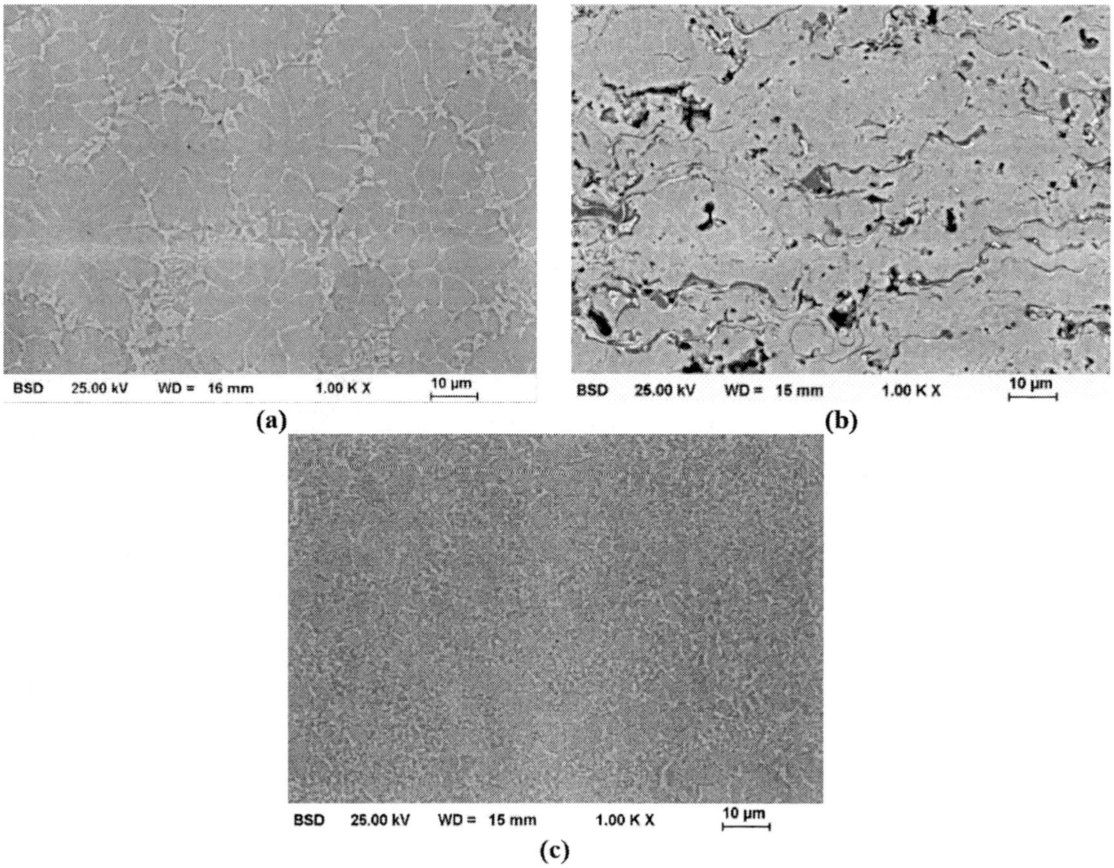

Figure 2. Microstructural SEM images (BSD) of $AlCoCrFeNiTi_{0.5}$: (a) casting; (b) HVOF coating and (c) SPS.

A coarse dendritic structure with a distinct material contrast is formed for the casting produced by arc-melting. The HVOF coating exhibits a lamellar structure, comprising oxide lamellae and pores. Within single spray particles material contrast and a dendritic structure can be observed. In comparison with the feedstock material no grain coarsening occurs due to the relatively low thermal input. Dense $AlCoCrFeNiTi_{0.5}$ alloys were produced powder metallurgically by SPS. No distinct coarsening of the dendritic structure occurs in comparison with the feedstock material.

The phase formation in dependence of the production route was determined by XRD, Figure 3.

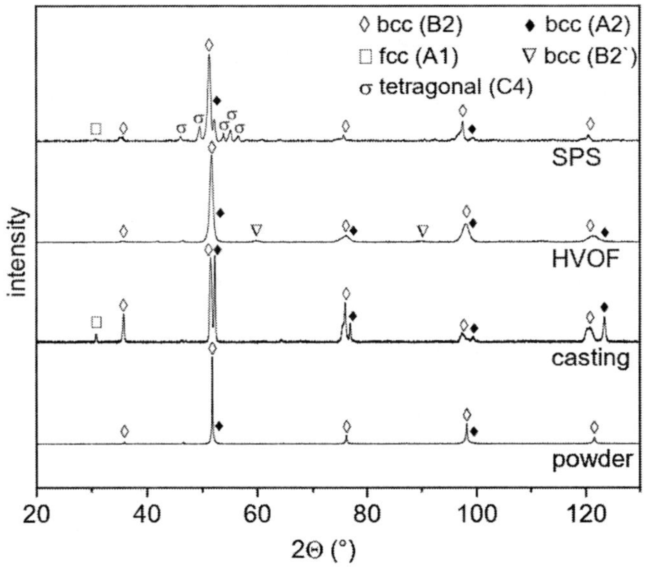

Figure 3. Diffractograms of $AlCoCrFeNiTi_{0.5}$ feedstock powder; casting; HVOF coating and powder metallurgically produced alloy (SPS).

The diffractogram of the feedstock powder exhibits main diffraction peaks of a chemically ordered bcc phase with B2 structure. Furthermore, a shoulder towards higher diffraction angles occurs at 51.8° and 98.2°, indicating the presence of a second minor phase. These peaks can be assigned to a chemically disordered bcc phase with A2 structure. The casting produced by arc-melting also shows intensity maxima of the bcc phases with B2 and A2 structure. An additional peak occurs at 30.9° indicating the presence of a minor phase with fcc structure. Further peaks of this phase overlap with the bcc phase with B2 structure. The diffractogram of the HVOF coating shows no distinct change in comparison to the feedstock powder. Additional minor diffraction peaks prove the formation of a further chemically ordered bcc phase with B2 structure with a distinctly changed lattice parameter. Powder metallurgical processing causes the formation of an additional tetragonal σ-phase. The formation of this phase in dependence of the manufacturing conditions for the alloy $AlCoCrFeNiTi_{0.5}$ has already been described in previous studies [13].

The effect of the microstructure and phase formation on the resulting properties was investigated by means of microhardness measurements in a first step, Figure 4.

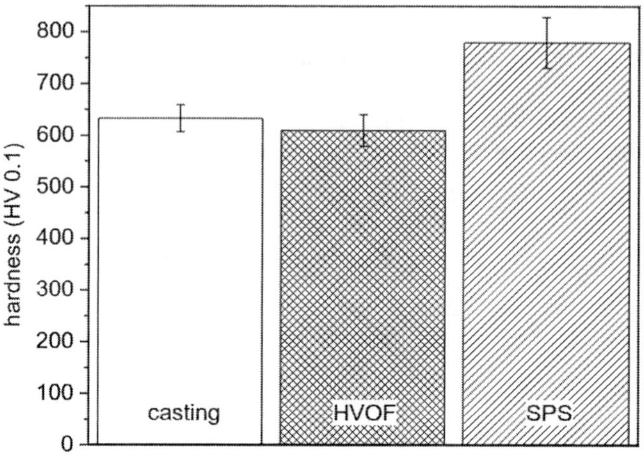

Figure 4. Microhardness HV 0.1 of $AlCoCrFeNiTi_{0.5}$ casting; HVOF coating and powder metallurgically produced alloy (SPS).

For the casted $AlCoCrFeNiTi_{0.5}$ alloy a hardness of 630 ± 30 HV 0.1 was measured. The coating produced by HVOF exhibits a similar hardness. The deviation is within the range of standard deviation. For the powder metallurgically produced alloy a distinct increase of hardness occurs. A value of 780 ± 50 HV 0.1 was measured. Phase analyses revealed the formation of an additional σ-phase with tetragonal structure. A significant hardening of HEAs by the formation of an additional σ-phase has been described in literature by Tsai et al. [14]. Furthermore, the average hardness can be influenced by the strong variation of grain size (Figure 2).

3.4. Wear behaviour

The wear behaviour was investigated under sliding, reciprocating and abrasive conditions. The results are summarised in Figure 5. A distinct influence of the production route on the wear resistance can be observed. The highest wear resistance under sliding wear conditions in ball-on-disk test was determined for the casting. Solely solid solutions with cubic structure were formed for this production route. Processing of the alloy by thermal spraying causes a higher wear depth and hence lower wear resistance. This behaviour can be explained by the formation of structural defects especially pores and oxides. The phase composition was not significantly altered in comparison to the casting. Despite the formation of a fine-grained microstructure and the absence of structural defects, the alloy produced by SPS exhibits an increased wear depth in comparison to the casting. Phase analyses revealed the formation of an additional σ-phase for this production route. This phase causes an increase in hardness. However, the fracture toughness and ductility are reduced [14, 15]. The impairment of mechanical properties has a negative influence on the wear resistance.

Higher wear depths were determined for the investigations under reciprocating conditions for all production routes. The results reveal a similar tendency. The highest wear resistance was achieved for the casting.

For the investigations under abrasive wear conditions in scratch test a less pronounced dependence on the production route occurs. The highest wear resistance was determined for the casting. Structural defects of the coating cause a slight increase of the wear depth. The additional σ-phase formed for the SPS sample causes no distinct impairment of wear resistance under abrasive conditions.

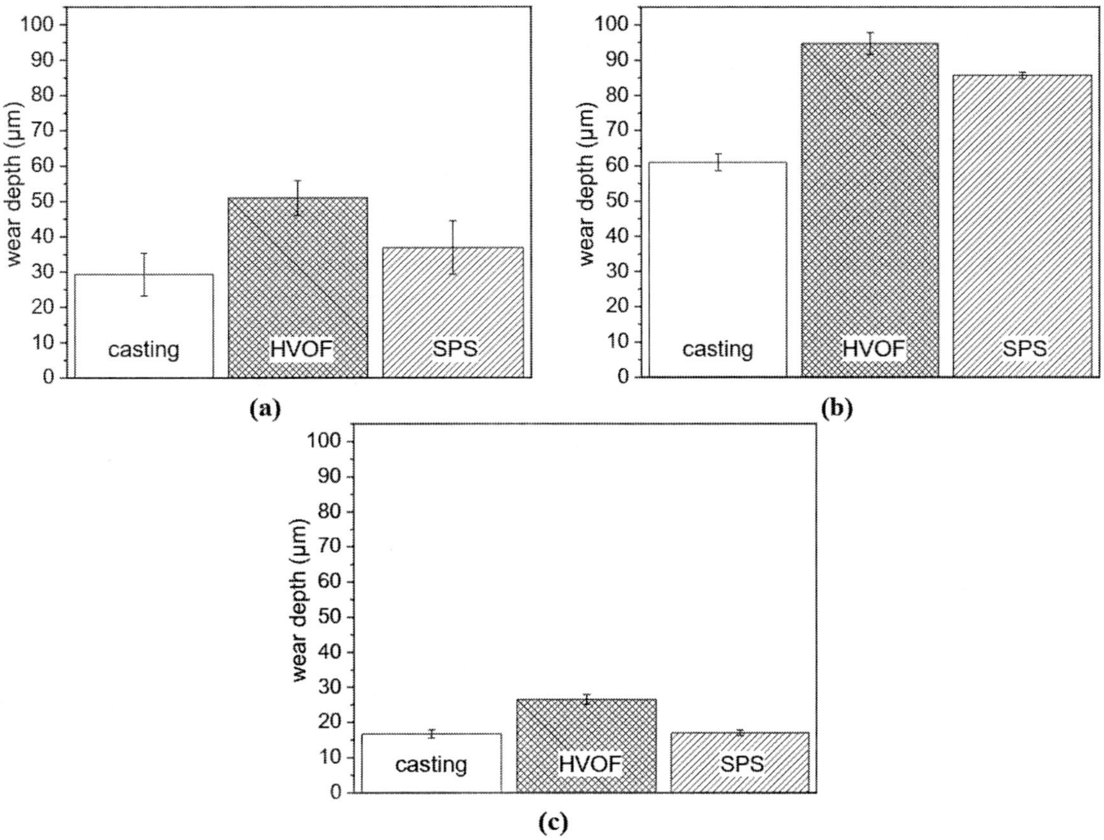

Figure 5. Results of wear investigations: (a) ball-on-disk test; (b) reciprocating wear test and (c) scratch test.

4. Summary and Conclusions

The influence of the production route on the microstructure, phase formation and wear behaviour was investigated in detail for the alloy $AlCoCrFeNiTi_{0.5}$. A two phase microstructure comprising two body centred cubic phases was formed for the feedstock powder. In comparison with the arc-melted alloy a minor face centred cubic phase is suppressed. Due to the increased cooling rate, a fine-grained microstructure is formed for the atomised powder. Processing of the powder by thermal spraying causes no major changes of the phase composition. However, typical structural features of thermal spray coatings were identified. Dense material can be produced by spark plasma sintering. No distinct grain coarsening occurs for both processes. However, due to the increased heat input in comparison to the thermal spray process, an additional σ-phase with a tetragonal structure is formed.

The σ-phase and the formation of a fine-grained microstructure cause an increase of microhardness in comparison to the casting. For the thermally sprayed coating no distinct change of hardness in comparison to the casting occurs.

A dependence of wear behaviour on structural features and additionally formed phases was observed, especially under sliding and reciprocating conditions. Despite the coarse-grained microstructure, the highest wear resistance was observed for the casted alloy comprising solely body centred cubic phases. The structural features formed for the coatings cause a reduction of wear resistance. Although these defects were not formed for the powder metallurgically processed alloy, a reduced wear resistance was determined in comparison with the casting. This behaviour is a result of the additional tetragonal σ-phase.

Suitable HEA compositions have to be identified in subsequent studies to avoid the formation of the σ-phase for all production routes. Thus, the equilibrium state has to be considered. For the improvement of coating properties, the content of porosity and oxides has to be further reduced.

Funding
This research was funded by the German Research Foundation (Deutsche Forschungsgemeinschaft/DFG), Grant No. La-1274/54-1.

References

[1] Gao M C, Yeh J-W, Liaw P K and Zhang Y 2016 *High-Entropy Alloys: Fundamentals and Applications, 1st edition* Springer International Publishing: Cham, Switzerland

[2] Zhang K and Fu Z 2012 Effects of annealing treatment on phase composition and microstructure of $CoCrFeNiTiAl_x$ high-entropy alloys *Intermetallics* **22(0)** 24–32 doi.org/10.1016/j.intermet.2011.10.010

[3] Li C, Li J C, Zhao M and Jiang Q 2010 Effect of aluminum contents on microstructure and properties of $Al_xCoCrFeNi$ alloys *J Alloy Comp* **504** 515-518 doi.org/10.1016/j.jallcom.2010.03.111

[4] Zhou Y J, Zhang Y, Wang YL and Chen GL 2007 Solid solution alloys of $AlCoCrFeNiTi_x$ with excellent room-temperature mechanical properties *Appl Phys Lett* **90(18)** 181904 doi.org/10.1063/1.2734517

[5] Löbel M, Lindner T, Mehner T and Lampke T 2018 Influence of Titanium on Microstructure, Phase Formation and Wear Behaviour of $AlCoCrFeNiTi_x$ High-Entropy Alloy *Entropy* **20(7)** 505 doi.org/10.3390/e20070505

[6] Tsai M-H and Yeh J-W 2014 High-Entropy Alloys: A Critical Review *Mater Res Lett* **2(3)** 107–23 doi.org/10.1080/21663831.2014.912690

[7] Olevsky E A and Dudina D V 2018 *Field-Assisted Sintering, Science and Applications* Springer International Publishing: Cham, Switzerland

[8] Oksa M, Turunen E, Suhonen T, Varis T and Hannula S-P 2011 Optimization and Characterization of High Velocity Oxy-fuel Sprayed Coatings: Techniques, Materials, and Applications *Coatings* **1(1)** 17–52 doi.org/10.3390/coatings1010017

[9] Chen L, Bobzin K, Zhou Z, Zhao L, Öte M and Königstein T et al. 2019 Wear behavior of HVOF-sprayed $Al_{0.6}TiCrFeCoNi$ high entropy alloy coatings at different temperatures *Surf Coat Tech* **358** 215–222 doi.org/10.1016/j.surfcoat.2018.11.052

[10] Löbel M, Lindner T, Mehner T and Lampke T 2017 Microstructure and Wear Resistance of AlCoCrFeNiTi High-Entropy Alloy Coatings Produced by HVOF *Coatings* **7(9)** 144 doi.org/10.3390/coatings7090144

[11] Ashok M, Ameey A, Murty B S, Berndt C. C, Kottada R V and Ang A S M 2020 Thermal Spray High-Entropy Alloy Coatings: A Review *J Therm Spray Tech* **29** 857-893 doi.org/10.1007/s11666-020-01047-0

[12] Löbel M, Lindner T, Pippig R and Lampke T 2019 High-Temperature Wear Behaviour of Spark Plasma Sintered $AlCoCrFeNiTi_{0.5}$ High-Entropy Alloy *Entropy* **21(6)** 582 doi.org/10.3390/e21060582

[13] Moravcik I, Cizek J, Gavendova P, Sheikh S, Guo S and Dlouhy I 2016 Effect of heat treatment on microstructure and mechanical properties of spark plasma sintered $AlCoCrFeNiTi_{0.5}$ high entropy alloy *Mater Lett* **174** 53–56 doi.org/10.1016/j.matlet.2016.03.077

[14] Tsai M-H, Tsai K-Y, Tsai C-W, Lee C, Juan C-C and Yeh J-W 2013 Criterion for Sigma Phase Formation in Cr- and V-Containing High-Entropy Alloys *Mater Res Lett* **1(4)** 207–212 doi.org/10.1080/21663831.2013.831382

[15] Tsai M-H, Chang K-C, Li J-H, Tsai R-C and Cheng A-H 2016 A second criterion for sigma phase formation in high-entropy alloys *Mater Res Lett* **4(2)** 90–95 doi.org/10.1080/21663831.2015.1121168

IOP Conf. Series: Materials Science and Engineering 1147 (2021) 012010 doi:10.1088/1757-899X/1147/1/012010

Electrical contacting of High-Velocity-Air-Fuel sprayed NiCr20 coatings by brazing

K Bobzin, W Wietheger*, H Heinemann, X Liao and C Vogels

RWTH Aachen University, Aachen / Surface Engineering Institute

*e-mail: wietheger@iot.rwth-aachen.de

Abstract. Thermally sprayed heater coatings can be used to control the surface temperature of parts and tools. A crucial point for such applications that has been neglected in the past is a reliable electrical contact with the power supply. One possible solution is the soldering or brazing of cables and cable connectors onto the heater coating. In this study, the possibility of brazing cable connectors to High-Velocity-Air-Fuel sprayed NiCr20 coatings was investigated to ensure electrical contact. The eutectic Ag-Cu filler was used for the vacuum brazing process. The insufficient wetting of the filler on the as-sprayed coating surface and the favored wetting of this filler on the copper cable connector led to low joint quality. This joint was improved by grinding the coating surface and changing the cable connector's material. Furthermore, measurements of the heater coatings' electrical resistance confirmed that electrical contacting by brazing is possible without significant influence on the heater coatings' properties.

1. Introduction

To prevent premature solidification of the melt and thus to enable the fabrication of thin-walled parts, in short cycle time by means of the die-casting process, increasing the surface temperature of the die-casting mold is sufficient. Among others, the concept of surface heating by means of a heater coating on the tool surface represents a time and energy-efficient solution, since the temperature of the interface between the tool and the melt and the heat exchange can be controlled directly. A typical structure of such a heating element is shown in figure 1. A bond coating is applied to the tool to ensure the following layers to adhere. For electrical insulation two ceramic coatings surround the heater coating.

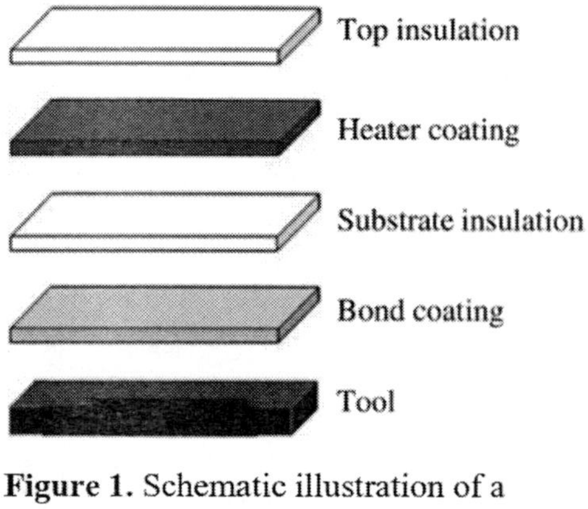

Figure 1. Schematic illustration of a coating structure of a heating element.

One important component of the system is the heater coating, as the surface heating is realized with it. Thermally sprayed heater coatings have been investigated in numerous studies. The heater coating development by atmospheric plasma spraying (APS) [1] and High-Velocity-Oxygen-Fuel (HVOF) [2] was examined, inter alia, by Prudenziati et al. For the heater coating different feedstock materials have been investigated in the literature, which can be divided into two categories. One category of feedstock materials are ceramics, which have been investigated by M. Floristán et al. by the application of TiO_2 heater coatings onto LiO_2-Al_2O_3-SiO_2 glass substrate [3] and S. Scheitz et al. by examining TiO_x and Cr_2O_3-TiO_2 as feedstock material [4]. The other investigated group of materials used in these applications are metallic heater coatings. FeCr-alloys, CrAl-alloys and NiCr-alloys have been investigated by Killinger et al. [5]. Michels et al. studied the properties of NiCr20 coatings [6]. The authors have investigated the heating behavior of heater coatings [7] and their temperature distribution [8] in case of plasma sprayed TiO_2/Cr_2O_3 before.

Although the general feasibility of thermally sprayed coatings as a heating element has been investigated in the literature, little attention has been payed to the electrical contact to the coating. In the papers referenced here, it is not mentioned nor described in detail, how the electrical contact has been realized. In the industry, this variant of tempering has not been used so far and therefore no contacting concept has been implemented yet. Since the electrical contact is crucial for the functionality of the heater coating system and must be carefully adjusted for the aimed application, it is necessary to develop suitable concepts. Therefore, a possible contacting of the heater coating by brazing a connector is investigated in this paper. Generally, literature on the brazing of additional parts onto thermally sprayed coatings is sparse. In most cases, the filler metals were used as a thermally sprayed coating e.g. Nicolaus et al. [9]. The behavior of a coating onto which a brazing alloy has been applied by thermal spraying has been examined before. Bobzin et al. [10] and Zhao et al. [11] applied brazing alloy Al12Si on aluminium substrates. The substrates were coated with the brazing alloy by cold spraying process. In this paper, it will be investigated whether the filler metal can be brazed onto a thermally sprayed High-Velocity-Air-Fuel (HVAF)-NiCr20 surface coating and a sufficient connection can be achieved.

2. Experimental and materials

2.1. Thermal spray process

The substrates are hot-working steel coupons made of X38CrMoV5-1. This steel was chosen because it is often used to produce modular die-casting tools. Before the thermal spraying processes, the substrates have been prepared by grit blasting with corundum with the fraction of F16. A bond coat made of NiCr20 and an insulation coating made of Al_2O_3 were first applied with APS, which is a common process for ceramic coatings, using a TriplexPro-210 gun (Oerlikon Metco Europe GmbH, Kelsterbach, Germany). The parameter sets used to apply these two coatings are listed in table 1. All parameters presented in this paper are based on previous studies performed at the Surface Engineering Institute, Aachen. During the deposition, the gun was mounted on a robot and traveled above the sample in a meandering path.

Table 1. Parameter set to apply the NiCr20 bond coating and the Al_2O_3 substrate insulation using APS.

Parameter	Unit	NiCr20 Bond coating	Al_2O_3 Insulation
Normal particle size distribution	[µm]	-53 +20	-45 +20
Flow rate of argon	[SLPM]	70	60
Flow rate of hydrogen	[SLPM]	0	6
Electrical current	[A]	400	450
Spray distance	[mm]	120	120
Nozzle diameter	[mm]	9	9
Traverse speed of the gun	[mm/s]	800	800
Offset of the meandering path	[mm]	5	5
Rotation speed of the conveyor disk	[%]	15	20
Cooling	[MPa]	0.5	0.5
Flowrate of carrier gas argon	[SLPM]	4.5	5.5
Passages	[-]	5	9

The feedstock material used to apply the heater coating is also NiCr20, same as the material used for the bond coating but in this case with a nominal particle size distribution of -25 + 5 µm. To apply the heater coating onto the APS-sprayed insulation coatings the HVAF process was used with an AK 7 Gun (Kermetico Inc, Benicia, USA) equipped with a shortened 3L nozzle. The parameter settings used to apply the heater coating in this study are listed in table 2. As in case of the bond coat and the substrate insulation, the gun was carried by a robot and traveled in a meandering path. The HVAF process, which is used to achieve a low oxidation, necessitates the selection finer particle size for the heater coating compared to the bond coat.

Table 2. Parameter set for applying the NiCr20 heater coating.

Parameter	Unit	NiCr20 Heater coating
Normal particle size distribution	[µm]	-25 +5
Pressure of compressed air	[MPa]	0.61
Pressure of propane	[MPa]	0.55
Flowrate of carrier gas nitrogen	[SLPM]	37.2
Spray distance	[mm]	250
Nozzle type	[-]	shortened 3L
Traverse speed of the gun	[mm/s]	1,000
Offset of the meandering path	[mm]	5
Rotation speed of the conveyor disk	[rpm]	3.5
Cooling	[MPa]	0.5
Passages	[-]	4

The samples can be divided into two groups, one with the smaller dimension of $30 \times 30 \times 8$ mm^3 and the other one with the greater dimension of $100 \times 18 \times 8$ mm^3. The samples with the smaller dimension were named with prefix "A" followed by the sample number. The aim of the experiments done with the samples of group "A" is to investigate the brazing ability of the coating and to identify the suitable pretreatment for the brazing process. During the deposition of the heater coating, a mask was mounted on the sample surface, which reduced the heater coating to a dimension of 20×15 mm^2. This is necessary to prevent any electrical contact between the heater coating and the substrate. To optimize the brazing process in the experimental examination, two samples of the described dimension were ground with 1,200# SiC abrasive paper before the brazing process. Figure 2 a) and b) shows the as-sprayed sample A1 and the corresponding height profile with the determined arithmetic average roughness Ra. The ground sample A3 is shown in figure 2 c) and d). The height profile and the corresponding roughness value Ra were measured using the Keyence VK-X200 (Keyence Corporation, Osaka, Japan).

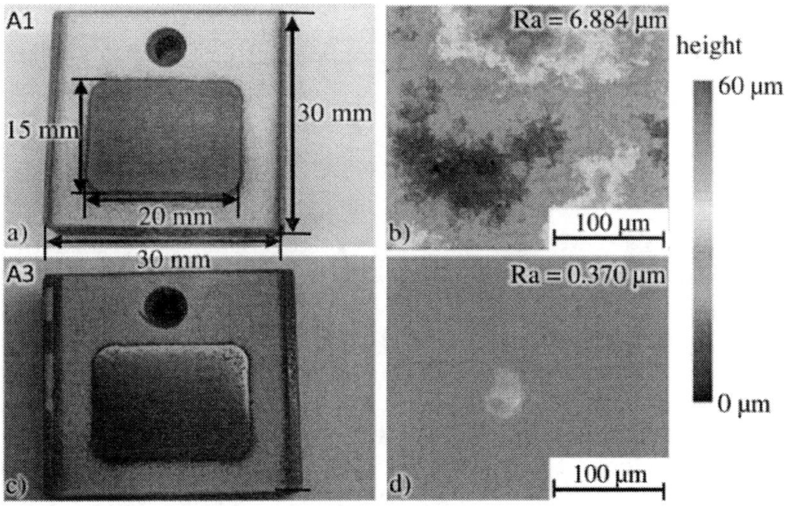

Figure 2. a) as-sprayed sample A1 with b) corresponding height profile and c) ground sample A3 with d) corresponding height profile.

2.2. Brazing process

In this study the joint of electrical contacting by brazing was examined. The choice of a suitable filler metal should be made under the consideration of the application environment. According to the literature of Long et al. [12] and Norwood et al. [13], the surface temperature of casting dies during high pressure die-casting of aluminum alloys can reaches as high as $T \approx 450°C - 500°C$. Based on this condition the eutectic AgCu alloy (72 wt.-% Ag, 28 wt.-% Cu) was chosen as a foil filler metal, with the dimension of 7×12 mm^2, to braze the electrical connector on the heater coating. The eutectic AgCu alloy, with its melting temperature of 779°C, is able to withstand the high temperatures of the die-casting process and the brazing temperature is low enough to keep the impact on the microstructure of the coating system as low as possible. The electrical contacts used in this study were common cable connectors, with and without copper cable attached, and a X5CrNi18-10 steel screw. The brazing process was carried out in vacuum furnace MOV 533T (PVA TePla Industrial Vacuum Systems GmbH, Wettenberg, Germany). The time-temperature setting of the brazing process is shown in figure 3, the process was established in a previous project for a similar filler metal. The peak temperature is above the tempering temperature of the chosen X38CrMoV5-1 substrate. However, the focus of this study was to investigate the realization of a suitable bonding of the electrical connector to the heater coating by brazing. The future areas of application for the presented heating coatings are injection moulding and die-casting. In these applications, it is not practical to treat the entire tool in a brazing process, as only small areas of these tools need to be connected. These connections can then be brazed using laser brazing or induction brazing. Therefore, in this first step only the bonding mechanism of the AgCu filler metal to the connector and the NiCr20 heater coating are of interest.

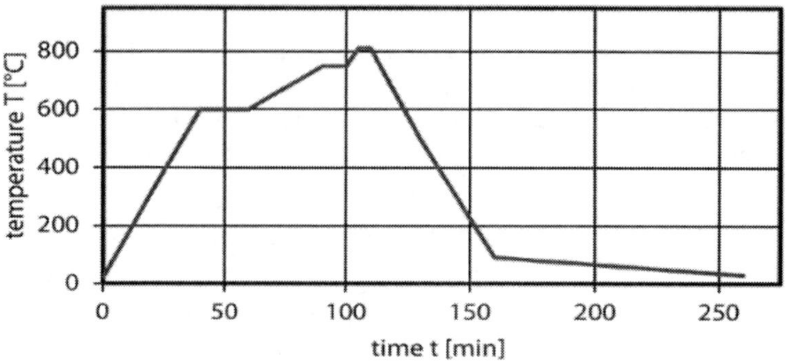

Figure 3. Time-temperature setting of the brazing process.

Three samples of the group "A" were fabricated in this study. Sample A1, remained in as-sprayed state and was brazed with a cable connector with a copper cable inside. The surface of the heater coating of A2 was slightly grounded with 1,200# SiC abrasive paper. As electrical contact for A2, a cable connector without a copper cable was used. In case of A3 the cable connector was replaced by a steel screw. The parameter settings of the brazing process remained the same as for all specimen.

After the brazing ability examinations, two samples of same type with the greater dimension were fabricated to obtain a statistical validation. These two samples are named with the prefix "B" followed by the sample number. The longer shape of the samples in group "B" enables the deposition of a heater coating with the shape of a conducting track with the help of a corresponding mask. The samples were produced in the same manner and with the same parameter in table 1 and table 2. An overview of the fabricated samples is given in table 3.

Table 3. Overview of the fabricated samples.

sample	preparation for brazing	electrical connector	sample shape
A1	as-sprayed state	Cu cable connector with copper cable	square brazing joint
A2	grounded with 1,200 SiC	Cu cable connector without copper cable	square brazing joint
A3	grounded with 1,200 SiC	X5CrNi18-10 steel screw	square brazing joint
B1.1	grounded with 1,200 SiC	X5CrNi18-10 steel screw	conducting track
B1.2	grounded with 1,200 SiC	X5CrNi18-10 steel screw	conducting track

Additionally, an area of the heater coating was covered with an Al_2O_3 top insulation, which was necessary for further examinations. This top insulation was applied with the APS process using the same parameter sets as taken for the bottom insulation. The conducting track has two circular areas at its ends that function as the contact surface. The contact surfaces of the heater coating were also ground and steel screws were brazed to the sample with the same parameter sets as used before. Figure 4 shows a schematic illustration of these samples. The influence of the brazing process on the heater coating was investigated by testing the electrical resistance of the conducting tracks on B1.1 and B1.2 using four wire method with TM-508A (Isothermal Technology, Southport, United Kingdom).

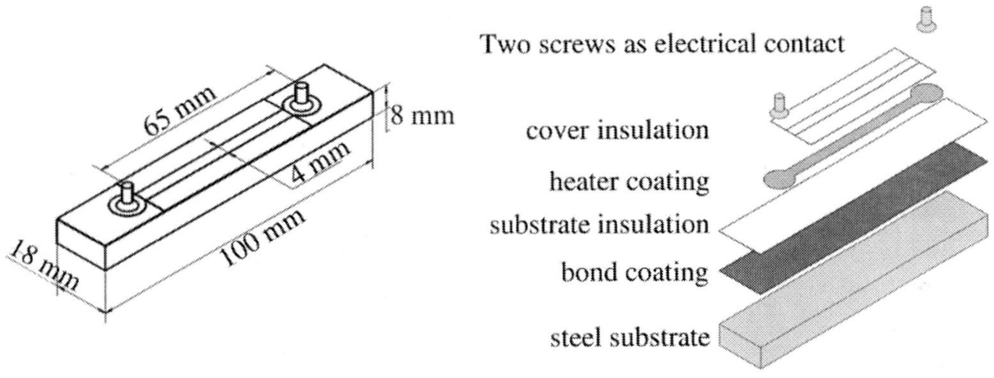

Figure 4. Schematic structure of samples B1.1 and B1.2.

3. Results and discussion

Figure 5 a) shows a closer view of sample A1 after the first brazing trial, in which the cable connector mounted with a copper cable was brazed onto the heater coating in the as-sprayed state. The area of the filler metal foil is still visible as a dark grey spot on the surface. Between the copper cables, a trace of the eutectic AgCu filler metal is visible. The capillary effect caused by the fine wires of the cable allows the filler metal to flow upwards along the cable connector, is one possible explanation for this phenomenon. Furthermore, a larger filler metal droplet on the surface of the heater coating is clearly noticeable. This droplet formation, as well as the visible gap between the cable connector and the coating, indicates poor wetting of the filler metal on the heater coating. This low wettability is probably caused by an oxide layer on the surface. Therefore, it must be concluded that this first brazing attempt was not successful.

The copper cable was removed from the cable connector for sample A2 to investigate whether the capillary effect caused by the copper wires was the main reason for the poor bonding between the cable connector and the heater coating. In addition, the surface of the heater coating was lightly ground with a 1200# SiC abrasive paper to remove the oxide layer. Figure 5 b) shows a closer look at the sample A2. A droplet formation on the surface of the heater coating can also be seen here, but it is not quite as pronounced as in the previous trial. The visible gap between cable connector and heater coating also indicates a weak connection of the two components. Therefore, the quality of the connection still remained insufficient despite the two chosen measures.

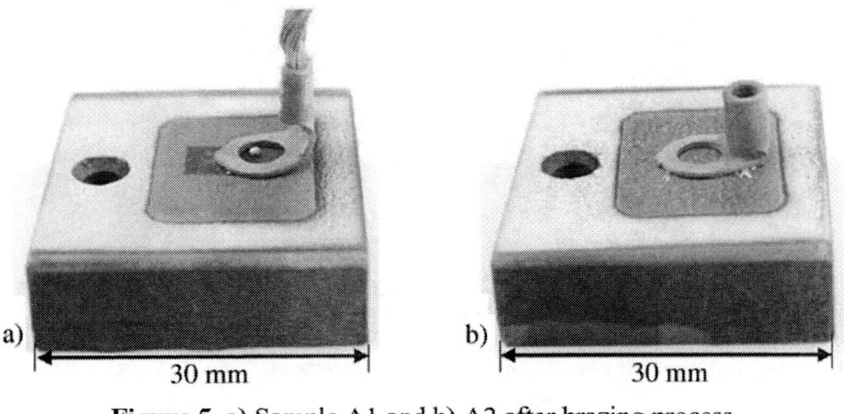

Figure 5. a) Sample A1 and b) A2 after brazing process.

To examine the underlying mechanisms in more detail, cross sections of the brazed position were prepared. The brazing gap of sample A2 was examined closely under the scanning electron microscope

(SEM), as shown in figure 6 a) and figure 6 b). Both images confirm that the filler metal only partially connects the cable connector with the heater coating. This poor connection is not sufficient to guarantee a functional electrical contact of the resistance heating element. The SEM images also indicate that most of the filler metal disappeared from the brazing gap even without the cable.

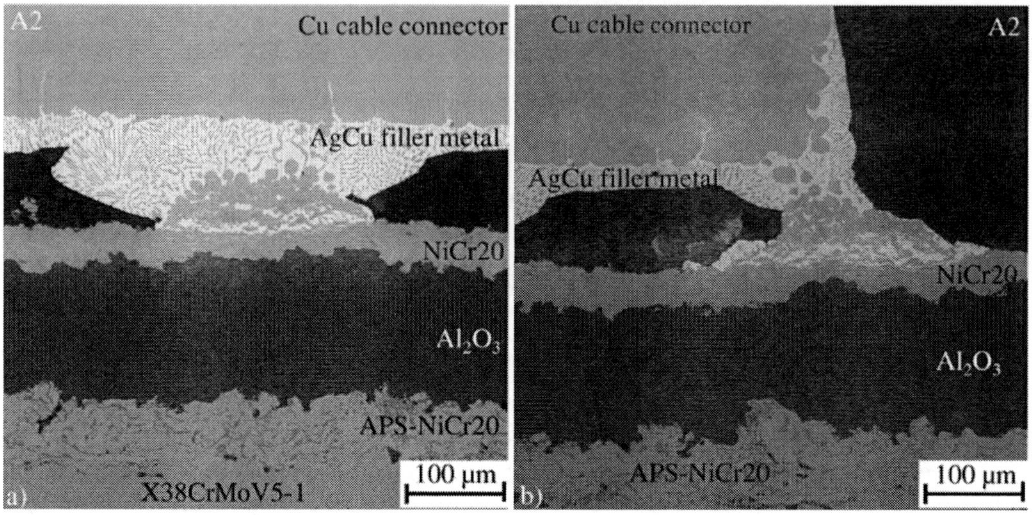

Figure 6. SEM images of the brazing gap of sample A2 in a) central position and b) side position.

Figure 7 shows another SEM image of the A2 sample, in which a bright film at the edge of the cable connector is visible. Apparently, the filler metal flowed upwards along the surface on the side of the cable connector.

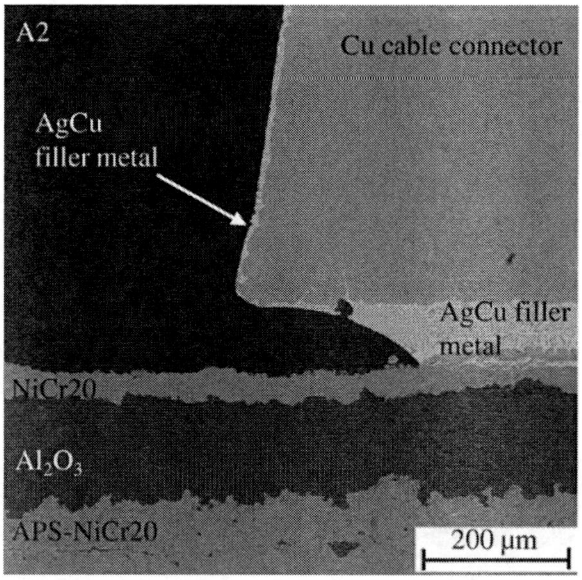

Figure 7. SEM image of sample A2 at the side of the cable connector.

The cylindrical section of the cable connector was cut and its surface was examined. In addition to the SEM images of the lateral surface of the cable connector, energy dispersive X-ray (EDX) measurements were carried out at selected positions to provide evidence of AgCu filler metal on the lateral surface of the cable connector. SEM images were taken at the middle section and the section far from the heater coating surface. Figure 8 shows two examples of the SEM images taken at these two sections. The EDX measurement points were marked with arrows on the images.

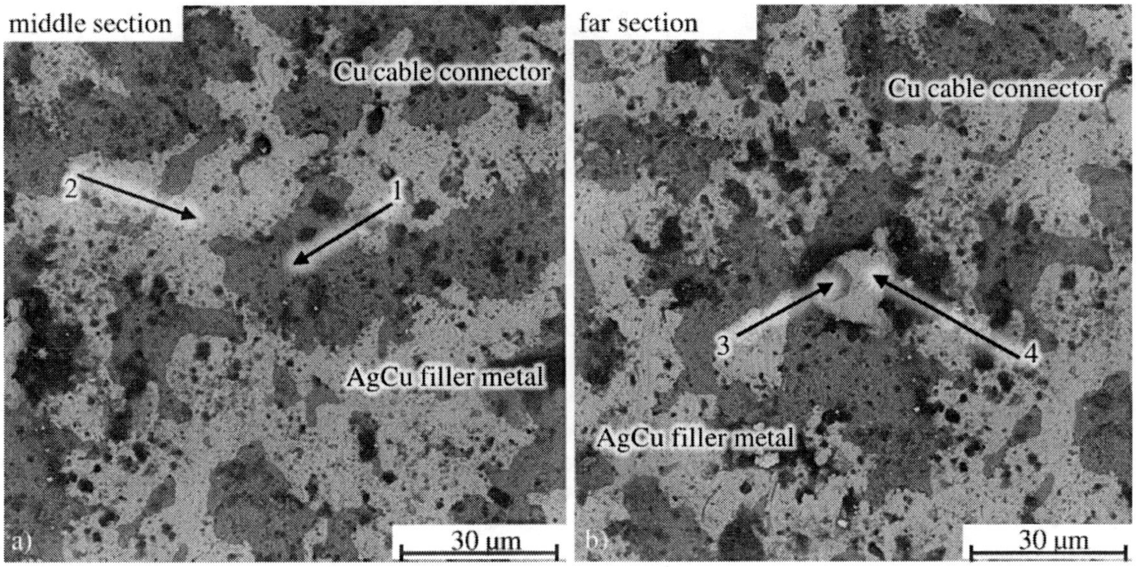

Figure 8. EDX analysis of A2 copper cable connector after brazing process at a) middle section and b) far section. Measurement points are marked with arrows.

Table 4 shows the results of the EDX analysis. It is evident that even at the far section silver can be detected on the surface of the cable connector. It can be concluded that even without the copper wire inside the cable connector the filler metal flowed up and along the surface of the cable connector due to its excellent wettability on the cooper connector. Thus, there is not enough filler in the gap between the cable connector and the coating to form a satisfactory connection. Therefore, neither grinding nor removing the copper cable from the cable connector are sufficient to achieve a satisfactory connection.

Table 4. EDX results of the side surface of the cable connector.

Position	Element [wt.-%]			
	Ag	Cu	Al	O
1	-	92.60	3.74	3.66
2	92.57	4.10	-	3.33
3	-	100	-	-
4	93.44	6.52	-	-

Based on the results obtained with A1 and A2, a steel screw was used as the connector for A3 to prevent the upward flow of the filler metal. On the other hand, the HVAF-sprayed NiCr20 heater coating was further ground with 1200# SiC abrasive paper until most of the coating surface appears glossy. The aim is the removal of most of the surface oxide to further increase the filler metal's wettability on the coating. Likewise, the lower arithmetic average roughness Ra of the ground sample, which is shown in figure 2 b) and d), should result to a better wettability. Figure 9 illustrates the new trial with sample A3. Neither droplet formation nor a gap between the screw and the coating is visible.

Figure 9. Sample A3 after brazing process.

The investigation of A3 indicates that the use of a steel screw and the thoroughly grinding of the heater coating's surface can ensure a good bonding between the screw and the coating. Therefore, this concept was used to fabricate the samples B1.1 and B1.2. A closer look at the brazing gap of sample B1.1 using SEM in figure 10 a) shows that the filler metal has completely filled the gap between the screw and the heater coating and has distributed itself homogeneously. The filler adheres well to both. The wetting of the screw and the coating surface therefore is satisfactory. The closer observation of the interface between the filler metal and the heater coating shows, that the filler metal was able to penetrate into the coating along the splat boundaries by capillary effect, see figure 10 b). Furthermore, EDX results, which are shown in table 5, confirm that silver can be detected in the gap. A solid bonding was formed.

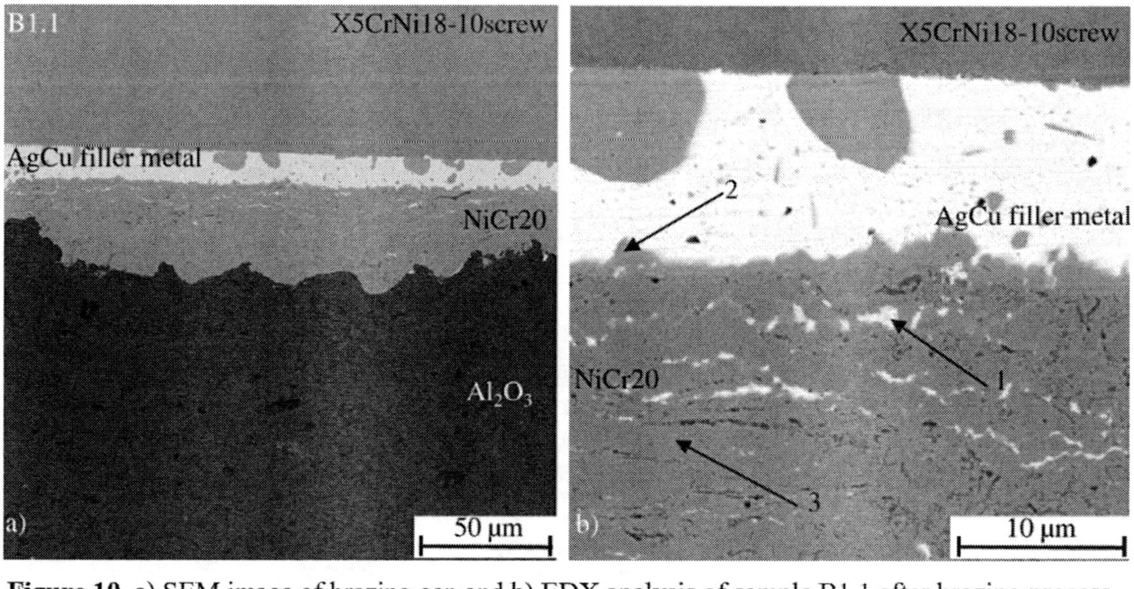

Figure 10. a) SEM image of brazing gap and b) EDX analysis of sample B1.1 after brazing process.

Table 5. EDX results of sample B1.1.

Position	Element [wt.-%]			
	Ag	Cu	Ni	Cr
1	90.80	5.15	4.05	-
2	3.79	80.55	15.66	-
3	-	-	81.45	18.55

Further cross section of B1.1 was also examined by SEM, see figure 11 a). Furthermore, the boundary layer of the coating is dissolved by the filler metal. A closer look at the boundary layer, as shown in figure 11 b), reveals that the filler metals nearly reaches the insulation coating. This diffusion process results in the formation of a solid bond.

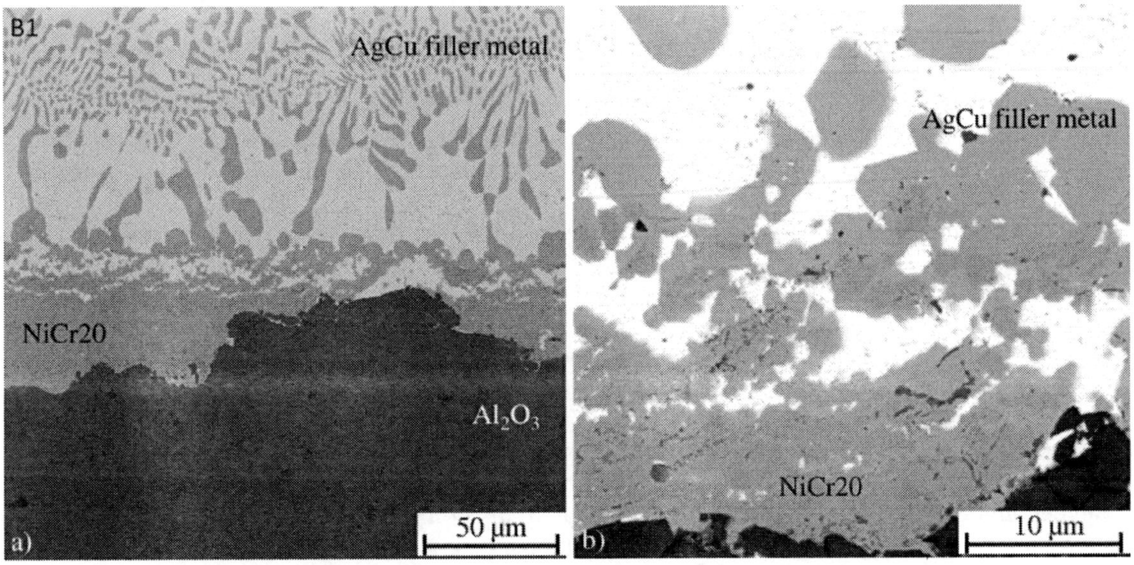

Figure 11. a) SEM image and b) closer SEM image of the B1.1 sample.

To investigate the influence of the brazing process on the functionality of the heater coating as a resistance heating element, the electrical resistance of samples B1.1 and B1.2 was measured before and after the brazing process. To verify the electrical resistance measurement both samples were fabricated similar. Table 6 shows that there are no significant differences in electrical resistance between measurements. Although the filler metal infiltrated the coating, this phenomenon only takes place in the connection area. The microstructure of the linear section of the conducting track should remain mostly uninfluenced, since the electrical resistance of the coating show almost no difference before and after the brazing process. Therefore, it can be assumed that the electrical contacting of the NiCr20 heater coating by brazing seems promisingly.

Table 6. Electrical resistance measurement.

Sample	Electrical resistance [mΩ]	
	Before Brazing	After brazing
B1.1	455	464
B1.2	534	534

4. Conclusions and Outlook

As the electrical contacting of the thermally sprayed heating elements has only been investigated to a very limited extent so far, it is necessary to examine the possibilities of contacting further. The aim of this study was to investigate the brazing of cable connectors onto a HVAF sprayed heater coating to realize an electrical contact. For this purpose, a multilayer resistance heating element was fabricated. A NiCr20 bond coating and a Al_2O_3 insulation coating were applied using the APS process. The subsequent NiCr20 heater coating was applied via HVAF process. An electrical connector was brazed onto the heater coating with a eutectic AgCu filler metal. The brazing gap and the bonding mechanism of the connector to the coating were analysed. The results can be summarized as follows:

- An as-sprayed coating surface has an oxide layer, which greatly reduces the wettability of the coating. The wettability can be increased by grinding the surface.
- Cable connectors made of copper are unsuitable for the studied case, since the filler metal's excellent wettability on copper leads to its upward flow during the brazing process and an insufficient bonding to the heater coating. Replacing the cable connector with a steel screw prevents this phenomenon and sufficient bonding can be achieved.
- If the coating surface is sufficiently wetted, the filler metal penetrates the heater coating and forms a firm solid bonding.
- The filler metal AgCu (72 wt.-% Ag, 28 wt.-% Cu) is a suitable material selection for the combination of a NiCr20 heater coating and a X5CrNi18-10 steel screw.
- The used brazing process has no significant influence on the electrical resistance of the HVAF sprayed NiCr20 heater coating.

After the coating surface was ground and a steel screw was used as an electrical connector, the brazing process enabled a satisfactory connection to the heater coating. In further studies, the developed contacting must be validated in heating cycle tests. Thereby it must be examined whether this contacting can withstand requirements and high heating rates. In future studies it must also be investigated if the presented combination can be bonded using laser or induction brazing. These processes would prevent an improper heat affection of the tools in the intended applications.

Acknowledgments

Gefördert durch die Deutsche Forschungsgemeinschaft (DFG) im Rahmen der Exzellenzstrategie des Bundes und der Länder – EXC-2023 Internet of Production – 390621612. Funded by the Deutsche Forschungsgemeinschaft (DFG, German Research Foundation) under Germany's Excellence Strategy – EXC-2023 Internet of Production – 390621612.

References

[1] Prudenziati M 2008 Development and the Implementation of High-Temperature Reliable Heaters in Plasma Spray Technology *J Therm Spray Tech* **17** 234–43
[2] Prudenziati M and Gualtieri M L 2008 Electrical Properties of Thermally Sprayed Ni- and Ni20Cr-Based Resistors *J Therm Spray Tech* **17** 385–94
[3] Floristán M, Fontarnau R, Killinger A and Gadow R 2010 Development of electrically conductive plasma sprayed coatings on glass ceramic substrates *Surface and Coatings Technology* **205** 1021–8
[4] Scheitz S, Toma F L, Berger L M, Puschmann R, Sauchuk V and Kusnezoff M 2011 Thermally sprayed multilayer ceramic heating elements *Thermal Spray Bul* **2** 88–92
[5] Killinger A and Gadow R 2006 Thermally Sprayed Coating Composites for Film Heating Devices *AST* **45** 1230–9
[6] Michels D, Hadeler J and Lienhard V J H 1998 High-Heat-Flux Resistance Heaters from VPS and HVOF Thermal Spraying *Experimental Heat Transfer* **11** 341–59
[7] Bobzin K, Wietheger W, Knoch M A and Schacht A 2020 Heating behaviour of plasma sprayed TiOx/Cr2O3 coatings for injection moulding *Surface and Coatings Technology* **399** 126199

[8] Bobzin K, Öte M, Knoch M A and Alkhasli I 2019 Temperature distribution on thermally sprayed heating conductor coatings *IOP Conf. Ser.: Mater. Sci. Eng.* **480** 12002
[9] Nicolaus M, Möhwald K and Bach F-W (eds) 2012 *A New Hybrid Process for Repair Brazing and Coating of Turbine Blades Proceedings from the International Thermal Spray Conference and Exposition*
[10] Bobzin K, Zhao L, Schlaefer T and Warda T 2010 Application of cold spraying for flux-free brazing of aluminium alloy 6060 *Front. Mech. Eng. China* **5** 256–60
[11] Zhao L, Bobzin K, Ernst F, Zwick J, Roesing J and Lugscheider E 2006 Feasibility Study of Brazing Aluminium Alloys Through Pre-Deposition of a Braze Alloy by Cold Spray Process *Adv. Eng. Mater.* **8** 751–3
[12] Long A, Thornhill D, Armstrong C and Watson D 2012 Predicting die life from die temperature for high pressure dies casting aluminium alloy *Applied Thermal Engineering* **44** 100–7
[13] Norwood A J, Dickens P M, Soar R C and Harris R A 2007 Surface temperature of tools during the high-pressure die casting of aluminium *Proceedings of the Institution of Mechanical Engineers, Part B: Journal of Engineering Manufacture* **221** 1659–64

Feasibility, processing properties and thermal behavior of Ni nanopastes produced by ultrasound-enhanced dispersing

B Sattler*, S Hausner and G Wagner

Professorship of Composites and Material Compounds, Institute of Materials Science and Engineering, Chemnitz University of Technology, D-09107 Chemnitz, Germany

*e-mail: benjamin.sattler@mb.tu-chemnitz.de

Abstract. Nickel nanoparticles are materials that can be used for a novel type of joining, so called nanojoining, as an alternative to conventional brazing. The physical characteristics of these nano-sized particles lead to a lower melting temperature compared to nickel bulk material and even lower than typical brazes, without the need for melting point depressants. Such elements like boron or silicon cause issues due to formation of brittle intermetallic phases. In order to be able to use Ni nanoparticles in practical applications, they have to be in forms such as suspensions, e.g pastes. In the present work, therefore, a series of Ni nanopastes is prepared and investigated with emphasis on the solvent systems. These suspensions consist of Ni nanoparticles (mean diameter 20 / 90 / 180 / 300 nm) and organic compounds. The examinations show the feasibility of producing such nanopastes using ultrasound-enhanced dispersing, as well as margins for adjusting their processing properties. In addition, first results of the thermal behavior of the Ni nanopastes, measured by simultaneous DSC/TG, are shown.

1. Introduction

In high-temperature applications, nickel base superalloys are heavily employed due to their high-temperature strength and great resistance to oxidation and creeping [1,2]. For joining of these alloys, a range of welding and brazing techniques are available, but there are some points to consider. For example, several Ni base superalloys show segregation issues during solidification when welded due to its high diffusion rate, which weakens it's strength [3]. According to manufactures, brazing processes in the range of 1200 °C negatively affect properties of single-crystal Ni base materials, due to recrystallization effects above 1100 °C. Brazes also limit the maximum operating temperature of joined components and can cause brittleness when intermetallic phase formation takes place at the centerline of the joint [4,5].

Therefore, this paper deals with viscous suspensions consisting of pure Ni nanoparticle and organic additions, which are intended to be used in nanojoining processes. Nanojoining is an alternative method and similar to brazing resp. soldering but possess, besides some challenges, specific advantages. Nanoparticle in general show lower melting and sintering temperatures compared to the corresponding bulk material due to their high specific surface area or high surface-to-volume ratio, respectively [6–8]. In comparison to brazing fillers no melting point depression elements like Si or B are needed. After Ni nanoparticles coalesce during the joining process, the resulting metal regain almost the properties of the bulk material. This is beneficial for joining purposes, because the process can be carried out at relatively low temperatures while the maximum operating temperature of the joint is not strongly affected [9,10]. In contrast to many previous papers dealing with low viscosity suspensions (e.g. nano inks) or

IOP Conf. Series: Materials Science and Engineering 1147 (2021) 012011 doi:10.1088/1757-899X/1147/1/012011

suspensions as a reagent in order to produce nanoparticles bottom-up [11–13], this work here presents results of paste-like, higher viscosity Ni nano suspensions.

Mixing Ni nanoparticles with solvents and stabilizers (dispersing agents) is necessary to cover their surfaces and to keep them separate from each other in order to avoid agglomeration before the actual joining process takes place [14,15]. Furthermore, for a reliable joining process handling, it is necessary to use nanoparticles in the dispersed form of a suspension, comparable to common brazing pastes.

2. Experimental and materials

2.1. Starting materials

For examination here, pure Ni nanoparticles in different mean diameters and several organic solvents and agents are used. In the following, Ni nanoparticles is abbreviated to “Ni-NP”. The mean diameters of the particles are 300, 180, 90 and 20 nm, delivered by manufacturer Amogreentech Co., LTD. (South Corea) and M K Impex Corp. (Canada), respectively. A summary of all organic materials is provided in Table 1.

Table 1. Organic compounds used for preparation of Ni nanopastes.

Compound	formula	CAS No.	Supplier
Polyethylene glycol 400 (PEG 400)	$C_{2n}H_{4n+2}O_{n+1}$	25322-68-3	Carl Roth GmbH *(GER)*
Polyethylene glycol 600 (PEG 600)	$C_{2n}H_{4n+2}O_{n+1}$	25322-68-3	Carl Roth GmbH *(GER)*
Terpineol, pure	$C_{10}H_{18}O$	8000-41-7	Carl Roth GmbH *(GER)*
Dihydroterpineol (DHT)	$C_{10}H_{20}O$	498-81-7	MuseChem *(USA)*
Ethyl cellulose (EC)	- *(more complex)*	9004-57-3	Carl Roth GmbH *(GER)*

The herein investigated nanopastes are based of two chemical systems. First, a mixture of Ni-NP and PEG, where PEG acts as solvent and stabilizers at the same time. PEG 400 is a macromolecular liquid at room temperature (20 °C) whereas PEG 600 is at it’s melting point, so appear as a pasty material. Both are miscible, so by combining them the viscosity of the resulting suspension is adjustable within a certain range. Second one is a mixture of Ni-NP, Terpineol or DHT as well as EC for the option to enhance viscosity. For tests of suspensions with EC, the EC must be solved in terpineol/DHT at approx. 85 °C in advance.

2.2. Measurement of viscosity characteristics of solvents

The viscosity range of solvent systems, which can be adjusted by their composition, is important for later applications of the nanopaste in terms of handling. In addition to detailed methods, simplified methods can also be used to determine the viscosity, such as using a viscosity cup or flow tests on a inclined plane [16]. Performing a detailed viscometer measurement is not reasonably for this work, because it would require large amounts of some limited cost-intensive starting materials and further, only a relative comparison is intended.

The measurement method used here is based on the principle of flow due to gravity. A rotatable mounting is used for this, into which sample bottles are inserted, each filled with 1 g of the solvent investigated. The mounting, including the filled bottles, is then tilted by 45° to horizontal position and turned back subsequent. This is recorded digitally and evaluated by using image analysis. The experimental design is shown in Figure 1.

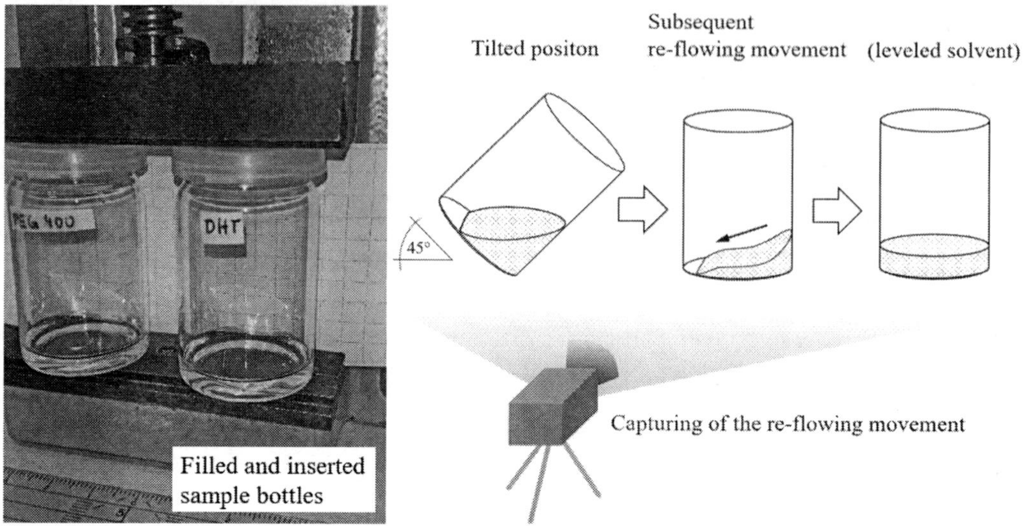

Figure 1. Sample mounting and measurement principle for evaluating viscosity characteristics of solvents.

The duration until the re-flow movement meets a certain criterion, is used as the measured value. Therefore, taking into account the flow front is inclined, a special measuring bar was defined which the flow front has completely to travel. This duration was taken as result. The duration for complete leveling of the solvent seems also as an obvious one, but preliminary tests in this case show an extreme spread of the measured values, since this non-linear process takes a long time towards the end, at least in the case of solvents with higher viscosity.

2.3. *Volatility of solvents*

The volatility of the solvents that keep the Ni-NP dispersed a suspension, has strong influence on their processing properties and storage capacity, i.e. on drying behavior. In order to obtain appropriate and reproducible results, all solvents were exposed to a defined ambient at elevated temperature to accelerate the process of volatilization. For this purpose, 1 g of each solvent was heated to 85 °C in an open sample tube in a water bath whereas the tubes do not touch the ground. After specified periods of time (between 0 to 30 minutes), the residual mass of the remaining solvent was weighed, so that the loss of mass could be determined as a function of time.

2.4. *Preparation of the Ni nanopastes*

In order to evaluate the maximum achievable metal content and to carry out thermal analysis, nanopastes were produced by first giving together precisely weighted amounts of Ni-NP and solvents together in a sample bottle. Then the components are mixtured by pulsed ultrasound-enhanced dispersing (USP 200, Hielscher-Ultraschall-Technologie) a several times while it's temperature is controlled by using a water bath. An important variable in this mixtures is the metal content of the nanopastes. Preliminary tests have shown that the metal content cannot be increased up to any value. If the content of solvent is too low, the wetting of Ni-NP is not sufficient to form a viscous suspension. Therefore, in a separate test series PEG 400 was added in small steps to a given content of Ni-NP. After each step, it was attempted to produce a suspension by using ultrasound-enhanced dispersing. If this was not possible, the solvent content was increased and so on. This was repeated with Ni-NP in all different mean sizes to evaluate the maximum achievable metal content.

For thermal analysis, three PEG based Ni nanopastes were produced, see table 2 for details. The investigation of the thermal behavior was carried out by a measuring system from *Netzsch*, model *STA 449 F1 Jupiter*. This device record heat flow (DSC) and the weight over temperature (TG, thermogravimetric analysis) simultaneous.

Table 2. Composition of Ni nanopastes used for thermal analysis.

	Ni-NP		solvent		
Nanopaste	Mean diameter	Weight	Type	Weight	Resulting metal content
Ni90_PEG400	90 nm	2518 mg	PEG 400	1037 mg	70.83 %
Ni90_PEG600	90 nm	2497 mg	PEG 600	1039 mg	70.62 %
Ni20_PEG400	20 nm	2500 mg	PEG 400	1781 mg	58.40 %

3. Results and discussion

3.1. Measurement of viscosity characteristics of solvents

All re-flow tests, like described in section 2, took place at room temperature (20 °C) and three measurements were carried out for each solvent, so 3 sequences were captured, evaluated and a mean result was formed. Results are shown in figure 2, please note EC additions are in weight percent and the 20:80 proportion is by weight too.

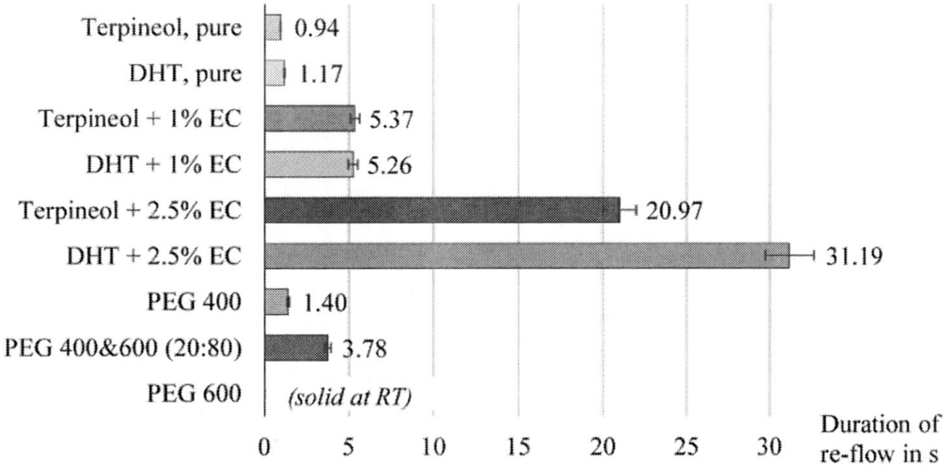

Figure 2. Results of the re-flow tests for evaluating viscosity characteristics of solvents, all solvent quantities refer to weight.

The re-flow durations are to be interpreted in this way, that a higher value corresponds to a higher viscosity of the liquid. Pure terpineol and pure dihydroterpineol (DHT), show the lowest values and behave very similarly. If only 1 wt.-% ethyl cellulose (EC) is added, the re-flow duration increases to more than four times in both cases. An addition of 2.5 wt.-% EC leads to a further large increase, which is even greater in case of DHT. The significantly more viscous properties of terpineol and DHT with additions of EC are also clearly noticeable during the experimental handling. In summary, the system [terpineol/DHT] + EC provides a good viscosity range that can be set to application desired values with relatively small amounts of EC.

Polyethylene glycol (PEG) 400 has a slightly higher viscosity than the terpineol or DHT without EC. At room temperature, PEG 600 is at the transition to the solid and does not flow. If PEG 400 and PEG 600 are mixed together, the melting point can be shifted or the viscosity of the mixture can be adjusted as desired. For example, a mixture of PEG 400 and PEG 600 in a ratio of 20:80 by weight shows a re-flow duration that is more than twice as long as pure PEG 400. Pure PEG 600 is also suitable for processing a usable suspension, but for dispersing with Ni nanoparticles, the temperature must increase above the melting point (e.g. 40° C) while mixing.

For a better assessment of the values, it should be stated that all solvents are more viscous then water at room temperature. In general, the behavior is strongly dependent on temperature, which was observed, for example, when terpineol or DHT are heated to 80 °C during the dissolution of EC. At this temperature, their viscosity is comparable to that of water, despite the EC content. The PEGs show a sensitive viscosity behavior particularly around their melting point.

3.2. Volatility of solvents

The investigation was carried out with all solvents of which the viscosity was characterized. The figure 3 shows the loss of mass after 5 / 10 / 20 / 30 minutes.

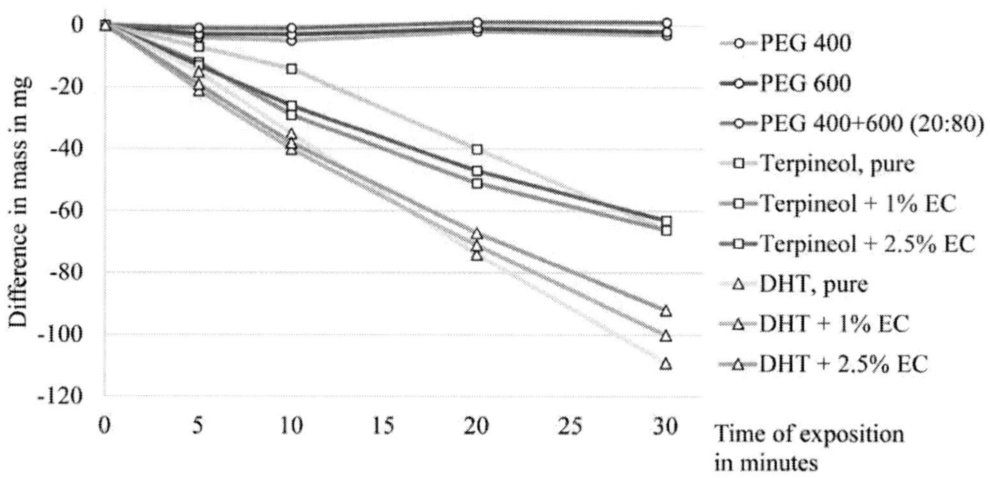

Figure 3. Volatility of solvents heated to 85 °C, measured as loss of mass over time.

It can be observed that the PEGs show almost no loss over the entire exposure duration of 30 minutes; this is true for pure PEGs as well as the 80:20 mixture (PEG 400: PEG 600). Terpineol solvents shows mass losses that are roughly linear over time. The same applies to DHT based solvents, but with a higher rate of loss. Pure DHT lost 109 mg of mass after 30 minutes, that is over 10% based on the initial mass of 1g. A significant influence of the EC on the volatility cannot be identified. The generally higher volatility of terpineol- and DHT-based solvents is also clearly seen in the processed form as Ni nanopaste during experimental handling. Small amounts of suspension of this kind dry up quickly, whereas PEG based suspensions are still stable after exposition to air for weeks. This behavior is an important aspect as it is crucial for the long-term viscosity properties and dispersion stability of nanopastes.

3.3. Maximum metal contents and thermal analysis of PEG based Ni nano suspensions

First, according to the method described in experimental, a evaluation of the maximum metal content for Ni nano suspensions with PEG 400 as solvent was carried out. The results are listed in table 3.

Table 3. Maximum metal contents in PEG 400 based Ni nano suspensions.

	Mean diameters of Ni-NP			
	300	180	90	20
Max. metal content in wt.%	89.2	85.2	78.2	~58.4

Compared to other publications [12,17,18] it turns out that relatively high metal contents are feasible, but it depends on the particle size. So, the smaller the mean diameters of the particles, the more solvent is necessary to obtain a suspension. In other words, the maximum realizable metal content decreases with decreasing particle sizes. The mixture which consists of 20 nm particles does not show any clear suspension forming behavior, since strong agglomerations were observed, so this value is uncertain.

The following diagrams in figure 4, 5 and 6 show initially examinations of thermal behavior of the Ni nanopastes (combined curves of DSC/TG). The results for the nanopastes Ni90_PEG400, Ni90_PEG600 and Ni20_PEG400 under an argon shield gas atmosphere at a heating of 10 K/min are presented here, further investigations are in progress.

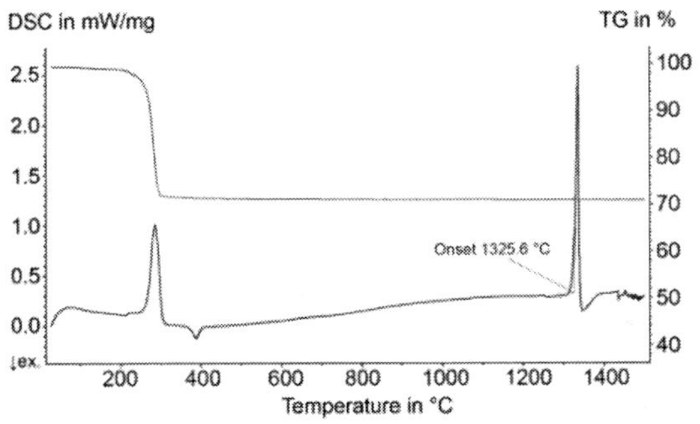

Figure 4. DSC / TG signal of nanopaste "Ni90_PEG400".

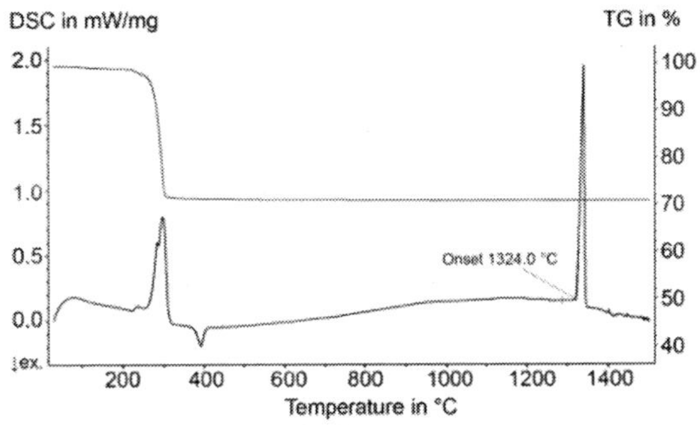

Figure 5. DSC / TG signal of nanopaste "Ni90_PEG600".

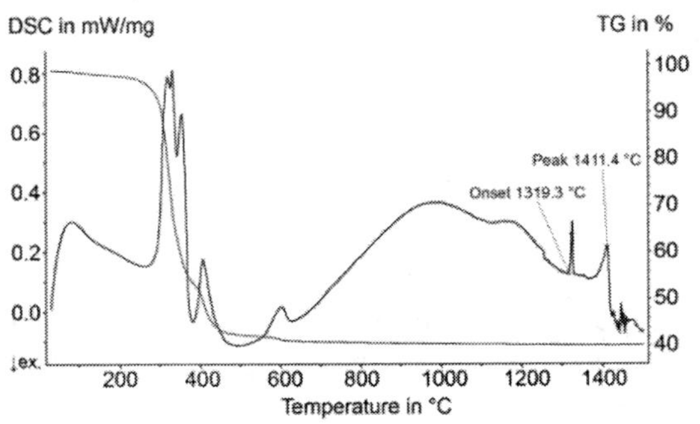

Figure 6. DSC / TG signal of nanopaste "Ni20_PEG400".

The TG signal in pastes Ni90_PEG400 and Ni90_PEG600 indicates an almost complete evaporation of the organic components of the suspension, since the mass after the end of the heating process corresponds almost exactly to the mass fraction of the given nickel (metal content). However, carbon residues from the solvent are to be expected. The evaporation extends over a temperature range of approx. 200 ... 300 °C in both cases and manifests itself in the DSC signal with a corresponding reaction. The weaker exothermic reactions at just below 400 °C cannot be easily interpreted and must be further investigated, but based of former investigations it is assumed to be a sinter reaction of the Ni-NP. At high temperatures, the DSC signals shows a clear endothermic reaction at around 1325 °C. This is not the melting reaction of pure nickel with it's significantly higher melting point at 1455 °C. Due to the assumed carbon residues, this might be the melting reaction of eutectic mass from the Nickel-Carbon binary system. The eutectic melting point in this system is calculated to be at 1326.4 °C according to Franke et al.[19]. Basically, carbon residues, that lowering the melting point, are not desired in later joining applications, but they do not have to occur there due to the higher diffusion activity in thin joining layers. Apart from that, other organic solvents do not show this behavior in DSC signals and instead indicate the melting temperature of bulk nickel [9].

The diagram of the nanopaste Ni20_PEG400 is different. Knowing the thermal characteristics of pastes Ni90_PEG400 and Ni90_PEG600 from their diagrams, a roughly similar behavior can be recognized, but overall it is not that clearly. First, the mass loss according to the TG signal does not match the solvent content of the paste. There is a discrepancy of about 18 wt.-%. Further, the heat flow signal indicates more distributed reactions, especially no strong reaction at around 1325 °C. It must be stated, that the 20nm Ni-NP show agglomerations, even after ultrasound-enhanced dispersing, so the resulting suspension is not that homogenous like the other ones. This might be a reason for the less defined thermal behavior in comparison to the suspensions consists of 90nm Ni-NP.

4. Conclusions

In this work, organic solvents systems and suspensions of Nickel nanoparticles are investigated. These so called nanopastes are intended to be used for alternative joining methods ("nanojoining") of Ni base superalloys. It is similar to brazing, but offers several advantages, most notably a significantly lower process temperature and, at the same time, a higher operating temperature of the joint. This is based on the special properties of nanoparticles, which, due to their very high surface-to-volume ratio, have a lower melting point compared to bulk material.

Investigations reveals the viscosity and volatility behavior of two different solvent systems. The first system, terpineol and dihydroterpineol (DHT) with ethyl cellulose (EC) additives provides a wide range of viscosity viscosities that can be tailored to the needs of an application. However, these solvents also show highly volatility during tests in a controlled ambient. Second system is based on polyethylene glycols (PEGs) and also offers a range of viscosity by mixing PEGs of different molecular weights, like PEG 400 and PEG 600 here in this study. Furthermore, it turns out that PEGs are almost non-volatile, even dispersed in a Ni nanopastes after weeks exposed to air.

Ni nanopastes have been successfully prepared using these solvent systems. Results are presented for PEG based nanopastes. The maximum metal content, which can be achieved in these suspensions, is examined. Results show a clearly dependence of the particles mean diameter. Also, results in thermal behavior using simultaneous DSC/TG measurement were shown.

Acknowledgments

This work was performed within the IGF project 20.980 B / DVS No. 07.3249 of the research association "Forschungsvereinigung Schweißen und verwandte Verfahren e. V. des DVS, Aachener Straße 172, 40223 Düsseldorf" which is, based of a resolution of the German Bundestag, promoted by the Federal Ministry of Economic Affairs and Energy via AiF within the framework of the program for the promotion of joint industrial research.

Financial support from the AiF is gratefully acknowledged. Further gratitude goes to the members of the project-accompanying committee and of DVS for their support of this as well as to the members at our institute.

IOP Conf. Series: Materials Science and Engineering 1147 (2021) 012011 doi:10.1088/1757-899X/1147/1/012011

References

[1] Madhusudhana R G, Srinivasa M C V, Srinivasa R K and Prasad R K 2009 Improvement of mechanical properties of Inconel 718 electron beam welds–influence of welding techniques and postweld heat treatment *The International Journal of Advanced Manufacturing Technology* **43** 671–680

[2] Amato K N, Gaytan S M, Murr L E, Martinez E, Shindo P W, Hernandez J, Collins S and Medina F 2012 Microstructures and mechanical behavior of Inconel 718 fabricated by selective laser melting *Acta Materialia* **60** 2229–2239

[3] Xiao H, Li S M, Xiao W J, Li Y Q, Cha L M, Mazumder J and Song L J 2017 Effects of laser modes on Nb segregation and Laves phase formation during laser additive manufacturing of nickel-based superalloy *Materials Letters* **188** 260–262

[4] Ghasemi A and Pouranvari M 2019 Intermetallic phase formation during brazing of a nickel alloy using a Ni–Cr–Si–Fe–B quinary filler alloy *Science and Technology of Welding and Joining* **24** 342–351

[5] Samani M S, Bahrami A and Karimzadeh F 2019 Microstructure and mechanical properties of transient liquid phase (TLP)-bonded Ni3Al intermetallic compounds *Materials Today Communications* **21** 100619

[6] Shyjumon I, Gopinadhan M, Ivanova O, Quaas M, Wulff H, Helm C A and Hippler R 2006 Structural deformation, melting point and lattice parameter studies of size selected silver clusters *The European Physical Journal D - Atomic, Molecular, Optical and Plasma Physics* **37** 409–415

[7] Delogu F 2005 Thermodynamics on the Nanoscale *The Journal of Physical Chemistry B* **109** 21938–21941

[8] Nanda K K 2009 Size-dependent melting of nanoparticles: Hundred years of thermodynamic model *Pramana* **72** 617–628

[9] Hausner S 2015 Potential von Nanosuspensionen zum Fügen bei niedrigen Temperaturen *Chemnitz University of Technology*

[10] Hausner S, Weis S, Wielage B and Wagner G 2016 Low temperature joining of copper by Ag nanopaste: correlation of mechanical properties and process parameters *Welding in the World* **60** 1277–1286

[11] Bala T, Gunning R D, Venkatesan M, Godsell J F, Roy S and Ryan K M 2009 Block copolymer mediated stabilization of sub-5 nm superparamagnetic nickel nanoparticles in an aqueous medium *Nanotechnology IOP Publishing* **20** 415603.

[12] Tseng W J and Chen C-N 2006 Dispersion and rheology of nickel nanoparticle inks *Journal of materials science* **41** 1213–1219

[13] Yin H and Chow G-M 2009 Effects of oleic acid surface coating on the properties of nickel ferrite nanoparticles/PLA composites *Journal of Biomedical Materials Research Part A* **91A** 331–341

[14] Nagasawa H, Maruyama M, Komatsu T, Isoda S and Kobayashi T 2002 Physical Characteristics of Stabilized Silver Nanoparticles Formed Using a New Thermal-Decomposition Method *Physica status solidi (a)* **191** 67–76

[15] Jiang H, Moon K-S, Lu J and Wong C P 2005 Conductivity enhancement of nano silver-filled conductive adhesives by particle surface functionalization *Journal of Electronic Materials* **34** 1432–1439

[16] Mezger T G 2016 *Das Rheologie Handbuch: 5., überarbeitete Auflage* (Hannover: Vincentz Network) pp 332

[17] Jay Deiner L, Piotrowski K A and Reitz T L 2013 Mechanisms of Fatty Acid and Triglyceride Dispersant Bonding in Non-Aqueous Dispersions of NiO *Journal of the American Ceramic Society* **96** 750–758

[18] Lee S, Paik U, Yoon S-M and Choi J-Y 2006 Dispersant-Ethyl Cellulose Binder Interactions at the Ni Particle-Dihydroterpineol Interface *Journal of the American Ceramic Society* **89** 3050–3055
[19] Scientific Group Thermodata Europe (SGTE) 2012 *Thermodynamic Properties of Inorganic Materials: Binary Systems and Ternary Systems from C-Cr-Fe to Cr-Fe-W* vol Subvolume C, ed Franke P and Seifert H J (Springer) pp 15

IOP Conf. Series: Materials Science and Engineering 1147 (2021) 012012 doi:10.1088/1757-899X/1147/1/012012

Shallow neural networks to predict glass transition, crystallization and liquidus temperature of iron-based metallic glasses

K Bobzin, W Wietheger and L M Johann*

Surface Engineering Institute (IOT), RWTH Aachen University

*e-mail: johann@iot.rwth-aachen.de

Abstract. In the designing of iron-based metallic glasses the prediction of the glass transition temperature T_g, crystallization temperature T_x and liquidus temperature T_l is of special interest. The determination of these temperatures allows not only conclusions about the glass forming ability by calculating glass formation criterions but also about the thermal stability of the alloy in the amorphous state. In the course of alloy development, T_g, T_x and T_l are usually determined by means of differential scanning calorimetry (DSC) on an amorphous sample produced, for instance, by melt spinning or copper casting techniques. The test procedures are time-consuming and cost-intensive. In the present work T_g, T_x and T_l of iron-based metallic glasses are predicted. For that purpose, shallow neural networks with a varying number of neurons are trained using Bayes regularization. The data set for training and testing consists of corresponding literature data. The chemical compositions of iron-based metallic glasses are used as inputs and T_g, T_x and T_l are the outputs. A threshold method is used for data balancing and limiting the number of inputs. Low Root Mean Squared Error (RMSE) and correspondingly high prediction accuracies were achieved during the testing.

1. Introduction

In 1960, Klement et al. [1] discovered non-crystalline structures during the rapid solidification of the metallic alloy $Au_{75}Si_{25}$. Due to the high cooling rate and the resulting rapid solidification beyond equilibrium, crystallization was suppressed. Since then, numerous metallic glasses with different main alloying elements, such as Zr, Fe, Mg, Pd, Pt have been found [2]. Extensive research has shown the great potential that lies in metallic glasses. The amorphous state lacks grain boundaries and enhances passivation behavior, which leads to a better corrosion resistance than in the crystalline state, as shown for $Cu_{50}Zr_{50}$ by Naka et al. [3]. Inoue et al. [4] investigated the yield strength of Fe-based metallic glasses $[(Fe_{1-x}Co_x)_{0.75}B_{0.2}Si_{0.05}]_{96}Nb_4$ ($x = 0$–0.5) and discovered an extremely high fracture strength of $\sigma_f = 3{,}900$–$4{,}250$ MPa and very high hardness. In order to achieve these properties, the metallic glass must be cooled faster than its critical cooling rate R_c. Cohen and Turnbull [5] assumed that any ionic or metallic melt could be converted to the amorphous state if it was sufficiently supercooled. The critical cooling rate necessary for the vitrification of the alloy, R_c, is also considered a parameter for measuring its glass forming ability (GFA). Thermodynamic approaches to determine the GFA, i.e. the critical cooling rate, led to the definition of various quantitative criteria, some of which are shown in Table 1. One of the first quantitative criterions, the reduced glass transition temperature, T_{rg} was proposed by Turnbull [6]. T_{rg} is defined as the ratio of glass transition temperature T_g and liquidus temperature T_l (Table 1). Turnbull suggested that liquids with a T_{rg} of 2/3 or higher would, if seed-free, crystallize only in a small temperature range, i.e. between T_l and T_g. Inoue proposed ΔT_x, the supercooled liquid region, as a criterion for the GFA [7]. Although the GFA is also dependent on the thermal stability of the alloy, which is well-indicated by ΔT_x [8], no strong correlation of ΔT_x with R_c was found [9, 10].

Table 1. Overview of some quantitative GFA criterions.

Parameter	Equation	Reference
T_{rg}	$T_{rg} = \frac{T_g}{T_l}$	[6]
γ	$\gamma = \frac{T_x}{T_g+T_l}$	[11]
γ_m	$\gamma_m = \frac{2T_x-T_g}{T_l}$	[12]
ΔT_x	$\Delta T_x = T_x-T_g$	[7]

Lu et al. [11] considered high ratios of crystallization temperature to liquidus temperature and crystallization temperature to glass transition temperature to be beneficial for the GFA. They proposed the γ-criterion [11], which showed a strong correlation with R_c of $R^2 = 0.91$ [12]. Further considerations led to the definition of the γ_m-criterion by Du et al. [12], which showed an even higher correlation with R_c at $R^2 = 0.931$. Among the criterions in Table 1 Guo et al. found that for various metallic glasses γ_m shows the highest correlation [10, 13]. As the definition of these criterions reveals, the determination of T_g, T_x and T_l is of essential importance in the development of metallic glasses. Not only may the GFA be estimated using these characteristic temperatures but also the thermal stability of the metallic glass, i.e. its resistance against crystallization at elevated temperatures.

2. Procedures

The procedures in the present paper were conducted using the commercial software MATLAB® 2020b. Neural networks were implemented using functions of the Statistics and Machine Learning Toolbox and the Deep Learning Toolbox.

2.1. Datamining and processing

Data which is considering the composition, T_g, T_x, T_l and heating rates of Fe-based metallic glasses is collected. The basis of this data set is Suryanarayana et al. [14] containing ~300 compositions. Further amorphous alloy compositions and the corresponding T_g, T_x, T_l are added by surveying respective literature [15–32] and another data set was provided in Johnson et al. [33]. Data points for ternary iron-based alloys are added from the Landolt Börnstein Phase Diagrams and Physical Properties of Nonequilibrium Alloys, Physical Properties of Ternary Amorphous Alloys series, Part 2 and 3 [34]. Multiple types of data bias can be observed in the resulting data set. A nearly bimodal distribution of the number of components is shown in Figure 1. Most of the amorphous alloys in the data set comprise of six components. The peak at ternary compositions originates from the Landolt-Börnstein data.

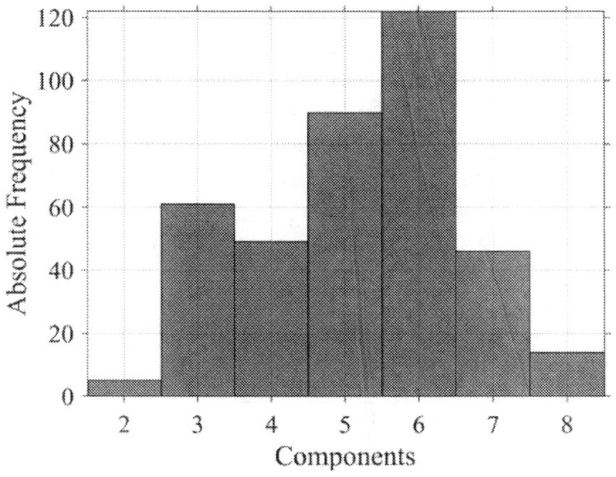

Figure 1. Histogram plot of the absolute frequency of components in the iron-based amorphous alloys data set.

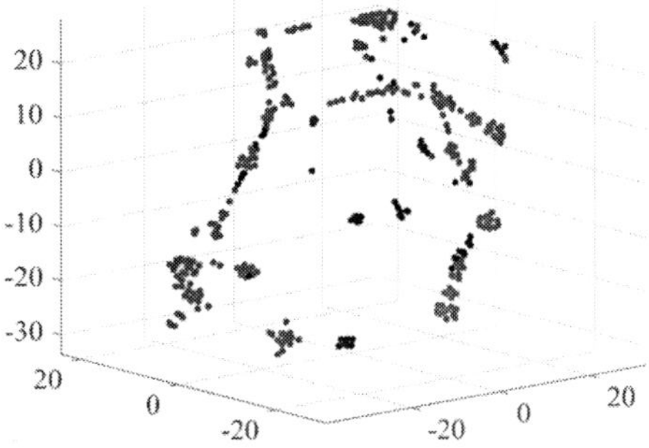

Figure 2. Three dimensional t-SNE plot of the input data at a perplexity of 19 and a learning rate of 250; alloys with a metalloid content of 15 atm.-% or higher marked in red, others black.

According to Inoue's rules more than three alloying elements are beneficial to the glass forming ability [7, 8], which may explain the observations in Figure 1. Most Fe-based amorphous alloys comprise of a metal-metalloid composition, wherein the metal content is ~80 at. % and the metalloid content is at ~20 at. % [2, 14]. The majority of the Fe-based metallic glasses considered here contain at least 15 at. % metalloids (Figure 2). The t-distributed stochastic neighbor embedding (t-SNE) algorithm, created by L. v. d. Maarten and G. Hinton [35], provides a visualization of high dimensional data which are frequently faced in machine learning problems. As to be seen in Figure 2, the data points reveal a somewhat clustered structure when the t-SNE algorithm is applied. This may well be interpreted as a representation of the alloy development in iron-based amorphous alloys. It is suspected that when an iron-based amorphous alloy composition is found, related compositions, i.e. with the same alloying elements and slight variations in concentration are tested and consequently related alloy compositions are found.

In order to prepare the data for a prediction model data processing must be conducted as well. Some alloying elements are only seldom added, e.g. Pt, Sc, As. The influence of these elements on T_g, T_x and T_l cannot be estimated reliably. To account for that a threshold is used to filter out compositions which contain rarely added elements. Thereby elements which are only contained in 3 % or less of all compositions in the data set are neglected.

Glass transition temperature and crystallization temperature are usually determined by differential scanning microscopy (DSC). However, both temperatures depend on the crystallization kinetics, the heating rate used to acquire the DSC data. This dependency has been thoroughly investigated by Kissinger [36] regarding peak temperature T_p of the reaction. The investigations of Mitrovic et al. [37] revealed for the iron-based metallic glass $Fe_{72}Al_5Ga_2P_{11}C_6B_4$ an increase in T_p from $T_{p,5} = 776$ K at $\beta = 5$ K/min up to $T_{p,80} = 810$ K at $\beta = 80$ K/min. As T_p increases, so does T_x, which is defined as the onset temperature of the respective crystallization peak. The data set exhibits a heating rate of $\beta = 40$ K/min for 143 data points and a heating rate of $\beta = 20$ K/min for 90 data points, covering most of the data set. In order to account for the different heating rates at the measured glass transition and crystallization temperatures, both are approximated to a standard heating rate of 40 K/min. The T_g and T_x values at $\beta = 40$ K/min of data points measured at differing heating rate are estimated using a logarithmic fit proposed by M. Lasocka [38] for $Te_{85}Ge_{15}$.

$$T_{g,x} = A_{g,x} + B_{g,x} * \ln(\beta) \quad (1)$$

T_x and T_g are increased by higher heating rates and the slope of the crystallization temperature curve is always larger than that of the glass transition temperature curve, $B_g < B_x$ [39]. For the Fe-based amorphous alloy $Fe_{77}C_5B_4(AlGa)_3(PSi)_{11}$ B_g was calculated by Ile et al. [40] to be 6.29. B_x is estimated based on multiple sources investigating the crystallization of Fe-based amorphous alloys [37, 41, 42]. The coefficient is calculated as the mean of the minimum and the maximum value of all fitting curves B_x; $B_x = 13.34$. The intercept $A_{g,x}$ is calculated for every T_g and T_x with its corresponding heating rate. The respective values at $\beta = 40$ K/min can thereafter be estimated. Data points without a heating rate are not approximated in any way, i.e. are assumed to be $\beta = 40$ K/min. More specifically, extremely low heating rates are only common when the crystallization kinetics are investigated, in which case a heating rate will obviously be given. Secondly the error due to heating rates which are not $\beta = 40$ K/min is expected to outweigh the benefits of additional correct data in terms of training and testing. To avoid high approximation errors in T_g and T_x, data points with a heating rate of less than $\beta = 10$ K/min are omitted.

2.2. Neural Network Design

Some approaches for predicting material properties by using neural networks (NN) have been reported in the literature. Cassar et al. [43] trained a multilayer perceptron on a data set of 55,000 inorganic glasses, aiming to predict T_g. Therein a neuron dropout routine as well as an early stopping routine have been implemented to prevent overfitting. In that work although a fairly low uncertainty was reached, some outliers remained during testing. These have been ascribed to measurement errors and faulty data in the large data set. Dewapriya et al. [44] used shallow neural networks (SNN) as well as deep neural networks (DNN) to predict the fracture stress of graphene samples under various processing conditions. Herein the largest SNN comprised of up to 61 parameters and the largest DNN comprised of 56,872,321 parameters respectively. It has been concluded that both, the approach taken with the SNNs and DNNs are in principle suitable to match the requirements. The authors highlighted the capabilities of the SNNs and suggest the implementation of SNN especially when facing a limited number of training samples.

SNN with a sigmoid function in the first layer and a linear function in a second layer may approximate most functions arbitrarily well [45]. For this work a SNN architecture is chosen comprising of two hidden layers and an output layer. A logistic-sigmoid activation function is used on the first layer and a linear rectifier on the second layer. The output layer applies a linear function. To reach the best network topology, the number of neurons in each hidden layer is varied from one up to ten neurons, leading to a total of 100 neural networks per regressand (T_g, T_x, T_l). In order to keep the networks with varying size comparable, each network topology is initiated for 100 times with biases at 0 and weights at random. A schematic illustration of the neural network design is depicted in Figure 3.

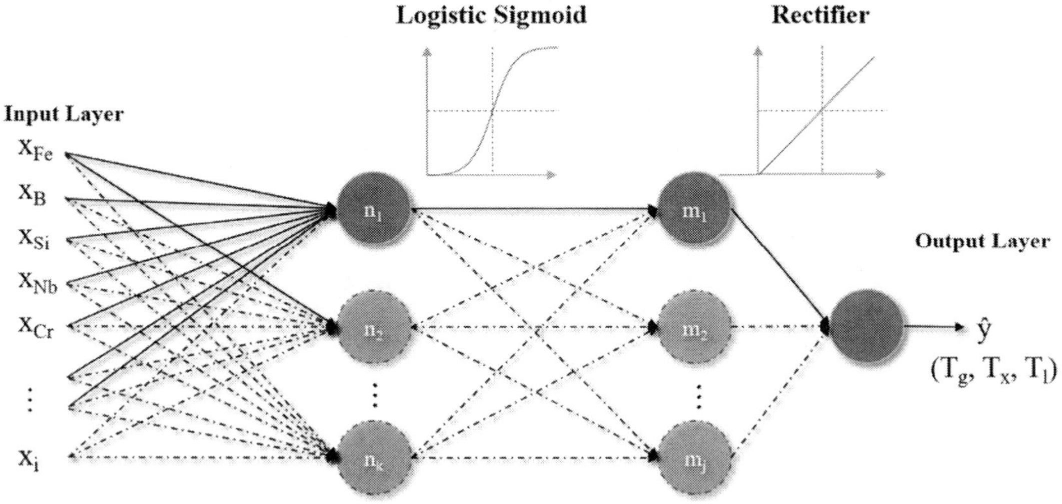

Figure 3. Schematic illustration of the neural network design with inputs x_i, neurons n_k in the first and neurons m_j in the second layer for the regressand $\hat{y}$.

2.3. Neural Network Training

Feed-forward neural networks for T_g, T_x and T_l are set up as described in subsection 2.2. The data set for the respective regressand is randomized and divided into a subset for testing and a subset for training. In order to ensure a sufficient amount of training data, the first contains 5 % of the data and is used for a cross-validation, the former comprises of the remaining 95 %. Bayesian regularization (BR) is applied to avoid overfitting and improve model generalization. In contrary to the early stopping method, BR does not require a validation subset. Hence, a higher proportion of the limited data available can be used for training. Moreover, BR training has proven to be able to fit complex relationships, yielding a robust model for shallow neural network predictions [46].

Each neural network topology is trained with random initial weights and initial biases set to zero for 100 initializations. The 100 neural networks of the set topology, e.g. 2 x 2 neurons, are then compared to the test set by mean squared error (MSE). The network with the lowest MSE during the cross-validation is chosen for the respective topology.

3. Results and discussion

3.1. Regressing on T_g

Ten by ten neural networks are found according to subsection 2.3 with the glass transition temperature set as regressand. After applying the aforementioned threshold, a total of 206 data points are used to train the neural networks. Each neural network is cross-validated using the test set (11 data points) and the RMSE is calculated.

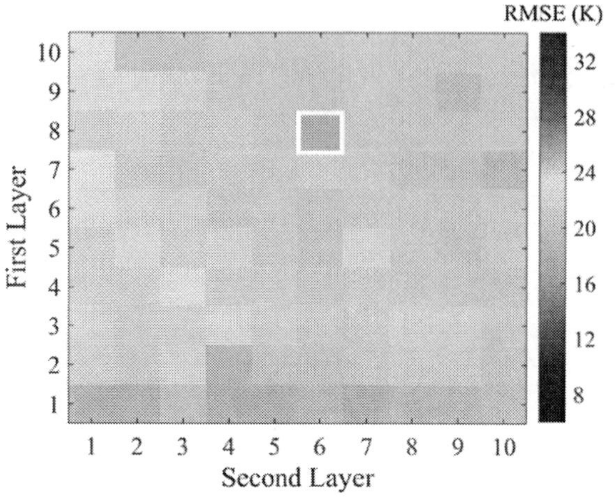

Figure 4. RMSE heatmap visualization of the ten by ten neural network topologies for the regressand T_g; best neural network marked white.

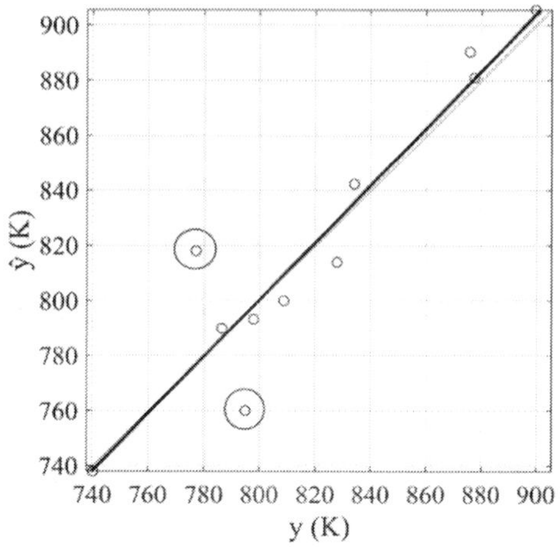

Figure 5. Regression plot of target values y and output values $\hat{y}$ for the neural network with a 8 x 6 structure; outliers marked red.

The RMSE of each neural network is illustrated in Figure 4. It can clearly be seen that a topology consisting of a single neuron in the first layer results in poor performance. This may well be explained by all the inputs passing through a single neuron which only returns a scalar, independent of the complexity of subsequent layers. Furthermore, for more than one neuron in the first layer, the performance improves successively from one to four neurons in the second layer. A topology comprising of less than three neurons in the second layer appears to yield suboptimal RMSE. The lowest RMSE of 17.89 K was achieved with eight neurons in the first layer and six neurons in the second layer. For this architecture 213 parameters (weights and biases) have been trained. The results of the cross-validation for this neural network are shown in Figure 5. Two major outliers (Figure 5, red) have been found, showing a very high deviation from the target of 40.91 K and 34.88 K respectively. The remaining nine test points have been predicted with a sufficiently high accuracy, i.e. at a deviation of less than 15 K.

3.2. Regressing on T_x

In analogy to subsection 3.1 ten by ten neural networks are determined according to subsection 2.3 with the crystallization temperature as a regressand. The threshold leaves a total of 222 data points in the training set. The 100 neural networks per topology initialized at random weights are cross-validated using a test set which consists of 12 data points. The resulting RMSE of the different topologies are visualized in Figure 6. Just as in Figure 4, any neural network topology containing a single neuron in the first layer will exhibit poor performance, i.e. at least two neurons in the first layer are required for decent prediction accuracy. Remarkably, the best performing network topology found for the prediction of T_g, comprising of an 8 x 6 structure, herein is an outlier. This may simply be a statistical artefact, i.e. none of the 100 initializations could be trained to a sufficient gradient minimum. A topology consisting of eight neurons in the first and the second layer, resulting in 241 trained parameters, yielded the lowest RMSE of 6.92 K. The difference between target and output in the test set did not exceed 15 K.

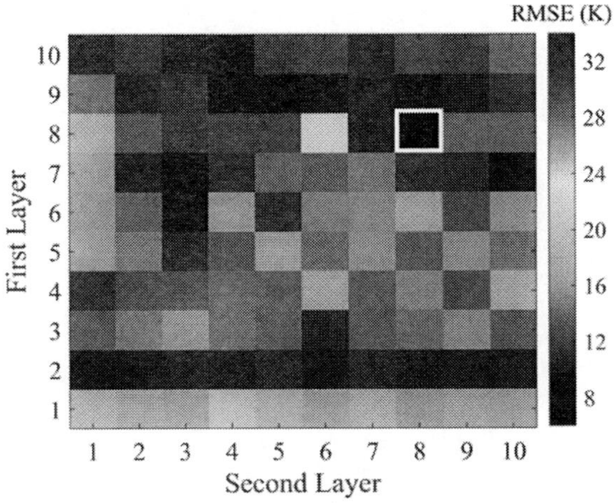

Figure 6. RMSE heatmap visualization of the 10 by 10 neural network topologies for the regressand T_x; best neural network marked white.

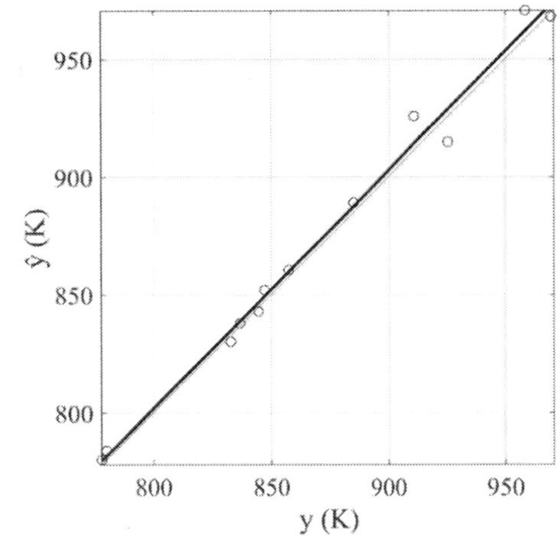

Figure 7. Regression plot of target values y and output values ŷ for the neural network with a 8 x 8 structure.

The regression plot in Figure 7 reveals the high accuracy for the cross-validation with the test set. Comparing Figure 6 and 4 illustrates that the training on T_g showed higher RMSE for most topologies. Whereas many topologies for the target being T_x reach a RMSE in the order of 10 K, T_g neural network topologies, to the contrary, exhibit much higher RMSE of about 20 K with few exceptions.

3.3. Regressing on T_l

As in subsection 3.1 and 3.2, ten by ten neural network topologies are investigated with the liquidus temperature as a regressand. After the 3 % threshold is applied, a total of 129 data points are available for training and seven data points for testing, i.e. the smallest subset among the three regressands.

As shown in Figure 8, any topology with one or two neurons in the first layer results in unsatisfactory performance, i.e. is not suited to fit the presented problem. The best performing neural network among the 100 initializations of ten by ten networks consisted of ten neurons in the first and six neurons in the second layer, attaining a RMSE of 16.81 K, which is similar to the lowest RMSE achieved in subsection 3.1. This neural network comprises of 253 parameters in total. The largest error identified in the test set was found to be 32.71 K, which is marked as an outlier in Figure 9.

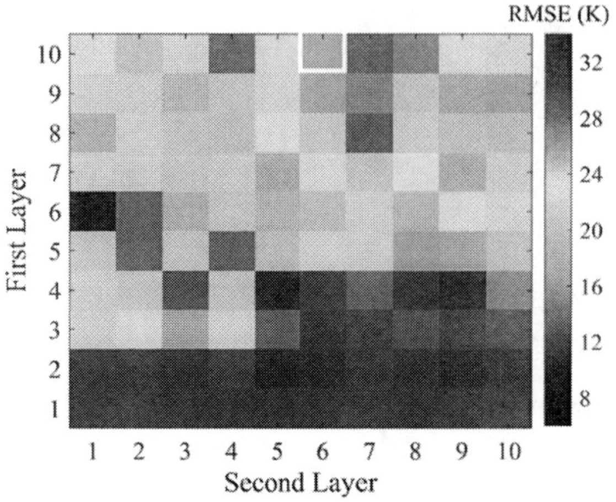

Figure 8. RMSE heatmap visualization of the 10 by 10 neural network topologies for the regressand T_l; best neural network marked white.

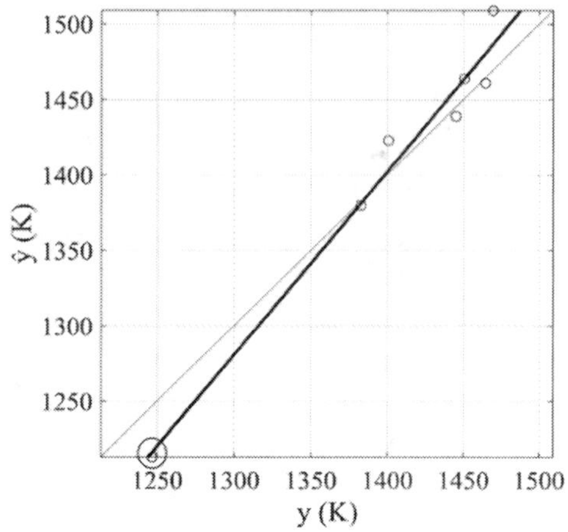

Figure 9. Regression plot of target values y and output values ŷ for the neural network with a 10 x 6 structure; outlier marked red.

3.4. Predictive Capabilities

Assuming a Gaussian distribution of errors and independent errors in the prediction of T_g, T_x and T_l, the uncertainties may be propagated through the equations of Table 1 using Gaussian uncertainty propagation. The equations used to calculate the resulting uncertainty in T_{rg}, γ, γ_m and ΔT_x are given in Table 2. For the uncertainty of the variables RMSE, mean absolute error (MAE) and maximum error (ME) of the respective variables T_g, T_x and T_l are used. Since ΔT_x assumes values near the predictive uncertainty of T_g and T_x, i.e. the mean ΔT_x in the data set is 56.03 K, propagated percentage uncertainties for ΔT_x go as high as 86.65 % for the maximum error in T_g and T_x.

Table 2. Uncertainties propagated through the GFA criterions. RMSE, MAE and ME are considered as respective uncertainty variables, denoted by u.

Parameter	Uncertainty Propagation Equation	Uncertainty variable used ($u_{T_{x,g,l}}$) RMSE	MAE	ME
T_{rg}	$u_{T_{rg}} = \sqrt{\left(\frac{\partial T_{rg}}{\partial T_x} * u_{T_x}\right)^2 + \left(\frac{\partial T_{rg}}{\partial T_g} * u_{T_g}\right)^2}$	2.48 %	1.80 %	5.5 %
γ	$u_\gamma = \sqrt{\left(\frac{\partial \gamma}{\partial T_x} * u_{T_x}\right)^2 + \left(\frac{\partial \gamma}{\partial T_g} * u_{T_g}\right)^2 + \left(\frac{\partial \gamma}{\partial T_l} * u_{T_l}\right)^2}$	1.52 %	1.14 %	3.26 %
γ_m	$u_{\gamma_m} = \sqrt{\left(\frac{\partial \gamma_m}{\partial T_x} * u_{T_x}\right)^2 + \left(\frac{\partial \gamma_m}{\partial T_g} * u_{T_g}\right)^2 + \left(\frac{\partial \gamma_m}{\partial T_l} * u_{T_l}\right)^2}$	2.73 %	2.01 %	5.97 %
ΔT_x	$u_{\Delta T_x} = \sqrt{\left(\frac{\partial \Delta T_x}{\partial T_x} * u_{T_x}\right)^2 + \left(\frac{\partial \Delta T_x}{\partial T_g} * u_{T_g}\right)^2}$	38.13 %	27.5 %	86.65 %

Hence no reliable predictions on ΔT_x can be made. Uncertainties of criterions that are also more closely related to the GFA may however be predicted with a sufficient accuracy. As can be seen in Table 2 the largest estimation error when calculating the GFA criterion γ_m is about 5.96 %, i.e. a deviation in γ_m of 5.96 % at most would be expected.

4. Conclusions

Neural networks have been designed and trained to predict T_g, T_x and T_l of Fe-based amorphous alloys. The prediction errors differ between the models, with T_x being estimated most accurately. Low RMSEs showed the potential of the approach taken in this work. The reason for the outliers seen in the prediction of T_g and T_l are yet to be found. Since the determination of T_g from the DSC-curve is more challenging and T_g may at times not be determined accurately, in contrary to T_x, higher deviations in the data set regarding T_g can be expected. It is presumed that the prediction accuracy in T_g and T_l may be vastly improved if more data is available and added to the data set. The neural networks created can however already augment the design of Fe-based amorphous alloys. If e.g. several supporting points of an alloy system are measured and added to the training data of the neural networks a decent prediction of T_g, T_x and T_l, subsequently the GFA of the alloys within the composition range can be expected. Hence, the approach presented by the authors is promising, especially in the computer-aided design of Fe-based amorphous alloys.

5. Supplementary Information

The MATLAB® script created for this work may be downloaded from https://git.rwth-aachen.de/johann/shallow-neural-networks-to-predict-glass-transition-crystallization-and-liquidus-temperature-of-iron-based-metallic-glasses.

Acknowledgments

The research presented in this work was funded by the Deutsche Forschungsgemeinschaft (DFG, German Research Foundation) within the project BO 1979/67-1 “Fundamental research on deposition of amorphous coatings on surfaces of inner geometries and evaluation of phase stability under tribological loading”. The authors greatly acknowledge the financial support by the DFG.

References

[1] Klement W, Willens R H and Duwez P 1960 Non-crystalline Structure in Solidified Gold–Silicon Alloys *Nature* **187** 869–70
[2] Takeuchi A and Inoue A Classification of Bulk Metallic Glasses by Atomic Size Difference, Heat of Mixing and Period of Constituent Elements and Its Application to Characterization of the Main Alloying Element
[3] Naka M, Hashimoto K and Masumoto T 1978 Corrosion behavior of amorphous and crystalline Cu50Ti50 and Cu50Zr50 alloys *Journal of Non-Crystalline Solids* **30** 29–36
[4] Inoue A, Shen B L and Chang C T 2004 Super-high strength of over 4000 MPa for Fe-based bulk glassy alloys in [(Fe1−xCox)0.75B0.2Si0.05]96Nb4 system *Acta Materialia* **52** 4093–9
[5] Cohen M H and Turnbull D 1961 Composition Requirements for Glass Formation in Metallic and Ionic Systems *Nature* **189** 131–2
[6] Turnbull D 1969 Under what conditions can a glass be formed? *Contemporary Physics* **10** 473–88
[7] Inoue A 1995 High Strength Bulk Amorphous Alloys with Low Critical Cooling Rates (*Overview*) *Mater. Trans., JIM* **36** 866–75
[8] Inoue A and Hashimoto K 2001 *Amorphous and Nanocrystalline Materials: Preparation, Properties, and Applications (Advances in Materials Research* vol 3*)* (Berlin, Heidelberg: Springer)
[9] Suryanarayana C, Seki I and Inoue A 2009 A critical analysis of the glass-forming ability of alloys *Journal of Non-Crystalline Solids* **355** 355–60
[10] Guo S, Lu Z P and Liu C T 2010 Identify the best glass forming ability criterion *Intermetallics* **18** 883–8

[11] Lu Z P and Liu C T 2002 A new glass-forming ability criterion for bulk metallic glasses *Acta Materialia* **50** 3501–12
[12] Du X H, Huang J C, Liu C T and Lu Z P 2007 New criterion of glass forming ability for bulk metallic glasses *Journal of Applied Physics* **101** 86108
[13] Guo S and Liu C T 2010 New glass forming ability criterion derived from cooling consideration *Intermetallics* **18** 2065–8
[14] Suryanarayana C and Inoue A 2013 Iron-based bulk metallic glasses *International Materials Reviews* **58** 131–66
[15] Pang SJ, Zhang T, Asami K and Inoue A 2002 Synthesis of Fe–Cr–Mo–C–B–P bulk metallic glasses with high corrosion resistance *Acta Materialia* **50** 489–97
[16] Kwon D H, Park E S, Huh M Y, Kim H J and Bae J C 2011 Wear behavior of Fe-based bulk metallic glass composites *Journal of Alloys and Compounds* **509** S105-S108
[17] Huang D, Li R, Huang L, Ji V and Zhang T 2011 Fretting wear behavior of bulk amorphous steel *Intermetallics* **19** 1385–9
[18] Stokłosa Z, Kwapuliński P, Rasek J, Badura G, Haneczok G, Pająk L and Lelątko L 2008 Structural relaxation, crystallization and improvement of magnetic properties in FeXSiB (X=Cr, Nb)-type amorphous alloys *Journal of Magnetism and Magnetic Materials* **320** e762-e765
[19] Imafuku M, Sato S, Koshiba H, Matsubara E and Inoue A 2000 Crystallization Behavior of Amorphous $Fe_{90-X}Nb_{10}B_X$ (X=10 and 30) Alloys *Mater. Trans., JIM* **41** 1526–9
[20] Jia C G, Pang J, Pan S P, Zhang Y J, Kim K B, Qin J Y and Wang W M 2019 Tailoring the corrosion behavior of Fe-based metallic glasses through inducing Nb-triggered netlike structure *Corrosion Science* **147** 94–107
[21] Farmer J C *et al* 2007 Corrosion resistance of thermally sprayed high-boron iron-based amorphous-metal coatings: Fe 49.7 Cr 17.7 Mn 1.9 Mo 7.4 W 1.6 B 15.2 C 3.8 Si 2.4 *J. Mater. Res.* **22** 2297–311
[22] Farmer J *et al* 2009 Iron-Based Amorphous Metals: High-Performance Corrosion-Resistant Material Development *Metall and Mat Trans A* **40** 1289–305
[23] Vavilova V V, Vygovskii D A, Kalinin Y E, Kovneristyi Y K and Palii N A 2004 Inelastic Properties of Amorphous and Nanocrystalline Fe–P–Mn–V Alloys *Inorganic Materials* **40** 815–21
[24] Joseph S, Shiflet G J, Ponnambalam V, Keppens V M, Taylor R and Petculescu G 2002 Synthesis and Properties of High-Manganese Iron-Based Bulk Amorphous Metals as Non-Ferromagnetic Amorphous Steel Alloys *MRS Proc.* **754** 1
[25] Makino A, Chang C, Kubota T and Inoue A 2009 Soft magnetic Fe–Si–B–P–C bulk metallic glasses without any glass-forming metal elements *Journal of Alloys and Compounds* **483** 616–9
[26] Wang A, Zhao C, Men H, He A, Chang C, Wang X and Li R-W 2015 Fe-based amorphous alloys for wide ribbon production with high Bs and outstanding amorphous forming ability *Journal of Alloys and Compounds* **630** 209–13
[27] Wang G, Huang Z, Xiao P and Zhu X 2016 Spraying of Fe-based amorphous coating with high corrosion resistance by HVAF *Journal of Manufacturing Processes* **22** 34–8
[28] Hirata A, Hirotsu Y, Amiya K, Nishiyama N and Inoue A 2008 Nanocrystallization of complex Fe23B6-type structure in glassy Fe–Co–B–Si–Nb alloy *Intermetallics* **16** 491–7
[29] Luborsky F and Walter J 1980 Preparation and properties of Fe-B-Si-C amorphous alloys *IEEE Trans. Magn.* **16** 572–4
[30] Mitera M, Masumoto T and Kazama N S 1979 Effect of silicon addition on the magnetic properties of Fe-B-C amorphous alloys *J. Appl. Phys.* **50** 7609
[31] Huang B, Perez R J, Crawford P J, Sharif A A, Nutt S R and Lavernia E J 1995 Mechanically induced crystallization of metglas Fe78B13Si9 during cryogenic high energy ball milling *Nanostructured Materials* **5** 545–53

[32] Kawazoe Y, Carow-Watamura U and Louzguine D V 2019 Magnetic, thermal and mechanical properties of B-Fe-Nb-Si alloys *Phase Diagrams and Physical Properties of Nonequilibrium Alloys: Subvolume C: Physical Properties of Multi-Component Amorphous Alloys, Part 2: Systems from B-C-Fe-Mo to B-Ni-Si-Ta (Condensed Matter* 37C2*)* ed Y Kawazoe and U Carow-Watamura 1st edn (Berlin: Springer Berlin; Springer) pp 247–9

[33] Johnson W L, Na J H and Demetriou M D 2016 Quantifying the origin of metallic glass formation *Nature communications* **7** 10313

[34] Carow-Watamura U, Louizguine D V, Takeuchi A, Kawazoe Y, Landolt H, Börnstein R and Martienssen W (eds) 2011 *Numerical data and functional relationships in science and technology: New series (Landolt-Börnstein - Group III Condensed Matter)* (Berlin: Springer)

[35] Laurens van der Maaten and Geoffrey Hinton 2008 Visualizing Data using t-SNE *Journal of Machine Learning Research* **9** 2579–605

[36] Kissinger H E Reaction Kinetics in Differential Thermal Analysis

[37] Mitrovic N, Roth S and Eckert J 2001 Kinetics of the glass-transition and crystallization process of Fe72−xNbxAl5Ga2P11C6B4 (x=0, 2) metallic glasses *Appl. Phys. Lett.* **78** 2145–7

[38] Lasocka M 1976 The effect of scanning rate on glass transition temperature of splat-cooled Te85Ge15 *Materials Science and Engineering* **23** 173–7

[39] Ichitsubo T, Matsubara E, Chen H S, Saida J, Yamamoto T and Nishiyama N 2006 Structural instability of metallic glasses under radio-frequency-ultrasonic perturbation and its correlation with glass-to-crystal transition of less-stable metallic glasses *The Journal of chemical physics* **125** 154502

[40] Ile D C, Révész Á, Grahl H, Eckert J, Crespo P, Marín P, Hernando A, Suriñach S, Muñoz J S and Baró M D 2004 Thermal stability and crystallization behavior of Fe77C5B4(AlGa)3(PSi)11 metallic glasses *Materials Science and Engineering: A* **375-377** 297–301

[41] Myung W-N, Yang S-J, Kim H-G and Masumoto T 1991 Glass transition and viscous flow behavior of amorphous Fe◘M◘P (M = Cr, V or Mo) alloys *Materials Science and Engineering: A* **133** 418–22

[42] Révész Á 2008 Crystallization kinetics and thermal stability of an amorphous Fe77C5B4Al2GaP9Si2 bulk metallic glass *J Therm Anal Calorim* **91** 879–84

[43] Cassar D R, Carvalho A CPLF de and Zanotto E D 2018 Predicting glass transition temperatures using neural networks *Acta Materialia* **159** 249–56

[44] Dewapriya MAN, Rajapakse RKND and Dias WPS 2020 Characterizing fracture stress of defective graphene samples using shallow and deep artificial neural networks *Carbon* **163** 425–40

[45] Hagan M T, Demuth H B and Beale M H 2002 *Neural network design* (Boulder, Colorado: Campus Publishing Service)

[46] Kayri M 2016 Predictive Abilities of Bayesian Regularization and Levenberg–Marquardt Algorithms in Artificial Neural Networks: A Comparative Empirical Study on Social Data *MCA* **21** 20

High-strength mechanical joints under combined mechanical and medial loads

D Wett[1,*], M Radanke[2], P Lebelt[1] and H Brauns[2]

[1] Institute for Corrosion Protection Dresden GmbH, Department of Corrosion Protection and Process Engineering, Gostritzer Straße 65, 01217 Dresden
[2] Fraunhofer Research Institution for Large Structures in Production Engineering IGP, Department of Testing Technology, Albert-Einstein-Straße 30, 18059 Rostock

*email: daniel.wett@iks-dresden.de

Abstract. This paper introduces a test method for joined hot-dip galvanized steelwork test specimens. The specimens joined by means of a lock bolt system were subjected to a simultaneously combined mechanical and medial loading. This was achieved by combining a vibration testing machine and a condensation chamber that was expanded by an electrolyte dosing unit. The electrolyte was dosed on one side of the specimen in the area around the lock bolt head. The zinc dissolution is locally accelerated in the area around the joining element possibly due to superimposed bimetal and crevice corrosion. In addition, the electrolyte enters the space between the joined steel sheets and initiates corrosion processes there, too. Besides the expected zinc ions, potassium ions were also detected in the collected electrolyte by means of elemental analysis. These are originally part of the friction-increasing coating.

1. Introduction

The ambition to increase efficiency in the field of wind energy leads to the construction of wind turbines with large hub heights. The most common tower concept are currently steel tube towers which are erected in segmental construction methods. However, in regard to large hub heights transporting the segments (diameter > 4,2 m) is a significant problem considering infrastructure [1]. In consequence, concrete-steel hybrid towers are used as an alternative on the one hand. On the other hand, design concepts based on a separated steelwork supporting structure, such as lattice towers or lengthwise separated steel tube towers are tested (cf. figure 1). These usually consist of bar profiles or shell profiles and are assembled on-site by means of mechanical joining technology, using high-tensile prestressed bolts or lockbolts, for example. The new construction methods reduce costs for transport and at the same time offer the possibility of saving material. Contrariwise the assembly costs on-site increase. The semi-finished parts are typically batch-galvanized [2], specifying the corrosion resistance by means of the zinc-film thicknesses. According to DIN EN 1090-2 [3] an alkali silicate zinc-dust coating is applied [4]. This leads to an increase in the coefficient of friction and a better corrosion protection for the slip-resistant prestressed joints. Typically, it is directly applied before joining and the joints themselves are not overcoated.

Figure 1. Examples for wind turbine supporting structures. Left: Lattice tower, Butzkies Stahlbau GmbH. Center: Lattice tower, Stieblich Hallenbau GmbH. Right: Steel tube tower separated lengthwise, SENVION S.A.

High-tensile prestressed bolts or lockbolt systems are used as joining elements. The bolts are usually hot-dip galvanized, as is the case with the semi-finished parts, and have an average film thickness of 50 µm in accordance with DIN EN 10684 [5]. Lockbolt joints have metallic coatings as well as organic coatings [6]. The durability of these protection systems is determined by means of laboratory test methods, such as the neutral salt spray test. Thus, for both corrosion protection systems sufficient durability for steelwork applications is theoretically verified. In regard to dynamically loaded constructions such as wind turbines the joints are executed as slip-resistant prestressed joints [7]. From an economic point of view it is crucial for such supporting structures to reach sufficient service life after construction. Due to increased loading a maximum of 20 years is the basis for calculation regarding wind turbines [8]. These service lives can only be accomplished if the joints resist the combined mechanical and medial loads on-site and a quantitative estimation of the effects of combined mechanical and medial loads can currently be done exclusively by means of mere conservative inspections of real constructions or prolonged natural weathering tests.

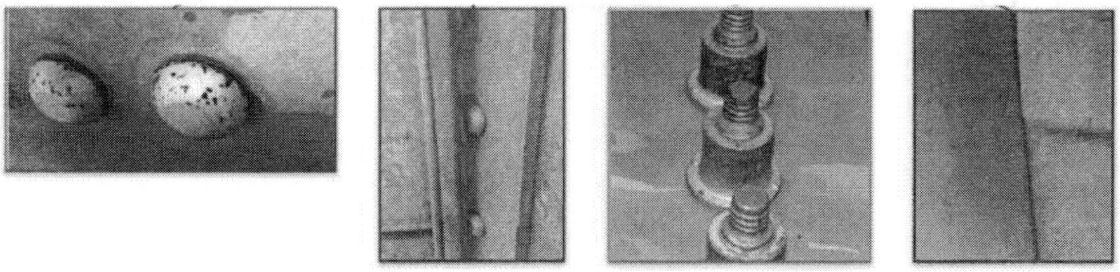

Figure 2. Effect of combined loads on mechanical joints after 5 years of operation [Source: construction inspection on 2016-10-25; wind turbine by the Butzkies Stahlbau GmbH in Neuenbrook]

The planning periods to be concluded from this (> 5 years) are not practice-oriented. For the estimation of the effect of combined mechanical and medial loads on the functional capability of mechanical joints in steel constructions there is currently no validated laboratory test method available. Construction inspections (as illustrated in figure 2) show, that despite nominally high corrosion protection durability (> 15 years) of the individual joint partners (semi-finished parts, anti-friction surface, joining element) already after a few years significant corrosion occurs on the assembled components. In accordance with the current state of the art no laboratory loading regimen on combined mechanical and medial loads for high-strength joints in steel constructions exists. Previously performed laboratory loading cycles refer to automotive and railway vehicle engineering, e.g. [9–13]. During these examinations three different test procedures with the following configurations had been carried out:

1. Pre-corrosion and subsequent mechanical loading
2. Alternating medial and mechanical loading
3. Simultaneous medial and mechanical loading

Only in test configuration 3 it is possible for gaps to occur, leading to forced electrolyte entry and, consequently, promoting corrosion processes on the joints in a practical manner. Until the electrolyte is dried this local continuous moisture load leads to significant corrosion increase [14]. It is to be emphasized that regarding simultaneous medial and mechanical loads a low test frequency shall be set, since the corrosion processes are dependent on time [11] and take place slowly. The consequences of test frequencies that are too high during combined loading would be corrosion influences that are too low. In addition, moisture entering crevices during test frequencies that are too high can lead to a modified tension distribution (local pressure increases due to electrolyte concentrations) in the contact area of the joint. As a consequence, test specimens prematurely fail during laboratory testing and, thus, real conditions cannot be reproduced.

This paper presents the first developing steps of a laboratory test method for simultaneous medial and mechanical loading of steelwork test specimens. This method is intended to help with the practice-oriented design of corrosion protection systems and their application for high-strength mechanical joints for supporting structures of wind turbines. During a test period of 3 weeks a (locally limited) corrosion corresponding to a duration of 5 years in corrosivity category C4 [15] is to be accomplished.

2. Experimental

2.1. Test chamber for combined loading

The combined mechanical and medial loading tests have been carried out in a test installation by the Institute for Corrosion Protection Dresden GmbH (IKS). The installation is a special construction consisting of a 100 kN horizontal vibration test device (DynaMess, Aachen) with hydrostatically supported test cylinder and a condensation-water chamber S450-S (Weisstechnik). In order to achieve the high corrosion requirements mentioned above, the installation had to be extended by a salt dosing unit. Since due to installation-specific reasons the application of salt spray is not possible, it was specified to drip the electrolyte onto the surface of the (inclined) test specimen by means of a cannula during the vibration test in condensation-water atmosphere. An approach with similar electrolyte placing is described in [16]. The build-up of the complete installation is given in figure 3. The electrolyte is conveyed onto the test specimen from a storage container outside the chamber through a tube and a cannula. The volumetric flow of the salt solution is controlled with a dosing valve. During the load tests the electrolyte solution dripping from the test specimen is conveyed into a collection container by means of a funnel.

2.2. Preparation of test specimens

The geometry of the test specimen is given in figure 4 and is following the standard DIN EN 1090-2 [3]. In the underlying research project these test specimens are used for the validation of the test system as well as for the determination of the failure behavior under combined loading. Due to the limited dimensions of the test chamber an implementation conforming to the standard to 100 % was not possible at the IKS.

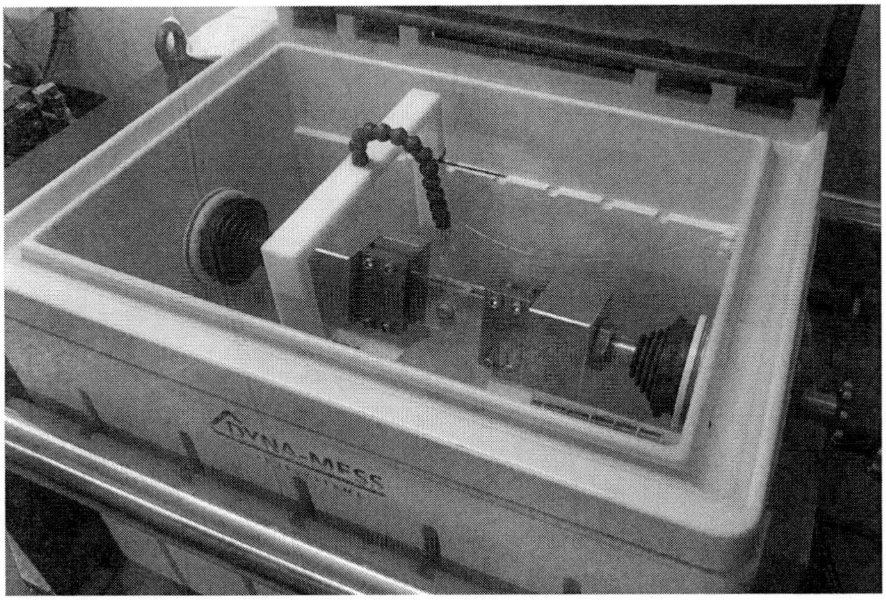

Figure 3. Test chamber of the IKS for the combined mechanical and medial loading (special construction consisting of 100 kN horizontal vibration tester, condensation-water chamber and electrolyte dosing unit) and alignment of the test specimen. [Image source: IKS]

The test specimens consist of 3 semi-finished parts with the substrate S355J2+N, which is hot-dip galvanized. The surfaces are mechanically pre-treated at the joints and have an alkali silicate zinc dust coating (Interzinc 697). The semi-finished parts are slip-resistant prestressed and are joined with a joining element. Latter consists of a lockbolts of the type MBT-DT16-35J45 (substrate non-galvanized steel + two-coat corrosion protection system) as well as a locking ring of the type MBTC-R16J46 (substrate hot-dip galvanized steel, single-coat corrosion protection system). The prestress was 116 kN. An industrial partner manufactured the semi-finished parts and applied the alkali silicate zinc-dust system. The preparation of the joining elements for the prestress determination (during mechanical loading) and the joining of all joints was done at the Fraunhofer IGP.

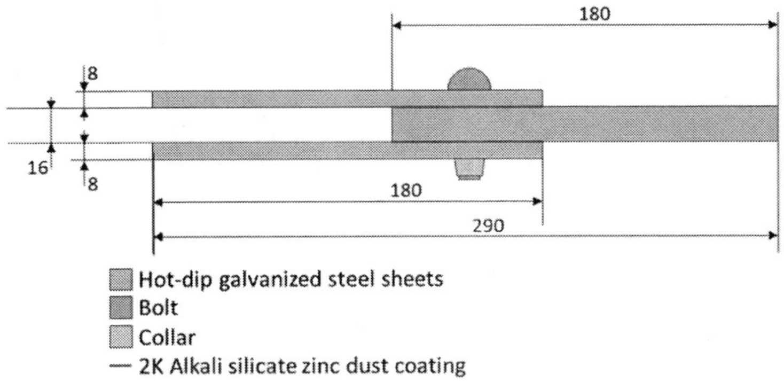

Figure 4. Reduced geometry (side view) of the test specimens following DIN EN ISO 1090-2. 3 semi-finished parts (80 mm with each) slip-resistant prestressed by a lockbolt system.

2.3. Parameters of the mechanical load

In 2015 the Fraunhofer IGP implemented a comprehensive monitoring of the developed maintenance-free wind turbine lattice tower [17]. The positions of the acceleration sensors and the used DMS bolts, serving the exact absorption of load spectra, are given in figure 5. The detailed analysis of the real mechanical load is used as the basis for the combined laboratory loading at IKS.

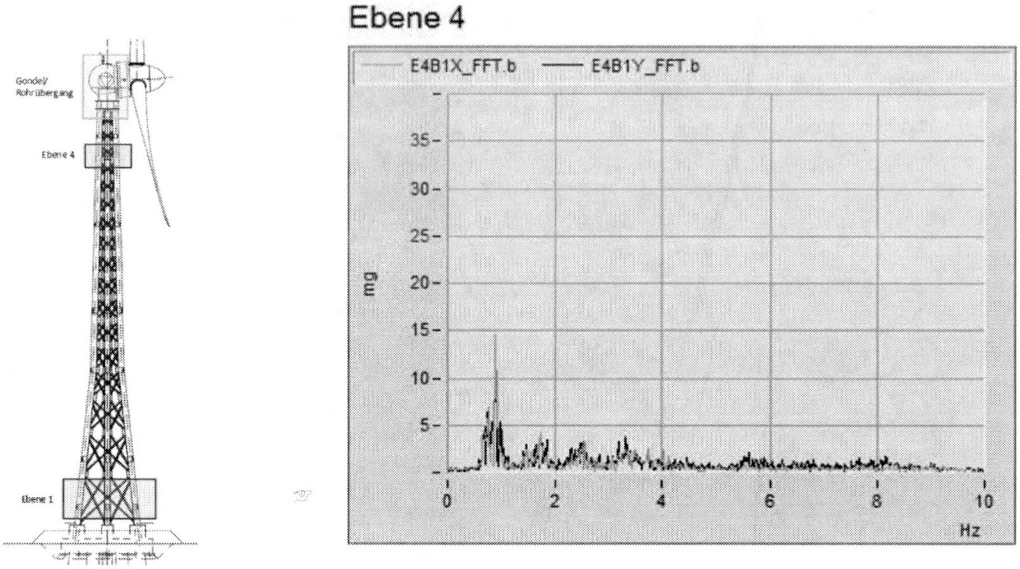

Figure 5. Left: Overview of the measuring levels on the lattice tower for comprehensive monitoring. Right: Measuring values of an acceleration sensor with main amplitudes around approx. 1 Hz.

The frequencies of the first bending vibrations of the installed acceleration sensors are roughly about 1 Hz. The frequency for the vibrating mechanical loading of the test specimens results from this. The other parameters of the mechanical load were estimated on the basis of studies on current calculation models [18] as well as publications on prestressed steel construction connections [19]. As amplitude of the mechanical load a highest load of 25 kN and an R-correlation of 0,1 has been specified [20]. Within the elaborated loading cycle of 3 weeks 10^6 load changes (LC) are pursued.

2.4. Parameters of the medial load

The atmospheric zinc removal over a period of 5 years in corrosivity category C4 can be estimated with the help of the correlations and calculation models, which are explained in more detail in [15, 21]. Depending on which model is used, the zinc removal is about 11,7 ±3,9 µm. In order to estimate the amount of salt solution to drip onto the test specimen in one location within 3 weeks, a strip-galvanized steel sheet (Z100; zinc film approx. 14 µm), inclined by 15° to the vertical, has been exposed to a condensation-water atmosphere (relative humidity ≈ 99 %, $\vartheta = 40 \pm 3$ °C) with a 5% by mass NaCl solution (the same is used for neutral salt spray tests). The electrolyte consumption was specified to 25 ± 1 ml per hour. After 24 hours the 14 µm thick zinc film was removed and red-rust formation occurred. Consequently, it was detected that the 24-hour loading is above the intended value. For this reason, at first, a total duration of salt exposure of 15 h was specified, corresponding to a local minimum zinc removal of 8,75 µm. This value lies at the bottom of the above-mentioned range. The salt has been dosed for 1 h during each of the 15 days of loading.

2.5. Combined loading

The mechanical and medial test parameters determined in *2.3.* and *2.4.* were now combined in a test program. It was specified that in each of the 3 test weeks the test specimens are exposed to condensation-water alternating climate ("AHT" in accordance with [22]; 8h relative humidity ≈ 100 % and 40 ±3°C and 16 h drying at 18-28°C and ambient humidity) for approx. 104 hours. During this time the vibration stress is applied simultaneously (f = 1 Hz; R = 0,1; F = 25 kN; approx. 374000 LC each week). In order to accomplish the required total LC (>1x10^6), an overall loading duration of 3 cycles (= 3 weeks) is required. The dosing of the NaCl solution was distributed to 3x5 days and was carried out in each case in the middle of the damp phases for one hour on the bolt side of the test specimen.

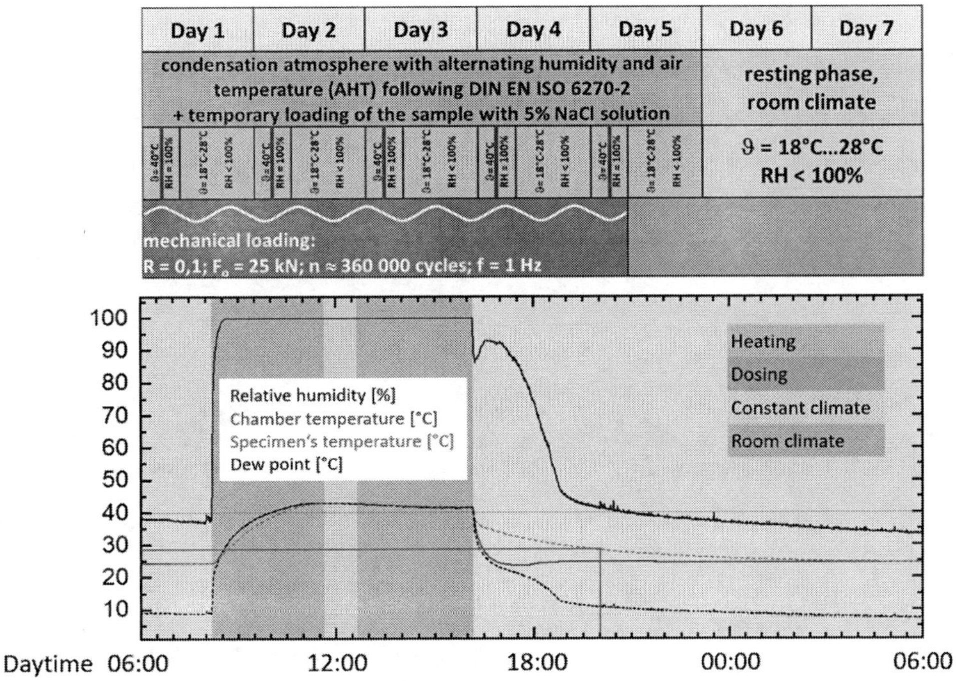

Figure 6. Upper: Sequence of the combined test program during a one-week cycle. Lower: Measured values of atmospheric temperature and humidity as well as specimen temperature during a day cycle.

The electrolyte dripping from the test specimen was conveyed into a collection container and filled up to a volume of 50 ml using deionized water. Subsequently, these solutions have been acidified and analyzed regarding their cation content.

A one-week loading cycle is given in the upper part of figure 6, the alignment of the test specimens can be taken from figure 3. The lower part of Figure 6 shows the development of the chamber temperature and humidity as well as the temperature of the sample over a 24-hour cycle. It has to be mentioned, that the sample's temperature profile does differ from that of the atmosphere. In the beginning the chamber heats up a little faster. The humidity rises very quick so that the dew point is reached after around 30 min. Returning to room clime after 8 h one can observe that the sample cools down very slowly and falls below the required temperature of 28°C after 4 hours.

2.6. Characterization methods

A digital camera Canon EOS 2000D and a flatbed scanner EPSON PERFECTION 4990 PHOTO were used for the macroscopic images. The light-microscopic images were recorded using an inverted light-microscope Zeiss Axio Observer.Z1M and a stereomicroscope Zeiss Discovery.V20. Further, the 3D laser-scanning microscope Keyence VK-X130 K was used. The FE-SEM images and EDX analyses were recorded with a field-emission electron microscope Zeiss ΣIGMA VP with 2 XFlash 5030 EDX detectors (Bruker Nano). For the chemical analyses a Varian 725-ES optical emission spectrometer was available.

3. Results and discussion

3.1. Condition of the test specimens and the joining elements after the end of stress application

Figure 7 exemplifies a test specimen prior to and after combined mechanical and medial loading for 3 cycles (= 3 weeks). The run-off marks of the salt solution are clearly visible, as well as the deposition and corrosion products formed in these locations. Only a rather small percentage of the sheet surface is affected by corrosion due to the locally limited electrolyte deposition; however, it is clearly visible that the drained off NaCl solution distributes over the underside of the test specimen and almost entirely wets it.

Figure 7. Condition of a test specimen prior to (left) and after (right) the end of the stress application. The electrolyte dripping from the test specimen distributes over the entire area of the underside. The joining element is free from red rust.

After severing the bolt, the test specimen was disassembled. It was observed that parts of the double layer bolt coating remained adhered to the galvanized steel surface of the sheet after disassembling the test specimen (figure 8 top right section). This was not the case for a merely joined but unloaded test specimen (not shown), meaning that the humidity during the test affects the adhering phenomenon. In the area around the removed bolt head, where the electrolyte flowed by, circular accumulations of corrosion products are clearly visible, which are mostly separated from one another. On the joining elements (figure 8 bottom section) macroscopic corrosion could not be found.

The cause for the heterogeneous corrosion around the bolt head is not clear yet. We suspect that sporadic coating defects in the edge area of the bolt head may result in local bimetal corrosion, which would lead to a locally limited but accelerated zinc dissolution.

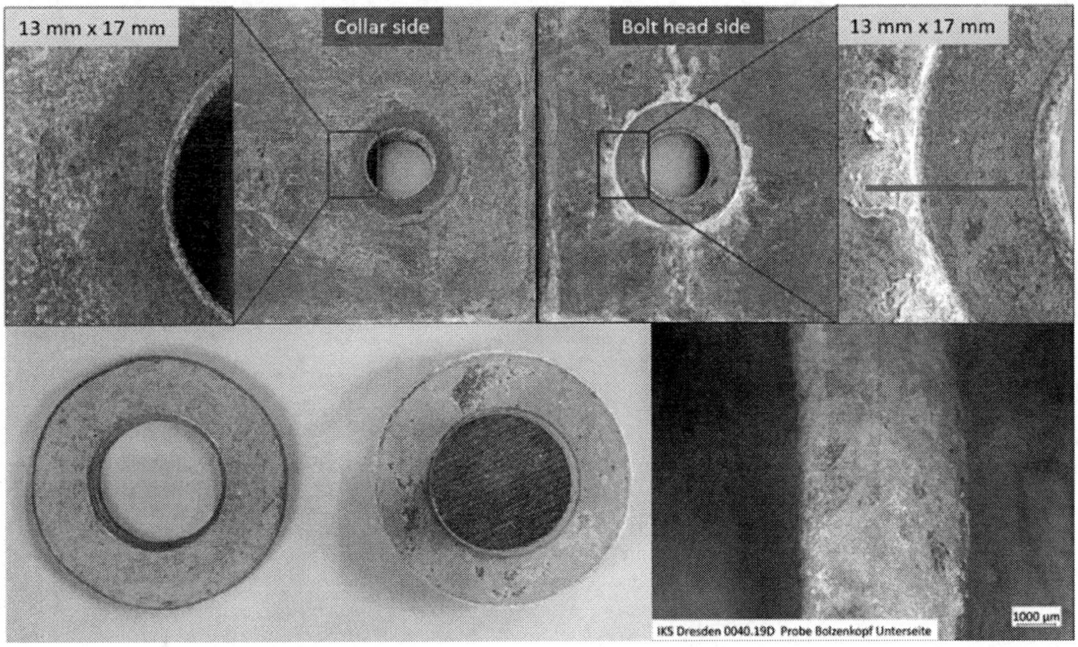

Figure 8. Top: Condition in the areas around the drilling after disassembling (Left: Side of the locking ring not directly loaded with electrolyte during the test. Moisture entry in the outer area. Right: Bolt-head side loaded with electrolyte during the test with partially adhering bolt-head coating and locally increased zinc corrosion). Bottom: Top view of both joining elements and side view of the bolt head.

Figure 9 shows such a point. In the area close to the edge the base coating (phosphorus-free, zinc-containing) is missing and the thickness of the top coating (phosphorus-containing, zinc-free) is rather small. The thin Zn and P-containing layer on the bolt surface corresponds to the zinc phosphating. The base coating typically contains certain corrosion protection pigments. If this coating is not present the corrosion protection efficiency is reduced.

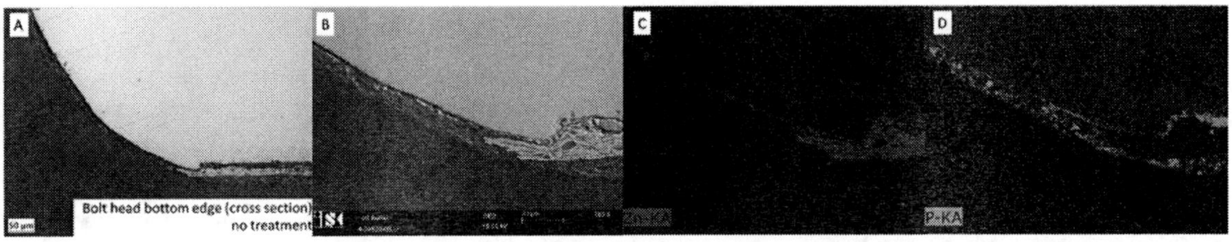

Figure 9. Inspection and analysis of the joining elements. (A, B) Separated locking ring flange and bolt after loading test. (C) Side view of the lower edge of a bolt head after loading test. (D) Cross-sectional view (light microscope) of the poorly covered edge of an unloaded bolt head. (E) FE-SEM image of the same position. (F) and (G) Mapping of the Zn-Kα and P-Kα signals.

It was then examined whether the bolt coating or the crevice were infiltrated by the electrolyte. For this purpose, a cross-section of a metal sheet with adhering bolt coating was analyzed in the FE-SEM-EDX after the loading test and disassembling. The results are shown in figure 10. From left to right the images are arranged in the following way: Outer area with the circular accumulations of corrosion products; Crevice area under the bolt; Crevice end with a little indent in the galvanization caused by the bolt; Area under the bolt. The qualitative EDX maps of the elements zinc, oxygen, chlorine and phosphorous are shown below the FE-SEM images.

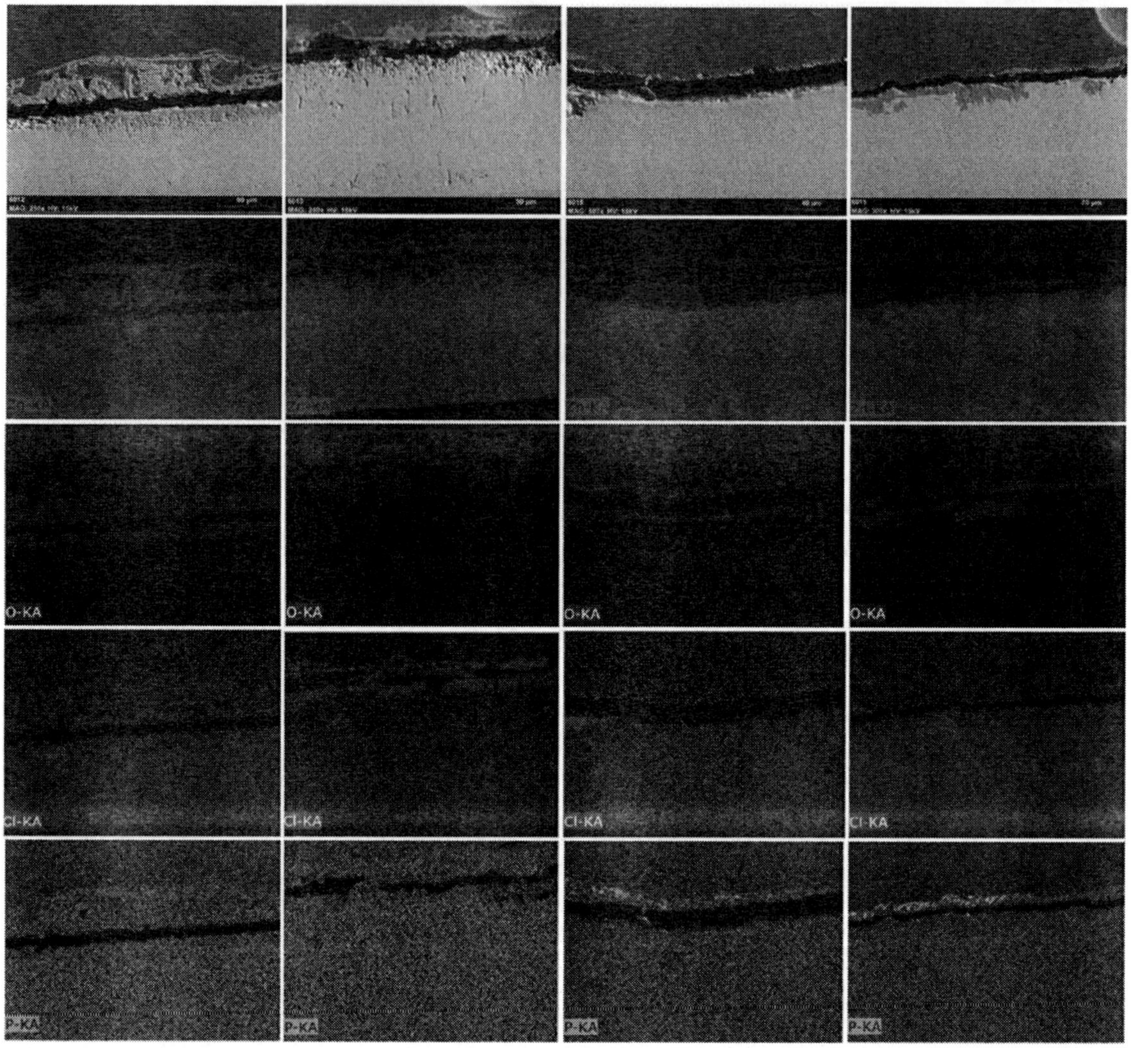

Figure 10. Cross section FE-SEM-EDX-Analyses in the region with increased zinc corrosion (cf. Figure 8). From left to right: Outer area; Crevice area; Crevice end with bolt indention; Area under the bolt. Fields of view [μm]: 435 x 300, 435 x 300, 210 x 145, 360 x 250, resp. Upper line: FE-SEM images (HV: 15 kV). Red, Blue, Purple and Green: Zn-Kα, O-Kα, Cl-Kα and P-Kα EDX intensities.

In the outer area mainly loose corrosion products were identified and, as one could expect, mainly zinc and oxygen were detected; the contrast in the maps of chlorine and phosphorous are very weak. In case of chlorine a very little intensification is noticeable in the region of the corrosion front. In the crevice, the chlorine map comes to a higher contrast meaning that Chlorine has penetrated in that zone. A noticeably attack on the zinc surface can also be seen in that images. In the transition area containing the bolt indention, no attack on the zinc surface was detectable and the chlorine signals drop to the background noise level. In the phosphorous map the signal intensities increase in the area corresponding to the adherent (phosphorous containing) top coat of the bolt head. In the inner area, the galvanizing was affected by weak corrosion in some places; However, chlorine could not be detected there. Some parts of the top coat were found again. In summary, we suggest that bimetal and crevice corrosion are superimposed in the area around the bolt head.

Since the locking ring side was not loaded by the electrolyte and was only exposed to humidity, we will discuss it only in a short way. Due to a typically occurring weak deformation of the locking ring flange during the bolting process a small crevice is formed, in which condensation water seeps into. This entry is limited to a depth of around 3,5 mm with the present flange width of approx. 7 mm. The opening angle of the crevice is approx. 1,5° (see figure 11). This value was determined by evaluating 2 profiles from 3D laser-scanning microscope images taken from opposite parts on the locking ring flange. Macroscopically, there was only a slight and facile corrosion of the galvanized layer detectable.

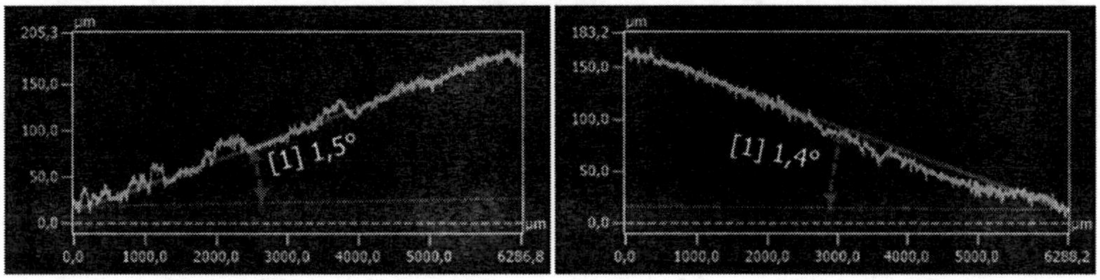

Figure 11. Inclination profiles of a locking ring flange determined by means of laser-scanning microscopy. Dimensions [µm]: Left: 6286,8 x 205,3, Right: 6288,2 x 183,2.

3.2. Condition of the inner interfaces and of the alkali silicate zinc-dust coating after the end of stress application

After disassembling the test specimens, it is noticeable that the electrolyte enters the spaces between the joined sheet stack (figure 12) starting from the bottom edge (framed in red) and rising at the sides (framed in blue). This occurs mainly due to the capillary effect of gap between the layers. In addition, the quickly drying alkali silicate zinc-dust coatings were applied by means of a brush, resulting in a wavelike texture. Due to this, small channels form in the intermediate areas, promoting entry of the salt solution. A further reason is the formation of cracks in the layers, since the reference value of 75 µm dry film thickness specified in the manufacturer's data sheet is only difficult to match under practical conditions. The resulting cracks should further increase the entry of the electrolyte.

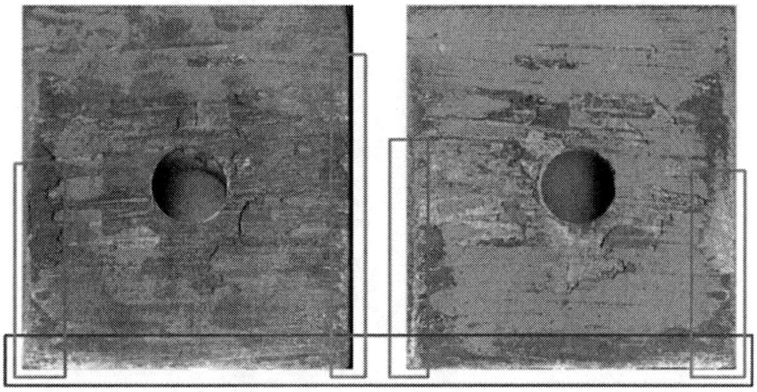

Figure 12. View of the inner contact surfaces of two sheets (left: middle sheet, right: outer sheet of the specimen) after 3 loading cycles. Parts of the electrolyte enter the space between the sheets. In the areas where the coating was partially pulled off, a dark discoloration occurs. These can also be found in the opposite areas with adherent coating.

Furthermore, it is noticeable that the coating pulls off of the outer sheets. The pulled-off parts largely adhere to the opposite side of the center sheet. In locations in which the coating pulls off, the galvanized surface often shows dark discoloration. Such a dark discoloration was analyzed by means of FE-SEM-

EDX, both in top view (figure 13) as well as on the cross-section (figure 14). From the analysis we can derive that an attack occurred on the galvanized zinc layer. A local oxygen content of approx. 27 percent by mass could be verified. Because the mass fraction of oxygen in pure zinc oxide is only approx. 20% and the concentration of other metals was found to be comparably low, the formation of zinc hydroxides appears plausible.

	C	O	Al	Si	Fe	Zn	Pb
	2,7	7,2	0,3	0,2	0,6	87,7	1,2
	3,8	26,9	0,2	0,9	0,6	61,1	0,6
	2,3	15,9	0,2	3,9	1,5	76,0	0,2

Figure 13. Left: FE-SEM-EDX analysis on a dark region with delaminated alkali silicate zinc-dust coating. The table below shows the results of the EDX-Quantification for 3 surface segments (marked by colors) given in percent by mass. Right: The elements O, Zn and Si are distributed very heterogeneously.

From the cross-section analysis in it was deduced that the attack on the zinc layer reaches down up to a depth of 20 µm (lower Zn Kα and increased O Kα signal intensities). On the other hand, the analysis finds that the attack can also occur beneath the applied alkali silicate zinc-dust coating. This is most likely possible because of the formation of little gaps due to a decreased adhesion between the coating and the surface of the galvanized steel.

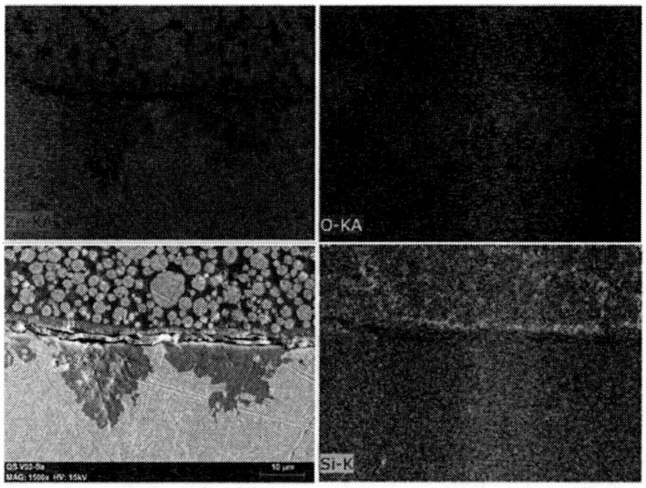

Figure 14. SEM-EDX images on the cross-section of an area, in which the zinc coating was attacked. The area with the sphere-shaped formations is the zinc-dust coating. Corrosion attack occurred beneath it, reaching in up to a depth of approx. 25 µm.

3.3. Element analysis of the collected electrolyte dripping from the test specimen

During loading a test specimen, the electrolyte was conveyed into a collection container by means of a funnel. The collected solution was filled up to a volume of 50 ml using deionized water and was acidified using 1 % nitric acid, in order to stabilize it. After completion of the test program all test specimens of the batch have been analyzed by means of ICP-OES [23]. An exemplified curve of the concentrations of the elements K, Zn and Fe is given in figure 15.

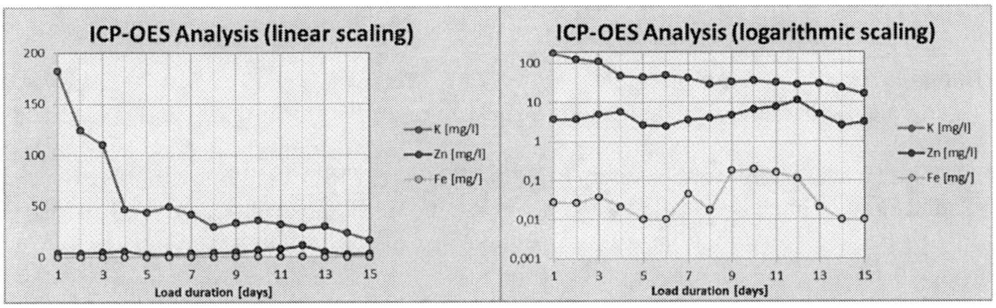

Figure 15. Results of the element analyses by means of ICP-OES.

It was found that a comparatively high release of potassium occurs at the beginning of the test series. This can either be caused by flushing of the excess potassium silicate or by its decomposition. Even on day 15 potassium has the highest concentration of the 3 diagrammed elements. The value of the potassium concentration decreases from initially approx. 180 mg/l to approx. 17 mg/l towards the end of the test. The discharge of zinc is almost constant, the concentration varies between 2 and 12 mg/l. The concentration values of iron are 2 magnitudes below that and are, in consequence, close to the detection limit. We consider the release of the potassium ions as another proof of the electrolyte activity in the spaces between the joined sheets.

4. Summary and conclusion

In this paper we describe the development of a test method allowing for combined mechanical and medial tests on practice-oriented, joined steelwork test specimens. This was accomplished by combining a vibration test in a closed chamber atmosphere with alternating humidity. During the damp phases an electrolyte solution was specifically placed on the area around the bolt head of the joining element. This method enables the detection of weak points in the joined structure, in the material selection (e.g., zinc coating, joining elements) as well as in the preparation process.

The first results indicate a heterogeneous corrosion in the area around the edges of the bolt head, if the electrolyte is dosed on the bolt head side of the sample. This behavior we attribute to a superposition of crevice corrosion with bimetallic corrosion due to local coating defects of the bolt head.

We further found out that the electrolyte enters the spaces between the joined sheet stack mainly caused by the capillary effect of the layers. The reason for that we see in the wavelike texture and the micro cracking coming from the brush application and the high coating thickness. We discovered that the penetrating electrolyte leads to a corrosion of the galvanized surface. Further evidence of electrolyte penetration was derived from the result of the chemical analysis of the electrolyte that dripped from the sample. From the beginning of the experiment potassium is released. This element originates from the zinc dust coating which is only applied to the inner surfaces of the galvanized steel sheets.

Future tasks will be the medial loading on a larger area on both sides of the test specimen (bolt head and locking ring) in order to enable a more realistic corrosive attack over the whole sample. Examinations are already conducted in this regard. Other focal points are the preparation and loading of test specimens, on which the alkali silicate zinc-dust coating has been applied by means of spraying and/or by rolling, and the determination of slip factors.

Acknowledgments
The research project "High-strength mechanical joints for supporting structures of wind turbines with large hub heights under combined mechanical and medial loads" (AiF-No. 24EWBR) of the European Research Association for Sheet Metal Working (EFB) has been funded by the AiF within the program for sponsorship by Industrial Joint Research (IGF) of the German Federal Ministry of Economic Affairs and Energy (BMWi) based on an enactment of the German Parliament. We are grateful to our partners from the industry to supply us with the semi-finished products and the joining elements.

References
[1] Bundesverband WindEnergie e. V. Turm und Mast https://www.wind-energie.de/themen/anlagentechnik/konstruktiver-aufbau/turm-und-mast/ (accessed 14 Sep 2020)
[2] DIN Deutsches Institut für Normung e. V. 2017 DIN EN ISO 14713-1: Zinküberzüge - Leitfäden und Empfehlungen zum Schutz von Eisen- und Stahlkonstruktionen vor Korrosion - Teil 1: Allgemeine Konstruktionsgrundsätze und Korrosionsbeständigkeit (Berlin: Beuth Verlag GmbH)
[3] DIN Deutsches Institut für Normung e. V. 2018 DIN EN 1090-2: Ausführung von Stahltragwerken und Aluminiumtragwerken - Teil 2: Technische Regeln für die Ausführung von Stahltragwerken (Berlin: Beuth Verlag GmbH)
[4] AkzoNobel 2015 Interzinc® 697 Anorganisches Zinksilicat: Produktdatenblatt (Ref: 2567) (AkzoNobel)
[5] DIN Deutsches Institut für Normung e. V. 2011 DIN EN ISO 10684: Verbindungselemente - Feuerverzinkung (Berlin: Beuth Verlag GmbH)
[6] Deutsches Institut für Bautechnik 2019 Allgemeine bauaufsichtliche Zulassung/ Allgemeine Bauartgenehmigung - Schließringbolzen ohne Abrissteil (Z-14.4-591) (Deutsches Institut für Bautechnik)
[7] Wittel H, Muhs D, Jannasch D and Voßiek J 2015 Roloff/Matek Maschinenelemente: Normung, Berechnung, Gestaltung 22nd edn (Wiesbaden: Springer Vieweg)
[8] Hau E 2014 Windkraftanlagen: Grundlagen, Technik, Einsatz, Wirtschaftlichkeit 5th edn (Berlin: Springer Vieweg)
[9] Voelkner W, Liebrecht F, Hahn O and Bangel M 1998 Untersuchungen zum Fügen organisch beschichteter Aluminiumbleche mittels Stanznieten *(EFB-Forschungsbericht* Nr. 112*)* (Hannover: EFB)
[10] Neugebauer R, Mauermann R and Grützner R 2012 Einfluss von kombinierter mechanisch-medialer Beanspruchung auf die Schwingfestigkeit von stanz- und blindgenieteten Mischverbindungen *(EFB-Forschungsbericht* Nr. 348*)* (Hannover: EFB)
[11] Hahnel W, Franke R, Donat B, Kaiser W-D and Friedrich S 2006 Ableitung eines zuverlässigen Tests für die Wirkung der Korrosion auf die Lebensdauer von mechanischen Fügeverbindungen im Schienenfahrzeugbau *(EFB-Forschungsbericht* Nr. 255*)* (Hannover: EFB)
[12] Voelkner W and Hahn O 1997 Fügen von Feinblechen mittels Durchsetzfügen - Kleben und Stanznieten - Kleben *(EFB-Forschungsbericht* Nr. 102*)* (Hannover: EFB)
[13] Thierry D, Vucko F, Luckeneder G, Weber B, Dosdat L, Bschorr T and Rother K 2016 Fatigue behavior of spot-welded joints in air and under corrosive environments: Part II: Fatigue under alternating and combined corrosion and fatigue load *Weld World* **60** 1231–45
[14] Seidel W W and Hahn F 2018 Werkstofftechnik: Werkstoffe - Eigenschaften - Prüfung - Anwendung: mit 389 Bildern sowie zahlreichen Tabellen, Beispielen, Übungen und Testaufgaben *(Lernbücher der Technik)* 11th edn (München: Hanser)
[15] DIN Deutsches Institut für Normung e. V. 2012 DIN EN ISO 9223: Korrosion von Metallen und Legierungen – Korrosivität von Atmosphären – Klassifizierung, Bestimmung und Abschätzung (Berlin: Beuth Verlag GmbH)

[16] Ziebermayr S, Fleischanderl M and Stellnberger K-H 12. und 2018 Tropfender Elektrolyt – Korrosion von zinkbasierten Korrosionsschutzüberzügen bei dynamischer Belastung mit Elektrolyt *3-Länder-Korrosionstagung (Empa-Akademie, Dübendorf, Schweiz, 12. und 13. April 2018)* ed R Bender and C Steinmetz (Frankfurt am Main: GfKORR - Gesellschaft für Korrosionsschutz e.V) pp 48–56
[17] Normen Fuchs and Froitzheim P 2015 Langzeitmonitoring der Strukturkomponenten und Schließringbolzenverbindungen in WEA Gittertürmen , doi: 10.13140/RG.2.2.27487.87201
[18] Flügge W and Denkert C 2018 Ausnutzung planmäßig vorspannbarer Verbindungselemente für dünne Klemmbereiche unter montagegerechten Bedingungen *(EFB-Forschungsbericht* Nr. 483*)* (Hannover: Europäische Forschungsgesellschaft für Blechverarbeitung e.V)
[19] Ebert A 2020 Ein Beitrag zur versuchsgestützten Ermittlung des Gleitwiderstandes vorgespannter Stahlbauverbindungen *Dissertation* Fakultät für Maschinenbau und Schiffstechnik, Universität Rostock
[20] Glienke R, Ebert A, Wanner M-C and Kupzok A 2015 Anforderungen an die mechanische Fügetechnik für Türme von WEA in Stahlbauweise für große Nabenhöhen *Stahlbau* **84** 556–70
[21] DIN Deutsches Institut für Normung e. V. 2012 DIN EN ISO 9224: Korrosion von Metallen und Legierungen – Korrosivität von Atmosphären – Anhaltswerte für die Korrosivitätskategorien (Berlin: Beuth Verlag GmbH)
[22] DIN Deutsches Institut für Normung e. V. 2018 DIN EN ISO 6270-2: Beschichtungsstoffe - Bestimmung der Beständigkeit gegen Feuchtigkeit - Teil 2: Kondensation (Beanspruchung in einer Klimakammer mit geheiztem Wasserbehälter) (Berlin: Beuth Verlag GmbH)
[23] DIN Deutsches Institut für Normung e. V. 2009 DIN EN ISO 11885: Wasserbeschaffenheit – Bestimmung von ausgewählten Elementen durch induktiv gekoppelte Plasma-Atom-Emissionsspektrometrie (ICP-OES) (Berlin: Beuth Verlag GmbH)

Effect of sample geometry on the macroscopic shear deformation of the titanium alloy Ti-10V-2Fe-3Al subjected to quasi-static and dynamic compression-shear loading

S Winter[1,2,*], M Scholze[1] and M F-X Wagner[1]

[1] Institute of Materials Science and Engineering, Materials Science Group, Chemnitz University of Technology, 09125 Chemnitz, Germany
[2] now at Fraunhofer Institute for Machine Tools and Forming Technology, IWU, 09126 Chemnitz, Germany

*e-mail: Sven.Winter@iwu.fraunhofer.de

Abstract. We investigate the evolution of shear bands in a Ti-10V-3Fe-2Al alloy with three different initial microstructures: solution-annealed β-titanium, β-titanium with primary α-phase precipitates as well as a condition with primary and secondary α-phase precipitates. Quasi-static and dynamic testing of compression-shear specimens with strain rates of 10^{-3} s^{-1} and 10^{2} s^{-1} is performed at room temperature. The influence of two different sample geometries on the macroscopic shearing tendency is investigated: samples with a cylindrical cross section and samples with a square cross section. The results of digital image correlation and the nominal stress-strain behavior under quasi-static and dynamic loading indicate a stronger localization tendency of the aged conditions with additional α-phase precipitates compared to the solution-annealed pure β-condition. Our results also demonstrate that compression-shear testing with samples with a square cross-section reduces facet loss, increases the resolution even at high strain rates and therefore allows to analyze strain distributions during shear band formation and propagation with higher accuracy.

1. Introduction

When subjected to dynamic loading, titanium alloys tend to fail spontaneously by the formation and growth of adiabatic shear bands [1,2]. This is primarily due to their low thermal conductivity [3]: High strain rates (>10^{2} s^{-1}) lead to a local temperature increase since the heat cannot be transferred quickly enough into the surrounding material. This effect results in local thermal softening which further leads to a concentration of deformation and finally to the formation of an adiabatic shear band [4–6]. The macroscopic deformation is therefore concentrated in relatively thin, microscopic regions, and microstructural evolution in these regions ultimately leads to spontaneous material failure by crack nucleation and growth [7,8].

The mechanical behavior of body-centered cubic (bcc) β-titanium alloys is strongly affected by microstructural parameters like β-grain size, morphology and volume fraction of primary and secondary hexagonal close-packed (hcp) α-phase precipitates [9–12]; the effect of these microstructural features on shear band formation and growth is not yet well understood. The present study is part of an ongoing experimental and theoretical investigation on the evolution of adiabatic shear bands in the β-titanium alloy Ti-10V-2Fe-3Al for different types of initial microstructures [13]: β-titanium with primary α-phase and β-titanium with primary and secondary α-phase precipitates. We carried out extensive compression-shear tests in a previous experimental study [14,15] to specifically characterize the influence of the different initial microstructural characteristics on the formation of adiabatic shear bands. Compression-shear specimens, first proposed by Meyer et al. [16], are geometrically derived from conventional cylindrical compression samples, but the top and lower surfaces are cut at a small angle of inclination with respect to the central axis. When loaded in compression, this special geometry leads to a superposition of shear stresses. While the resulting stress state is inhomogeneous and thus not easy to interpret in terms of nominal stress-strain curves [17], compression-shear specimens have the advantage of allowing to test a material for its tendency to exhibit localized (shear) deformation. Compression-shear testing has been successfully used to characterize localized deformation in functional materials [18,19] where complex stress-states can considerably affect phase transitions [20,21]; this experimental approach is, moreover, specifically well-suited to analyze the shear tendency and further the formation of adiabatic shear bands and subsequent material failure in structural materials at high strain rates. To evaluate sample deformation during compression-shear experiments with respect to the shear tendency, the local strain fields can be documented and analyzed by means of digital image correlation (DIC). In experiments with high strain rates, however, the very short testing time and the three-dimensional measurement of the surface typically lead to increased facet failures during the measurements. The missing facets have to be interpolated numerically, which effectively leads to a reduction of the local resolution of the DIC measurements. This effect limits the general analysis possibilities and especially affects the determination of the shear tendency.

In the present study, modified compression-shear specimens with a square cross-section are used in an attempt to significantly increase the resolution in dynamic tests and reduce facet failures. As with the more common cylindrical compression-shear specimens, the inclination is also 6°. The advantage is that the surface to be measured is completely flat. For this reason, it is possible to work with a simpler experimental technique using only one camera for the two-dimensional evaluation of the strain field (instead of two cameras needed in a 3D setup for cylindrical specimens). We show that the shear zone is homogeneous in the entire cross section of the sample and that, by using the optimized specimens, the formation of shear bands can be documented in-situ on the surface of the specimens during mechanical testing using digital image correlation. The DIC results are further compared with those of cylindrical specimens.

2. Experimental and materials

A metastable β-titanium alloy of commercial purity was received in a forged and solution annealed condition from *Otto Fuchs KG* (Meinerzhagen, Germany) as cylindrical billets with a diameter of 52 mm. The chemical composition of the alloy Ti-10V-2Fe-3Al (Ti-10-2-3) was determined by spark spectral analysis as 10.3V-1.71Fe-3.15Al-0.01C-(bal.) Ti (in wt.-%). Three different heat treatment strategies were used to modify the microstructure of the titanium alloy, conducted in a convection furnace with Argon atmosphere to prevent oxidation. The specific heat treatment sequences are shown schematically in Figure 1. In a first step, the material was solution annealed at 830 °C above the β-transus temperature (792 °C) for 30 min and water-quenched to room temperature to achieve a fully recrystallized microstructure in a β-phase condition without additional α-precipitates or orthorhombic martensite. Using additional aging treatments, modified microstructures with additional α-phase precipitates was produced: The alloy was annealed at 700 °C for 60 min followed by air cooling, which resulted in precipitation of the primary α-phase within the β-grains. In a third step, the material was annealed at 500 °C for 480 min followed again by air cooling, which led to the formation of additional fine, secondary α-phase precipitates. Consequently, the three microstructural conditions used for investigations in this study are: (i) a pure β-condition (**SA(β)**), (ii) a condition with additional primary α-phase precipitates (**A(α_p)**), and (iii) a condition with additional primary and secondary α-precipitates (**A(α_{p+s})**). Microstructural investigations were performed by optical and scanning electron microscopy (SEM) to document the effects of heat treatment. For the SEM investigations, a *Zeiss* NEON40EsB field emission SEM in QBSD (quadrant back scatter diffraction) mode with an acceleration voltage of 15 kV was used to visualize the material contrast between the α- and β-phases.

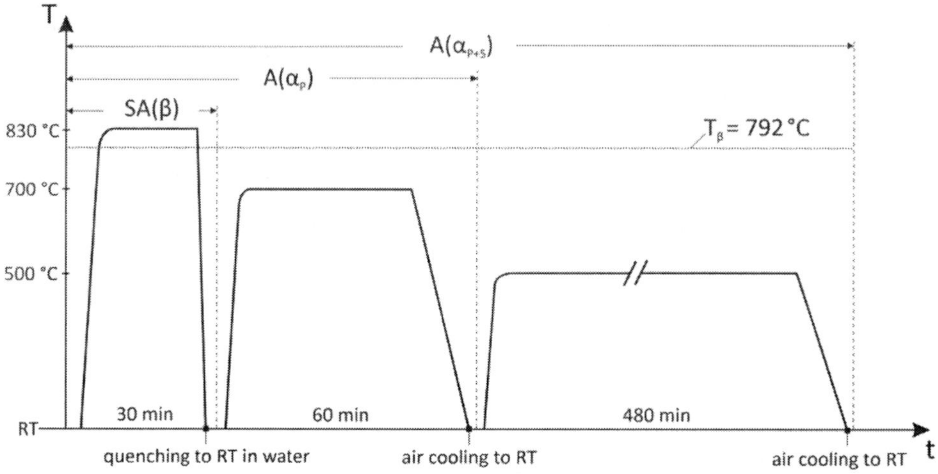

Figure 1. Schematic representation of the heat treatment sequences used in the present study. Solution annealing and subsequent aging results in three different conditions: solution annealed - SA(β), aged to precipitate primary α-phase - A(α_p), aged to form primary and secondary α-phase precipitates - A(α_{p+s}).

Compression-shear testing was performed to document and compare the shear tendency of the three microstructural material conditions. In contrast to simple compression testing, the compression-shear samples used in the present study have an angle of inclination of 6° with respect to the loading direction. In addition to the more common cylindrical samples, a new type of "inclined cube" (simply referred to as "cubic" in the remainder of this paper) compression-shear specimen was used. The geometries of both sample types are illustrated in Figure 2. The cylinders with a diameter of 9 mm were cut by spark erosion from the titanium billets; their front faces were ground under an angle of inclination of 6° to obtain a height of 9 mm. The samples with a square cross section of 8 mm x 8 mm were cut directly by spark erosion, producing an inclination angle of 6° and a height of 8 mm. The effective cross section of approx. 64 mm^2 was almost the same in both sample types. Both types of samples were used for testing at two different strain rates: Quasi-static tests (initial strain rate of 10^{-3} s^{-1}) were performed in a *Zwick/Roell* UPM 1475 universal testing machine with a maximum force of 100 kN. An additional compression fixture with an inductive sensor was used to measure displacements, excluding measurement error related to elastic deformation of the testing device. The dynamic experiments (initial strain rates of 220 s^{-1} and 250 s^{-1} for the cylindrical and cubic samples, respectively) were performed in a drop tower with a drop mass of 600 kg. All samples were coated with stochastic black and white patterns and surface displacements were recorded simultaneously for all quasi-static and dynamic experiments by DIC to estimate the magnitude of the shearing tendency for the three different microstructural conditions. A single camera was oriented perpendicular to the surface of all cubic samples (2D DIC measurements). For all cylindrical samples, a 3D DIC system with two high-speed cameras of the type SA5 from *Photron* were used. The ARAMIS software by *GOM/Zeiss* was used for subsequent analysis of the strain fields.

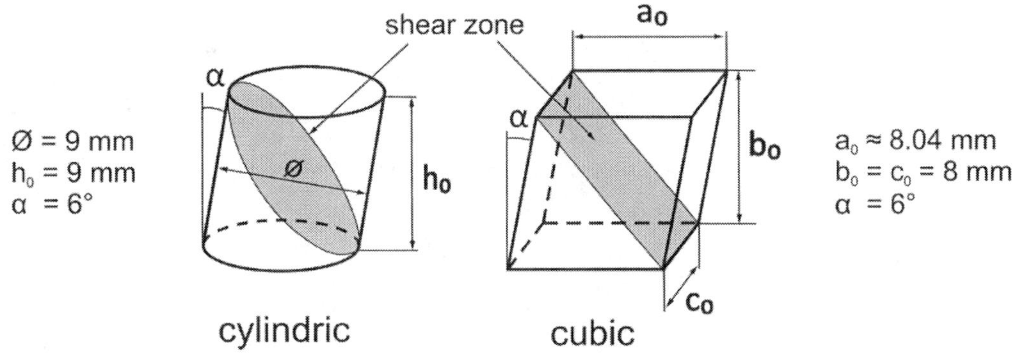

Figure 2. Two different sample geometries (cylindric and cubic) were used for both quasi static and dynamic compression-shear tests.

3. Results and discussion

3.1. Microstructural characteristics after different heat treatments

We first discuss the influence of the different heat treatments on the microstructure of the alloy Ti-10V-2Fe-3Al. Figure 3 shows optical and scanning electron micrographs of three differently heat-treated material conditions. All material conditions are characterized by a coarse-grained microstructure with globular grains and an average grain size of about 300 µm. Pure β-grains without precipitates are observed after solution annealing, as shown in the optical micrograph of material condition SA(β) (Figure 3a). Aging was conducted for additional precipitation of the α-phase. A significantly darker appearance in the optical micrographs, primarily due to the precipitation of primary α, is visible after etching of both aged conditions (Figures 3b&c). It is not possible to directly detect any differences between the aged conditions $A(\alpha_p)$ and $A(\alpha_{p+s})$ by optical microscopy. Higher magnifications are necessary to detect size, type, morphology, or volume content of α-phases. The SEM micrographs (Figures 3d&e) show lamellar primary α-phases with lengths between 1 and 4 µm and a volume fraction of about 40% in both aged conditions. A typical microstructural orientation between bcc β-grains and hcp α-phases is observed, which is well-known as Burgers orientation relationship [22]. Two-step aging as expected leads to the additional precipitation of lamellar secondary α-phase, with lamella sizes of approximately 100 nm and a volume fraction of about 15% (Figure 3e). Furthermore, a regular arrangement of secondary α-needles, parallel to primary α precipitates, is observed in the SEM micrograph. The microscopic characterization therefore confirms that three distinctly different material conditions were produced by the different heat treatments, which subsequently allowed to study the effect of different initial microstructures on shear band formation during compression-shear testing.

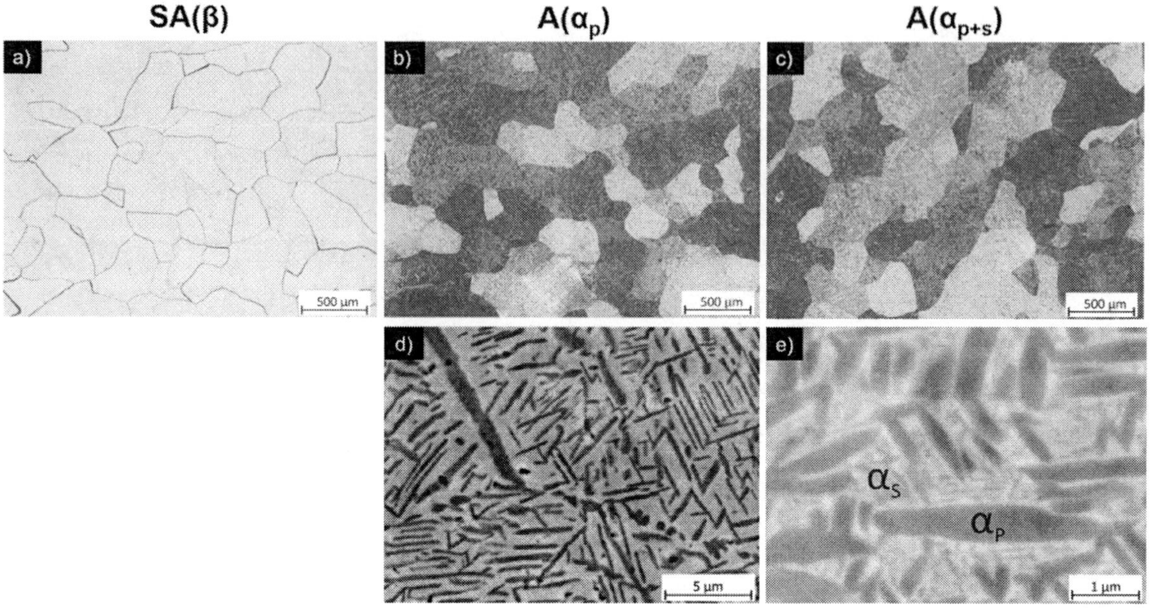

Figure 3. Microstructures of Ti-10V-2Fe-3Al after different heat treatments. (a) Optical micrograph after solution annealing (830 °C for 30 min) – condition SA(β), (b) optical micrograph after one-step aging (700 °C for 60 min) – condition A(α_p), (c) optical micrograph after two-step aging (700 °C for 60 min and 500 °C for 480 min) – condition A(α_{p+s}). Scanning electron micrographs indicate differences between both aged conditions: (d) One-step aging leads to a precipitation of primary α-phase precipitates within the β-grains. (e) Two-step aging leads to the additional formation of secondary α-phase precipitates.

3.2. Investigation of the shear tendency by DIC

In compression-shear samples, the angle of inclination of 6° leads to a complex stress state during nominally uniaxial compression; therefore, the stress-strain behavior can only be estimated by considering the compressive load component. Nevertheless, for a first-order, quantitative comparison of the mechanical behavior of the three different material conditions, nominal engineering stress-strain curves were calculated from the uniaxial load-displacement data recorded during compression-shear testing.

We first characterize the local strain fields at different stages of deformation as reflected by the nominal stress-strain data. With the help of DIC, we investigate to what extent optical measuring methods can be used to analyze shear band formation. Two different strain rates (quasi-static and dynamic) of both cylindrical and cubic compression-shear specimens are compared. Strain fields obtained by DIC are evaluated at 2%, 5% and 10% of axial compression. In all data sets discussed below, in case of a failure earlier than 10% (which occurred primarily when using the cubic specimens due to stress concentrations in the corners), strain data recorded immediately prior to failure was analyzed. Preliminary investigations showed that considering individual strain components in X- or Y-direction (where Y is the loading direction) are not sufficient for an unambiguous characterization of shear banding in compression-shear samples because the specimens deform in both directions during the test. To fully appreciate the multi-axial deformation of the samples, we instead calculate and visualize the major strain (determined at each measurement point from the 2D surfaces strain tensor information).

Figure 4 shows the behavior of the cylindrical compression-shear specimens under quasi-static loading. The solution annealed condition SA(β) exhibits a stronger strain hardening compared to both aged conditions A(α_p) and A(α_{p+s}) but the corresponding nominal stress of SA(β) are at a lower level for any given strain. Already at 2% plastic (compressive) strain, an initial localization of shear deformation can be seen in the DIC strain fields for all material conditions. The shear zone – the region where strains are increased compared to adjacent, less deformed regions – can be observed clearly in both aged states up to a plastic deformation of 10%. In contrast, the solution-annealed condition shows a somewhat more homogeneous deformation of the sample. With increasing plastic strains, several individual bands occur in the SA(β) samples, which are located either in the shear plane, parallel to the shear plane or perpendicular to the shear plane. A slightly stronger localization tendency with higher local strains under quasi-static loading of the two aged samples, compared to the solution-annealed samples is observed.

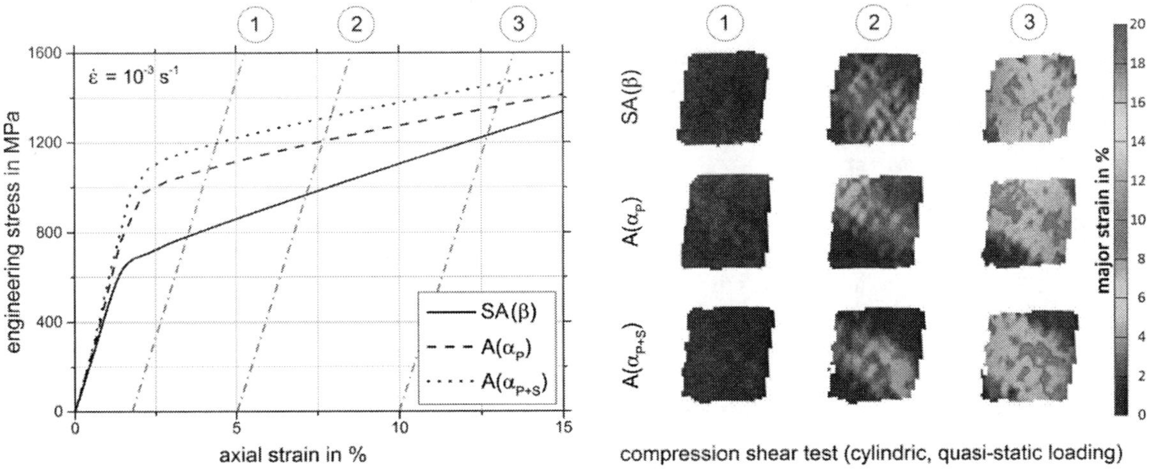

Figure 4. Nominal stress-strain curves of Ti-10-2-3 after different heat treatments under quasi-static loading of cylindrical compression-shear specimens. The corresponding strain fields (DIC) are shown at 2%, 5% and 10% nominal axial compressive strains, respectively. In the three material conditions, the formation of a shear zone with increased major strain values can be observed in the DIC data. In the aged conditions A(α_p) and A(α_{p+s}), this zone is narrower than in the solution annealed condition SA(β), where several parallel bands in and perpendicular to the shear plane are formed.

Figure 5 shows the nominal stress-strain behavior of the cylindrical compression-shear specimens under dynamic loading in the drop tower and representative DIC strain fields. Similar to the results of the quasi-static compression-shear tests, the SA(β) condition shows strain hardening until failure. In contrast, both aged conditions A(α_p) and A(α_{p+s}) are characterized by a significantly lower strain hardening rate after reaching the yield point. After 5% axial compression a significant softening is noticeable for both aged samples. It is likely that from this moment on shear band initiation and subsequent growth occurred, which counteracted the material's strain hardening due to local thermal softening [23,24]. In the DIC strain maps, the formation of a broad shear band region is confirmed. Compared to the quasi-static tests, the deformation in all three conditions is much more concentrated in the shear plane, and this effect becomes more intense with increasing axial strains. The shear zone of the SA(β) samples again is wider and, similar to the quasi-static tests, the main deformation appears to be localized in several bands. In the aged samples, a concentration of the main deformation in the "corner regions" indicates subsequent fracture initiation sites.

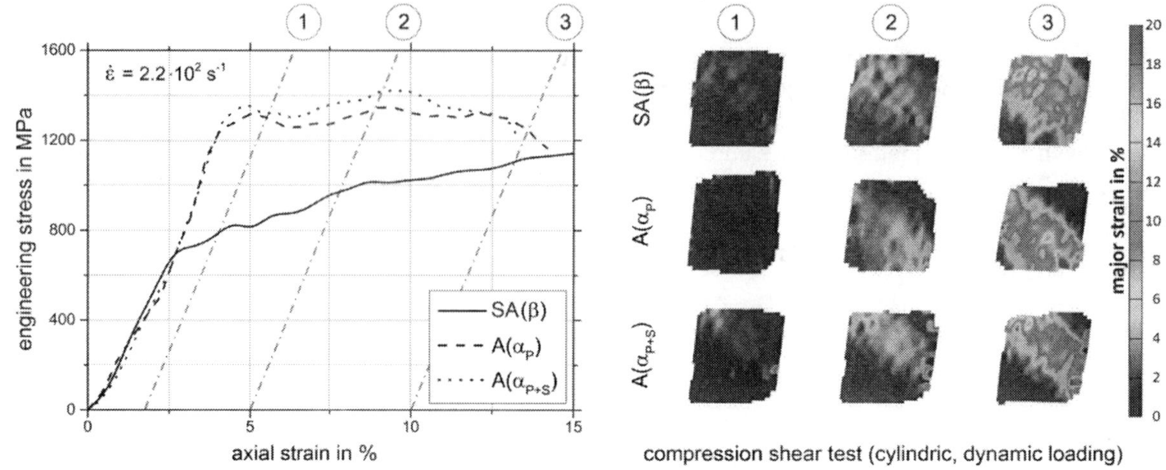

Figure 5. Nominal stress-strain curves of Ti-10-2-3 with different heat treatments under dynamic loading of cylindrical compression-shear specimens. The corresponding strain fields (DIC) are shown at 2%, 5% and 10% axial compressive strains. All material conditions show the formation of a shear zone with high major strain values. SA(β) forms deformation bands in the shear zone and perpendicular to it, similar to the quasi-static tests. The shear zone is slightly wider than in the aged samples. For A(α_p) and A(α_{p+s}), softening occurs, starting at about 5% axial strain. This indicates shear band initiation and propagation, which ultimately leads to material failure.

As a consequence of the high impact energy during drop tower tests, the DIC camera system is influenced by vibrations transmitted through the drop tower foundation into the ground. Especially in 3D DIC with two cameras, oriented and calibrated in a certain angle to each other, these vibrations can slightly change the camera positions. This affects the analysis of strain fields and may lead to the appearance of image correlation errors in individual image areas, which makes it necessary to subsequently interpolate facets, ultimately diminishing the resolution of strain measurements in the shear band regions. In order to address these issues, we additionally studied the deformation behavior of the cubic compression-shear specimens. The key advantage of this experimental approach is that the flat surface perpendicular to the shear plane of the specimen can be directly used for the DIC measurements. Most importantly, DIC can then be performed with only one camera in 2D mode. This leads to a reduced error rate and therefore to a significantly higher number of facets that can be analyzed and, accordingly, a higher resolution. The goal of the results presented below was to study whether the experimental observation and subsequent strain analysis of localized shear deformation processes can thus be simplified and improved.

Figure 6 summarizes the stress-compression behavior of cubic compression-shear samples under quasi-static loading. The nominal stress-strain curves are quite similar to the corresponding results determined from the cylindrical compression-shear specimens. In the solution annealed condition SA(β), the strain field clearly shows a local increase of major strain values; this localization becomes more distinct with increasing total deformation in the shear plane. Again, due to the quasi-static strain rate, a splitting of the deformation into several bands can be observed on the specimen's surface already at low nominal strain values of about 2% (and very clearly from 5%). In the cubic specimens, shear bands occur parallel and perpendicular to the shear plane (see Figure 2) on the entire surface. One possible explanation for the observed grid-like structure in the strain field is the relatively large grain size of 300 μm: It is likely that individual grains that are well oriented relative to the maximum shear stress are plastically deformed earlier than adjacent grains. This leads to the formation of more strongly deformed areas that can be directly detected in the DIC strain field and thus to the observed lattice-like patterns in the image correlation data. In both aged conditions, the formation of a "forging cross" (i.e., an X-shaped strain profile typically associated with uniaxial compressive loading), dominates the surface strain fields. A slight grid-like structure is also visible. Due a force limitation of 100 kN of the testing machine, both aged conditions A(α_p) and A(α_{p+s}) could not be deformed until fracture. Starting at a strain of 5%, shear localization in the shear zone also becomes evident.

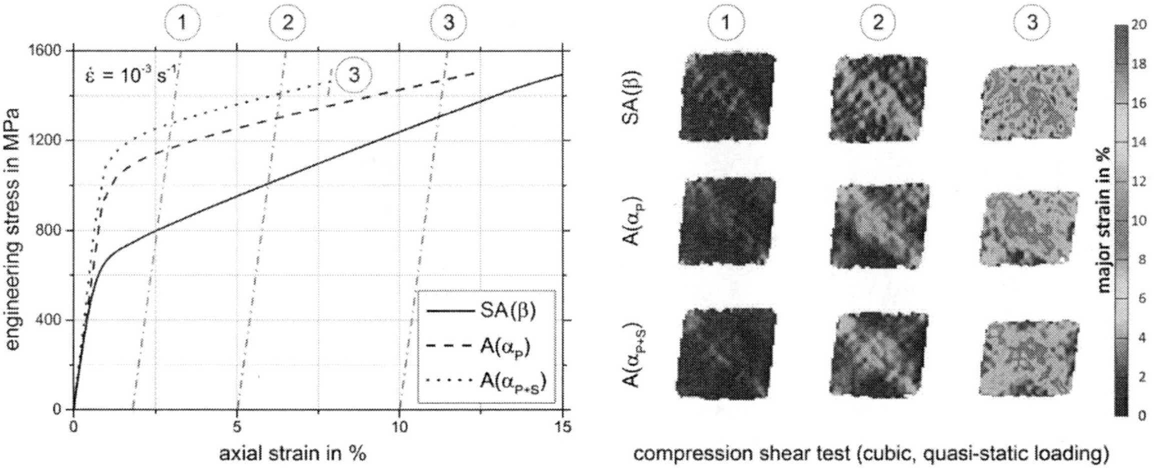

Figure 6. Nominal stress-strain curves of Ti-10-2-3 with different heat treatments under quasi-static loading of cubic compression-shear specimens. The corresponding strain fields (DIC) are shown at 2%, 5% and 10%. All material conditions show the formation of a shear zone with increased major strains.

Figure 7 shows the nominal stress-strain behavior of cubic compression-shear specimens under dynamic loading. Due to premature material failure, a high variation of the maximum achievable nominal strains was observed in the aged samples. This behavior is related to the sample geometry: Due to the small radii, local stress peaks occur in the corners of the sample. These corner regions are starting points for shear band initiation and, in some experiments, for early initiation of fracture. The representative stress-strain curves of the conditions A(α_p) and A(α_{p+s}) shown in Figure 7 were selected from experiments where fracture occurred relatively late. For all material conditions, the strain fields show the formation of a well-defined, sharp shear zone with high major strain values, in which also material failure occurs in the later stages of deformation. These shear band regions are concentrated in the shear plane, and they are already formed at small strains. The solution annealed material condition also exhibits a more pronounced deformation in the regions adjacent to the shear band region. This is due to the fact that the SA(β) condition shows a significantly higher amount of strain hardening and thus also reaches a higher total compressive strain. Consequently, localized shear deformation occurs much later compared to the aged material conditions.

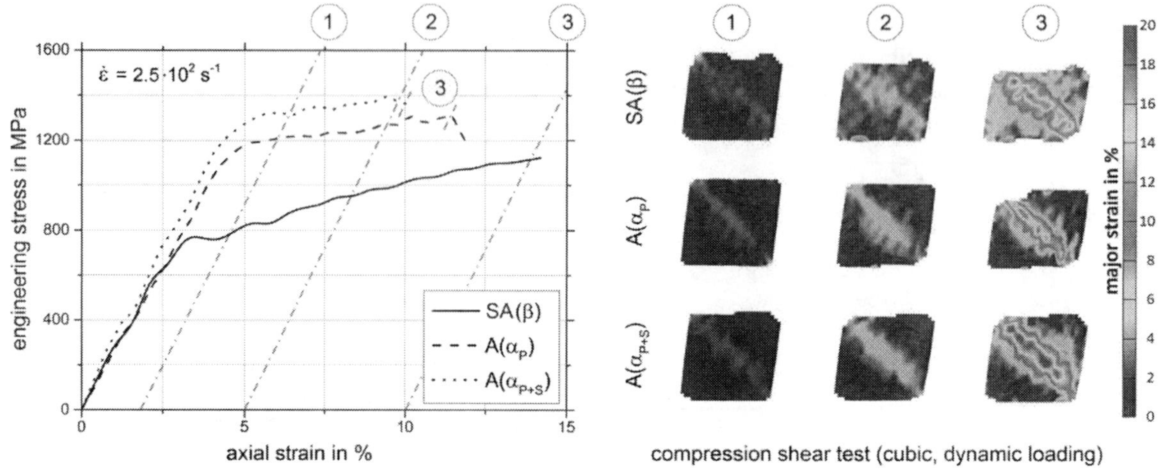

Figure 7. Nominal stress-strain curves of Ti-10-2-3 with different heat treatments under dynamic loading of cubic compression-shear specimens. The corresponding strain fields (DIC) are shown at 2%, 5% and 10% axial compressive strains, respectively. All material conditions form a distinct shear zone with high major strains during dynamic compression. The material condition SA(β) is characterized by significantly higher strain hardening than the aged conditions and therefore reaches a higher total deformation before shear localization sets in.

The flat surfaces, the possibility to perform more stable 2D DIC measurements, and the resulting reduced loss of facets even in dynamically tested samples, make the cubic compression-shear specimens particularly useful for an in-depth analysis of local strain fields during shear banding. One possible method of evaluation is shown as an example in Figure 8: In order to evaluate and compare the shearing tendency of the three differently heat-treated material conditions, we consider strain maps that were recorded immediately prior to material failure during dynamic loading. The improved resolution allows to visualize the major strain distribution on the planar surface of the cubic sample with high accuracy. Moreover, we plot major strain values along the linear paths (red lines; length 10 mm) highlighted in each DIC strain map. It is possible to carefully evaluate the widths of the shear zones of the different material conditions – and thus to analyze their shearing tendencies under dynamic loading conditions. In all material conditions, strongly localized deformation in the shear zone is observed. There is no trace of grid-like patterns in the strain fields as in the quasi-statically tested specimens. The deformation therefore clearly predominantly takes place in the shear plane at increased strain rates. Only the solution annealed material condition SA(β) exhibits a somewhat increased deformation in the remaining parts of the sample volume, confirming the observations discussed above in relation to Figure 7. An evaluation of the width of the shear zones was performed at 1% below the respective maximum of the major strain. The resulting values are 675 µm for SA(β), 400 µm for A(α_p) and 220 µm for A(α_{p+s}). These results agree well with the qualitative observations of the different shear zones: The solution annealed condition shows a significantly wider shear zone than both aged conditions and the deformation is generally more widely distributed over the entire sample volume. With A(α_p) and A(α_{p+s}), the tendency for shear localization is considerably more pronounced. These results are fully in line with our previous investigations presented in [13]. The two-stage aged specimens with additional secondary α-phase show a very high shear tendency and are also characterized by the highest local major strain values of all tested conditions.

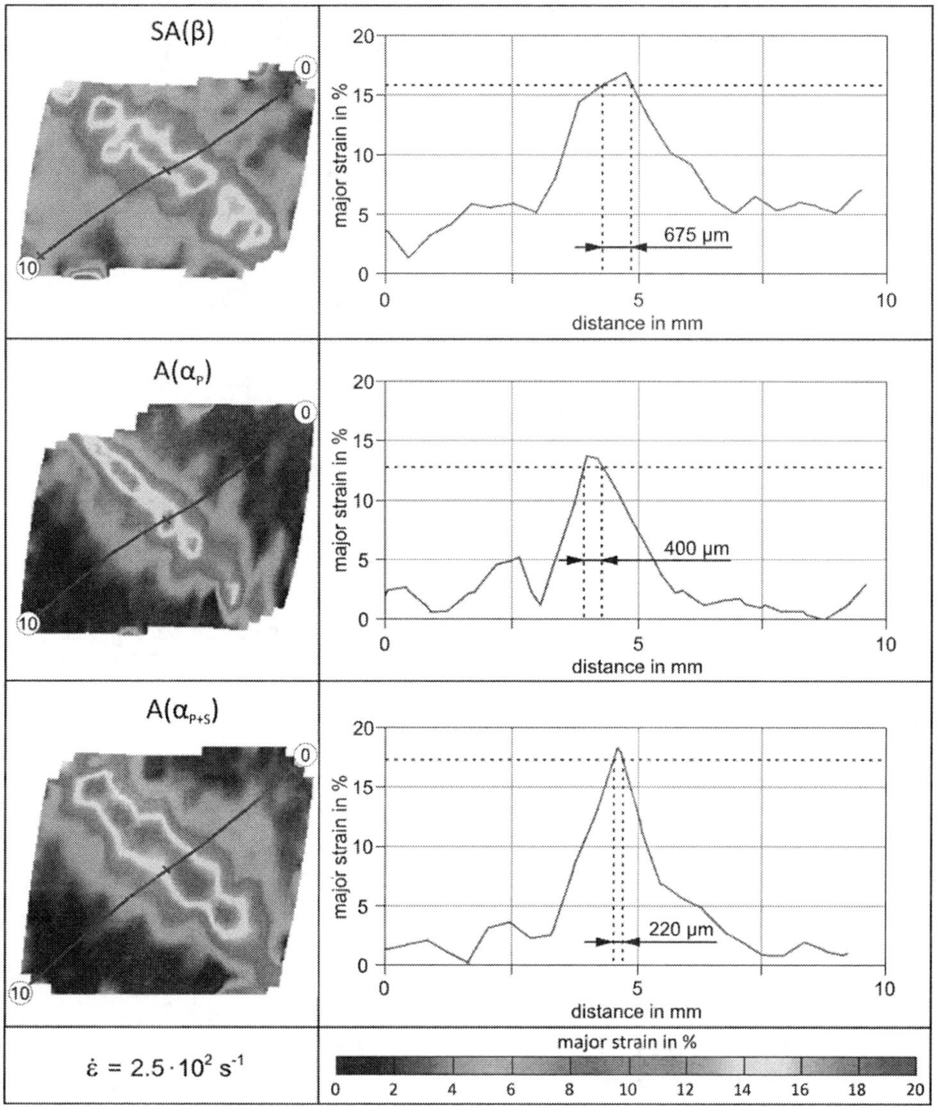

Figure 8. Analysis of the shear zones of the three heat-treated conditions under dynamic compressive shear loading of cubic compression-shear specimens. Additionally, the strain fields of the digital image correlation are shown. The corresponding diagrams show major strain values along a 10 mm long section (red lines, 0 – 10 mm, indicated in the DIC strain maps on the left). There is a clear localization of the shear deformation with differently pronounced peaks in all three material conditions. The solution annealed material condition SA(β) shows a slightly wider shear zone and the entire cross section of the sample exhibits more deformation than the aged samples. A(α_p) and A(α_{p+s}) show a strong localization of the major strain in narrower areas close to the shear zones.

4. Summary and conclusions

In the present study, we have investigated the influence of two different sample geometries and three different heat treatment conditions on the macroscopic shearing tendency of the titanium alloy Ti-10V-2Fe-3Al. Compression-shear tests with cylindric and cubic samples have been performed under quasi-static and dynamic loading in combination with digital image correlation. The results of the digital image correlation measurements and the nominal stress-strain behavior under quasi-static and dynamic loading clearly show that there is a stronger localization tendency in the aged material conditions with primary as well as with primary and secondary α-phase precipitates compared to a solution-annealed β-titanium condition. Our results again demonstrate that the mechanical behavior of the titanium alloy is significantly influenced by the superposition of compressive and shear stresses. Using the cubic sample geometry, a strong localization tendency can be detected by digital image correlation with high accuracy even under dynamic loading conditions. The less accurate representation of 3D deformation fields on the cylindrical compression-shear specimens under dynamic loading is avoided by using the cubic specimen shape. The two-dimensional strain mapping significantly increases the effective resolution capacity while reducing facet loss, providing an opportunity for more detailed investigations on adiabatic shear banding.

References

[1] Meyers M A and Pak H-R 1986 Observation of an adiabatic shear band in titanium by high-voltage transmission electron microscopy *Acta Metall.* **34** 2493–9

[2] Meyers M A, Subhash G, Kad B K and Prasad L 1994 Evolution of microstructure and shear band formation in α-hcp titanium *Mech. Mater.* **17** 175–93

[3] Machai C and Biermann D 2011 Machining of a hollow shaft made of β-titanium Ti-10V-2Fe-3Al *Proc. - 2011 IEEE Int. Symp. Assem. Manuf. ISAM 2011* 1–6

[4] Yang Y and Wang B F 2006 Dynamic recrystallization in adiabatic shear band in α-titanium *Mater. Lett.* **60** 2198–202

[5] Meyers M A and Wittman C L 1990 Effect of metallurgical parameters on shear band formation in low-carbon (approx. 0.20 wt pct) steels. *Metall. Trans. A, Phys. Metall. Mater. Sci.* **21A** 3153–64

[6] Nesterenko V F, Meyers M A and Wright T W 1998 Self-organization in the initiation of adiabatic shear bands *Acta Mater.* **46** 327–40

[7] Bai Y and Dodd B 2012 *Adiabatic Shear Localization* (Pergamon Press)

[8] Zener C and Hollomon J H 1944 Effect of Strain Rate Upon Plastic Flow of Steel *J. Appl. Phys.* **15** 22–32

[9] Jackson M, Jones N G, Dye D and Dashwood R J 2009 Effect of initial microstructure on plastic flow behaviour during isothermal forging of Ti-10V-2Fe-3Al *Mater. Sci. Eng. A* **501** 248–54

[10] He L, Dehghan-Manshadi A and Dippenaar R J 2012 The evolution of microstructure of Ti-6Al-4V alloy during concurrent hot deformation and phase transformation *Mater. Sci. Eng. A* **549** 163–7

[11] Chen W, Sun Q, Xiao L and Sun J 2012 Thermal stability of bulk nanocrystalline Ti-10V-2Fe-3Al alloy *Mater. Sci. Eng. A* **536** 223–30

[12] Winter S and Wagner M F-X 2016 On the effect of incremental forming on alpha phase precipitation and mechanical behavior of beta-Ti-10V-2Fe-3Al *IOP Conf. Ser. Mater. Sci. Eng.* **118** 012026

[13] Winter S, Pfeiffer S, Bergelt T and Wagner M F-X 2019 Finite element simulations on the relation of microstructural characteristics and the formation of different types of adiabatic shear bands in a β-titanium alloy *IOP Conf. Ser. Mater. Sci. Eng.* **480** 012022

[14] Winter S, Schmitz F, Clausmeyer T, Tekkaya A E and Wagner M F-X 2017 High temperature and dynamic testing of AHSS for an analytical description of the adiabatic cutting process *IOP Conf. Ser. Mater. Sci. Eng.* **181** 012026

[15] Schmitz F, Winter S, Clausmeyer T, Wagner M F X and Tekkaya A E 2020 Adiabatic blanking of advanced high-strength steels *CIRP Ann.* **69** 269–72

[16] Meyer L W, Staskewitsch E and Burblies A 1994 Adiabatic shear failure under biaxial dynamic compression/shear loading *Mech. Mater.* **17** 203–14

[17] Pfeiffer S, Frint P and Wagner M F-X 2017 Analysis of the complex stress state during early loading in cylindrical compression-shear specimens *IOP Conf. Ser. Mater. Sci. Eng.* **181** 012025

[18] Elibol C and Wagner M F-X 2015 Investigation of the stress-induced martensitic transformation in pseudoelastic NiTi under uniaxial tension, compression and compression–shear *Mater. Sci. Eng. A* **621** 76–81

[19] Elibol C and Wagner M F-X 2015 Strain rate effects on the localization of the stress-induced martensitic transformation in pseudoelastic NiTi under uniaxial tension, compression and compression–shear *Mater. Sci. Eng. A* **643** 194–202

[20] Wagner M F X and Eggeler G 2006 Stress and strain states in a pseudoelastic wire subjected to bending rotation *Mech. Mater.* **38** 1012–25

[21] Pfetzing-Micklich J, Wagner M F X, Zarnetta R, Frenzel J, Eggeler G, Markaki A E, Wheeler J and Clyne T W 2010 Nanoindentation of a pseudoelastic NiTiFe shape memory alloy *Adv. Eng. Mater.* **12** 13–9

[22] Leyens C and Peters M 2002 *Titan und Titanlegierungen* (Weinheim: Wiley-VCH)

[23] Dormeval R 1987 *The adiabatic shear phenomena. Materials at high strain rates.* ed T Z Blasynski (New York: Elsevier)

[24] Wright T W 2002 *The physics and mathematics of ASB* (New York: Cambridge University Press)

IOP Conf. Series: Materials Science and Engineering 1147 (2021) 012015 doi:10.1088/1757-899X/1147/1/012015

Chemical structure of amino-terminated alkyl silanes influencing the strength of aluminum-polyamide joints

A Dittes*[1], I Scharf, L Lehmann[1], E Saborowski[1], T Mehner[1] and T Lampke[1]

[1]Chemnitz University of Technology, Materials and Surface Engineering Group, Erfenschlager Straße 73, 09125 Chemnitz, Germany

*e-mail: axel.dittes@mb.tu-chemnitz.de

Abstract. Organo-functional silane coupling agents are widely used to promote adhesion in between inorganic and organic materials. Amino-functional silanes can improve bonding with polyamide. Aiming on a mechanically performant aluminum-polyamide joints, the effect of six amino silanes of different chemical structure, namely number and type of amine groups and alkyl spacers lengths, on the joint strength is investigated by means of lap-shear testing. Higher shear strengths are found along with a more pronounced capability of the amine group to form hydrogen bonds with polyamide. The results show that an additional amine group within the organo-functional group can increase joint strength, whereas long alkyl spacers reduce the observed joint strength. It is shown that high lap-shear strengths, in maximum about 15 MPa for N-(2-aminoethyl)-3-aminopropyltrimethoxysilane are achieved and that the high reproducibility can be assured when using right processing routines.

1. Introduction

For reasons of lightweight construction, dissimilar materials are increasingly combined into load-bearing components. Well-known applications based on fiber-reinforced plastics combined with metal alloys into polymer-metal hybrids (PMH) are numerously established in aircrafts [1]. The design of the material's interface is a key aspect and numerous methods are already applied for thermosetting resin-based PMH [2]. In order to open up broader areas of application outside of aviation, ongoing efforts are conducted to reduce production times and costs. Hereto, a promising approach is replacing thermosetting by thermoplastic resins [3]. Among others, particularly polyamide is of great interest, as it combines good processing properties with high availability and beneficial mechanical properties. Hence, appropriate interface-design techniques for polyamide-based PMH are demanded. Several studies outline the potential of e.g. laser structuring [4, 5] or grit blasting [6, 7] of the metal surface to achieve high joint strengths by mechanical interlocking with the polymer. Besides, organo-functional silane coupling agents are suitable to improve adhesion with polyamides and are already established in commercial glass fiber sizing [8]. The composition of such silane molecules is typically made up of a silicon central atom with three attached alkoxy groups and an organo-functional group. Whereas the alkoxy groups can form covalent bonds with metal or mineral substrates through hydrolysis and condensation reactions, the chemical structure of the organo-functional group determines the achievable adhesion with the polymer [9, 10]. In this context, silanes that bear amine groups which are located at the end of the organo-functional group, so-called amino-terminated alkyl silanes, are used to increase adhesion with polyamide [8, 11, 12, 13]. However, there are various amino-terminated alkyl silanes that differ from each other in the type (primary, secondary, tertiary) and number of amines, and alkyl group lengths with respect to the structure of the organo-functional group. It is reported, e.g. for basalt fiber - reinforced polyamide 6,6 and for talcum polyamide 6 composites, that adhesion with polyamide is influenced by this [12, 13]. In this context, this contribution deals with six amino-terminated

IOP Conf. Series: Materials Science and Engineering 1147 (2021) 012015 doi:10.1088/1757-899X/1147/1/012015

alkyl silanes. The influence of the chemical structure of the organo-functional group on the joint strength of aluminum polyamide joints is investigated. Hereto, aluminum sheets are coated by the silanes and joined with polyamide into lap-shear specimens by heat-conductive hot pressing. Based on this, the achieved adhesion is evaluated by means of lap-shear testing.

2. Material and methods

2.1. Materials and applied silane coupling agents

Lap-shear testing of polyamide 6-aluminum specimens is carried out to study the influence of the chemical structure of the amino-functional group of silane coupling agents on the achievable bond strength. For the specimens production, polyamide 6 sheets of type Ultramid B40 from Co. BASF were shaped into the dimensions of 80 mm × 25 mm × 4 mm by water-jet cutting. Cold-rolled sheets of the aluminum alloy EN AW 6082-T6 (supplier: Co. Gemmel Metalle) in the dimensions of 100 mm × 25 mm × 3 mm were used. For the silane application on the aluminum sheets, sodium hydroxide, ethanol, and xylene (Co. Merck) were utilized. The tested silane coupling agents are listed in table 1. The given abbreviation contains the number of amino groups per silane molecule (Mono or Di). For the three silanes with one amine, the amine type (primary, secondary, or tertiary amine) is given. For the three silanes with two amines (primary and secondary amine each), the lengths of the alkyl spacers (i.e. methyl, ethyl, propyl, hexyl, undecyl) are provided. Thereby, the first abbreviated name (e.g. "Prop") refers to the alkyl group in between the silicon atom and the secondary amine group, whereas the second abbreviated name belongs to the alkyl group in between the two amine groups.

Table 1: Abbreviations and key information regarding the investigated amino-functional silanes.

Silane, supplier, purity and CAS number	Abbreviation	Chemical structure
(N,N-Dimethyl-3-aminopropyl)trimethoxysilane Co. ABCR; 97 %; CAS 2530-86-1	Mono-tert	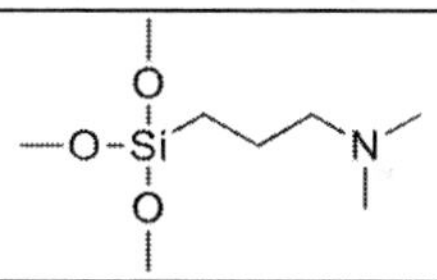
N-Methylaminopropyltrimethoxysilane Co. ABCR; 97 %; CAS 3069-25-8	Mono-sec	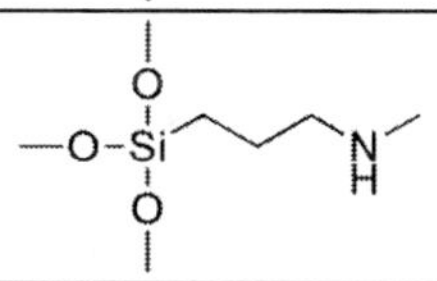
3-Aminopropyltrimethoxysilane Co. Merck; 98 %; CAS 13822-56-5	Mono-prim	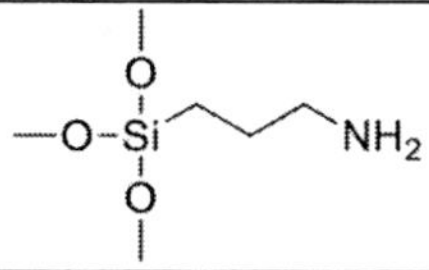
N-(2-aminoethyl)-3-aminopropyltrimethoxysilane Co. Merck; 97 %; CAS 1760-24-3	Di-Prop/Eth	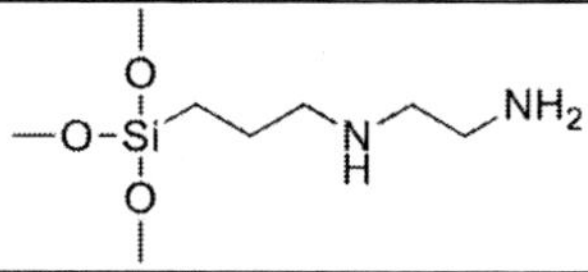
N-(6-aminohexyl)aminomethyltriethoxysilane Co. Gelest; 92 %; CAS 15129-36-9	Di-Meth/Hex	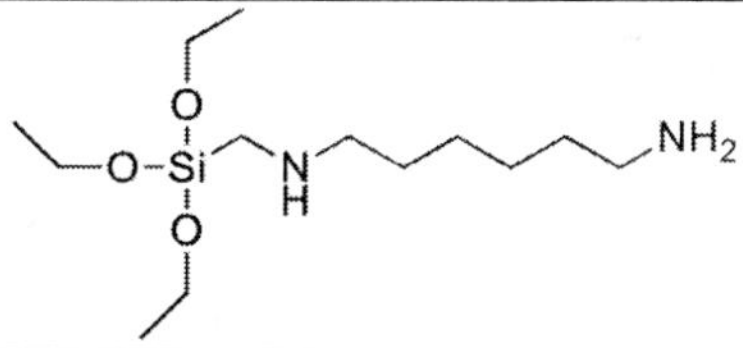
N-(2-aminoethyl)-11-aminoundecyltrimethoxysilane Co. Gelest; 95 %; CAS 121772-92-7	Di-Unde/Eth	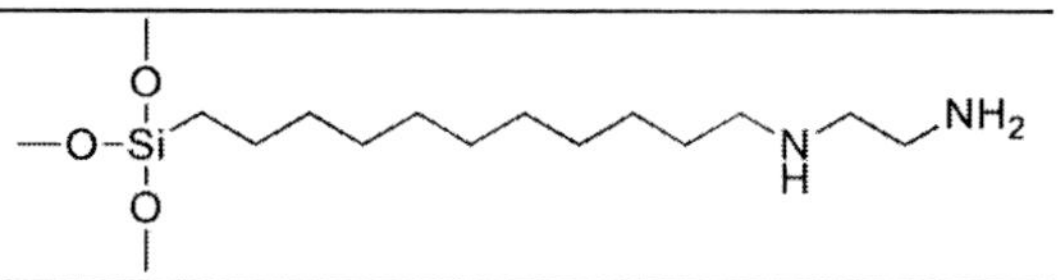

2.2. Preparation and mechanical testing of PMH specimens

For silane functionalization, the aluminum strips are pretreated by ultrasonically assisted degreasing in ethanol for five minutes, pickling in 3 % sodium-hydroxide solution at 50 °C for one minute. This is followed by thorough rinsing with deionized water and immersion in boiling deionized water for 30 min to ensure surface hydroxylation as precondition to the formation of covalent bonds with the silane. Subsequent to a 30 min air drying step at ambient temperature, five aluminum strips each are functionalized by the above-listed silane coupling agents. Hereto, the sheets are immersed in a mixture of anhydrous xylene and silane at a molar ratio of 25 to 1 at a solution temperature of 130 °C for 30 minutes. Afterwards, intense washing steps with ethanol and deionized water were performed to remove weakly bonded, residual silane molecules. The polyamide 6 sheets are pretreated by ultrasonically assisted degreasing in ethanol for five minutes and drying at 70 °C until mass equilibrium is achieved.

Based on the prepared components, lap-shear specimens (see figure 1) with an overlap length of 5 mm and overlap width of 25 mm are produced by heat-conductive hot pressing using a constant pressure of 0.2 MPa, which was applied to the joining zone. A maximum temperature of 235 °C ensures complete interfacial melting of the polyamide. A complete description of the hot-pressing process in provided in [6]. Thereafter, the specimens are conditioned using a climate chamber (VC 4018, Co. Fötsch Industrietechnik).

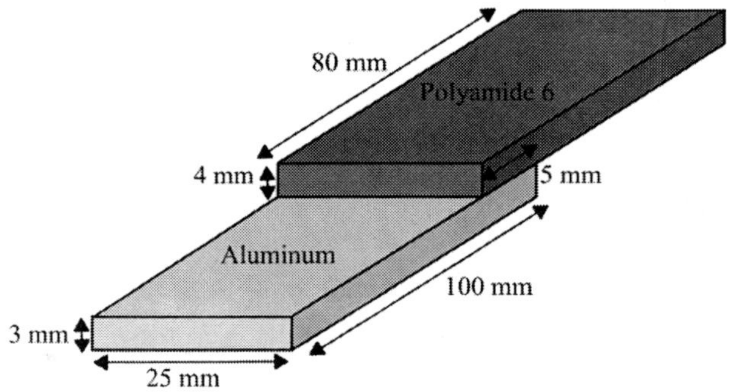

Figure 1: Geometry of the produced PMH specimens.

Accelerated conditioning (70 °C and 62 % relative humidity) to European standard climate is conducted according to ISO 1110 [14] until the sample mass reaches an equilibrium state. Subsequently, mechanical testing of the specimens is performed using a Z020 (Co. Zwick/Roell) testing machine equipped with a 20 kN load cell according to test standard DIN EN ISO 1465 [15]. A test speed of 1 mm/min is applied and all tests were conducted until failure.

3. Results and discussion

In order to study the influence of the chemical structure of the organo-functional group of amino silane coupling agents on the bond strength of aluminum-polyamide joints, lap-shear tests were conducted. It has to be mentioned, that the silane-coating process of the aluminum specimens was initially performed without subsequent intensive washing steps with deionized water and ethanol. Thereby, low lap-shear strengths with high scatter of the results were observed for all the tested silanes. This can be related to the ability of the amine group to form hydrogen bonds with the different functional groups of other amino-silane molecules [16] or with hydroxides attached to the aluminum surface. In consequence, the interfacial zone is partially blocked with weakly connected silane molecules instead of silane molecules that are connected to each other by strong covalent oxane bonds and with the aluminum surface. Hence, the ability of the interface to transfer high mechanical loads is lowered.

After introducing the aforementioned washing steps, significantly higher lap-shear strengths at lower scatter were determined for the tested amino silanes (see figure 2). For all specimens tested, interfacial failure is observed without apparently visible polyamide residues at the fractured aluminum surface. Mechanical tests of two reference conditions, one of them prepared by just degreasing the aluminum and the other one by boiling water treatment of the aluminum without subsequent silanization show both negligible joints strength. Hence, it is concluded that all the tested silane coupling agents sufficiently improve adhesion towards polyamide 6.

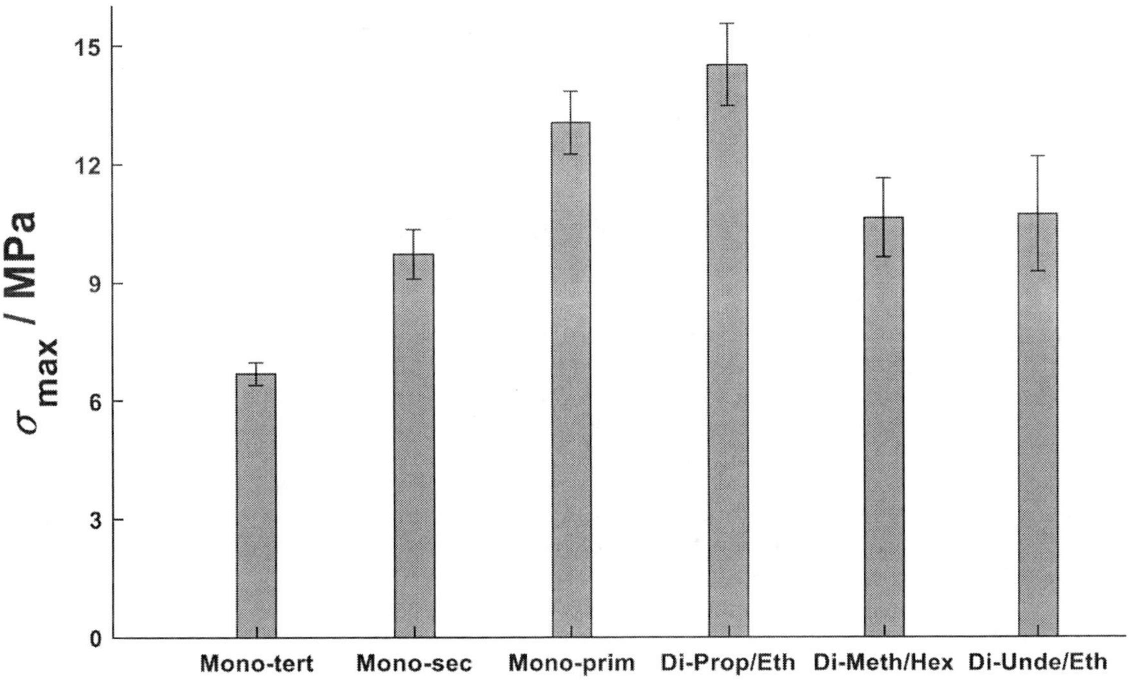

Figure 2: Maximum shear strength (σ_{max}) according to DIN EN ISO 1465 of aluminum EN AW-6082/PA 6 joints for different amino-silane coupling agents that are used for the interfacial design.

The chemical structure of the three mono amino-silanes is similar, except for the type (tertiary, secondary, primary) of amine group that is located at the chain end of the organo-functional group. Hereby, the lowest joint strength is observed for the silane bearing the tertiary amine group (Mono-tert), higher strength is achieved with the silane with secondary amine group (Mono-sec), and the silane with primary amine (Mono-prim) provides the highest strength. This can be directly related to the capability of the individual amine types to form hydrogen bonds with the amide group of polyamide 6 that is highest for primary amine and lowest for tertiary amine. Figure 3 illustrates possible options for hydrogen bonding between the amine and amide group. Note that possible hydrogen bonding between polyamide 6 and alkoxy, silanol, siloxane and aluminum hydroxide groups are not depicted. Primary and secondary amine can form all hydrogen bonding options depicted in figure 3 (a-e), but hydrogen bonds formed by secondary group with the amide group are less strong, since the partial charge of the involved hydrogen is less pronounced compared with primary amine. The tertiary amine can only form hydrogen bonding options (b, d-e) with the amide group. Due to the basicity increasing from primary to tertiary amines, the tendency to form ammonium salts is higher, giving rise hydrogens boning options (d-e) with the amine group. However, bonds arising from ammonium salts are weak compared to hydrogen bonds between amine and amide groups (a-c), as the partial charge of the hydrogen of the ammonium group is small in relation to hydrogen of amine groups.

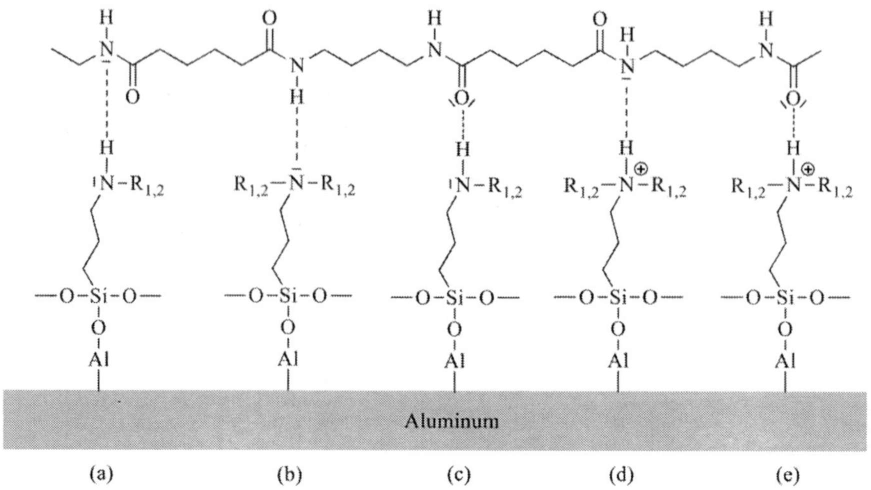

Figure 3: Simplified scheme of selected options for hydrogen bonding in between polyamide 6 and amine groups of specific type: Mono-prim (a, b, c: $R_1 = H$); Mono-sec (a, c: $R_1 = CH_3$; b: $R_1 = H$, $R_2 = CH_3$) and Mono-tert (b, $R_2 = CH_3$) and their corresponding ammonium salts (d, e).

For the three silanes that contain two amine groups (Di-Prop/Eth, Di-Meth/Hex, Di-Unde/Eth), namely a primary amine located at the end of the chain of the organo-functional group and a secondary amine positioned within the organo-functional group, significantly different joint strengths are recorded. This shows that not only the number and type of amine groups have an impact on the shear strength of the tested aluminum-polyamide specimens, but also the lengths of the alkyl spacers and the position of the amine groups within the organo-functional group have to be considered. Hereby, the diamino silane Di-Prop/Eth shows an increased strength that is the highest of all tested silanes, compared with the silane Mono-prim. This is in accordance with other studies that applied the same silanes Di-Prop/Eth and Mono-prim but for other material configurations [12, 13]. The increase in strength in between Mono-prim and Di-Prop/Eth can be explained by the additional amine within Di-Prop/Eth that can form an extra hydrogen bond with the amide group of polyamide 6.

Compared to that, the diamino silanes Di-Meth/Hex and Di-Unde/Eth, even though they bear two amine groups each, provide lower shear strength than mono-prim and Di-Prop/Eth. In comparison to Di-Prop/Eth, the silanes Di-Meth/Hex and Di-Unde/Eth contain longer alkyl spacers in total. Consequently, the interfacial functional density of amine groups is lower for Di-Meth/Hex and Di-Unde/Eth in relation to the overall space blocked by the silanes. This may explain the lower adhesion observed during lap-shear testing. This finding is in contrast to another study [12] that reported a higher interfacial strength for Di-Unde/Eth in comparison to Di-Prop/Eth and related this effect to more pronounced van-der-Waals interaction of the longer alkyl chains with polyamide 6,6 and chain entanglement between the polyamide 6,6 chains and the organo-functional group of Di-Unde/Eth.

In this context, no significant difference in joint strength can be observed in between Di-Meth-Hex and Di-Unde/Eth, even though Di-Unde/Eth contains significantly longer alkyl spacers. This might be related to the position of the secondary amine in Di-Unde/Eth that is next to the end of the organo-functional group and is thereby sterically better accessible for the polyamide 6 chain in order to form hydrogen bonds. In contrast, the secondary amine in Di-Meth/Hex is next to the central silicon atom and thus sterically hindered.

4. Conclusions

This contribution presented an experimental study on the influence of the chemical structure of amino silane coupling agents on the lap-shear strength of aluminum-polyamide 6 hybrids. Hereto, aluminum strips were coated by different amino silanes, joined to polyamide and tested by lap-shear tests. Appropriate coating procedures, including washing steps after the actual silane application are found to increase the achieved joint strengths and to enable high reproducibility. The results further show that the amine type (primary, secondary, tertiary), the number of amine groups (one, two), the length of the alkyl linkers (methyl, ethyl, propyl, hexyl, undecyl) and the position of the amine groups with respect to the organo-functional group impacts the achieved joint strength. It is concluded that high adhesion in aluminum-polyamide 6 joints is achieved if the organo-functional group contains (i) a primary amine chain-end group, (ii) multiple amine groups and (iii) short alkyl spacers. Based on this, the predominant mechanism of adhesion is concluded to be hydrogen bonding in between the amine groups and the amide groups of polyamide 6. The highest lap-shear strengths are determined when using N-(6-aminohexyl)aminomethyltriethoxysilane with about 15 MPa. This is sufficiently high in comparison to competitive methods [6], which are used for the interfacial of aluminum-polyamide 6 hybrids. Furthermore, the conducted investigations show the potential of the different silanes as constituent in sol-gel solutions, which can combine high adhesion with corrosion protection for aluminum substrates in combination with fiber-reinforced polyamide 6 composites [11].

Acknowledgements
This research was supported by the German Research Foundation (DFG) within the Priority Program 1712, project LA 1274/32-2.

References

[1] Petrie E M 2008 Adhesives for the Assembly of Aircraft Structures and Components *Metal Finishing* **106** 26-31
[2] Sinmazçelik T, Avcu E, M. Ö. Bora M Ö and Çoban O 2011 A review: Fibre metal laminates, background, bonding types and applied test methods *Materials and Design* **32** 3671-3685
[3] Zopp C, Dittes A, Nestler D, Scharf I, Kroll L and Lampke T 2019 Quasi-static and fatigue bending behavior of a continuous fiber-reinforced thermoplastic/metal laminate *Composites Part B* **174** 107043
[4] Saborowski E, Steinert P, Dittes A, Lindner T, Schubert A and Lampke T 2020 Introducing Fractal Dimension for Interlaminar Shear and Tensile Strength Assessment of Mechanically Interlocked Polymer–Metal Interfaces *Materials* **13** 2171
[5] Amend P, Pfindel S and Schmidt M 2013 Thermal Joining of Thermoplastic Metal Hybrids by Means of Mono- and Polychromatic Radiation *Physics Procedia* **41** 98–105
[6] Saborowski E, Dittes A, Steinert P, Lindner T, Scharf I, Schubert A and Lampke T 2019 Effect of metal surface topography on the interlaminar shear and tensile strength of aluminum / polyamid 6 polymer-metal-hybrids *Materials* **12** 2963
[7] Bergmann J P and Stambke M 2012 Potential of Laser-manufactured Polymer-metal hybrid Joints *Physics Procedia* **39** 84-91
[8] Thomason J L 2019 Glass fibre sizing: A review *Composites Part A* **127**
[9] Plueddemann E 1991 *Silane coupling agents. 2nd ed.* New York and London: Plenum Press
[10] Harding P H and Berg J C 1998 The Adhesion Promotion Mechanism of Organofunctional Silanes *Journal of Applied Polymer Science* **67** 1025-1033
[11] Kießling R, Ihlemann J, Riemer M, Drossel W-G, Dittes A, Scharf I, Lampke T, Sharafiev S, Pouya M and Wagner M-X 2018 A process and load adjusted coating system for metallic inserts in hybrid composites *Production engineering* **12** 249-257
[12] Yu S, Oh K H, Hwang J Y and Honga S H 2019 The effect of amino-silane coupling agents having different molecular structures on the mechanical properties of basalt fiber-reinforced polyamide 6,6 composites *Composites Part B* **163** 511-521
[13] Wiebeck H, Borrelly D F, Xavier C, Santos P S, Asciutti S A and Correa M P 1998 The effect of silane coupling agents on a composite polyamid-6/talc *Brazilian journal of chemical engineering* **15** 406-409
[14] DIN EN ISO 1110 1997 Plastics - polyamides - accelerated conditioning of test specimens., Beuth Verlag.
[15] DIN EN 1465 2009 Adhesives - determination of tensile-lap-shear strength of bonded assemblies., Beuth Verlag.

[16] Zhu M, Lerum M Z and Chen W 2012 How to prepare reproducible, homogenous, and hydrolytically stable aminosilane-derived layers on silica *Langmuir* **28** 416-423

IOP Conf. Series: Materials Science and Engineering 1147 (2021) 012016 doi:10.1088/1757-899X/1147/1/012016

Effect of local reinforcement particle distribution on mechanical properties of in-situ TiB/TiC_{1-x} particle reinforced Ti-5Al-5Mo-5V-3Cr – comparison between model and experiment

S Grützner[1,*], L Krüger[1] and I Schneider[2]

[1] Institute of Materials Engineering, TU Bergakademie Freiberg, Gustav-Zeuner-Str. 5, 09599 Freiberg, Germany
[2] Bundeswehr Research Institute for Materials, Fuels and Lubricants (WIWeB), Institutsweg 1, 85435 Erding, Germany

*e-mail: Steffen.Gruetzner@iwt.tu-freiberg.de

Abstract. In-situ TiB and TiC_{1-x} particle-reinforced titanium matrix composites (TMCs) based on a near-β Ti-5Al-5Mo-5V-3Cr alloy (Ti-5553) reacting chemically with 4 vol.-% B_4C were processed by spark plasma sintering (SPS). Different local distributions of the reinforcement particles formed within the composite were adjusted by varying the conditions during powder preparation. The measured Young's modulus, hardness and compressive yield strength of the composites increased, as the reinforcement particles were distributed more homogeneously within the matrix. In addition, Young's modulus and hardness were modelled considering the hybrid TiB-whisker and TiC-particle reinforcement applying the rule of hybrid mixtures (RoHM) and the Fu-method, respectively. The compressive yield strength was modelled in accordance with the summation of the particular effective strengthening mechanisms and the Clyne-method. Measured values of the Young's modulus and the hardness were overestimated by the modelling as the clustering of the reinforcement particles increased. The compressive yield strength of the composites is modelled appropriately using the Clyne-method. However, pronounced agglomeration of the reinforcing particles within the matrix resulted in an overestimation of the measured values.

1. Introduction

Owing to the low density of titanium, ceramic particle reinforced titanium matrix composites (TMCs) provide a combination of outstanding specific properties, such as high specific strength, specific stiffness, and hardness with heat- and/or corrosion resistance [1–4]. However, the strength of these composites is limited by the comparatively low strength of the pure titanium matrix. A considerable increase in strength of the composite would be possible, if the pure titanium matrix was replaced by a high strength titanium alloy. In this context, the high strength near-β alloy Ti-5Al-5Mo-5V-Cr (Ti-5553) appears to be a suitable matrix material. In Ti-5553, the high-temperature modification (β-phase) is stable up to ambient temperature. The high strength results from the solution strengthening effect of the alloying elements in addition with a specific heat treatment including solution treatment and aging (STA) leading to a β-microstructure with small dispersed α-precipitations [5–10].

According to the literature it is well established that most effective strengthening in particle reinforced TMCs is achieved by titanium carbide particles TiC_p and whisker-shaped titanium boride TiB_w

[11–18]. By means of powder metallurgy processes it is possible to create such ceramic reinforcement particles *in-situ* within the matrix during the TMC-synthesis. The advantage of this in-situ method is that the reinforcement's surfaces are free of contaminations like oxygen or nitrogen. Consequently, a stronger bonding of the particle/matrix interfaces is facilitated [11]. Titanium exhibits high chemical reactivities with boron and carbon. Hence, B_4C-powder can be used as reactive powder for the in-situ formation process of the TiC- and TiB-reinforcing particles together with titanium [16]. The solid-state reaction behavior is based on a variety of complex diffusion-controlled processes between the titanium- and the B_4C-particles occurring at elevated temperatures up to 700–1200 °C (sometimes even higher). Detailed information concerning the thermodynamics and the kinetics of the solid-state reaction between pure titanium and B_4C are given in references [19–24]. Among others, the spark plasma sintering is a common fabrication technique for such TMCs. Unlike described in the literature, own studies have demonstrated the necessity of considering the real stoichiometrically composition of the TiC_{1-x} particles formed during the in-situ reaction with B_4C, see reaction equation (1) below [25].

$$(x+6)Ti + B_4C \rightarrow xTi + 4TiB + 2TiC_{0.5} \quad (1)$$

Furthermore, in the case of Ti-5553 the solid-state reaction behavior is strongly influenced by the matrix alloying elements, resulting in a more sluggish reaction kinetics compared to the reaction with pure titanium [26]. In addition, it could be shown that the largest diameter of completely transformable B_4C clusters was limited to ~ 10 µm, if the sintering process was conducted at 1200 °C with a dwell time of 15 minutes. Consequently, the powder preparation conditions strongly influence the final degree of solid-state reaction between titanium/Ti-5553 and B_4C by controlling the formation or prevention of undesirably large B_4C clusters in the initial powder mixtures [25].

Beside the degree of B_4C conversion, the powder preparation conditions significantly affect the local distribution of the in-situ formed reinforcement particles within the TMCs as well. Low energy powder milling in conjunction with the sluggish reaction kinetics cause the in-situ formation of reinforcement particles predominantly in the vicinity of the former matrix particle surfaces. This particle clustering corresponds to an inhomogeneous local distribution of the reinforcement particles [25, 26].

Generally, the simultaneous use of two different ceramic phases (TiC_p and TiB_w) within the metal matrix is designated as *hybrid particle reinforcement*. However, the mechanical properties of such composites, e.g. hybrid Young's modulus, hybrid hardness and hybrid strength can hardly be predicted by material models typically including only one single reinforcement phase, its phase fraction and sometimes its morphology. Relevant models in this context concerning particle reinforced composites are the Hashin-Shtrikman-model [27, 28] or the Halpin-Tsai-model [11] (Young's modulus), the Rice-model [29] or the Halpin-Tsai-model [30] (hardness) and the summation of the contributions of particular strengthening mechanism [31–35] and the Clyne-method [31, 36, 37] (strength). Fu et al. introduce two approaches, which enable the prediction of the Young's modulus for hybrid particle/short-fiber reinforced polymer matrix composites: The *Rule of Hybrid Mixtures* (RoHM) and a modified laminate analogy approach – hereinafter referred to as the *Fu-method* [38]. The *RoHM* describes the hybrid Young's modulus of the composite E_c resulting from the two separate subsystems particle/matrix and short-fiber/matrix, see Figure 1. E_c is calculated according to the RoHM in the following way

$$E_c = E_{c1}V_{c1} + E_{c2}V_{c2} \quad (2)$$

E_{c1}, E_{c2} and V_{c1}, V_{c2} are the Young's moduli and the relative hybrid volume fractions of the systems particle/matrix and short-fiber/matrix, respectively. E_{c1} and E_{c2} have to be calculated applying suitable models for the particular subsystems, each including only one single reinforcement phase. Note that the total reinforcement volume fraction $V_t = V_p + V_f$ (V_p and V_f are volume fractions of the particles and short-fibers, respectively) should be used as reinforcement volume fraction for calculation of E_{c1} and

E_{c2}. Finally, the relative hybrid volume fractions in equation (2) are defined as $V_{c1} = V_p/V_t$ and $V_{c2} = V_f/V_t$.

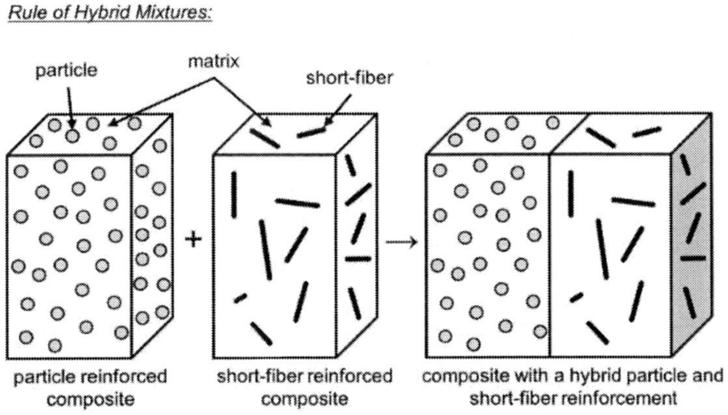

Figure 1. Application of the Rule of Hybrid Mixtures (RoHM) to a particle and short-fiber reinforced composite (schematically) according to Fu et al. [38].

The *Fu-method* is a two-step procedure. In a first step, the particle reinforced matrix is defined as a new phase – referred to as the *effective matrix*. Within the model, this effective matrix is reinforced by the short-fibers in a second step, see Figure 2. Both steps allow the application of material models for composites with a single reinforcement phase. In contrast to the RoHM, the calculation of the Young's modulus of the effective matrix $E_{m,eff}$ in the system particle/matrix is based on the relative volume fractions of the matrix $V_{m,rel}$ and the reinforcing particles $V_{p,rel}$. These are defined as $V_{m,rel} = V_m/(V_m + V_p)$ and $V_{p,rel} = V_p/(V_m + V_p)$, where V_m and V_p represent the volume fractions of the matrix and the reinforcing particles, respectively.

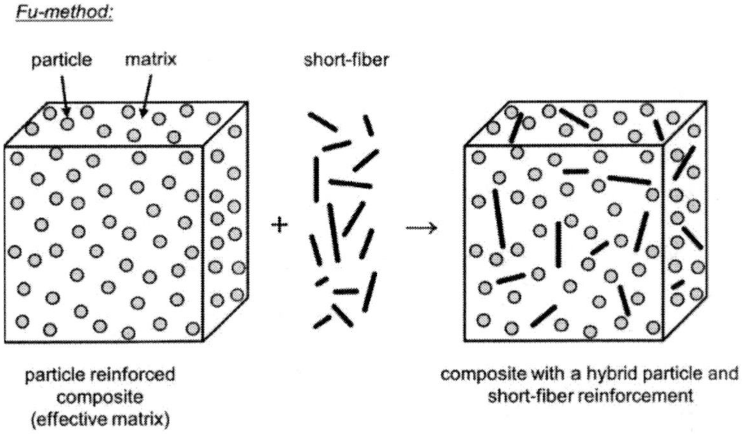

Figure 2. Application of the Fu-method to a particle and short-fiber reinforced composite (schematically) according to Fu et al. [38].

The present study analyses the influence of the local reinforcement particle distribution within hybrid in-situ TiB/TiC particle-reinforced Ti-5Al-5Mo-5V-3Cr matrix composites on selected mechanical properties. Especially, with respect to improve the predictability of these properties, first investigations focused on the comparison between the measured and the modelled values for the composite's hybrid Young's modulus, hardness and compressive yield strength. Suitable model approaches, which consider the hybrid particle reinforcement were applied to the TMC property calculations. Deviations between experiment and model were discussed including the role of the local distribution of the reinforcing particles within the TMCs.

2. Experimental and materials

In the present study, a Ti-5Al-5Mo-5V-3Cr powder (Ti-5553, TLS Technik GmbH & Co Spezialpulver KG, Germany) exhibiting a spherical particle morphology served as the matrix material. B_4C (abcr GmbH, Germany) with a significantly lower particle size was used as a reactive powder for the in-situ reinforcement of the matrix. The chemical compositions and particle size distributions of the initial powders are given in Table 1 and Table 2.

Table 1. Measured chemical composition of the initial powders.

Ti-5553								
Composition	Ti	Al	Mo	V	Cr	Fe	O	N
(wt.-%)	bal.	5.22	5.05	5.10	2.83	0.31	0.13	< 0.01
B_4C								
Composition	B	C	N	O	Si	Al	Fe	
(wt.-%)	bal.	21.90	0.20	1.40	0.11	0.01	< 0.01	

Table 2. Particle size distribution of the initial powders measured by laser diffraction particle size analysis.

Values (μm)	d_{10}	d_{50}	d_{90}
Ti-5553	22.59	33.36	47.83
B_4C	0.51	1.80	1.96

Powder mixtures of Ti-5553 and 4 vol.-% B_4C were prepared under high energy milling conditions to break up the large B_4C clusters existing in the initial B_4C-powder, see Table 3. The powders were either milled in a planetary ball mill (Pulverisette 6, Fritsch GmbH, Germany) at ambient conditions or by vibration milling in a cryogenic mill (CryoMill, Retsch GmbH, Germany), while the grinding bowl was permanently cooled from the outside by gaseous nitrogen.

Table 3. Preparation conditions of the initial powder mixtures and TMC designation of the consolidated samples.

Method	Ball material	Ball diameter	BPR	Milling time	Rotating Speed/ Frequency	TMC Designation
Planetary ball milling	Cr-steel (hardened)	10 mm	5:1	4 h	250 rpm	TMC-A
Cryogenic milling	Cr-steel (hardened)	10 mm	5:1	0.5 h	30 Hz	TMC-B

The powder mixtures were consolidated by SPS in an FCT-HP D 25/2-2 apparatus (FCT Systeme GmbH, Germany). Cylindrical samples with a diameter of 20 mm and a final height of 6 mm were sintered under vacuum atmosphere. The powders were heated up to 1200 °C at a heating rate of 100 K/min and kept at this temperature for 15 minutes. Parallel to the heating, the pressure was increased from 16 MPa at 450 °C to 51 MPa at 1200 °C and maintained constant during the dwell time. Parallel to the uncontrolled cooling, the pressure was reduced to 16 MPa within 15 minutes.

The density of the samples was measured using the Archimedes method in distilled water. For the microstructural investigations, samples were mechanically ground and polished. This was followed by chemical-mechanical polishing and subsequent etching with Kroll's reagent for 20–40 seconds. Matrix grain size determination and size measurements of the reinforcement particles required etching with Kroll's reagent for a shorter time and subsequently thermal etching according to the procedure proposed by Vander Voort [39]. Phase analysis was performed by X-ray diffraction as described in an earlier study [25].

The Young's modulus was determined by measurement of the longitudinal and transversal wave velocities, employing the acoustical logging method on the faceground compact samples using a Hitachi V-1565 Oscilloscope (Hitachi, Japan) in combination with a Panametrics Model 5800 computer-controlled pulser/receiver (Panametrics GmbH, Germany). Due to the narrow sample diameter, the running periods of the acoustic sound waves were measured three times with an accuracy of at least 0.01 µs at the same position at each sample. A micrometer screw was used to determine the sample thickness between the faceground surfaces with an accuracy of 0.001 mm at three different sample positions. Finally, the wave velocities for Young's modulus calculation were derived from the mean values of the measured sound wave running periods and the mean sample thickness. The Vickers hardness was determined from at least five indentations introduced with an indentation load of 294.2 N. For the compression tests, cylindrical samples with equal diameters and heights of 4 mm were cut from the sintered bodies by means of electrical discharge machining. Quasi-static compression tests at a strain rate of 10^{-3}/s were carried out at room temperature using an MTS 810 servo-hydraulic testing machine (MTS, USA). The compressive strain was calculated from the actual compression of the specimen, measured by a displacement gage. At least three compression tests were carried out for each composite to improve the statistical validation of the measured values.

3. Results and discussion

Figure 3 shows a comparison of the initial powder mixtures and the microstructures of the sintered samples for each milling condition, respectively. Especially, during the high energy ball milling the matrix particles undergo a strong plastic deformation, which is associated by pressing the smaller B_4C-particles into the matrix particle surfaces. Moreover this, coarse Ti-5553/B_4C particle agglomerates were formed through an increasing cold welding of the deformed matrix particles. In contrast, less plastic deformation of the matrix particles is observed during the cryogenic milling. Nevertheless, the high energy impact successfully avoids the formation of coarse B_4C particle agglomerates. However, the individual B_4C particles rather adhere to the matrix particle surfaces than they were pressed into them.

The different morphologies of the initial powder mixtures strongly effect the local distribution of the reinforcing particles within the bulk TMCs. Obviously, the reinforcement particles in TMC-B were predominantly located in the vicinity of the former matrix particle surfaces, while less reinforcements were formed inside the matrix particle volume. This appearance is in agreement with the findings for a comparable powder mixture based on Ti-5553 and B_4C presented in an earlier study [26]. The microstructure of the TMC-A sample, shows a more homogeneously distribution of the reinforcement particles. This can be ascribed to the repeated deformation, fracture and cold welding of the matrix particles during powder preparation resulting in the formation of agglomerates with embedded B_4C particles.

A comparison of the mechanical properties between TMC-A and TMC-B is limited to the requirement of approximately equal phase contents with respect to the in-situ formed TiB- and $TiC_{0.5}$-particles within the TMC samples. In other words, the completely solid-state reaction of the B_4C particles during sintering has to be ensured, thus avoiding superimposed effects of different particle contents. Both composites exhibit almost the same density, see Table 4. This and the absence of unreacted B_4C clusters within the microstructure prove high degrees of B_4C conversion in the TMCs. The measured TiB- and $TiC_{0.5}$ phase fractions only vary in the precision of the XRD-method resulting in almost identical total reinforcement particle volume fractions, designated as $\Sigma(TiB + TiC_{0.5})$.

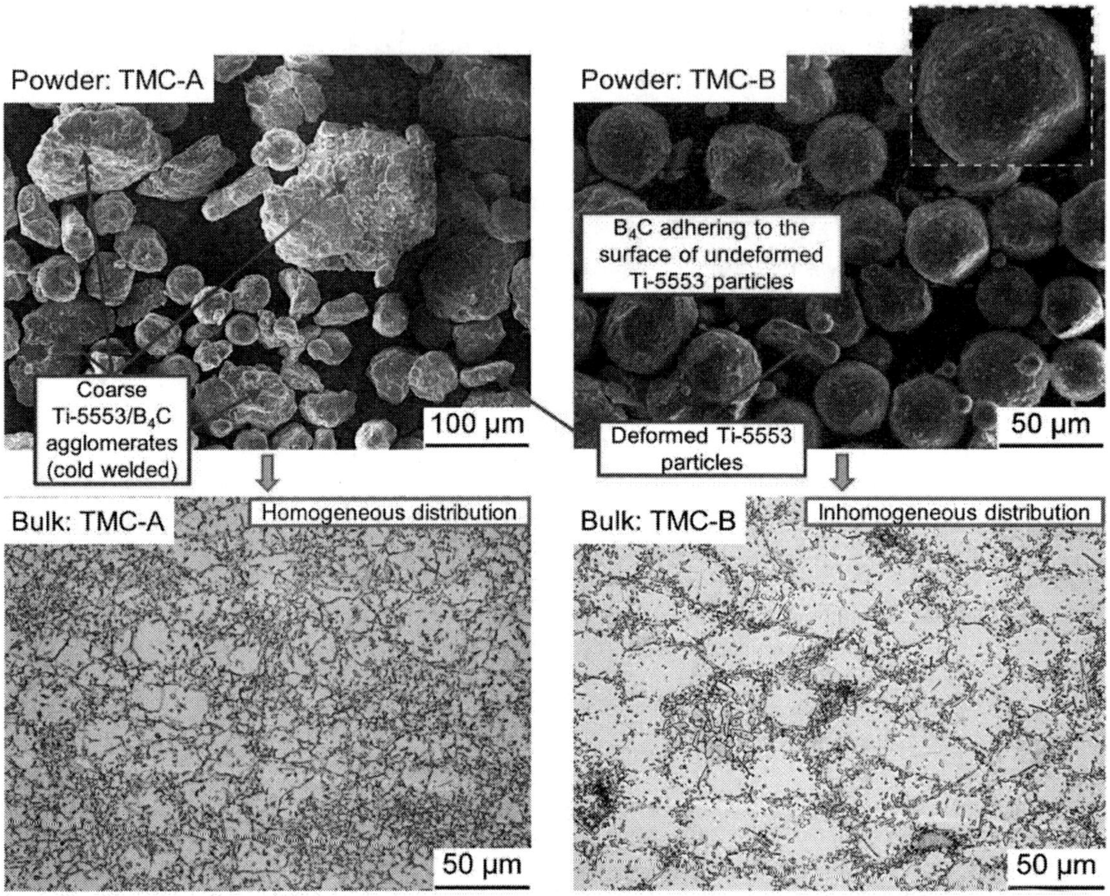

Figure 3. TMC-A and TMC-B – morphology of the initial powder mixtures (SEM-SE) and optical micrographs of the sintered composites showing different local distributions of the reinforcement particles.

Table 4. Density and phase composition of the TMCs.

TMC	Density (g/cm^3)	Volume phase fractions (vol.-%) Ti-bcc	Ti-hcp	TiB	$TiC_{0.5}$	$\Sigma(TiB + TiC_{0.5})$
TMC-A	4.66	74	9	11	6	17
TMC-B	4.65	84	–	9	7	16

The measured phase fraction of α-Ti (Ti-hcp) in TMC-A, may have resulted from multiple effects connected to the powder milling and sintering processes. Due to the high chemical affinity of titanium, oxygen and nitrogen, the repeated deformation and fracture of the matrix particles during the powder preparation in air atmosphere caused their contamination through interstitial solid solutioning, see Table

5. Oxygen and nitrogen have a strong α-phase stabilizing effect on titanium. The influence of carbon on the phase stability in titanium is similar to that of oxygen and nitrogen. However, solid solution of carbon within the matrix particles occurs first during the sintering process as a consequence of the B_4C-decomposition. Finally, an earlier study proved that the in-situ reaction with B_4C effected a significant change in the chemical composition within the initial Ti-5553 particles, in particular leading to an increase of the aluminum concentration [25]. Next to the interstitials mentioned before, aluminum acts as a strong α-stabilizing element in titanium alloys as well. However, the matrix in TMC-B consisted of β-phase, only. Consequently, the β-phase destabilizing effects associated to the sintering process seem to have a much lower efficiency, than the powder contamination with oxygen and nitrogen.

Table 5. Oxygen- and nitrogen-concentrations within the Ti-5553 powder and the initial powder mixtures measured by hot gas extraction.

	Concentrations (wt.-%)	
	oxygen	nitrogen
Powder: Ti-5553	0.130	< 0.010
Powder: TMC-A	0.503	0.250
Powder. TMC-B	0.320	0.024

The measured microstructural parameters of the TMCs are given in Table 6. Identically sintered samples based on the unreinforced Ti-5553 powder reveal a β-phase grain size d_0 of 229 μm. Hence, the β-phase grain size d_1 within both composites was decreased in order of two magnitudes. The TiB-whiskers were characterized by their length l, diameter d and aspect ratio S. Characteristically microstructure parameters of the plate-shaped $TiC_{0.5}$ particles are represented by their width L_2, thickness t and aspect ratio $A = 2L_1/t$ including the particle length $2L_1$ according to the definition of Nardone and Prewo [40].

Table 6. Measured microstructural parameters of the TMCs.

	β-matrix		TiB			$TiC_{0.5}$	
	d_1	l	d	S	L_2	t	A
TMC	(μm)	(μm)	(μm)	–	(μm)	(μm)	–
TMC-A	4.8	15.4	1.3	13	2.4	1.6	1.7
TMC-B	7.2	13.1	1.4	10	2.5	1.6	1.6

The TMC's mechanical properties Young's modulus, hardness und compressive yield strength increase with a more homogeneously distribution of the reinforcement particles, see Table 7. A Comparison of the compressive behavior is given by the compressive stress-strain curves in Figure 4.

Table 7. Measured hybrid Young's modulus E_c, hybrid hardness H_c and hybrid compressive yield strength at 0.2 % plastic deformation σ_c of the TMCs.

	E_c	H_c	σ_c
TMC	(GPa)	(HV 30)	(MPa)
TMC-A	125	527 ± 13	1546 ± 14
TMC-B	110	377 ± 8	1245 ± 19

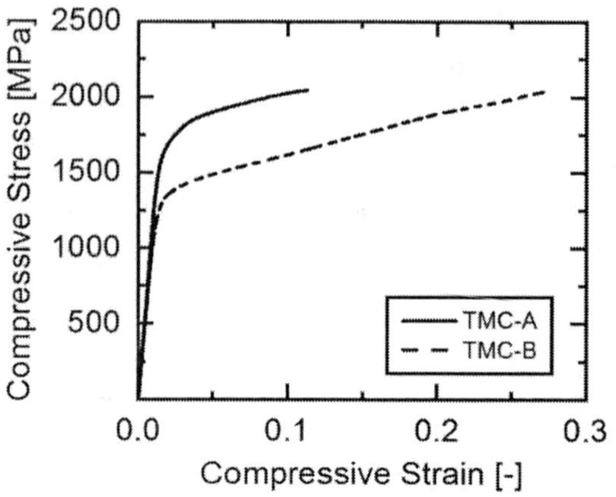

Figure 4. Typical compressive stress–strain curves of TMCs.

Modelling of TMC's mechanical properties was based on measured phase compositions and microstructure parameters. The partial destabilization of the β-phase in the matrix of TMC-A was considered for the matrix property calculations by applying a linear rule of mixture using the relative volume fractions of the α- and β-phase, respectively. This also includes the Hall-Petch strengthening effect due to the reduced β-grain size within both composites. Mechanical properties of the individual phases were taken from the literature (if available) or had to be determined in additional experiments. Input variables for the hybrid Young's moduli and the hybrid hardness modelling are summarized in Table 8.

Table 8. Input variables for modelling of the hybrid Young's modulus and hybrid hardness.

Phase	E (GPa)	Hardness (HV 30)	References
Ti-bcc (Ti-5553)	86	$\mathrm{HV} = 278HV + \frac{0.22HV\sqrt{m}}{\sqrt{d}}$	measured
Ti-hcp (Ti-5553)	110	390	[10, 41]
TiB	450	1800	[42, 43]
$TiC_{0.5}$	265	2000	[43, 44]

The Young's moduli and the hardness of the subsystems TiC_p/matrix and TiB_w/matrix (RoHM) as well as the effective matrix and their reinforcement by TiB_w (Fu-method) were calculated according to the Halpin-Tsai equations [11, 30, 45], thus considering the microstructural morphology of the individual reinforcement phases. Another fact that could not be ignored was the misorientation between TiB-whiskers and the load direction. The Halpin-Tsai equations are applicable just for short-fibres and whiskers, that are aligned parallel to the load line. Ryu et al. found, that the whisker orientation influence can be expressed in form of an effective aspect ratio $S_{eff,F(\theta)}$ considering the probability density function of misaligned whiskers $F(\theta)$, where θ is the misorientation angle [46]. $S_{eff,F(\theta)}$ is calculated as below:

$$S_{eff,F(\theta)} = \int_{0^\circ}^{90^\circ} \left[S\cos^2(\theta) + \left(\frac{3\pi-4}{3\pi}\right)\left(1+\frac{1}{S}\right)\cos^2(\theta) \right] F(\theta)d\theta \quad (3)$$

IOP Conf. Series: Materials Science and Engineering 1147 (2021) 012016 doi:10.1088/1757-899X/1147/1/012016

For randomly aligned whiskers – like in the present TMCs – the probability density function is reduced to $F(\theta) = 2/\pi$ [47]. A comparison between measured and modelled values is given in Table 9. Measured values of E_c and H_c were overestimated by the RoHM and the Fu-method, respectively. However, this overestimation increases significantly, as the local distribution of the reinforcement particles within the matrix becomes more inhomogeneous.

Table 9. Comparison of the measured and the modelled hybrid Young's moduli E_c and hybrid hardness H_c (differences between model and experiment expressed as $\Delta_{Mod-Exp}$).

	Hybrid Young's modulus E_c (GPa)			Hybrid hardness H_c (HV 30)		
	Experiment	RoHM	Fu-method	Experiment	RoHM	Fu-method
TMC		($\Delta_{Mod-Exp}$)	($\Delta_{Mod-Exp}$)		($\Delta_{Mod-Exp}$)	($\Delta_{Mod-Exp}$)
TMC-A	125	128 (+3)	129 (+4)	527 ± 13	561 (+34)	562 (+35)
TMC-B	110	119 (+9)	121 (+11)	377 ± 8	520 (+143)	522 (+145)

A great plenty of effective strengthening mechanism could be identified in the present composites. This includes the grain boundary strengthening (Hall-Petch-effect) of the β-matrix grains $\Delta\sigma_{H-P}$, the solid solution strengthening by interstitials (oxygen, nitrogen and carbon) $\Delta\sigma_{ss}$, the load-transfer strengthening $\Delta\sigma_{L-T}$ and the Orowan strengthening $\Delta\sigma_{OR}$ of both, carbides and whiskers. In TMC-A the β→α phase transformation resulted in an increased matrix strength σ_{m0} by the amount of $\Delta\sigma_{Phase}$. A detailed description of the single calculation steps in determining the strengthening contributions is given by Grützner in [48]. The modelled values are listed in Table 10.

Table 10. Modelled strengthening contributions identified in the TMCs.

	$\Delta\sigma_{Phase}$	σ_{m0}	$\Delta\sigma_{H-P}$	$\Delta\sigma_{ss}$	$\Delta\sigma_{L-T(TiB)}$	$\Delta\sigma_{L-T(TiC)}$	$\Delta\sigma_{OR(TiB)}$	$\Delta\sigma_{OR(TiC)}$
TMC					(MPa)			
TMC-A	45	965	451	178	358	28	14	12
TMC-B	0	920	400	114	237	28	12	14

As mentioned in the introduction, two distinct approaches are suggested in the literature, which enable the strength modelling of particle reinforced composites. One is the summation-approach, where the single contributions of particular strengthening mechanism are added to σ_{m0}. According to the Clyne-method the final strength of a composite is calculated by summing the root of squares of all the individual strengthening contributions, as below

$$\sigma_c = \sigma_{m0} + \sqrt{\sum_i \Delta\sigma_i^2} = \sigma_{m0} + \sqrt{\Delta\sigma_{H-P}^2 + \Delta\sigma_{L-T}^2 + \Delta\sigma_{ss}^2 + \Delta\sigma_{OR}^2 + \cdots} \quad (4)$$

Table 11 shows a comparison of the modelled hybrid yield strengths for the composites with the measured values. Yield strengths were significantly overestimated by the summation approach. Hence, this approach is unsuitable for the current composites, which is associated with the superposition of the individual strengthening effects [31, 37]. These superposition effects are considered by the Clyne approach. However, precision decreases again, if the reinforcing particles were distributed more inhomogeneously within the matrix, see TMC-B in Table 11.

Table 11. Comparison of the measured and the modelled hybrid compressive yield strength at 0.2 % plastic deformation σ_c of the TMCs (differences between model and experiment expressed as $\Delta_{Mod-Exp}$).

TMC	Experiment	Hybrid yield strength σ_c (MPa) Summation approach ($\Delta_{Mod-Exp}$)	Clyne approach ($\Delta_{Mod-Exp}$)
TMC-A	1546 ± 14	2033 (487)	1585 (39)
TMC-B	1245 ± 19	1725 (480)	1400 (155)

4. Conclusions

Mechanical properties of hybrid TiB/TiC particle-reinforced Ti-5Al-5Mo-5V-3Cr matrix composites were strongly influenced by the local reinforcement particle distribution. Despite almost identical total reinforcement particle volume fractions, the composite exhibiting a more homogeneously reinforcement particle distribution within the matrix reveals higher measured values of Young's modulus, hardness and compressive yield strength. Furthermore, these mechanical properties were modelled taking into account the phase composition as well as microstructural parameters and the chemical composition of individual phases. The rule of hybrid mixtures (RoHM) and the Fu-method – both considering the hybrid TiB-whisker and TiC-particle reinforcement – were used for modelling of the Young's modulus and the hardness, respectively. Concerning the compressive yield strength of the composite, it was necessary to identify the particular effective strengthening mechanisms. The strengthening contributions of these mechanism represent input parameters for the compressive yield strength modelling in accordance with the summation approach and the Clyne-approach. With pronounced agglomeration of the reinforcement particles the measured values of the hybrid Young's modulus and the hybrid hardness were overestimated by the RoHM and the Fu-method. The hybrid compressive yield strength of the composites was modelled more appropriately using the Clyne-approach. However, with increasing clustering of the reinforcement particles the measured values of the compressive yield strength were overestimated by the Clyne-approach as well.

Acknowledgments

The authors would like to thank the Bundeswehr Research Institute for Materials, Fuels and Lubricants (WIWeB) for its financial support of this project.

References

[1] Gorsse S and Miracle D B 2003 Mechanical properties of Ti-6Al-4V/TiB composites with randomly oriented and aligned TiB reinforcements *Acta Mater.* **51** 2427–42
[2] Yang Z F, Lu W J, Xu D, Qin J N and Zhang D 2006 In situ synthesis of hybrid and multiple-dimensioned titanium matrix composites *J. Alloys Compd.* **419** 76–80
[3] Yanbin L, Yong L, Huiping T, Bin W and Bin L 2011 Fabrication and mechanical properties of in situ TiC/Ti metal matrix composites *J. Alloys Compd.* **509** 3592–601
[4] Huang L J, Geng L, Xu H Y and Peng H X 2011 In situ TiC particles reinforced Ti6Al4V matrix composite with a network reinforcement architecture *Mat. Sci. Eng. A* **528** 2859–62
[5] Nag S, Banerjee R, Srinivasan R, Hwang J Y, Harper M and Fraser H L 2009 ω-Assisted nucleation and growth of α precipitates in the Ti–5Al–5Mo–5V–3Cr–0.5Fe β titanium alloy *Acta Mater.* **57** 2136–47
[6] Veeck S, Lee D, Boyer R and Briggs R 2004 The castability of Ti-5553 alloy *Adv. Mater. Process.* **162** 47–9

[7] Cotton J D, Hellenbrand P J, Bryan D J, Bayha T D, Leder M and Levin I 2014 The Effect of Hydrogen on the Fracture Toughness of Ti-5Mo-5V-5Al-3Cr *TMS 2014 143rd Annual Meeting & Exhibition, Annual Meeting Supplemental Proceedings* ed The Minerals, Metals & Materials Society 1st edn (Springer International Publishing)

[8] Fanning J C 2005 Properties of TIMETAL 555 (Ti-5Al-5Mo-5V-3Cr-0.6Fe) *J. Mater. Eng. Perform.* **14** 788–91

[9] Nyakana S L, Fanning J C and Boyer R R 2005 Quick reference guide for β titanium alloys in the 00s *J. Mater. Eng. Perform.* **14** 799–811

[10] Wagner V, Baili M, Dessein G and Lallement D 2010 Experimental characterization of behavior laws for titanium alloys: application to Ti5553 *Key Eng. Mater.* **446** 147–55

[11] Tjong S C and Mai Y W 2008 Processing-structure-property aspects of particulate- and whisker-reinforced titanium matrix composites *Compos. Sci. Technol.* **68** 583–601

[12] Li G, Li D, Liu Y, Wang Q and Guan S Microstructure and Mechanical Properties of Ti-35V-15Cr-0.05C Nonburning Titanium Alloy *J. Mater. Sci. Technol.* **14** 411–4

[13] Ma Z Y, Tjong S C and Gen L 2000 In-situ Ti-TiB metal-matrix composite prepared by a reactive pressing process *Scripta Mater.* **42** 367–73

[14] Soboyejo W O, Lederich R J and Sastry SML 1994 Mechanical behavior of damage tolerant TiB whisker-reinforced in situ titanium matrix composites *Acta Metall. Mater.* **42** 2579–91

[15] Ni D R, Geng L, Zhang J and Zheng Z Z 2006 Effect of B_4C particle size on microstructure of in situ titanium matrix composites prepared by reactive processing of Ti-B_4C system *Scripta Mater.* **55** 429–32

[16] Feng H, Zhou Y, Jia D, Meng Q and Rao J 2006 Growth mechanism of in situ TiB whiskers in spark plasma sintered TiB/Ti metal matrix composites *Cryst. Growth Des.* **6** 1626–30

[17] Carman A, Zhang L C, Ivasishin O M, Savvakin D G, Matviychuk M V and Pereloma E V 2011 Role of alloying elements in microstructure evolution and alloying elements behaviour during sintering of a near-β titanium alloy *Mat. Sci. Eng. A* **528** 1686–93

[18] Vadayar K S, Rani S D and Prasad V V 2013 Microstructural and mechanical characteristics of in-situ titanium metal matrix composites *Int. J. Theor. Appl. Res. Mech. Eng.* **2** 12–6

[19] Zhang X, Lü W, Zhang D, Wu R, Bian Y and Fang P 1999 In situ technique for synthesizing (TiB+TiC)/Ti composites *Scripta Mater.* **41** 39–46

[20] Brodkin D, Kalidindi S R, Barsoum M W and Zavaliangos A 1996 Microstructural evolution during transient plastic phase processing of titanium carbide-titanium boride composites *J. Am. Ceram. Soc.* **79** 1945–52

[21] Mogilevsky P, Gutmanas E Y, Gotman I and Telle R 1995 Reactive formation of coatings at boron carbide interface with Ti and Cr powders *J. Eur. Ceram. Soc.* **15** 527–35

[22] Zhao H and Cheng Y-B 1999 Formation of TiB_2-TiC composites by reactive sintering *Ceram. Int.* **25** 353–8

[23] Vallauri D, Atías Adrián I C and Chrysanthou A 2008 TiC-TiB_2 composites: A review of phase relationships, processing and properties *J. Eur. Ceram. Soc.* **28** 1697–713

[24] Jia L, Wang X, Chen B, Imai H, Li S, Lu Z and Kondoh K 2016 Microstructural evolution and competitive reaction behavior of Ti-B_4C system under solid-state sintering *J. Alloys Compd.* **687** 1004–11

[25] Grützner S, Krüger L, Schimpf C, Radajewski M and Schneider I 2018 Microstructure and mechanical properties of in-situ TiB/TiC particle-reinforced Ti-5Al-5Mo-5V-3Cr composites synthesized by spark plasma sintering *Metall. Mater. Trans. A* **49** 5671-5682

[26] Grützner S, Krüger L, Radajewski M and Schneider I 2018 Characterization of in-situ TiB/TiC particle-reinforced Ti-5Al-5Mo-5V-3Cr matrix composites synthesized by solid-state reaction with B_4C and graphite through SPS *Metals* **8** 377

[27] Zhang Q, Wu G, Chen G, Jiang L and Luan B 2003 The thermal expansion and mechanical properties of high reinforcement content SiCp/Al composites fabricated by squeeze casting technology *Compos. Part A Appl. Sci. Manuf.* **34** 1023–7

[28] Geiger A L and Jackson M 1989 Low expansion MMCs boost avionics *Adv. Mater. Process.* **7** 23–30
[29] Rice R W, McMillan P W and Stryjak A J 1979 Internal stress dependence of the hardness of crystallized glasses *J. Mater. Sci.* **14** 2768–72
[30] Goyal R K, Tiwari A N and Negi Y S 2008 Microhardness of PEEK/ceramic micro- and nano-composites: Correlation with Halpin-Tsai model *Mat. Sci. Eng. A* **491** 230–6
[31] Sanaty-Zadeh A 2012 Comparison between current models for the strength of particulate-reinforced metal matrix nanocomposites with emphasis on consideration of Hall–Petch effect *Mat. Sci. Eng. A* **531** 112–8
[32] Jia L, Chen B, Li S, Imai H, Takahashi M and Kondoh K 2014 Stability of strengthening effect of in situ formed TiCp and TiBw on the elevated temperature strength of (TiCp+TiBw)/Ti composites *J. Alloys Compd.* **614** 29–34
[33] Callister W D and Rethwisch D G 2014 *Materials science and engineering: An introduction* 9th edn (Hoboken, NJ: Wiley)
[34] Ozerov M, Klimova M, Stepanov N and Zherebtsov S 2018 Microstructure evolution of a Ti/TiB metal-matrix composite during high temperature deformation *Mater. Phys. Mech.* **38** 54–63
[35] Ozerov M, Klimova M, Sokolovsky V, Stepanov N, Popov A, Boldin M and Zherebtsov S 2019 Evolution of microstructure and mechanical properties of Ti/TiB metal-matrix composite during isothermal multiaxial forging *J. Alloys Compd.* **770** 840–8
[36] Hull D and Clyne T W 1996 *An Introduction to composite materials* (Cambridge: Cambridge University Press)
[37] Casati R and Vedani M 2014 Metal matrix composites reinforced by nano-particles—A review *Metals* **4** 65–83
[38] Fu S Y, Xu G and Mai Y W 2002 On the elastic modulus of hybrid particle/short-fiber/polymer composites *Compos. Part B Eng.* **33** 291–9
[39] Vander Voort G 2015 *Metallographic preparation of titanium and its alloys (Tech Notes* 3*)*
[40] Nardone V C and Prewo K M 1986 On the strength of discontinuous silicon carbide reinforced aluminum composites *Scr. Metall.* **20** 43–8
[41] Opini V C, Salvador C A, Campo K N, Lopes E S, Chaves R R and Caram R 2016 α phase precipitation and mechanical properties of Nb-modified Ti-5553 alloy *Mat. Sci. Eng. A* **670** 112–21
[42] Madtha S, Lee C and Ravi Chandran K S 2008 Physical and mechanical properties of nanostructured titanium boride (TiB) ceramic *J. Am. Ceram. Soc.* **91** 1319–21
[43] Ravi Chandran K S and Panda K B 2002 Titanium composites with TiB whiskers *Adv. Mater. Process.* **160** 59–62
[44] Samsonov G V, Naumenko V Y and Okhremchuk L N 1971 Herstellung und Eigenschaften von Karbiden der Übergangsmetalle in ihren Homogenitätsbereichen *Phys. Stat. Sol. (a)* **6** 201–11
[45] Leblanc J L 2010 *Filled polymers – Science and industrial applications* (Boca Raton: CRC Press)
[46] Ryu H J, Cha S I and Hong S H 2003 Generalized shear-lag model for load transfer in SiC/Al metal-matrix composites *J. Mater. Res.* **18** 2851–8
[47] Fukuda H and Chou T W 1982 A probabilistic theory of the strength of short-fibre composites with variable fibre length and orientation *J. Mater. Sci.* **17** 1003–11
[48] Grützner S 2020 *Festphasenreaktionsverhalten bei der SPS-Synthese von in-situ TiB/TiC-partikelverstärktem Ti-5Al-5Mo-5V-3Cr (Freiberger Forschungshefte. B, Werkstofftechnologie* vol 388*)* 1st edn (Freiberg: Technische Universität Bergakademie Freiberg)

IOP Conf. Series: Materials Science and Engineering 1147 (2021) 012017 doi:10.1088/1757-899X/1147/1/012017

The influence of direct powder rolling parameters on the properties of aluminium strip

K Kunene[1], C P Bemont*[1], L A Cornish[2], A Dittes[3], T Grund[3], P Sinclair[1] and T Lampke[3]

[1] University of KwaZulu-Natal, Discipline of Mechanical Engineering, Durban, 4001, South Africa
[2] DSI-NRF Centre of Excellence in Strong Materials; School of Chemical and Metallurgical Engineering, University of the Witwatersrand, Johannesburg, South Africa
[3] Chemnitz University of Technology, Materials and Surface Engineering Group, 09125 Chemnitz, Germany

*e-mail: bemontc@ukzn.ac.za

Abstract. Direct powder rolling (DPR) can be used to create metallic sheet at a reduced cost compared to traditional methods. The process is particularly relevant for metals that are beneficiated to a powder form. The DPR process, however, is not yet optimised and the material properties corresponding to DPR sheet do not compare well with sheet produced from rolled metallic billet. This study explores the effect of DPR input parameters on DPR aluminium sheet material properties. Roll speed, roll gap, sintering time and sintering temperature are shown to have an impact on the density, microstructure, hardness and ductility of DPR aluminium strip. It is shown that a low roll speed and a small roll gap produced the most dense strip. A roll gap of 0.7 mm and roll speed of 0.366 rpm produced green strip with a density of 90% of cast aluminium while a roll gap of 0.8 mm and roll speed of 0.73 rpm produced green strip with a density of 79.6% of cast aluminium. In addition, a Vickers hardness analysis and a three-point bending test show that increasing the sintering temperature of the green DPR strip will produce softer aluminium strips with higher ductility, as expected.

1. Introduction

Powder metallurgy has been identified as a process capable of reducing beneficiation costs of metal products by removing some of its process complexities and reducing the required energy input. Direct powder rolling (DPR) is a powder metallurgical beneficiation process used to create sheet metal by compacting metallic powders using a rolling mill and further densifying the product by sintering. This process, however, is yet to be optimised and sheet metal produced using this method has lower density, strength, ductility and formability when compared to sheets produced conventionally despite having good tensile elongation [1]. Optimising this process can therefore have a major impact on the metal working industry by making formally exotic materials readily available for applications that would have previously used cheaper material such as steel. This is especially of interest for titanium, which is among Earth's most abundant metals although is not often used in industry due to the high costs associated with its beneficiation [2].

This study is focused on the investigation of the effects of varying the input parameters of aluminium DPR only. These input parameters are to be manipulated in order to improve the product's mechanical properties such as density, formability and strength. It is believed that the findings from this research can be adapted to be used effectively for other materials such as titanium. Aluminium's relatively low cost, low ultimate tensile strength and low melting point require less time and resources to perform multiple DPR tests making it an ideal material for DPR research. DPR process parameters known to impact the properties are powder morphology, roll size, roll speed, roll surface roughness, roll gap, sintering time and sintering temperature, although how these parameters can be optimised to create competitive properties is not well understood [3]. This study explores the effects that roll speed, roll gap, sintering time and sintering temperature have on the density, hardness and bending strength of DPR strips.

Roll speed refers to the rotational speed of the rolling mill rolls. It has been found that at high rolling speeds, the strip produced is of inferior quality, particularly of poor density when compared to strips produced at lower speeds. The reduction in density due to roll speed can be attributed to the shorter time in which each particle of powder mass is subjected to pressure. Additionally, as the roll speed increases, the velocity of the air escaping the powder does not increase proportionally, resulting in lower density [4]. It has also been implied that an increase in roll speed will affect the position of the nip angle negatively, however, no experimental evidence exists that can support this. It has been found that when powder is slowly rolled at a given roll gap, the density and thickness of green (not sintered) strip remain constant up to a certain speed named the transition speed. Above this speed these properties decrease as the roll speed increases due to the factors listed above [5, 6].

For a given powder and roll combination, a decrease in roll gap increases the roll separation force, thus increasing the strip density. This occurs because at high pressure, more particle deformation is present and there is an increase in contact points within the powder mass [7]. Increasing the roll gap increases the DPR sheet thickness, but the density of the sheet decreases resulting in only a moderate increase in mass flowrate. This produces low density and consequently low ductility and low strength green sheets [4, 8]. The density will continue to increase with a decrease in roll gap until the roll gap approaches a point where the roll gap is close to the powder particle size [9].

The green strip produced from DPR is porous with only the pressure induced cold welds holding the strip together. Sintering is a heat treatment process, lowering the surface energy of the powder particles, allowing the cold welds to form necks [10]. This improves the physical and mechanical properties of green compacted metal powders. Mechanisms such as vapour transport and surface diffusion, transport material from the surface to form the neck. These mechanisms do not change the distance between the centres of the particles, thus they contribute little to densification, however, a small increase in density is observed for an increase in temperature [8]. An increase in sintering time will also increase the density of aluminium strip until grain growth begins resulting in a decrease in strip density [10].

Additionally, sintering relieves the strain hardening occurring during cold working and significantly increases the ductility, reducing the hardness of the strip [11]. In order to achieve higher physical and mechanical properties of the powder metallurgy product, the sintered strip can be cold rolled multiple times, annealing the product between compactions to relieve cold working stresses applied during rolling [8].

2. Experimental apparatus and methodology

2.1. Powder material properties

The aluminium powder used to create aluminium strips in this research was purchased from Insimbi Alloys Suppliers Pty (Ltd) in South Africa. The powder has a particle size distribution of 75 MIC-90.04% and its angular morphology is depicted in the scanning electron microscopy (SEM) image in Figure 1. The powder is 98.7% pure aluminium by weight.

IOP Conf. Series: Materials Science and Engineering 1147 (2021) 012017 doi:10.1088/1757-899X/1147/1/012017

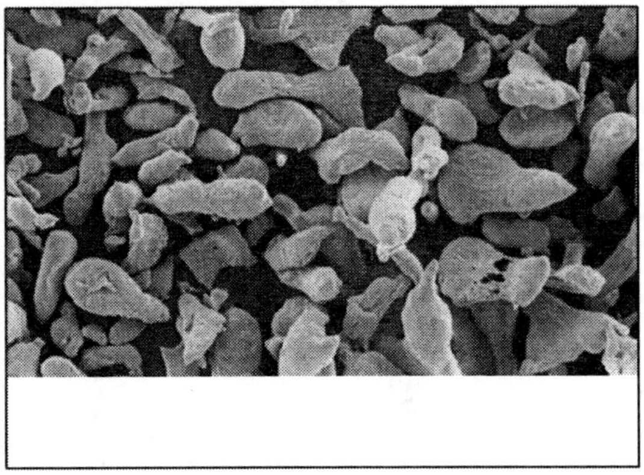

Figure 1. SEM-SE image showing the morphology of the aluminium powder.

2.2. Direct powder rolling parameters

A vertical rolling mill consisting of a powder feeder and 200 mm diameter rolls was used to directly powder roll the aluminium into continuous green strips. The surface roughness of the rolls was not measured but remained constant throughout experimentation. The rolling mill has adjustable roll speed and roll gap, allowing study of the effects of varying these two rolling parameters in DPR. The continuous strips were trimmed into test specimens, seen in Figure 2, using a metal file.

Green DPR strips were produced initially using a roll speed of 0.366 rpm and roll gaps of 0.7 mm, 0.75 mm and 0.8 mm. The roll speed was then doubled to 0.73 rpm and strips were produced for the same roll gaps. Strips produced with roll gaps less than 0.7 mm were not analysed due to severe cracking occurring when rolled below this limit. Green strips rolled above a 0.8 mm roll gap had very little density and crumbled when handled.

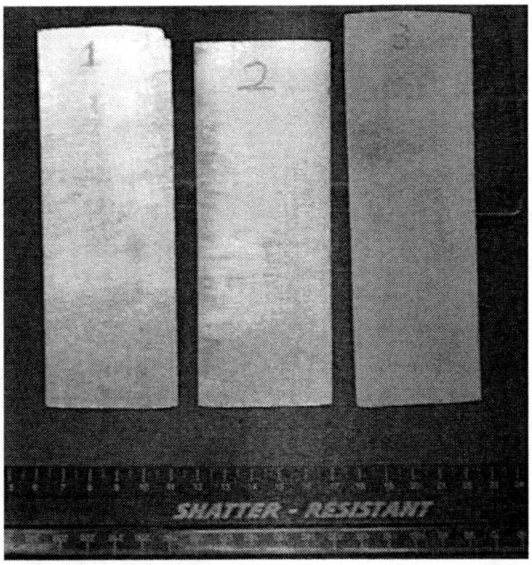

Figure 2. Trimmed green aluminium strips produced by DPR process.

2.3. Sintering parameters

Sintering was performed by heating the green aluminium strips using a Brother XD-1600A inert gas muffle furnace in three heating phases. The first phase involved heating the furnace at a constant rate of 10 K per minute until the peak temperature was reached. The temperature was then held constant at its peak for a predetermined time period. The third phase involved allowing the furnace to cool to room temperature at 10 K per minute.

In order to reduce the amount of oxidation occurring during sintering, all sintering occurred in an argon gas environment. This environment was achieved by pulling a vacuum of 0.05 MPa from the furnace before refilling it with 99.9 % pure argon gas. This process was conducted a total of five times before the gas was allowed to run at three litres per minute continuously. The three peak temperatures investigated along with their sintering hold times can be observed in Table 1.

Table 1. Sintering process parameters.

Sintering peak temperature (°C)	Phase 1 Heating time (min)	Phase 2 Peak temperature time (min)	Phase 3 Cooling time (min)	Total sintering time (min)
500	50	60	50	160
550	55	45	55	155
580	58	30	58	146

2.4. Post sintering densification parameters

Once sintered, the properties of the aluminium strips were further manipulated by re-rolling the strips and annealing the products. All re-rolling was performed at 0.366 rpm on strips initially rolled at 0.366 rpm. Re-rolling was conducted at a roll gap 0.5 mm less than each strip experienced during DPR, densifying the strips by further cold working them. Note that 0.5 mm was the maximum roll gap change that did not result in excessive cracking of the strips. Re-rolled strips were then annealed to further increase their density and release the internal stresses that occurred due to cold working during re-rolling. Annealing times and temperatures matched those experienced during sintering for each strip and therefore follow the procedures outlined in Table 1. Strips that have been re-rolled and annealed will henceforth be labelled, densified strips.

3. Experimental analysis and results

3.1. Density analysis

The manipulation of DPR and sintering parameters largely focuses on increasing the density of the final product, consequently improving its mechanical properties [7]. The aim is to achieve a final density that is close to that of wrought aluminium at 2.7 g/cm^3.

Table 2 depicts the impacts that a change of roll gap and a change of roll speed have on the density of green DPR strips. It is observed that a smaller roll gap produces strip with a higher density. It is also clear that the slower roll speed of 0.366 rpm results in a higher strip density than 0.73 rpm. This means that a roll speed of 0.73 is above Tundermann and Singer's transition speed and thus, too high for this powder and roll combination [5]. It is unknown whether a 0.366 rpm roll speed is above or below this transition speed. More experimentation needs to be completed in order to find the transition roll speed for time efficient DPR of the highest density for this powder and roll combination.

Table 2. Green aluminium strip density.

	0.366 rpm roll speed		0.73 rpm roll speed	
Roll gap (mm)	Density (g/cm^3)	(% of cast Al.)	Density (g/cm^3)	(% of cast Al.)
0.7	2.43	90	2.32	85.9
0.75	2.36	87.4	2.25	83.3
0.8	2.28	84.4	2.15	79.6

Following initial direct powder rolling, sintering of the green strips further increased their density. Table 3 represents the densities for direct powder rolled aluminium, rolled at 0.366 rpm, when green and when sintered at 500 °C, 550 °C and 580 °C respectively. The sintering process is in accordance with the procedure outlined in Table 1. Percentage change 1 uses the density of wrought aluminium, 2.7 g/cm^3, as the denominator while Percentage change 2 uses the measured green density as the denominator.

The results from these tables show that density is not greatly improved during sintering with a maximum density change from green of only 0.41% which was achieved by sintering 0.7 mm roll gap strips to 580 °C. The tables show that higher green densities result in a higher density change during sintering. In addition, a higher density is achieved by sintering at a higher temperature. Strips sintered to 550 °C and 580 °C, however, show very similar average density change values. A conclusion drawn from this is that high densities can be achieved by sintering strips at lower temperatures for a longer time.

Table 3. Sintered aluminium strip density.

Roll gap (mm)	Green density (g/cm^3)	Sintered density (g/cm^3)	Density increase 1 (% of wrought)	Density increase 2 (% of green)
Sintered (500 °C) aluminium strip density:				
0.7	2.40	2.406	0.222	0.250
0.75	2.37	2.373	0.111	0.127
0.8	2.28	2.282	0.074	0.088
Average			0.136	0.155
Sintered (550 °C) aluminium strip density:				
0.7	2.45	2.458	0.296	0.327
0.75	2.37	2.376	0.222	0.253
0.8	2.3	2.306	0.222	0.261
Average			0.247	0.280
Sintered (580 °C) aluminium strip density:				
0.7	2.44	2.45	0.370	0.410
0.75	2.38	2.386	0.222	0.252
0.8	2.28	2.284	0.148	0.175
Average			0.247	0.279

Further density analyses were conducted on densified strips after which they were rolled and sintered, were re-rolled and annealed according to the procedure outlined in Section 0. Table 4 presents density data for strips of varying roll and re-rolling gaps when green and when densified at 500 °C, 550 °C and 580 °C respectively.

The density results for densified DPR strips no longer show a significant increase with an increase in temperature, with the average density increases all showing very similar results. This implies that the maximum increase in density that can be achieved during sintering has been reached and the leading source of density increase during densification is due to re-rolling. Further re-rolling steps should be explored to achieve even denser strips. Strips previously rolled at a 0.75 mm roll gap and re-rolled at a 0.7 mm roll gap consistently produced the highest total increase in density.

Table 4. Sintered, re-rolled and annealed aluminium strip density.

Roll and re-roll gap (mm)	Green density (g/cm^3)	Densified density (g/cm^3)	Density increase 1 (% of wrought)	Density increase 2 (% of green)
Sintered (500 °C), re-rolled and annealed aluminium strip density:				
0.7 and 0.65	2.452	2.468	0.593	0.653
0.75 and 0.7	2.394	2.412	0.667	0.752
0.8 and 0.75	2.304	2.318	0.519	0.608
Average			0.593	0.671
Sintered (550 °C), re-rolled and annealed aluminium strip density:				
0.7 and 0.65	2.412	2.426	0.519	0.580
0.75 and 0.7	2.334	2.35	0.593	0.686
0.8 and 0.75	2.294	2.308	0.519	0.610
Average			0.543	0.625
Sintered (580 °C), re-rolled and annealed aluminium strip density:				
0.7 and 0.65	2.424	2.44	0.593	0.660
0.75 and 0.7	2.352	2.372	0.741	0.850
0.8 and 0.75	2.298	2.312	0.519	0.609
Average			0.617	0.707

3.2. Optical image analysis

A Nikon Eclipse MA200 was used to view the impact that sintering temperature has on the microstructures of aluminium strips. The strips represented were rolled at 0.366 rpm with a roll gap of 0.75 mm although the results were consistent for all three roll gaps. Figure 3 shows a green aluminium DPR strip, Figure 4, a strip sintered to 500 °C, Figure 5, a strip sintered to 550 °C and Figure 6, a strip sintered to 580 °C. Figure 7 shows an optical image of a strip that has been re-rerolled with a roll gap of 0.7 mm roll gap and annealed at 580 °C. The images show a small decrease in pore size with an increase in sintering temperature and no grain growth when compared to the green strip.

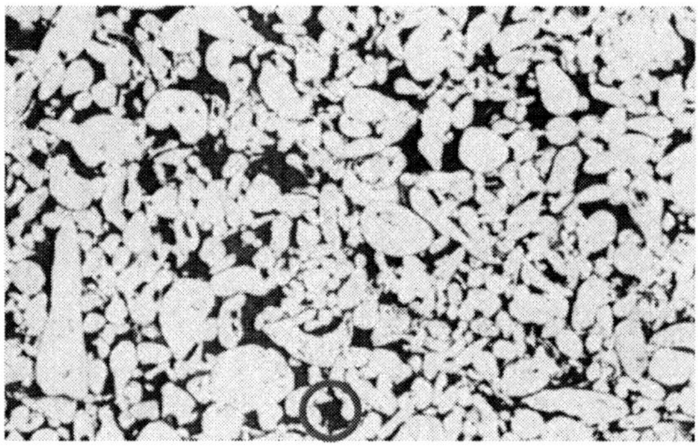

Figure 3. Optical image of a green DPR aluminium strip (pores depicted black, one is exemplarily marked by blue circle).

IOP Conf. Series: Materials Science and Engineering 1147 (2021) 012017 doi:10.1088/1757-899X/1147/1/012017

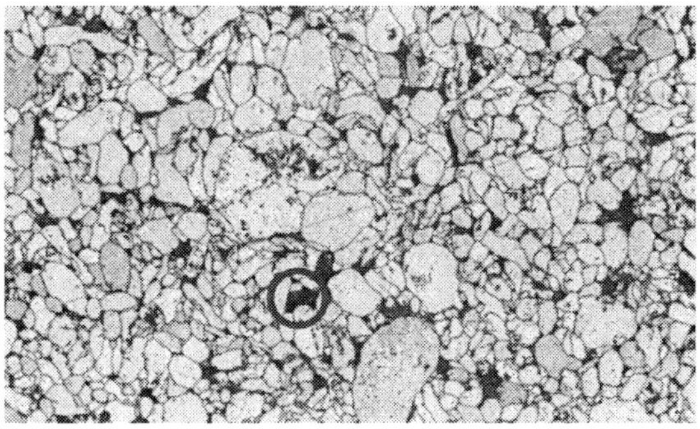

Figure 4. Optical image of DPR aluminium strip sintered at 500 °C (pores depicted black, one exemplarily marked by blue circle).

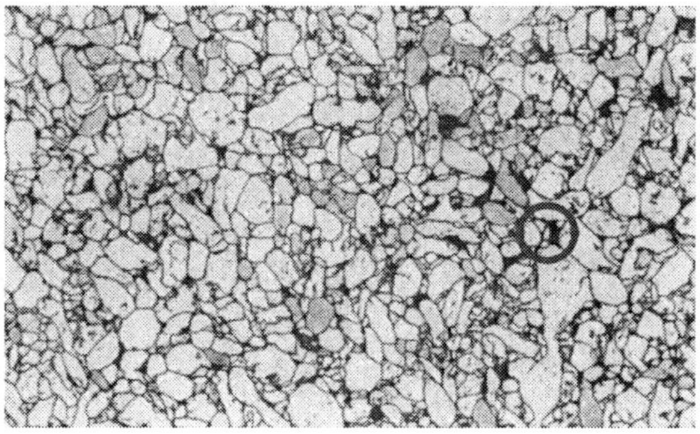

Figure 5. Optical image of DPR aluminium strip sintered at 550 °C (pores depicted black, one is exemplarily marked by blue circle).

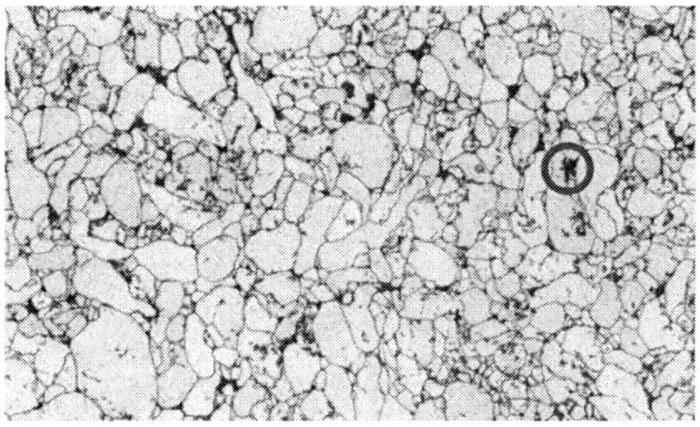

Figure 6. Optical image of DPR aluminium strip sintered at 580 °C (pores depicted black, one is exemplarily marked by blue circle).

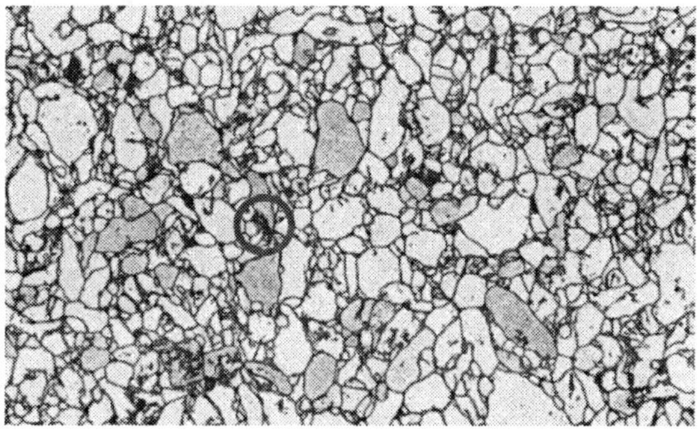

Figure 7. Optical image of DPR aluminium strip sintered at 580 °C, rerolled at 0.7 mm roll gap and re-sintered at 580 °C (pores depicted black, one exemplarily marked by blue circle).

3.3. SEM analysis
The effect that roll gap has on the microstructure on the surface of sintered DPR strips was analysed using a Zeiss Ultra plus FE scanning electron microscope. DPR aluminium strips were rolled at three different roll gaps, all sintered at 580 °C and characterized by SEM. Figure 8 shows a strip rolled at a roll gap of 0.7 mm, Figure 9 a roll gap of 0.75 mm and Figure 10 a roll gap of 0.8 mm. Smaller pores and larger pressure welds coupled with neck formation is evident in the images of the strips rolled at smaller roll gaps.

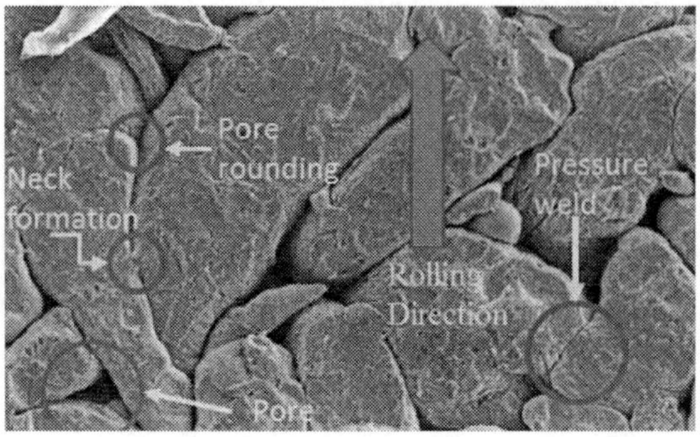

Figure 8. SEM-SE image of an aluminium strip surface after being rolled at 0.7 mm and sintered at 580 °C.

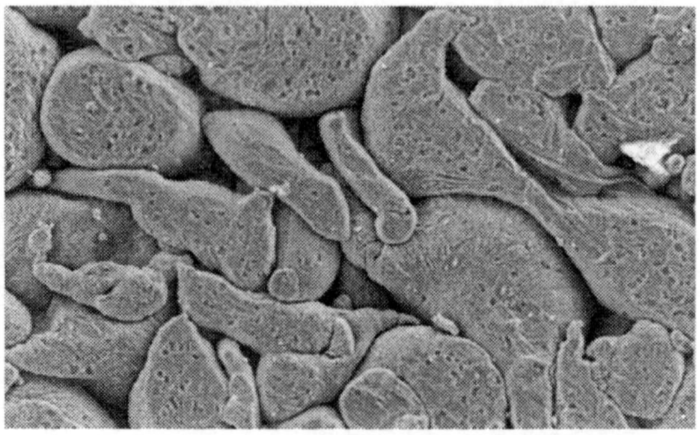

Figure 9. SEM-SE image of an aluminium strip surface after being rolled at 0.75 mm and sintered at 580 °C.

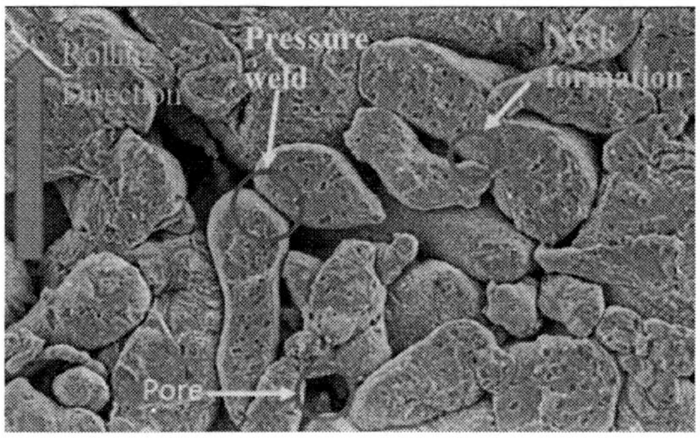

Figure 10. SEM-SE image of an aluminium strip surface after being rolled at 0.8 mm and sintered at 580 °C.

Densified strips that had been re-rolled and annealed were then analysed using the SEM. The following figures show strips that have been direct powder rolled, sintered and densified according to the procedure in Section 0. Figure 11 shows a densified strip originally rolled at a roll gap of 0.7 mm, Figure 12 a strip originally rolled at a 0.75 mm roll gap and Figure 13, a strip originally rolled at a 0.8 mm roll gap. All strips were sintered and annealed at 580 °C according to Table 1. These images show a clearer reduction in pore size than non-densified strips with densified strips originally rolled at 0.7 mm roll gap showing almost no surface pores. The pressure welds and necks have grown so much that in some cases, particles can no longer be identified as separate.

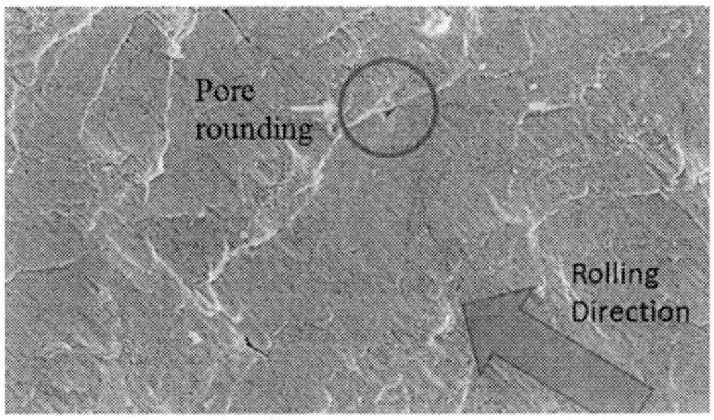

Figure 11. SEM-SE image of an aluminium strip surface after being rolled at 0.7 mm and sintered at 580 °C, then re-rolled at 0.65 mm roll gap then re-sintered at 580 °C.

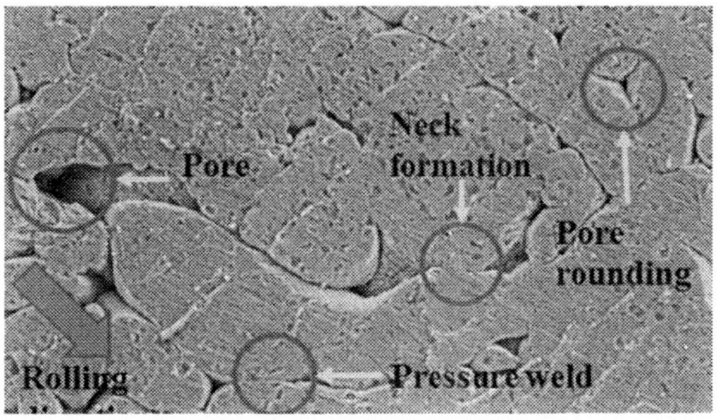

Figure 12. SEM-SE image of an aluminium strip surface after being rolled at 0.75 mm and sintered at 580 °C, then re-rolled at 0.7 mm roll gap then re-sintered at 580 °C.

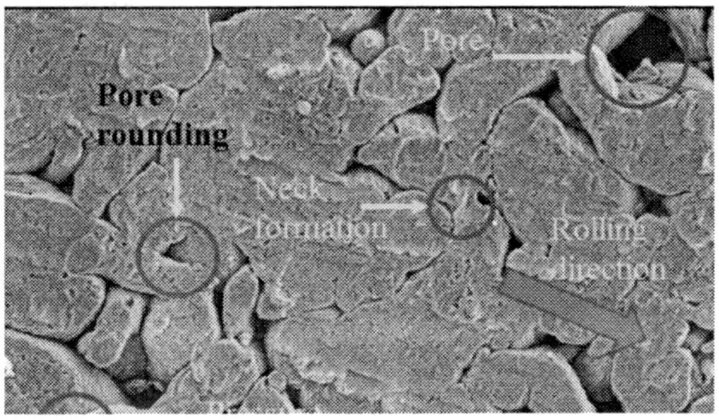

Figure 13. SEM-SE image of an aluminium strip surface after being rolled at 0.8 mm and sintered at 580 °C, then re-rolled at 0.75 mm roll gap then re-sintered at 580 °C.

3.4. Micro hardness analysis

The Vickers micro hardness of the aluminium samples was investigated by an applied force of 1 kgf using an Emco Test Durascan. Prior to indentation, the samples were mounted in bakelite resin with their rolled surface exposed and polished. The top three rows of data in Table 5 show the Vickers hardness for aluminium DPR strips rolled at 0.366 rpm with varying roll gaps and sintering temperatures. The bottom three rows of Table 5 show the Vickers hardness for aluminium DPR strips that have been sintered, and densified according to the procedure explained in Section 0. Each value in the table was achieved by averaging two readings taking from different places on each of three strips making a total of six independent readings per value.

In aluminium DPR, high hardness values are not desired because increased hardness is associated with decreased ductility. High ductility is a desired quality in aluminium DPR allowing higher levels of formability of the aluminium sheet.

Table 5 outlines that similar to density, hardness increases with decreasing roll gap. This is due to the higher pressure the rolls exert on the aluminium powder resulting in a higher level of cold working of the material. The table shows that this hardening is relieved during sintering and higher sintering temperatures resulting in a greater reduction of hardness. The values of hardness show that there is a limit to the reduction of hardness that can be achieved during annealing with this asymptotic value being close to 33 HV.

Table 5. Micro hardness of DPR aluminium strips.

Roll gap (mm)	Green (HV)	Sintered 500 °C (HV)	Sintered 550 °C (HV)	Sintered 580 °C (HV)
0.7	57 ± 3.10	52 ± 2.45	48 ± 1.2	44 ± 1.05
0.75	49 ± 2.03	40 ± 1.64	36 ± 2.51	34 ± 2.80
0.8	44 ± 2.30	38 ± 3.22	35 ± 2.34	33 ± 3.05
0.7 and 0.65		40 ± 2.36	37 ± 2.45	35 ± 2.74
0.75 and 0.7		40 ± 2.30	38 ± 3.00	37 ± 3.10
0.8 and 0.75		39 ± 2.21	36 ± 1.59	35 1.73

3.5. Bending analysis

A common disadvantage of powder metallurgical products is their poor strength when a bending load is applied. It is therefore of importance to prioritise maximising strip strength in bending in order to produce sheet that can easily be formed. In agreement with DPR literature, the specimens being tested displayed low resistance to bending and were thus analysed adapted from 3-point bending analysis standards using a bearing distance of 120 mm. The results obtained can be used for comparative analysis within this study only. The maximum deflection of the specimens was measured while 50-gram weights were incrementally loaded until failure of the specimen.

The bending tests were all conducted using aluminium strip that was rolled at 0.366 rpm. The effect that roll gap and sintering temperature have on the maximum deflection are of interest in order to maximise strip ductility. Figure 14 is a bar graph showing the maximum deflection that each strip experienced before failure and Figure 15 shows the maximum loads each strip experienced before failure. In contrast to strip density, strip ductility is largely affected by sintering while roll gap has a lesser effect on ductility. A high sintering temperature resulted in the maximum deflection of strip sintered at 580 °C to more than double the deflection of a green strip.

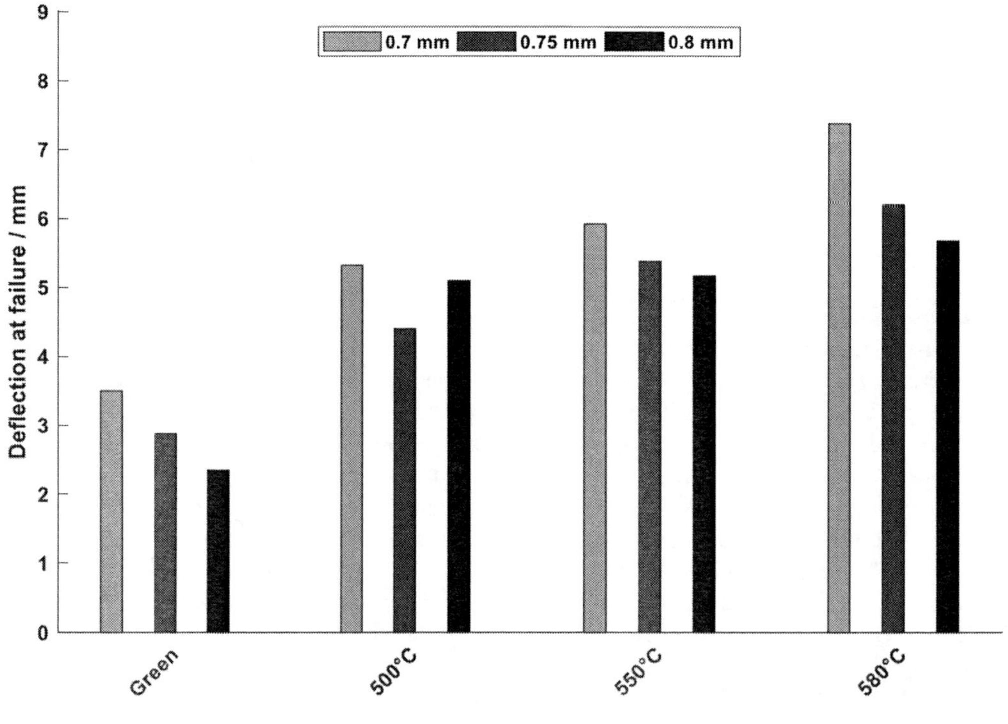

Figure 14. Bar graph representing aluminium strip maximum deflection at failure.

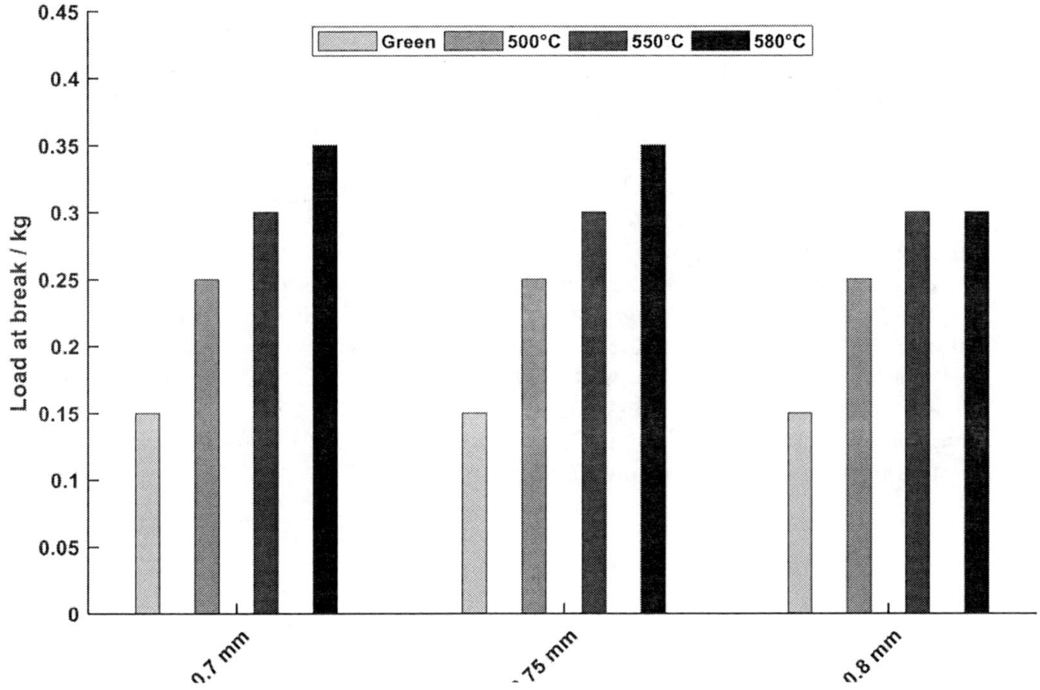

Figure 15. Bar graph representing aluminium strip maximum load at failure.

4. Conclusions

In this study, aluminium DPR strips were successfully created with varying roll speed, roll gap, sintering temperature and sintering time. The effects that each parameter had on the density, microstructure, hardness and ductility were explored. It can be concluded that the density is largely dependent on the rolling conditions with smaller roll gaps and slower roll speeds produces denser strip. Although only a small difference was recorded, higher sintering temperatures produced strip with higher density. The density can be further improved by re-rolling and annealing the strip. Furthermore, larger pressure welds are evident in strips rolled at smaller roll gaps. Re-rolling and annealing the strips produces much larger pressure welds and necks. The hardness of green strip increases with a decrease in roll gap. The cold working resulting in these pressure welds are relieved by sintering. Higher sintering temperatures had a larger influence in relieving this strain hardening. Ductility analyses in the form of a three-point bend test proved that the largest influence on ductility is the sintering conditions. High sintering temperatures produced significantly higher strip ductility.

References

[1] Peter W H, Muth T, Chen W, Yamamoto Y, Jolly B, Stone N A, Cantin G M D, Barnes J, Paliwal M, Smith R, Capone J, Liby A, Williams J and Blue C 2012 Titanium sheet fabricated from powder for industrial application *Journal of Materials* **64** 566-571

[2] Cantin G M D, Kean P L, Stone N A, Wilson R, Gibson M A, Yousuff M, Ritchie D and Rajakumar R 2011 Innovative consolidation of titanium and titanium alloy powders by direct rolling *Powder Metallurgy* **54** 188-192

[3] Johanson J R 1965 A rolling theory for granular solids *Journal of applied mechanics* **32** 842-848

[4] Evans P E and Smith G C 1959 The compaction of metal powders by rolling. II-An examination of the compaction process *Powder Metallurgy* **2** 26-44

[5] Tundermann J H and Singer A R E 1968 The flow of iron powder during roll compaction *Powder Metallurgy* **11** 261–294
[6] Dube R K 1990 Metal strip via roll compaction and related powder metallurgy routes *International Materials Reviews* **35** 253-292
[7] Delavari M, Salarvand A, Rahi A and Shahri F 2011 The effect of powder metallurgy process parameters on mechanical properties of micro and nano-iron powder *International Journal of Engineering, Science and Technology* **3** 86-94
[8] Worn D K and Perks R P 1959 Production of pure nickel strip by the direct powder rolling process *Powder Metallurgy* **2** 45-71
[9] Bindhumadhavan G, Seville J P K, Adams M J, Greenwood R W and Fitzpatrick S 2005 Roll compaction of a pharmaceutical excipient: Experimental validation of rolling theory for granular solids *Chemical Engineering Science* **60** 3891-3897
[10] Gürbüz M, Şenel M C and Koç E 2018 The effect of sintering time, temperature, and graphene addition on the hardness and microstructure of aluminum composites *Journal of Composite Materials* **52** 553-563
[11] Sturgeon G M, Jackson G, Barker V and Sykes G M H 1968 The production of stainless-steel strip from powder *Powder Metallurgy* **11** 314-329

Influence of the aluminum die casting process on PEEK layered CFRP in the manufacturing of a hybrid connection

A Marx[1], A Schmid[2,3], T Haubold[2], L Pursche[2], P Schiebel[1] and A Herrmann[1]

[1] Faserinstitut Bremen e.V., Am Biologischen Garten 2, 28359 Bremen, Germany
[2] Fraunhofer-Institut für Fertigungstechnik und Angewandte Materialforschung IFAM, Wiener Str. 12, 28359 Bremen, Germany
[3] University of Bremen, Bibliothekstr. 1, 28359 Bremen, Germany

*e-mail: amarx@faserinstitut.de

Abstract. Lightweight design is used as a key technology in a variety of fields to achieve significant weight reduction. Due to its low density in combination with high strength, CFRP and aluminum are predestinated lightweight materials. However, the joining of this material combination is still performed using common technologies like adhesive or mechanical joining. A major challenge at the joining of this material combination is the corrosion behavior. Therefore a novel approach of a hybrid connection between CFRP and aluminum was developed by using the HPDC process of aluminum. During efficient HPDC process, a CFRP structure is recast with aluminum (AlSi10MnMg). To resist the shortly appearing temperature of 700 °C from the liquid aluminum melt, the CFRP is coated with a layer of PEEK. This leads to a material transition, which spatial separates CFRP and aluminum and a hybrid connection with tensile shear strengths up to 22 MPa. Within this work, the influence of the several processing steps on the PEEK layer and the structure of the CFRP was investigated. DSC and TGA results show no significant change on thermal properties of the PEEK. Imaging investigations show the appearance of heat induced volatilizations in the PEEK layer of the hybrid specimens after the HPDC process.

1. Introduction

Lightweight applications are becoming more popular in serval industries, such as the transportation sector as they contribute to emissions reduction [1]. Due to their low weight in combination with high strength, aluminum and carbon fiber reinforced plastics (CFRP) are predestinated lightweight materials. Especially the combination in type of multi-material design enables both materials to be used due to their advantages [2]. However, the joining of this material combination is still accomplished by common joining techniques such as adhesive or mechanical joining [3]. This results in serval disadvantages, for example, extra process times, additional weight caused by the joining elements and a decrease of the strength because of drilling holes in the CFRP structure [4].

To overcome these problems, the focus of research investigations lies on new innovative joining technologies that do not increase the weight of the material combination by adding mechanical joining elements. A common approach is the use of the welding technology, as the partial melting of the thermoplastic results in a connection between the materials [5]. Besides the complex requirements regarding mechanical and thermal properties of the different materials, the occurrence of galvanic corrosion between CFRP and aluminum has to be considered [6]. Within the joining process, actions have to be taken to prevent the development of corrosion.

In this work a novel approach of combining thermoplastic CFRP and aluminum into an intrinsic hybrid composite is investigated. In the primary shaping process of high pressure die casting (HPDC), a CFRP structure is recast with aluminum. Essential for the connection is a decoupling layer, which avoids corrosion and protects the CFRP from high process temperature during the HPDC process. Polyetheretherketon (PEEK) was identified as suitable material due to its stability of high temperatures. In a first approach by [7], hybrid single lap specimens were manufactured with a PEEK layer thickness of 250 μm. It was demonstrated that the tensile shear strength of the hybrid connection increase if quenching and heat treatment is carried out on the specimens after the HPDC process [7]. Within the further development of the hybrid connection, unpublished results show a tensile shear strength up to 22 MPa. Besides single lap specimens, the implementation of CFRP bracket structure into the HPDC of aluminum was shown [8]. A method to describe the failure behavior of the hybrid composite is the use of a cohesive zone model in the finite element method [9].

Therefore, the aim of the presented work is the identification of potential thermal damage in the PEEK layer during the different process steps by the use of differential scanning calorimetry (DSC) and thermogravimetry analysis (TGA). Imaging measurements via computer tomography (CT) and micrographs are carried out to investigate the PEEK layer and the CFRP regarding process induced delaminations and pores.

2. Experimental and material

2.1. Materials

For the manufacturing of the hybrid specimens, the organo sheet Tenax-E TPCL PEEK 4-40-HTA40 consisting of 5 layers carbon fiber fabric [0°/90°|+/- 45°|0°/90°|+/-45°|0°/90°] with thermoplastic PEEK matrix and a thickness of 1.55 mm was used. To prepare the CFRP structure for the HPDC process, a PEEK coat was applied on the CFRP. Therefore, the PEEK film Victrex aptiv 1000 with a thickness of 250 μm was used.

2.2. PEEK coating

The CFRP organo sheets and the PEEK film were dried for 24 h at 120 °C. The application of the PEEK coating on the organo sheet was executed on a Rucks KV 228 press in an isoforming process, which is optimized for thermoplastic composites. Two separated heated zones allow a fast heating of the composite until the melt temperature of the thermoplastic is reached. A transfer of the mold to the colder zone ensures a defined cooling rate. This characterizes the isoforming process as time and energy efficient [10].

For this work, the organo sheets with PEEK film were placed in a compression tooling and heated up under the pressure of 4.2 bar until thermocouples inside the tooling viewed a temperature of 350 °C. At this temperature, the transfer to the colder zone was carried out and the cooling occurred under the pressure of 4.2 bar. Afterwards the coated specimens were trimmed to their designated dimension of 100 x 40 mm. In this approach, the whole specimens were coated with a PEEK layer. The Process parameters and specimens are shown in Figure 1.

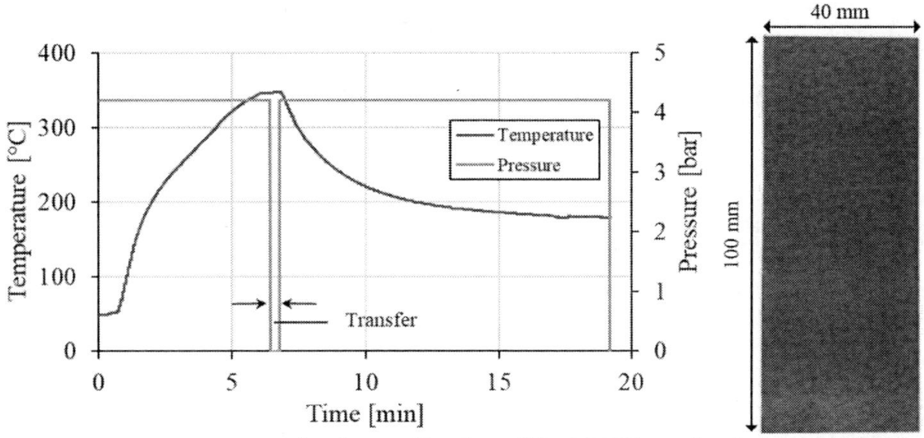

Figure 1. Process parameters for the application of the PEEK coating left, CFRP laminate coated with PEEK on the right.

2.3. *Aluminum high pressure die casting process*

The HPDC process of the aluminum was carried out at the Fraunhofer IFAM on a cold-chamber die casting machine Bühler type SC N/66. The CFRP specimens were dried for 24 h at 120 °C. The aluminum alloy Silafont 36 (AlSi10MnMg) was used. During the HPDC process, the aluminum melt had a temperature of 700 °C and was pressed with a piston speed of 2.35 m/s and a hold pressure of 1000 bar into the 150 °C hot mold. The closing time of the mold was 10 s. An aluminum body with a length of 100 mm, a width of 40 mm and a thickness of 10 mm was casted and simultaneously joined to the CFRP in an overlap area of 40 x 40 mm. Immediately after the HPDC process the hybrid specimens were quenched in water (temperature 20-30 °C), a heat treatment, inspired by T5 treatment for metals, was carried out at 190 °C for 2 h. A manufactured single lap hybrid specimen is shown in Figure 2.

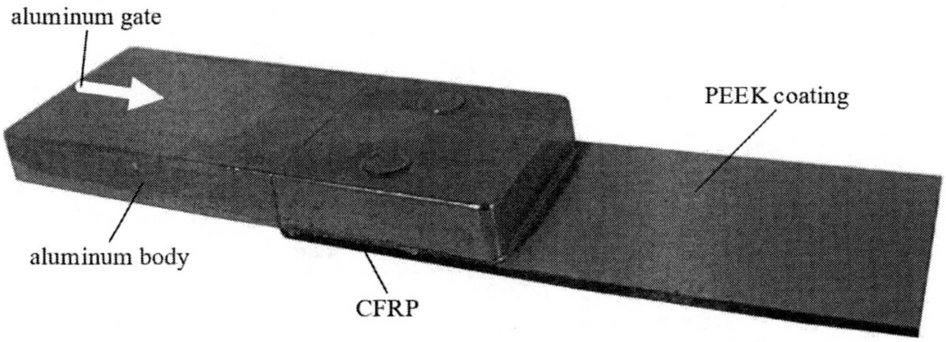

Figure 2. Manufactured single lap hybrid specimen.

2.4. *Analysis methods*

Thermally analysis with DSC and TGA were carried out to investigate the thermal influence on the PEEK interface layer during the different process steps and to identify potential thermal damage. The TGA determines the weight loss of the PEEK over a range of temperature to predict the thermal stability and to gain information about the thermal decomposition and degradation behavior [11]. The TGA measurements were carried out with a Q5000 from TA Instruments with a temperature range from 25 °C to 800 °C and a heating rate of 10 K/min in nitrogen and ambient atmosphere. The samples for the TGA were prepared out of the PEEK layer with scalpel and had a weight between 3 and 6 mg.

DSC analysis reveal the melting temperature and enthalpy of fusion of the PEEK, which leads to the degree of crystallinity. According to [12], the thermal history of PEEK causes an impact on the crystal structure. The DSC tests were performed at a DSC Discovery 250 from TA Instruments with a heating rate of 10 K/min and nitrogen atmosphere. The samples for the DSC analysis after the HPDC process were prepared from the middle of the fractural interface of the CFRP side after tensile shear tests. Therefore, they do not represent a direct contact between PEEK and aluminum melt. The sample weight was between 6 and 7 mg.

The process induced delaminations and pores in the CFRP laminate and PEEK layer were investigated by CT scans on a GE Phoenix v/tome/x m with a 240 kV tube, micrographs of the cross section and pictures of the fractural surface in the interface after adhesive failure at tensile shear tests.

3. Results and discussion

3.1. DSC

The used PEEK film Victrex aptiv 1000 reveals an enthalpy of fusion of 40.9 J/g and a melting temperature of 342.1 °C. According to [12], the enthalpy of fusion from complete crystalline PEEK is 130 J/g. This results in a degree of crystallinity of 31.5 %. After the application of the PEEK film via isoforming process, the degree of crystallinity increases to 34.1 %. According to [10] the degree of crystallinity of the PEEK is effected by the cooling rate. The slower the cooling rate, the higher the degree of crystallinity. The slow cooling rate in the isoforming process could lead to an increase of the crystallinity of 3 % compared to the initial PEEK film. After the HPDC process with a high cooling rate, the determined degree of crystallinity decreased to 29.8 %. An explanation for the lower degree of crystallinity after HPDC process is the performed quenching and heat treatment. A heating above the glass transition temperature leads to a crystallization of the PEEK [13]. However the annealing temperature of 190 °C was not high enough to lead to a higher degree of crystallinity [14]. According to [15] the properties of PEEK are highly dependent on the degree of crystallinity. A low degree of crystallinity decreases the mechanical properties of PEEK [16]. Hence, as there is no significant difference between the degree of crystallinity in the initial PEEK film and the PEEK after the HPDC process, it can be assumed that no major change in the properties of the PEEK occurred. The determined melting temperatures of the PEEK slightly decreases over the different process steps from 342.1 °C to 340.7 °C. The results are shown in Table 1.

Table 1. Results of the DSC analysis of the PEEK layer at different process steps.

Process step	Melting Temperature (°C)	Enthalpy of fusion (J/g)	Degree of Crystallinity (%)
Initial film	342.1	40.9	31.5
After application	341.9	44.4	34.1
After HPDC	340.7	38.7	29.8

The graphs of the DSC analysis are shown in Figure 3. All specimens show a similar heat flow behavior at the glass transition temperature and melting temperature. Since the curves are very similar, no indication of a thermal damage caused by the short contact time with the 700 °C aluminum melt is visible. The thermal properties of the PEEK remain unchanged during the different process steps. Since the mechanical properties of PEEK rise with an increasing crystallinity, a heat treatment adapted to PEEK should be investigated to increase the mechanical properties of the hybrid connection [14].

Since the DSC analysis identified no thermal damage on the PEEK, additional investigation such as X-ray photoelectron spectroscopy (XPS) and infrared spectroscopy (IR) are recommended to determine changes of PEEK direct at the joining zone with the aluminum.

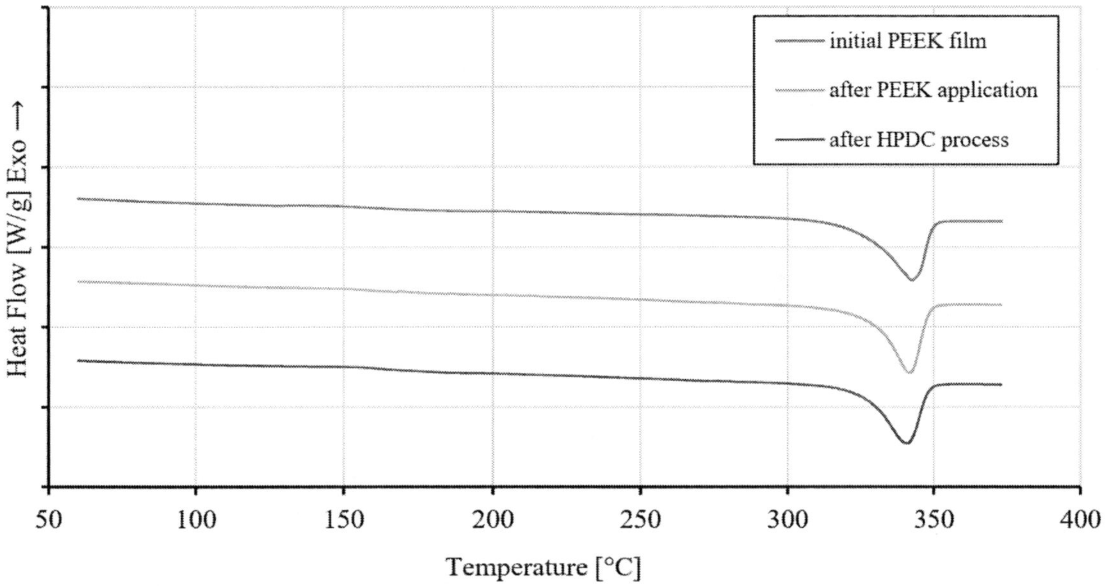

Figure 3. DSC analysis of the PEEK layer at a heating rate of 10 K/min after different process steps.

3.2. TGA

The decomposition of PEEK is described in two steps. During the first decomposition step a random chain scissions of ether and ketone occur and result in a fast and rapid weight loss of about 45 % of the mass, followed by a second step with a slower decomposition, in which under air almost the whole mass is volatized [17].

The start of thermal decomposition of the initial PEEK film Victrex aptiv 1000 was identified at about 530 °C under ambient atmosphere and 550 °C under nitrogen atmosphere see Figure 4. A rapid weight loss occurs until 600 °C. As described by [17] the decomposition under nitrogen and ambient atmosphere differs in the second step. Compared to the initial PEEK film, the decomposition of the PEEK, after the HPDC process, starts at 470 °C under ambient atmosphere and 540 °C under nitrogen atmosphere. Except for the difference in the initiation of the decomposition, the behavior under nitrogen is very similar for both specimens. Under ambient atmosphere, the decomposition of PEEK after the HPDC occurs at slightly lower temperatures.

Investigations from [15] show a similar behavior of thermally stressed PEEK at wear testing. A faster decomposition of PEEK occurs, because of the damage of linear chain segments and the reduction of active sides for pyrolytiy reaction. The weight loss in the first decomposition step indicates the quantity of the decomposition of the linear structure inside the PEEK [15]. Due to the fact, that just small differences in the decomposition behavior of the PEEK specimens appears, it can be assumed that a slight decomposition occurred during the HPDC process.

A TGA of the PEEK layer after the application in the isoforming process was renounced, because the process temperatures of 350 °C are below the investigated decomposition temperatures of PEEK.

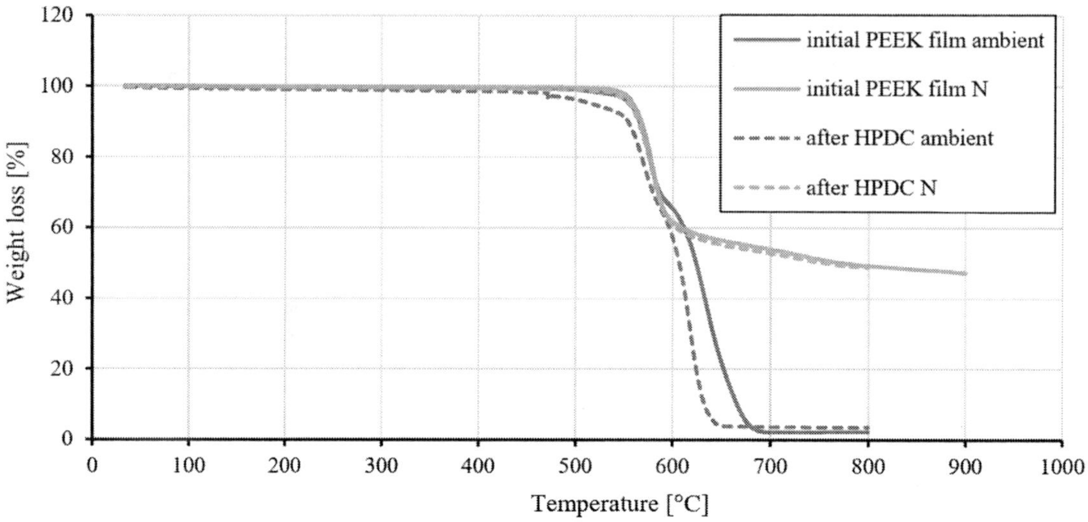

Figure 4. TGA analysis of the PEEK layer at a heating rate of 10 K/min under ambient and nitrogen atmosphere.

3.3. Micrographs

To investigate the quality of the PEEK coated CFRP specimens, micrographs of the cross sectional area were prepared, see Figure 5. The coated PEEK film is homogenously bonded with the PEEK matrix of the organo sheet. The thickness of the resulting PEEK layer was measured on the micrographs and reveal an average PEEK coating thickness of 294 µm. Furthermore, the micrographs indicate the appearance of single pores inside the CFRP. Due to the fact, that processing temperature at the PEEK coating were below the investigated decomposition temperature, volatilization on the basis of decomposition gases is excluded.

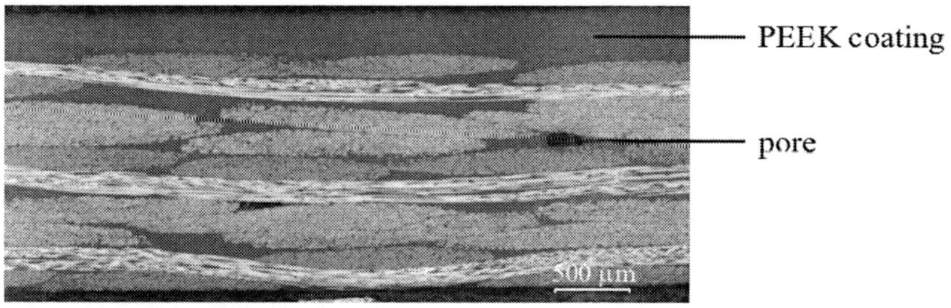

Figure 5. Mircograph in the cross sectional area of the PEEK coated CFRP before the HPDC process.

3.4. CT Scans

Micrographs just display a single section of a specimen, therefore CT scans were conducted to investigate imperfection like pores and delamination in the whole CFRP specimen. The CT scan, shown in Figure 6, displays partial single pores inside the CFRP after the PEEK coating in the isoforming process. The results agree with the investigated pores in the micrograph.

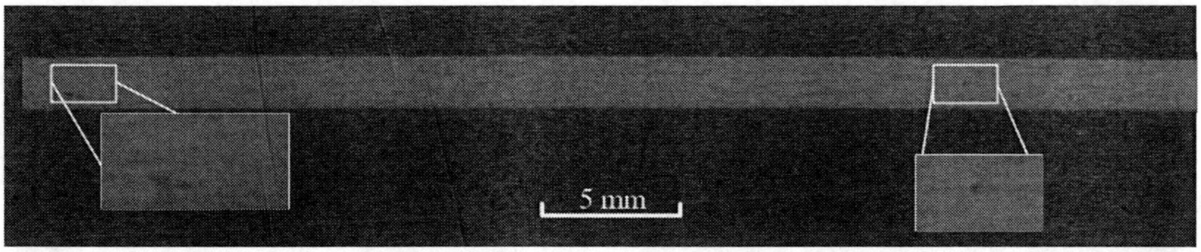

Figure 6. CT image of the CFRP specimen with applied PEEK coating, the pores are highlighted in red.

To show the influence of the HPDC process on the structural integrity of the PEEK coated CFRP, the same specimen as shown in Figure 6 was investigated after the HPDC process and heat treatment. Therefore, the interface area between the PEEK coated CFRP and aluminum was investigated, shown in Figure 7. It becomes obvious, that massive pores occur in the PEEK interface between aluminum and CFRP. Besides the interface no pores or delamination are visible in the CFRP. During the HPDC process the PEEK layer is for a low time span exposed to the melted aluminum with a temperature of 700 °C. According to [18] the volatile decomposition products of PEEK in this temperature range are phenol, benzene, CO and CO_2. The CT scans indicate that a volatilization of these gas components in consequence of the heat during the HPDC process occurred in the interface.

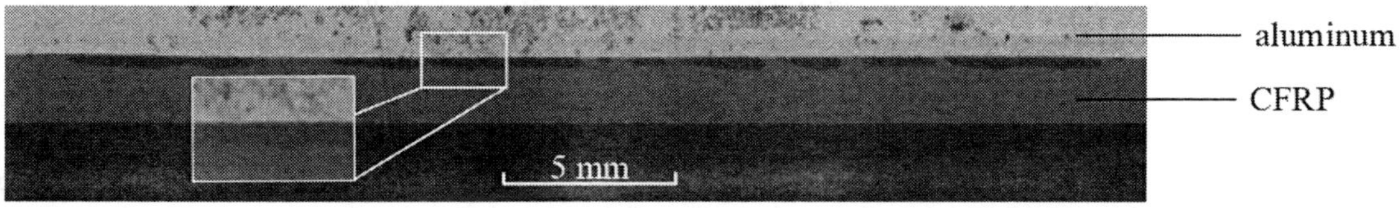

Figure 7. CT image of the interface between CFRP and aluminum after the HPDC process, pores are highlighted in red.

3.5. Fractural surface analysis

Conducted mechanical test reveal a tensile shear strength of 15.6 MPa, related to the casted area of 1,600 mm², for the presented series of hybrid specimens. A picture of the fractural surface area is shown in Figure 8. A mainly adhesive failure occurs between the PEEK interface and the aluminum, with thin remains of PEEK on the aluminum surface. The occurring pores are a consequence of the volatilization of gas products during decomposition in the PEEK interface. The specimen shows the biggest pores in the middle, the area assumed to achieve the highest thermal input during the HPDC process. A black value analysis reveals that 38 % of the surface form the PEEK interface consists of pores. The occurring pores reduces the connection surface and thus the strength of the connection. Nevertheless, tensile shear test of specimens without PEEK coating show a strength of just 6.4 MPa. Despite the pores, the spatial separation between the CFRP and aluminum through the PEEK layer is visible and indicates no direct contact between the CFRP and aluminum. Hybrid specimens with different HPDC process parameters and heat treatments achieve tensile shear strengths up to 22 MPa and show smaller pores in the fractural surface area.

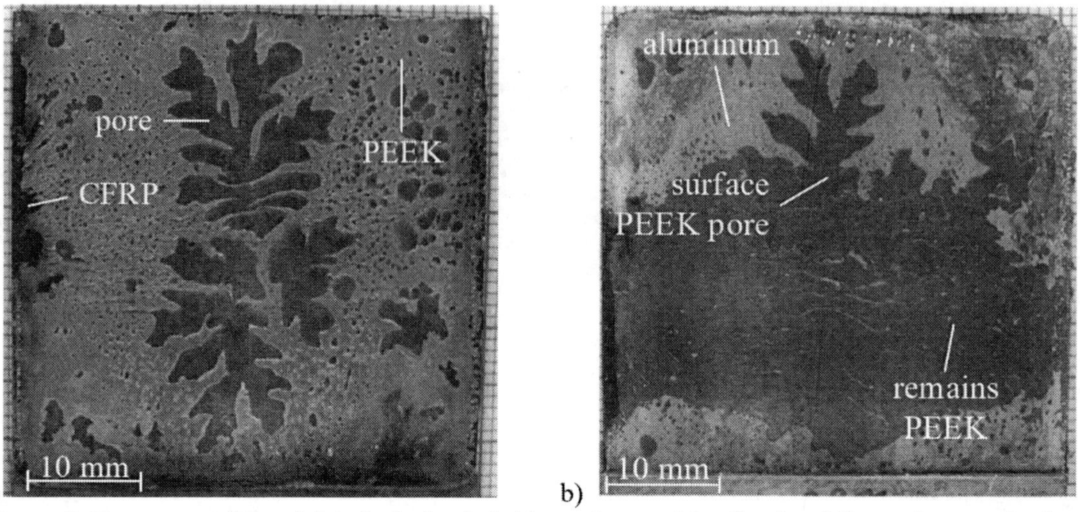

Figure 8. Fracture surface of the single lap hybrid specimen with adhesive failure after tensile shear tests, a) CFRP side with the PEEK interface, b) aluminum side.

4. Conclusion

In the presented work, the TGA and DSC analysis reveal no significant thermal degradation within the investigated PEEK layer due the HPDC process. However, the imaging investigation showed pores in the interface which are affiliated to the volatilization of the PEEK due to the high temperatures of the HPDC process. Further research needs to be conducted to investigate the interface between PEEK and aluminum with chemical analysis such as XPS or IR. The determined degree of crystallinity inside the PEEK layer is slightly lower than in the initial PEEK film. A more suitable PEEK annealing treatment of the hybrid specimens could increase the degree of crystallinity and strength.

Acknowledgments

Funded by the Deutsche Forschungsgemeinschaft (DFG, German Research Foundation)
– HE 2574/36-2

References

[1] Friedrich H E 2013 *Leichtbau in der Fahrzeugtechnik* ed H E Friedrich (Wiesbaden: Springer Fachmedien Wiesbaden)

[2] Kroll L and Nestler D 2019 Verbundwerkstoffe und Werkstoffverbunde *Technologiefusion für multifunktionale Leichtbaustrukturen* ed L Kroll (Berlin: Springer Vieweg) pp 11–22

[3] Zhang D, Zhang Q, Fan X and Zhao S 2019 Review on Joining Process of Carbon Fiber-reinforced Polymer and Metal: Applications and Outlook *Xiyou Jinshu Cailiao Yu Gongcheng/Rare Met. Mater. Eng.* **48** 44–54

[4] Amancio-Filho S T and Blaga L A 2018 *Joining of polymer-metal hybrid structures: Principles and applications* ed S T Amancio-Filho and L-A Blaga (JohnWiley & Sons)

[5] Pramanik A, Basak A K, Dong Y, Sarker P K, Uddin M S, Littlefair G, Dixit A R and Chattopadhyaya S 2017 Joining of carbon fibre reinforced polymer (CFRP) composites and aluminium alloys – A review *Compos. Part A Appl. Sci. Manuf.* **101** 1–29

[6] Wu X, Sun J, Wang J, Jiang Y and Li J 2019 Investigation on galvanic corrosion behaviors of CFRPs and aluminum alloys systems for automotive applications *Mater. Corros.* **70** 1036–43

[7] Schmid A, Arnaut K, Clausen J, Koerdt M, Struss A, Wöstmann F J and Busse M 2017 Intrinsic aluminum CFRP hybrid composites produced in high pressure die casting with polymer based decoupling layer *Key Eng. Mater.* **742 KEM** 197–204

[8] Schmid A, Arnaut K, Clausen J, Koerdt M, Struß A, Woestmann F-J and Busse M 2018 Process concepts for the manufacturing of hybrid composites made from aluminum and CFRP with a polymer-based decoupling layer *3. Int. Konf. Hybrid Mater. Struct. 2018. Proc.* 216–22
[9] Struß A, Schmid A, Ebrahimi A, Jablonski F and Busse M 2020 Description of the Boundary Layer Behavior of an Aluminum–Carbon-Fiber-Reinforced Polymer Hybrid Compound Using a Cohesive Zone Model *J. Fail. Anal. Prev.* **20** 930–5
[10] Koerdt M 2020 *Development and analysis of the processing of hybrid textiles into endless-fibre-reinforced thermoplastic composites* (Universität Bremen)
[11] Day M, Cooney J D and Wiles D M 1989 The thermal stability of poly(aryl-ether–ether-ketone) as assessed by thermogravimetry *J. Appl. Polym. Sci.* **38** 323–37
[12] Wu Z, Zheng Y Bin, Yu H X, Seki M and Yosomiya R 1988 Effect of thermal history on crystallization behavior of polyetheretherketone studied by differential scanning calorimetry *Die Angew. Makromol. Chemie* **164** 21–34
[13] Cebe P 1988 Annealing study of poly (etheretherketone) *J. Mater. Sci.* **23** 3721–31
[14] Regis M, Bellare A, Pascolini T and Bracco P 2017 Characterization of thermally annealed PEEK and CFR-PEEK composites: Structure-properties relationships *Polym. Degrad. Stab.* **136** 121–30
[15] Zhang M Q, Lu Z P and Friedrich K 1997 Thermal analysis of the wear debris of polyetheretherketone *Tribol. Int.* **30** 103–11
[16] Gao S L and Kim J K 2002 Correlation among crystalline morphology of PEEK, interface bond strength, and in-plane mechanical properties of carbon/PEEK composites *J. Appl. Polym. Sci.* **84** 1155–67
[17] Patel P, Hull T R, McCabe R W, Flath D, Grasmeder J and Percy M 2010 Mechanism of thermal decomposition of poly(ether ether ketone) (PEEK) from a review of decomposition studies *Polym. Degrad. Stab.* **95** 709–18
[18] Tsai C J, Perng L H and Ling Y C 1997 A study of thermal degradation of poly(aryl-ether-ether-ketone) using stepwise pyrolysis/gas chromatography/mass spectrometry *Rapid Commun. Mass Spectrom.* **11** 1987–95

Prediction of residual-stress depth profiles in turning of EN AW-2017 based on in-process measurements of machining forces and temperatures

T Mehner*[,1], T Junge[2], A Schubert[2] and T Lampke[1]

[1] Materials and Surface Engineering Group, Institute of Materials Science and Engineering, Chemnitz University of Technology, 09107 Chemnitz, Germany
[2] Professorship Micromanufacturing Technology, Chemnitz University of Technology, 09107 Chemnitz, Germany

*e-mail: thomas.mehner@mb.tu-chemnitz.de

Abstract. The surface integrity of parts is strongly impacted by the surface-layer properties, which are modified by machining processes. In particular, it is advantageous if the finish machining process generates a resilient residual-stress state without additional post-treatment. Thus, this paper describes relationships between the forces and temperature which are measured in-situ/during the process and the residual-stress profile for the turning of the aluminum alloy EN AW-2017. The residual-stress depth profiles are measured by X-ray diffraction after electrochemical removal of material by means of jet-electrochemical machining. The characteristic features of the residual-stress profile (value and depth under the surface of the local minimum of the residual stress) are determined and modeled using multiple regression. The predictions of the models are validated by test samples. An excellent agreement between experiments and the model is achieved. Thus, the models can be applied to predict the expected residual-stress profiles during the machining process, which allows for an in-process adjustment of the machining parameters in order to generate an advantageous residual-stress state.

1. Introduction

Machining of parts affects the surface-layer properties. In particular, the residual stresses are impacted by the mechanical and thermal loads during machining. As a result, a characteristic residual-stress state occurs, which strongly influences the stress-corrosion cracking and fatigue behaviors, as well as the overall surface integrity. Thus, in addition to the generation of the final geometry, the finish machining process (e.g. turning) can be applied to tailor the residual-stress state, which could improve these very important properties. In general, compressive residual stresses are considered to be beneficial. While the residual-stress state at the surface is important, the residual-stress depth profile has to be considered as well. Crack growth can be initiated below the surface in regions with sufficiently high tensile residual stresses. However, the residual-stress profiles caused by machining processes were rarely considered and mostly investigated for steels. Only a small number of studies deal with other materials like titanium or aluminum. For the modeling of these profiles, different modeling approaches were used.

Jacobus et al. [1] developed a numerical model for the prediction of residual-stress profiles. The model includes thermomechanical coupling, plastic and frictional heating, convection, conduction, thermal softening and strain hardening. In turning processes of the steel AISI 4340 (1.6563), the model was calibrated and tested. In spite of the limited set of machining parameters (two feeds and two depths of cut) used and the temperature measured optically in 5 mm circumferential distance behind the turning tool, a good agreement between the model and validation tests is achieved. Hua et al. [2] studied hard turning of the steel AISI 52100 particularly in regards to residual-stress profiles. They implemented a hardness-based flow-stress model into an elastic-viscoplastic finite-element model. The numerical investigation of different cutting-tool geometries allowed for a prediction of the residual-stress profiles, which were validated by face-turning experiments. Based on the good agreement between the experiments and the model, the conclusion was drawn that a high hardness of the material, a large hone radius as well as high feed rates cause higher compressive residual stresses at greater depths under the surface. The authors recommend that a low feed rate and high workpiece hardness should be used in order to keep both cutting-edge temperature and cutting force low and obtain higher compressive residual stress at the same time. Zhang et al. [3] modeled the residual-stress profile of a finish hard-turning process of AISI 1053 and AISI 1070 steels using neural networks and a regression model. The researchers included the type of cooling method, tool parameters like material, geometry, and edge preparation, as well as tool flank wear and cutting parameters. The model that is based on a neural network agreed well with the experimentally measured residual-stress profiles. Guo and Anurag [4] performed a finite-element modeling of the residual-stress profiles in hard turning of AISI 52100 steel. A coupled thermal-mechanical model without explicit chip formation was used to simulate the machining temperature and the material deformation for the prediction of the residual-stress profiles, and it was found that the plowed depth is the most important factor that causes the characteristic residual-stress profiles for the hard-turning process. Ordás et al. [5] investigated residual-stress profiles after hard turning of AISI D2 steel with tools of different wear states. They found sub-surface compressive residual stresses whose magnitudes are significantly lower, and depth is much shallower when worn tools were used instead of new ones. Thus, a strong impact of the tool wear state (i.e. the contact conditions during machining as well) is expected. Hence, the contact conditions have to be considered if residual-stress profiles are to be predicted.

Residual-stress profiles of turned titanium have been investigated as well. For example, for the orthogonal turning of Ti-6Al-4V, Pan et al. [6] successfully predicted the residual-stress profile with good agreement using an analytical model. In the model, grain-size evolution and dynamic recrystallization are included.

For aluminum alloys, processes like milling, shot peening, and grinding have been investigated with a focus on residual-stress profiles [7–10], but no such investigations could be found for the turning of aluminum alloys. In this paper, the residual-stress depth profiles of samples machined by turning with different parameters are determined, and their characteristic features (value and depth of the minimum of the residual-stress profile) are predicted. For the first time, the prediction is based on a multiple regression model that uses the in-process temperature and forces as input values instead of machining parameters. Thus, the effects of tool wear do not need to be considered separately.

2. Experimental and materials

The material EN AW-2017 T4511 (extruded rods, supplier BIKAR-ALUMINIUM GmbH, Germany) was machined by turning on a precision lathe of the type PD 32 (Spinner AG, Switzerland). The rotationally symmetric specimens had a total length of 80 mm and were stepped. For a length of 60 mm, the diameter was 37 mm, and the remainder had the diameter of 25 mm. The shorter side was utilized for clamping using a dead-length collet chuck. In the finish machining experiments, uncoated cemented-carbide indexable inserts were used. The tools of the type CCGT 09T304-AZ (Mitsubishi Materials, Japan) were characterized by a tool-included angle of 80°, a clearance angle of 7°, and a rake angle of 26° near the cutting edge. In combination with the tool holder used, the nominal tool cutting-edge angle was 95°. During pre-machining, the feed (0.1 mm), the depth of cut (0.5 mm), and the cutting speed (150 m/min) were kept constant on a machining length of 50 mm while keeping a cylindrical section of 10 mm as the cold junction for the electrical brush contact for the in-process measurement of the temperature. After pre-machining, the specimens had a diameter of 31 mm. In the finish machining experiments, two different depths of cut (a_p/mm = 0.4, 1.2) and five different feeds (f/mm = 0.04, 0.08,

0.12, 0.16, 0.20) at a constant cutting speed of 300 m/min were used. In addition, samples with three different cutting speeds (v_c/(m min^{-1}) = 175, 425, 550) were machined with f = 0.12 mm and a_p = 0.4 mm. In total, 16 different tools were used for the machining of three samples per each set of parameters. One of these three samples was then selected for further investigations. The forces (cutting force F_c, passive force F_p, feed force F_f) were measured during machining by a three-axis dynamometer type 9257A (Kistler Group, Switzerland), which was connected to a charge amplifier type 5070A (Kistler Group, Switzerland). The in-process temperature was determined using a tool–workpiece thermocouple. The rotating workpiece was brought into electrical contact with a copper-graphite brush at the cold end of the workpiece (= cold junction). The thermoelectric voltage between the cold junction and the machining area (= hot junction) was recorded by a digital multimeter DAQ6510 with an integrated multiplexer card type 7700 (Tektronix, USA). The Seebeck coefficients of the workpiece and the tool were used to determine the temperature. Additional details are described by the authors in [11, 12]. Residual stresses were measured by X-ray diffraction (XRD) using a D8 Discover (Bruker, USA) with Co Kα radiation (point focus, irradiated area of approximately 0.7 mm in diameter) by application of the $\sin^2\psi$ method. The {311} peaks of Al (elastic constants of the {311} lattice planes: Young's modulus $E^{\{311\}}$ = 69 GPa, Poisson's ratio $\upsilon^{\{311\}}$ = 0.35 [13]) were measured along three rotation directions (0°, 45°, 90°) each in the positive and negative tilt direction using nine tilt angles ψ that are equally spaced on the $\sin^2\psi$ axis from 0.0 to 0.8. In order to measure the depth profiles of the residual stresses, the material was removed in steps by electrochemical machining (ECM) using an electrolyte jet ($NaNO_3$ electrolyte with a mass fraction of 30 % $NaNO_3$, working gap 300 µm, voltage 56 V, 100 µm inner diameter of the nozzle, nozzle velocity 250–800 µm/s depending on the intended removal depth) on an area of about 5 mm × 5 mm [14, 15]. This process was chosen as it does not cause heating or mechanical deformation, and the removal of material by ECM mainly influences the residual stress in the normal direction, which cannot be measured by XRD in the used configuration. Thus, only a minor impact on the measured residual-stress profiles is expected. The intended removal steps were 10 µm, 25 µm, 45 µm, 70 µm, and 100 µm. The depths were measured by a stylus instrument T 4000 (Hommelwerke GmbH, Germany). The residual stresses were determined at the same position after each removal step. The residual-stress profiles $\sigma(x)$ were fitted using a sinusoidal decay function based on [16]:

$$\sigma(x) = C \cdot \exp(-A \cdot \omega_0 \cdot x) \cdot \cos(\omega_d \cdot x + \varphi) + \sigma_0 \quad (1)$$

with

$$\omega_0 = \frac{\omega_d}{\sqrt{1 - A^2}}. \quad (2)$$

C is the amplitude, A the damping coefficient, ω_0 the undamped frequency, ω_d the damped frequency, φ the phase angle, and σ_0 a residual-stress offset. σ_0 serves the purpose of improving the quality of the fit, but it is usually fixed at zero. The absolute value of the depth under surface is denoted with x. Based on these fits, the depth and value of the residual stress at the minimum of the residual-stress profile were determined.

Correlations between the forces, the temperature, the depth below the surface, and the value of the local minimum of the residual stresses were determined by multiple regression using the analysis of variance (ANOVA) with a square model that includes interactions between the parameters. The parameters (i.e., F_c, F_p, F_f, and T) were normalized in order to limit their values to the interval [-1, 1]. The modeling was done on ten samples (hereafter called "calibration samples") that were machined with constant v_c. The models were tested on four "test samples", which were machined using different v_c (table 1 in section 3). The model equations are independent of the machining parameters. This serves the purpose of validating it for varying cutting speeds.

3. Results and discussion

The forces and temperature that were measured in process during machining are summarized in table 1. N is the number of machining experiments which had been conducted with the same tool previous to the machining of the sample. Various different parameters have been used for these previous machining processes.

Table 1. Results of the in-process measurements of the forces and temperature during machining with different process parameters. The samples labeled as "test samples" will be used for the verification of the multiple regression model.

f (mm)	a_p (mm)	v_c (m/min)	N	F_c (N)	F_p (N)	F_f (N)	T (K)
				Calibration samples			
0.04	0.4	300	6	20.6	5.1	8.7	386.7
0.08	0.4	300	6	39.9	7.0	13.5	390.1
0.12	0.4	300	6	46.9	4.7	9.0	388.6
0.16	0.4	300	6	53.7	7.6	7.5	456.6
0.20	0.4	300	6	68.0	8.5	9.3	407.0
0.04	1.2	300	5	55.5	10.3	25.0	415.3
0.08	1.2	300	2	94.8	18.3	36.4	454.7
0.12	1.2	300	2	130.4	20.7	36.4	455.8
0.16	1.2	300	4	163.2	26.1	43.6	477.2
0.20	1.2	300	4	196.1	29.6	42.2	430.0
				Test samples			
0.12	0.4	175	0	46.4	8.1	10.7	384.0
0.12	0.4	300	0	45.7	6.2	8.97	425.1
0.12	0.4	425	0	42.2	4.5	6.5	452.1
0.12	0.4	550	0	44.7	4.1	6.7	479.7

The temperature actually reached and the required forces depend strongly on the contact conditions between the tool and the specimen, which are influenced by built-up edge formation and tool wear. For various tools, cutting-edge breakage occurred, which strongly affected the contact conditions [12]. Thus, the regression model will be applied using the forces and the temperature instead of the machining parameters. In addition, the forces and the temperature are the more relevant parameters as they affect the development of the residual stresses. Outside of some deviations that are caused by the different tool contact conditions due to built-up edge formation and tool wear, the forces rise with increasing feed and depth of cut. The cutting speed influences the forces only slightly, but the temperature is severely impacted. The slight reduction of the forces is caused by the increased temperature that improves the machinability of aluminum (e.g. decreased flow stress with rising temperature).

A typical profile of the area of ECM material removal is shown in figure 1. While not even in their entire, the middle portions of the profiles, in which the XRD measurements were done, were sufficiently even for all samples.

IOP Conf. Series: Materials Science and Engineering 1147 (2021) 012019 doi:10.1088/1757-899X/1147/1/012019

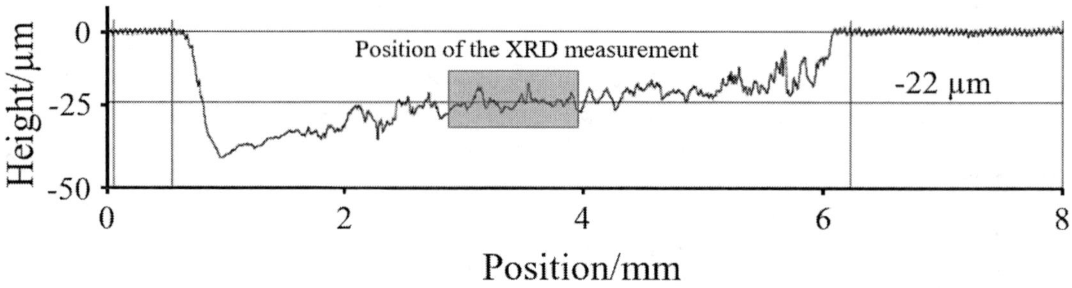

Figure 1. Typical profile after ECM material removal measured along the axial direction. In the center of the area, the residual stresses were measured. This part was flat for all the samples and removal steps. The profile that is shown corresponds to the second removal step with a depth in the center of -22 μm relative to the surface.

The fitting function (equation (2)) describes the course of all residual-stress profiles well. At the surface, tensile residual stresses can be measured for all samples except for low feeds (0.04 mm and 0.08 mm) in the axial direction. At the first removal step (around 10–15 μm), a steep decrease of the residual-stress value can be observed – compressive residual stresses are present in all samples in the axial direction. In the tangential direction, the tensile residual stress is reduced, but no compressive residual stresses are generated for most of the samples. At higher depths after its minimum, the value of the residual stress increases only slightly (typically about 10–30 MPa). The depth of the minimum in the residual-stress curve and the value of the residual stress in the axial direction are shown in figure 2.

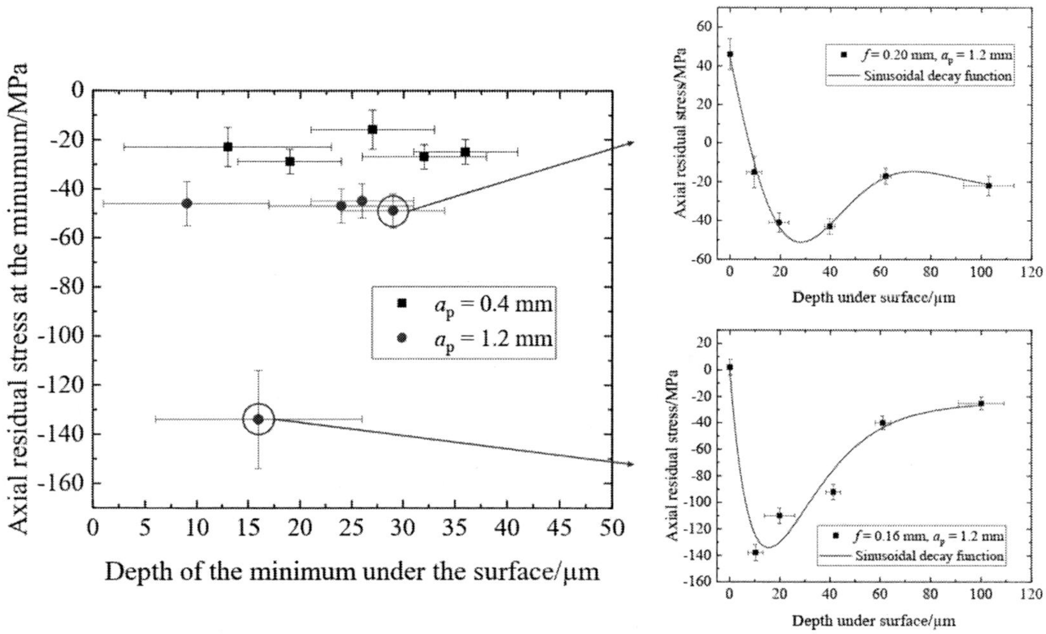

Figure 2. Value and depth under the surface of the residual stresses in the axial direction (left) and corresponding residual-stress depth profiles (right).

In the axial direction, the residual-stress value at the minimum is almost constant for each of the depths of cut (about -24 MPa for $a_p = 0.4$ mm and -47 MPa for $a_p = 1.2$ mm). Thus, the larger depth of cut increases the compressive residual-stress value by more than 20 MPa. However, the depth of the minimum decreases steadily with increasing feed. It should be noted that the sample that was produced with a_p = 1.2 mm and f = 0.16 mm has significantly different residual stresses. This is most likely caused by a local anomalous increase of the forces during machining. The compressive residual stresses below the surface at the curve's minimum are about 100 MPa lower than expected from the tendencies that can be observed in the other samples. As it can be seen in figure 2 (bottom right), this value is not caused by an outlier, as high compressive residual stresses were measured in three depths between 15 µm and 40 µm. The values in figure 2 (left) of this sample could be caused by a local difference at the measurement position in that particular sample. Thus, this sample is not considered in the further analysis. The results of the tangential direction are shown in figure 3.

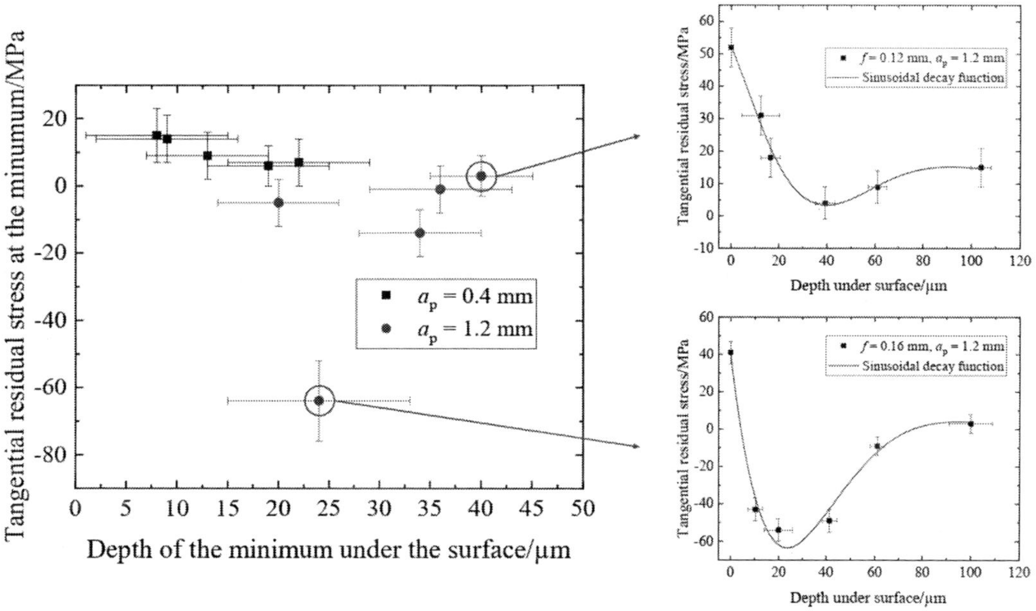

Figure 3. Value and depth under the surface of the residual stresses in the tangential direction (left) and corresponding residual-stress depth profiles (right).

As with the axial direction, the tangential direction shows almost constant values of the residual stress (about 10 MPa for a_p = 0.4 mm and -4 MPa for a_p = 1.2 mm). The larger depth of cut causes the residual stress at the minimum to become smaller or even slightly compressive. The same sample described above, which was produced with a_p = 1.2 mm and f = 0.16 mm, deviates from these trends as it was observed for the axial direction.

For both directions, the residual stress rapidly approaches values below 10 MPa, i.e. almost free of residual stresses, for depths above 60–80 µm – despite the presence of high tensile residual stresses at the surface in the tangential direction. No clear correlations between the results in figures 2 and 3 and the feed could be identified. This is caused by the differences in the tool-wear state, which varied by machining processes. Thus, the forces and the temperature are considered in the following.

Using multiple linear regression and ANOVA, the results from the calibration samples were modeled. The normalization of the parameters as well as the relevant coefficients of the underlying equations are shown in tables 2 and 3.

Table 2. Normalization of the parameters that are used in the multiple-regression model.

Term	Normalization
F_c	$(F_c - 108.36\ \text{N})/87.77\ \text{N}$
F_p	$(F_p - 17.18\ \text{N})/12.44\ \text{N}$
F_f	$(F_f - 24.85\ \text{N})/17.32\ \text{N}$
T	$(T - 421.69\ \text{K})/34.95\ \text{K}$

Table 3. Results of the multiple-regression model. The p-value corresponds to the error probability of the related term.

Term	Coefficient	p-value
Axial residual stresses at the minimum		
F_c	-17.34	< 0.001
F_f^2	23.00	< 0.001
Constant	-54.92	
Tangential residual stress at the minimum		
T	11.42	0.006
F_p	-16.57	< 0.001
$T \cdot F_c$	16.89	0.006
F_f^2	9.48	0.012
Constant	-12.81	
Depth of the minimum (axial)		
$T \cdot F_c$	22.22	0.176
$T \cdot F_f$	-14.62	0.144
$F_p \cdot F_f$	-58.77	0.057
F_f^2	74.55	0.026
Constant	7.62	
Depth of the minimum (tangential)		
F_p	14.21	< 0.001
$F_p \cdot F_c$	-14.59	0.003
Constant	35.07	

The experimental and calculated results are compared in figures 4 and 5.

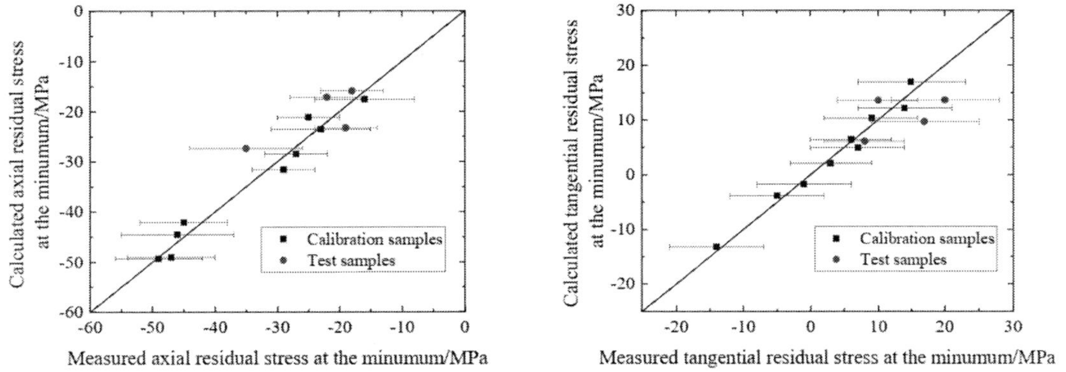

Figure 4. Comparison of the measured and calculated residual-stress values at the minimum of the residual-stress profile in the axial (left, $R^2 = 0.967$) and tangential (right, $R^2 = 0.975$) directions. The modeling was done using the calibration samples only. The lines correspond to $y = x$.

The residual-stress values at the minimum of the residual-stress profile can be fitted very well using only three (axial direction) or five (tangential direction) terms. The R^2 values are very high (> 0.960) and the error probabilities p very low (<< 0.05). For the axial direction, only F_c and F_f have a significant impact on the residual stresses. When the cutting force is above 108.36 N (table 2), the residual-stress value at the minimum becomes smaller (or more compressive), whereas an increase of the feed force always increases the tensile residual stresses. In the tangential direction, F_p, F_f and the temperature as well as its interaction with F_c have an influence on the residual-stress values. As for the axial direction, an increase in the feed force causes more tensile residual stresses. Only temperatures above 421.69 K would increase the compressive residual stresses, but due to the interaction with F_c, the interaction term of almost all machining processes is positive. The effects of both terms combined lead to the conclusion that only high temperatures and low cutting forces would increase the compressive residual stress. This is not a typical combination of these two parameters, which leads to the conclusion that high temperatures have a negative impact on the residual stress at the minimum of the depth profile. High passive forces favor compressive residual stresses.

The model shows a very good agreement with the experimental data. When the test samples are taken into account, their deviations from the line $y = x$ are small. Thus, both model equations can be used to predict the residual stresses at the minimum of the residual-stress profile with high accuracy. The calculated depth of the minimum of the residual stresses is shown in figure 5 in comparison to the experimental values.

The depth of the minimum of the residual stresses can be calculated with the models with high accuracy. The R^2 in the axial direction is the lowest of the four models, but the value is still high (0.874). For this model, the depth increases with increasing feed force. There are three relevant interactions for this direction, namely the interactions between temperature and cutting force, temperature and feed force, and passive and feed forces. Thus, the resulting depth of the minimum is a complex interaction between all parameters. The p-values are comparably high for the temperature-related interactions (> 0.10), but these interactions are necessary in order to predict the depth values of the test samples. In the tangential direction, the p-values are very low (<< 0.05). Only the passive force and its interaction with the cutting force influences the depth of the minimum. High passive forces (> 17.18 N) in combination with low cutting forces (< 108.36 N) would increase this depth, but this combination cannot be realized by the parameter sets that have been investigated in this study.

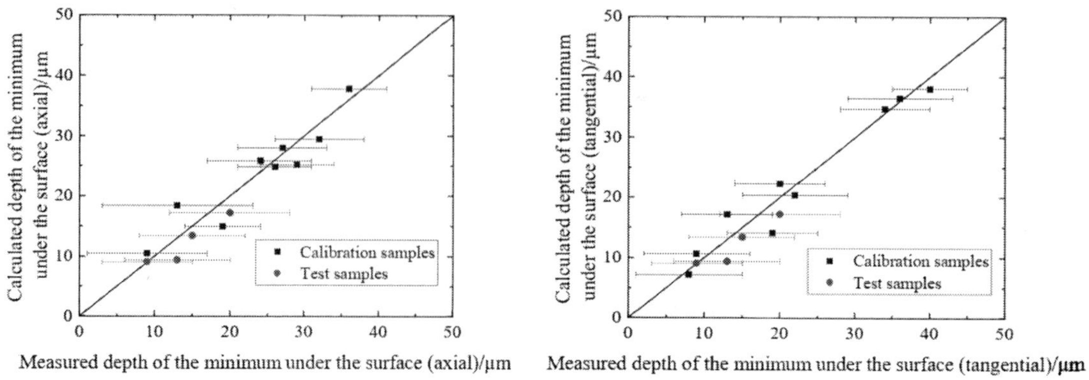

Figure 5. Comparison of the measured and calculated depths of the minimum of the residual-stress profile in the axial (left, $R^2 = 0.874$) and tangential (right, $R^2 = 0.949$) directions. The modeling was done using the calibration samples only. The lines correspond to $y = x$.

For both of the models in figure 5, the test samples show an excellent agreement. Thus, the general validity of all four models was shown by using test samples that were machined with a variation of v_c. No such variation was done for the calibration samples. Hence for the modeling, the approach to use F_c, F_p, F_f, and T, which were measured in process, was proven to be applicable to predict the relevant features of the residual-stress profiles in the turning of EN AW-2017.

4. Conclusions

The characteristic features of the residual-stress profiles generated by turning of the aluminum alloy EN AW-2017 were determined by XRD after stepwise removal of material through the application of jet electrochemical machining. On the surface, tensile residual stresses were present in most of the samples regardless of the machining parameters used. The residual stresses decline quickly and a typical local minimum was measured. At the minimum, mostly tensile residual stress were measured in the tangential direction, whereas compressive residual stresses were detected in the axial direction.

The multiple-regression models, which use the parameters measured in process, describe the experimental observations with excellent agreement, which is also true for the test samples. Thus, the models allow for the prediction of the aspired-to temperature and forces during the turning process in order to achieve an advantageous residual-stress profile (e.g. for improved fatigue strength). Due to the inclusion of test samples with varying cutting speed, the models were validated for a wide range of processing parameters. Hence, the approach to model the in-process temperature and forces (instead of processing parameters) has been proven to be suitable as the impacts of the tool wear on the residual-stress profile are implicitly included in these models. Time-consuming measurements of residual-stress profiles can be omitted. The models allow for an in-process control of the expected residual-stress state. Thus, despite tool wear, the turning process can be used to adjust the surface integrity robustly during machining.

Acknowledgments

The scientific work has been supported by the German Research Foundation (DFG) within the research priority program SPP 2086 (SCHU 1484/26-1, LA 1274/49-1) grant number 401805994. The authors thank the DFG for this funding and intensive technical support.

The support of Franz Pfaffendorf (ECM), Marc Pügner (XRD measurements), Claudia Albero Rojas (depth measurements), and Morgan Uland (English proofreading) are gratefully acknowledged.

References

[1] Jacobus K, DeVor R E, Kapoor S G and Peascoe R A 2001 Predictive Model for the Full Biaxial Surface and Subsurface Residual Stress Profiles from Turning *J. Manuf. Sci. Eng.* **123** 537–546

[2] Hua J, Shivpuri R, Cheng X, Bedekar V, Matsumoto Y, Hashimoto F and Watkins T R 2005 Effect of feed rate, workpiece hardness and cutting edge on subsurface residual stress in the hard turning of bearing steel using chamfer+hone cutting edge geometry *Mater. Sci. Eng.* A **394** 238–248

[3] Zhang J Y, Liang S Y, Zhang G and Yen D 2006 Modeling of residual stress profile in finish hard turning *Mater. Manuf. Process.* **21** 39–45

[4] Guo Y B and Anurag S 2009 Finite element modeling of residual stress profile patterns in hard turning *Powder Diffraction* **24** S22–S25

[5] Ordás N, Penalva M L, Fernández J and García-Rosales C 2003 Residual stresses in tool steel due to hard-turning *J. Appl. Crystallogr.* **36** 1135–1143

[6] Pan Z, Shih D S, Garmestani H and Liang S Y 2019 Residual stress prediction for turning of Ti-6Al-4V considering the microstructure evolution *J. Eng. Manuf.* **233** 109–117

[7] Denkena B, Boehnke D and de León L. 2008 Machining induced residual stress in structural aluminum parts *Prod. Eng. Res. Devel.* **2** 247–253

[8] Žagar S and Grum J 2011 Surface Integrity after Mechanical Hardening of Various Aluminium Alloys *Strojniški vestnik – J. Mechan. Eng.* **57** 334–344

[9] Žagar S and Grum J 2015 Roughness, residual stresses and pitting corrosion effect on shot peened AA 7075 *Tehnički vjesnik* **22** 1589–1595

[10] Žagar S and Šturm R 2019 Influence of Grinding on Residual Stress Measurements of Shot Peened Aluminium Alloy AA7075 *Strojniški vestnik – J. Mechan. Eng.* **65** 401–409

[11] Junge T, Liborius H, Mehner T, Nestler A, Schubert A and Lampke T 2020 Measurement system based on the Seebeck effect for the determination of temperature and tool wear during turning of aluminum alloys *Procedia CIRP* **93** 1435–1441

[12] Junge T, Nestler A and Schubert A 2021 Application Potential of Thermoelectric Signals for Temperature Monitoring in Turning of Aluminum Alloys *Production at the leading edge of technology, Proceedings of the 10th Congress of the German Academic Association for Production Technology (WGP)* 235–245

[13] Eigenmann B and Macherauch E 1996 Röntgenographische Untersuchung von Spannungszuständen in Werkstoffen (Teil 3) *Mat.-wiss. u. Werkstofftech.* **27** 426–437 (in German)

[14] Hackert M, Meichsner G and Schubert A 2009 Generating plane and microstructured surfaces applying Jet Electrochemical Machining *Proceedings of Fifth International Symposium on Electrochemical Machining Technology, Fraunhofer IKTS Dresden*, ed A Michaelis, M Schneider (Stuttgart: Fraunhofer Verlag) p 51

[15] Hackert-Oschätzchen M, Meichsner G, Zinecker M, Martin A and Schubert A 2012 Micro Machining with Continuous Electrolytic Free Jet *Precision Eng.* **36** 612–619

[16] Ulutan D, Arisoy Y M, Özel T and Mears L 2014 Empirical modeling of residual stress profile in machining nickelbased superalloys using the sinusoidal decay function Procedia CIRP 13 365–370

Influence of metal matrix powder size on the tensile strength of a SiC_p/AlSi7Mg0,6 composite produced by field assisted sintering technique

R Pippig*, S J Hirsch, T Grund and T Lampke
Chemnitz University of Technology, Chair of Materials and Surface Engineering
*e-mail: robert.pippig@mb.tu-chemnitz.de

Abstract. In the present study, 35 vol% SiC_p/AlSi7Mg0,6 composites were prepared using field assisted sintering technique in order to investigate the effect of different particle fractions and size distributions of the AlSi7Mg matrix powder on the tensile properties of the produced composite material. In most usecases the size of the reinforcement phase is given by the application and is only variable within narrow limits ($\leq$ 20 µm particle size in this work). On the other hand, there is potential for optimization of the matrix powder. In this investigation, fine (d_{50} = 25 µm), coarse (d_{50} = 52 µm), bimodal (50 wt% of fine + 50 wt% of coarse, d_{50} = 36 µm) and as received (d_{50} = 40 µm) aluminum powder was used as the matrix powder. Using fine matrix powder has improved yield strength by 5 % and ultimate tensile strength by 7 % compared to the as received condition. This is largely due to the lower porosity of the composite produced under the use of the fine matrix powder ((0.07 ± 0.04) %) in contrast to the composite using the as received aluminum powder ((0.62 ± 0.35) %). At the same time, the consumed heating energy of the composite was decreased by almost a third when using the fine matrix powder in comparison to the use of the as-received matrix powder. This paper presents results of an optimization approach for mechanical properties of aluminum matrix composites without any changes of the sintering parameters.

1. Introduction

Due to their widely adjustable properties, aluminum matrix composites (AMCs) promise high application potential in various industries. A wide range of applications is currently not given due to their complex and costly production. Therefore, different fabrication routes are the subject of current research. Field assisted sintering technique (FAST), often referred to as spark plasma sintering (SPS) in the literature, allows the production of various materials, among others ceramics [1], metals and their alloys [2], composites [3, 4] and even high entropy alloys [5]. It is not surprising that this promising process also seems to have potential in the production of AMCs [6, 7]. Almost pore-free AMCs with a homogeneously distributed reinforcement phase can be fabricated [8]. Therefore, both the matrix material and the reinforcing phase are mixed in powder form before the sintering process. One parameter to influence production and application properties is the powder size, both of the matrix and the reinforcement phase. The reinforcement particle size is mainly given by the application: nanometer range for improvements in strength properties [6, 9, 10] and micrometer range for the improvement of wear resistance [11, 12]. If the application is regarded as given, the variation of the matrix remains. Obviously, the smaller

Table 1. Chemical composition of the used metal powder according to DIN EN 1706 in wt%.

Si	Fe	Cu	Mn	Mg	Zn	Ti	others	Al
6.50...7.50	0.19	0.05	0.10	0.45...0.70	0.07	0.25	0.10	balance

Table 2. The used fractions of metal matrix powder and the SiC powder as the reinforcement phase provided by sieving. The combination of this description of the fraction and AMC is used in the text as an abbreviation for the corresponding mixed composites, e. g. "fine AMC".

fraction	d_{10} (µm)	d_{50} (µm)	d_{90} (µm)
fine	9	25	41
coarse	35	52	80
bimodal	15	36	60
as received	23	40	64
SiC	12	21	33

the powder size, the higher the packing density and the amount of contact points between particles which leads to a better diffusion and neck forming in sintering processes [13]. But on the other hand, aluminum is a strong passivating metal: The smaller the powder size the larger the oxidized surface area which leads to a worser sintering result. In addition to that, diffusion is by far not the only compacting mechanism during field assisted sintering [14]. Among others, "particle rearrangement" and localized deformation are to be mentioned. Both effects have positive impact on the sintering process and are more pronounced with coarser particles [2].

In this paper, the influence of the metal matrix particle fraction and size distribution on the mechanical properties of an AMC produced by field assissted sintering is investigated.

2. Materials and methods

Inert gas atomized powder "AlSi7Mg0,6 < 0,063 mm" by ECKA Granules Germany GmbH (for chemical composition see table 1) was used as the matrix material. The powder was sieved into two different fractions (see table 2), which were examined for their characteristic diameters by laser diffraction analysis using a "Cilas 920" device. Together with the bimodal fraction (mixture of 50 wt% of fine and 50 wt% of coarse fraction) and the as received condition, four fractions and size distributions were examined. SiC-Powder ($d_{50} = 21$ µm) was used as a reinforcement phase. The four different matrix powder fractions were mixed with the SiC powder using a tumbler mixer to a composite powder with 35 vol% reinforcement content, each. According to DIN EN ISO 3953, the bulk density of all the powder mixtures were determined.

The powder mixtures were compacted using field assisted sintering by a "SPS KCE FCT-HP D 25-SI" (FCT Systeme GmbH, Germany) in order to produce AMC variants. Three samples of each AMC variant with a diameter of 40 mm and a height of 8 mm were prepared in a random order. The used sintering tools (die and punchs) were made out of graphite. In addition to that, a release agent in form of a 0.3 mm graphite foil was used between die and punchs. To minimize reactions of both the sample and the tool with the atmosphere during sintering, the recipient was flushed with argon and evacuated to a low vacuum (< 1 mbar) twice. The sintering was performed with a constant current using the following manufacturing parameters: a sintering temperature of 520 °C, a constant pressure of 50 MPa, a heating rate of 100 K min^{-1} and a holding

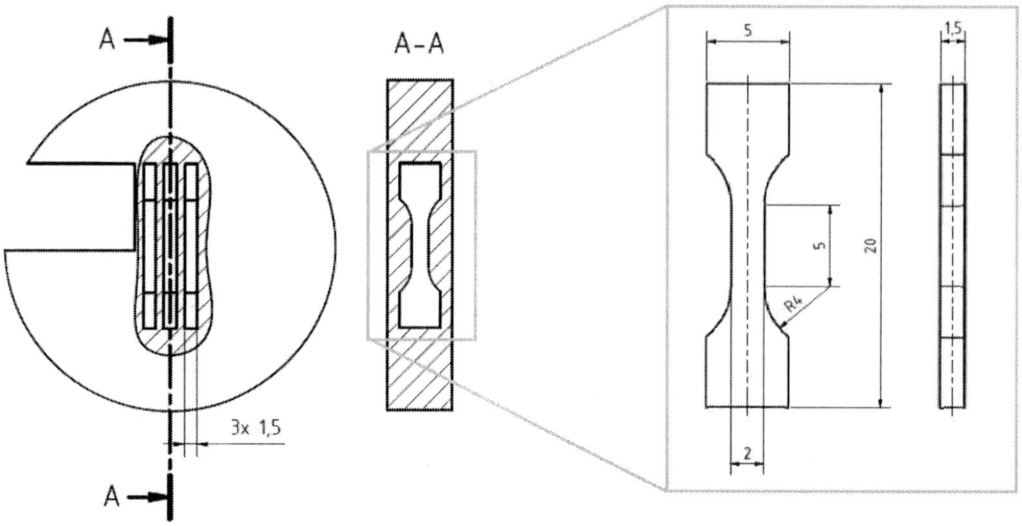

Figure 1. Principle sketch of sampling from the sintered pellet. Three tensile specimens were taken from the center of the pellet and perpendicular to the compression direction. Also the taking of the metallographic sample can be seen.

time of 5 min. The cooling to room temperature was uncontrolled with a resulting cooling rate between (50 and 100) $K\,min^{-1}$. By integrating the consumed heating power over time and corrected by the amount of energy used for a cylce in the empty sintering device, the heating energy per sample was calculated. The density of all samples was estimated by Archimedes´ principle using a Mettler-Toledo laboratory balance. From this the difference to the theoretical density could be determined. Porosity was determined optically using gray value correlation. According to figure 1, one specimen for metallographic investigations and three tensile specimens were eroded from the sintered samples. Cross-sections for the metallographic investigations were prepared by standard procedures. In total there were nine tensile specimens tested per condition. Tensile tests were performed with a "Zwick Allround-Line 20 kN" (ZwickRoell GmbH & Co. KG, Germany) at room temperature using an initial strain rate of $10^{-3}\,s^{-1}$. The strain was measured by gray value correlation using a "GOM Aramis" system. The toughness was calculated by integrating the stress-strain curve until failure.

3. Results and discussion

Figure 2 (a) shows the bulk density of the powder fractions. Obviously, this is highest with the fine AMC and lowest with the coarse AMC. The other two fractions fall in between and do not differ significantly from each other. A high bulk density means many contact points between the powder particles, which in turn promotes diffusion between the particles and leads to a high number of sinter bridges. A lower bulk density in turn allows intensive particle rearrangement and considerably larger local deformations, especially at the first stages of the sintering process. Furthermore, it should be noted that with decreasing particle size the total particle surface increases quadratically. In the case of a passivating material (e. g. the alloy considered here), this also means a significantly higher amount of oxide layers within the bulk material. Oxide layer could impede diffusion during the sintering process and may have a negative effect on microstructure formation.

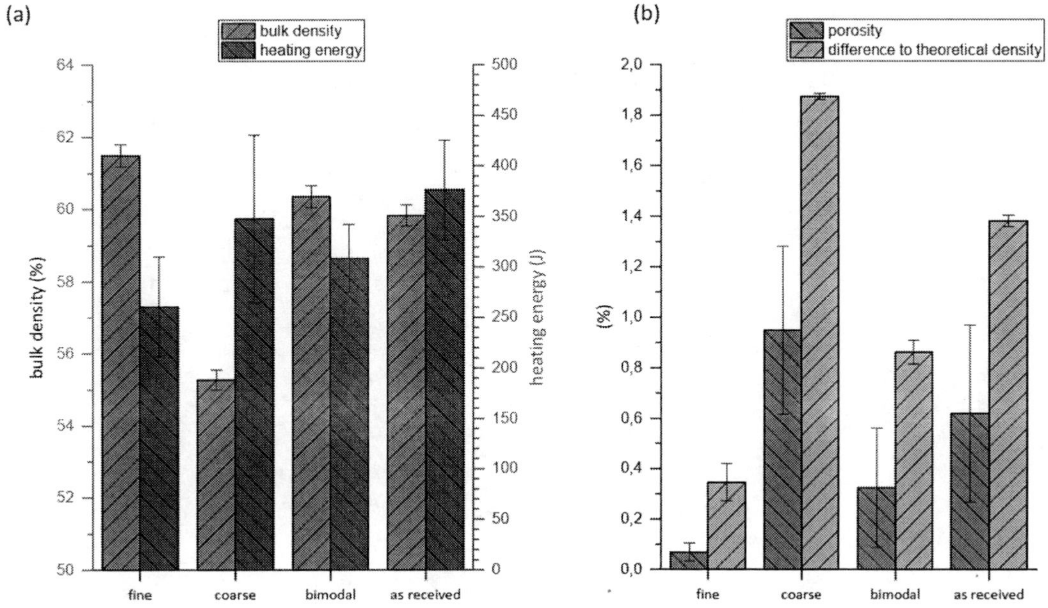

Figure 2. (a) Bulk density of the powder fractions in relation to the theoretical density of a pore-free-composite with the given composition. Furthermore, the consumed heating energy of the samples during sintering process, corrected by the amount of heating energy for the sintering apparatus, determined in an empty run. (b) Porosity of the compacted samples determined by microscopic investigations and by Archimedes principle using a laboratory balance, respectively.

The energy consumed by the samples during the sintering process is shown in Figure 2 (a). The as received AMC consumed a heating energy of (376 ± 49) J, which is on a similar level to the coarse AMC that consumed (348 ± 83) J. A significant reduction in energy can be observed in both the fine composite and the bimodal AMC. The bimodel AMC consumes an energy of (308 ± 84) J, which means a reduction of about 20 % compared to the as received AMC. In fact, the fine AMC consumes only an energy of (261 ± 49) J, which means a reduction of almost one third (31 %) compared to the as received AMC. This is probably due to the fact that there are significantly more contact points between the powder particles, especially at the beginning of the sintering process, which leads to a better conversion of the electric current into joule heat. Therefore, this temperature-controlled process requires less electrical power to reach and maintain the set temperature.

Figure 3 shows the metallographic cuttings of the four different AMC conditions. For the fine metall matrix powder there are an almost pore-free microstructure with a homogenous dispersion of the reinforcement phase. Almost no clusters of SiC-particles are visible. In contrast, the AMC with the coarse metall matrix powder shows huge clusters of the reinforcement phase and clearly visible porosity. This is due to the missing content of fines. There are no fines which could penetrate into the clusters evolved during sintering process. The results are large voids inside the SiC-clusters. This conclusion is supported by the other two AMCs. Despite there are a less homogenous distribution of SiC-particles inside the bimodal AMC with a tendency to clustering, there are no significiant voids visible. This is most likely due to the mixed in fines of this composite. The bimodal AMC and the as received AMC have a similar average powder size,

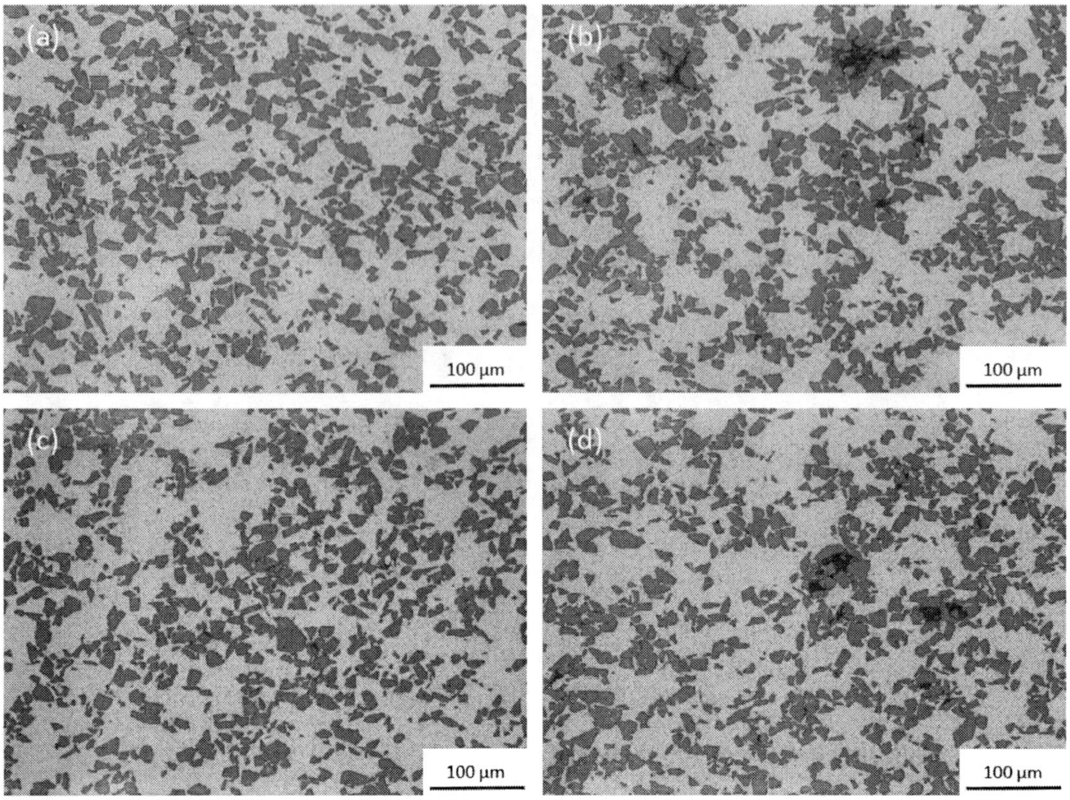

Figure 3. Micrographs of the different AMCs produced: (a) fine AMC, (b) coarse AMC, (c) bimodal AMC and (d) as received AMC. The bright phase is the aluminium alloy, the dark phase is the SiC and the black areas are pores.

but the content of fine fractions in the as received composite is much smaller. As a consequence, in this samples voids can be found similar to the coarse AMC. It seems, that not the average powder size of the metal matrix powder, but a critical amount of fine fraction influences the occurence of voids.

The qualitative microstructure observations are confirmed in the difference to theoretical density $\Delta\rho_{th}$ and the porosity. The fine AMC has a $\Delta\rho_{th}$ of (0.35 ± 0.07) %, only a porosity of (0.07 ± 0.04) % can be identified. This means a nearly non-porous material. The coarse AMC, on the other hand, shows the highest $\Delta\rho_{th}$ of (1.87 ± 0.02) %, porosity measured of (0.95 ± 0.33) %, as suspected when looking at the microstructure. A $\Delta\rho_{th}$ of (0.86 ± 0.05) %, porosity of (0.32 ± 0.24) %, is detected for the bimodal AMC, which allows a practical application of this composite. This porosity level is in the range of gravity casting, a potential competitive method. The bimodal AMC is also less porous than the AMC produced with the as received powder. The latter shows a $\Delta\rho_{th}$ of (1.38 ± 0.02) %, optically measured of (0.62 ± 0.35) %.

It is noticeable, that the $\Delta\rho_{th}$ calculated by Archimedes´ principle is higher than the porosity determined by optical microscopy for all samples. This is due to the density measurement of the whole sample by laboratory balance. In contrast, the microscopic determination is performed at some cross sections of the sample, not the sample as a whole. The higher standard deviation supports this conclusion. So, it is assumed and in accordance with other authors (e. g. [15]), that the Archimedes´ principle is more reliable than microscopic determination for absolute values.

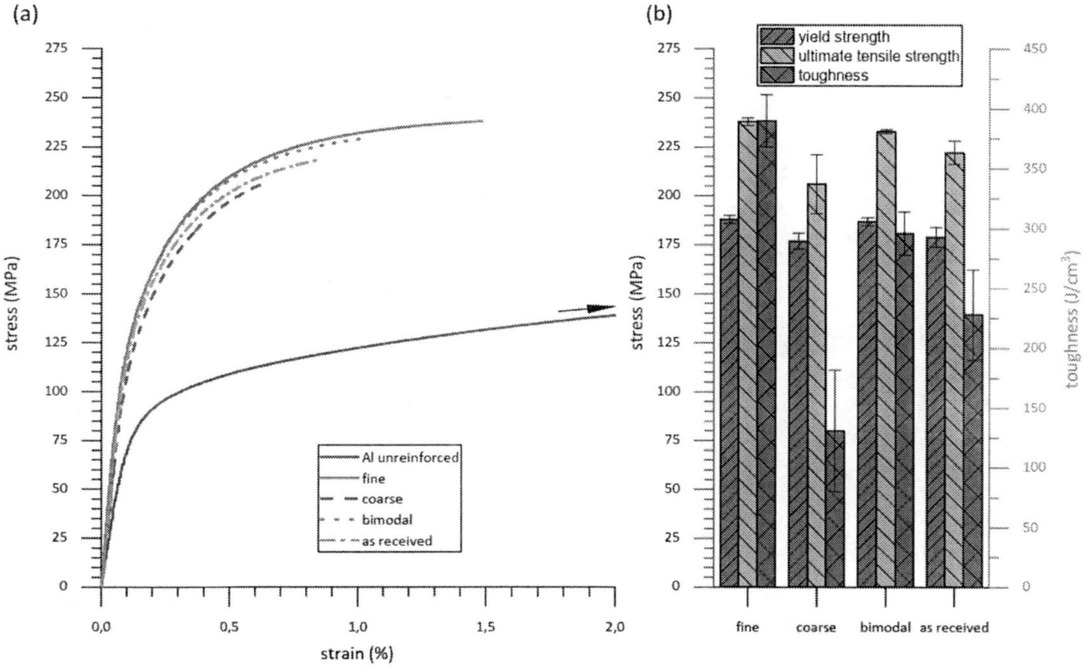

Figure 4. (a) Representative engineering stress-strain curves of the four different AMCs and the same matrix alloy without reinforcement with initial strain rate of $10^{-3}\,s^{-1}$ until uniform elongation. (b) Yield strength, ultimate tensile strength and toughness of the four different AMC conditions. The vertical stress-axis in (a) and (b) have the same scale.

However, it should also be mentioned that $\Delta\rho_{th}$ determined by Archimedes´ principle is related to the theoretical density of the composite. In practicle use, this can be subjected to slight fluctuations due to the mixing process. In relation, both measurement methods are consistent with each other.

Representative stress-strain curves of each composite are shown in Figure 4 (a) and the yield strength (YS), ultimate tensile strength (UTS) and toughness are shown in Figure 4 (b). The as received AMC shows a YS of (179 ± 5) MPa and an UTS of (222 ± 6) MPa, while the toughness was calculated on $(228 \pm 38)\,\mathrm{J\,cm^{-3}}$. The fine AMC shows both, the best strength and toughness. Out of all tensile specimens a YS of (188 ± 2) MPa (increase of 5 %) and an UTS of (238 ± 2) MPa (increase of 7 %) was measured with a toughness of $(390 \pm 22)\,\mathrm{J\,cm^{-3}}$ (70 % more than the as received AMC). The bimodal AMC shows a similar curve shape to the fine AMC, which results in similar strength values, namely a YS of (187 ± 2) MPa and an UTS of (233 ± 1) MPa, respectively. The lower toughness of $(296 \pm 18)\,\mathrm{J\,cm^{-3}}$ is apparently due to the lower uniform elongation. In contrast to the composites with fine content, the coarse AMC shows the worst strength and toughness values with a YS of (177 ± 4) MPa, an UTS of (206 ± 15) MPa and a toughness of $(131 \pm 51)\,\mathrm{J\,cm^{-3}}$. In particular, the toughness is over 40 % less then the as received AMC. Also worth mentioning is the greater fluctuation of the measured values between the parallel samples, especially for the UTS and the toughness. This is probably due to the pores and clusters, described at the microscopic observations, which act as inner notches.

Thus, the $\Delta\rho_{th}$ and the porosity of the samples were reflected one-to-one in the mechanical properties. A composite that is as dense as possible is therefore obtained by the presence of a certain amount of fines. It is therefore clear that the composite sintered exclusively from a very fine fraction has the highest density and the best mechanical properties. A significant influence due to the increasing proportion of oxide films with smaller particles does not seem to exist in this order of magnitude (micro-scale). On the other hand, adding a fine fraction to a coarse fraction has a very strong effect, as the bimodal composite proves. Compared to the as received AMC (and of course to the coarse AMC), the density and thus the tensile strength was significantly improved. This circumstance possibly allows the utilization of fine and coarse fractions by sintering, which are discarded in other manufacturing processes, e. g. in thermal spraying. An effect that coarse particles can have a positive impact on the FAST process (see [2]) could not be determined for the SiC_p/AlSi7Mg0,6 composites examined in this work.

4. Conclusions

SiC_p/AlSi7Mg0,6 composites with different powder fractions and particle sizes as a matrix, fine ($d_{50} = 25\,\mu m$), coarse ($d_{50} = 52\,\mu m$), bimodal (50 wt% of fine + 50 wt% of coarse, $d_{50} = 36\,\mu m$) and as received ($d_{50} = 40\,\mu m$), were fabricated by field assisted sintering technique and examined for their microstructure and mechanical properties. The following key aspects can be summarized from this study.

(i) Heating energy consumed by the fine AMC was over 30 % less than the as received AMC. Both, coarse and bidmodal AMC showed no analyzable differences to the as received AMC.

(ii) The fine AMC shows a nearly pore-free microstructure with a homogenous distribution of the reinforcement phase, whereas the coarse AMC shows the highest porosity with partly hugh clusters of SiC particles. Bimodal powder fraction can improve density in comparison to the as received powder fraction.

(iii) Using the fine aluminum powder as a matrix increases the yield strength about 5 % and the ultimate tensile strength about 7 % compared to the as received powder. The toughness was increased by about 70 %. The bimodal specimen achieves similar strength values, but with a considerably lower toughness. The coarse AMC had the worst mechanical properties, especially tensile strength (decrease of 7 %) and toughness (decrease over 40 %) are significantly lower than in the as received composite.

Acknowledgments

The authors would like to thank the technical staff of the institute for the production of the tensile specimens and the support in metallographic preparation.

References

[1] Wang L, Pouchly V, Maca K, Shen Z and Xiong Y 2015 *J. Asian Ceram. Soc.* **3** 183–187
[2] Diouf S and Molinari A 2012 *Powder Technol.* **221** 220–227
[3] Sulima I, Kowalik R and Hyjek P 2016 *J. Alloys Compd.* **688** 1195–1205
[4] Park K, Park J and Kwon H 2017 *Mater. Sci. Eng. A* **691** 8–15
[5] Löbel M, Lindner T, Pippig R and Lampke T 2019 *Entropy* **21** 582
[6] Vintila R, Charest A, Drew R and Brochu M 2011 *Mater. Sci. Eng. A* **528** 4395–4407
[7] Jiang H, Xu Z, Xiu Z, Jiang L, Gou H, Zhou C and Wu G 2018 *J. Alloys Compd.* **763** 822–834
[8] Ghasali E, Pakseresht A, Rahbari A, Eslami-shahed H, Alizadeh M and Ebadzadeh T 2016 *J. Alloys Compd.* **666** 366–371
[9] Casati R and Vedani M 2014 *Metals* **4** 65–83
[10] Amirkhanlou S, Ji S, Zhang Y, Watson D and Fan Z 2017 *J. Alloys Compd.* **694** 313–324
[11] Kan W H, Ye Z J, Zhu Y, Bhatia V K, Dolman K, Lucey T, Tang X, Proust G and Cairney J 2016 *Mater. Charact.* **119** 65–74
[12] Alaneme K K, Fajemisin A V and Maledi N B 2019 *J. Mater. Sci. Technol.* **8** 670–682
[13] Cheng Y, Cui Z, Cheng L, Gong D and Wang W 2017 *Adv. Powder Technol.* **28** 1129–1135
[14] Monchoux J P 2019 *Spark Plasma Sintering of Materials* ed Cavaliere P (Cham: Springer International Publishing) pp 93–115
[15] Spierings A, Schneider M and Eggenberger R 2011 *Rapid Prototyp. J.* **17** 380–386

Suitability of roughness parameters for the interlaminar strength prediction of mechanically interlocked polymer-metal-interfaces

E Saborowski[1,*], P Steinert[2], A Dittes[1], T Lindner[1], A Schubert[2] and T Lampke[1]

[1] Materials and Surface Engineering Group, Faculty of Mechanical Engineering, Chemnitz University of Technology, Erfenschlager Straße 73, D-09125 Chemnitz, Germany
[2] Micromanufacturing Technology Group, Faculty of Mechanical Engineering, Chemnitz University of Technology, Reichenhainer Straße 70, D-09126 Chemnitz, Germany

[*]e-mail: erik.saborowski@mb.tu-chemnitz.de

Abstract. The interlaminar strength of mechanically interlocked polymer-metal-interfaces strongly depends on the metal surface topography. Hence, this contribution assesses standardized surface roughness parameters as well as fractal dimension for interlaminar strength prediction of EN AW-6082/polyamide 6 polymer-metal-hybrids. Seven different metal surface topographies generated by mechanical blasting, alkaline etching, thermal spraying as well as laser structuring are investigated. Interlaminar strength testing is carried out with regard to tensile and shear strength using butt-bonded hollow cylinder specimens. The surface roughness profiles are recorded with a tactile stylus profiler. The fractal dimension of the metal surface is derived from cross-sectional images. The measured interlaminar strength values show good correlation to the fractal dimension, moderate correlation to the root mean square slope of the surface roughness profile and no correlation to surface roughness height parameters.

1. Introduction

Thermoplastic polymer-metal-hybrids (PMH) offer great potential for automotive mass production due to their quick processability. If reinforcing fibers are added to the polymer, they also provide high specific strength and stiffness. Thermal contact joining using micro-scale mechanical interlocking is a common approach for connecting both dissimilar material groups. Thereby, the thermoplastic polymer is used as hot melt adhesive as it infiltrates and interlocks the roughness features of the metal surface. For joining PMH, metal and polymer are put together under pressure, while the contact area between both materials is heated until the polymer melts. The necessary thermal energy can e.g. be generated by direct laser heating [1], indirect laser heating [2, 3, 4], induction heating [5, 6], ultrasonic oscillations [7, 8], friction [9] or heat conduction [10, 11, 12]. The achievable adhesion depends mainly on the surface topography of the metallic joining partner. Therefore, the precise micro-structuring of surfaces offers promising possibilities for increasing the interlaminar strength of dissimilar material joints. Possible surface structuring techniques are mechanical blasting [2, 13, 12, 14], laser structuring [3, 8, 15, 16, 17, 18], thermal spraying [6, 16, 19, 20] and etching [12, 21].

According to the state of the art, strength-enhancing properties of the surface structure are the structure density and the aspect ratio of the roughness features [15, 17, 22, 23] as well as the presence of undercuts [21, 22] and sub-structures [18]. Consequently, an appropriate measure for predicting the interlaminar strength needs to consider these specific properties. Standardized roughness height parameters like *Ra*, *Rz* and *Rc* do not consider any of these aspects and therefore, no considerable accordance to the interlaminar strength was found in various studies [2, 13, 21]. On the other hand, the root mean square slope $R\Delta q$ showed good accordance to the shear strength of steel-bone cement joints [23] since this measure is directly connected to structure density and aspect ratio of the roughness features. However, undercuts cannot be recorded using tactile surface roughness measurement. Another approach for roughness description is the fractal dimension. The surface topography can e.g. be derived from cross-sectional images, whereby undercuts and sub-structures are taken into consideration as far as the image resolution allows. A study investigating the adhesion of thermally sprayed ceramic coatings on blasted steel substrates showed good correlation between fractal dimension and bonding strength [24].

This contribution is the continuation of our previous work [12, 16] on the relation between surface topography and interlaminar strength of mechanically interlocked PMH. The aim of this work is the assessment of the suitability of standardized surface roughness parameters and comparison to the fractal dimension [16] for predicting the interlaminar strength. To this end, seven different surface topographies are applied to EN AW-6082 aluminium alloy specimens. The structured surfaces are joined to polyamide 6 (PA6) by heat conduction hot pressing. Interlaminar strength values are determined using butt-bonded hollow cylinder specimens, since this method allows interlaminar shear and tensile strength testing with one single specimen geometry [11, 25].

2. Experimental and materials

2.1. Materials

For all experiments, hollow cylinders made up from EN AW-6082 aluminum alloy were used as metal part. Hollow cylinders made up from extruded Ultramid® B3 PA6 (BASF, Ludwigshafen, Germany) were used as polymer part. Table 1 shows the relevant material data.

Table 1. Material properties of the used metal and polymer, supplier specifications.

	EN AW-6082	PA6 (humid condition)
Density [kg/m³]	2.7	1.14
Young's modulus [MPa]	70000	1800
Poisson's ratio [-]	0.34	-
Yield strength [MPa]	260	60
Ultimate strength [MPa]	310	-
Elongation to failure [%]	7	200
Melting temperature [K]	933	496
Thermal expansion coefficient [10^{-6}/K]	23.4	70
Thermal conductance [W/(m·K)]	170 - 220	0.23
Specific heat [J/(kg·K)]	898	1700

2.2. Surface pretreatment

In total, seven different surface topographies were applied to the aluminum surface using grit blasting (F16), grit blasting and additional etching (F16-NaOH), thermal spraying (NiAl5) and laser structuring (deterministically distributed pin structures: L-8, L-11, L-14; stochastically distributed pin structures: L-Pin). Figure 1 shows scanning electron (SE) microscopy topography images as well as cross-sectional back scattering detector (BSD) images of the generated surfaces.

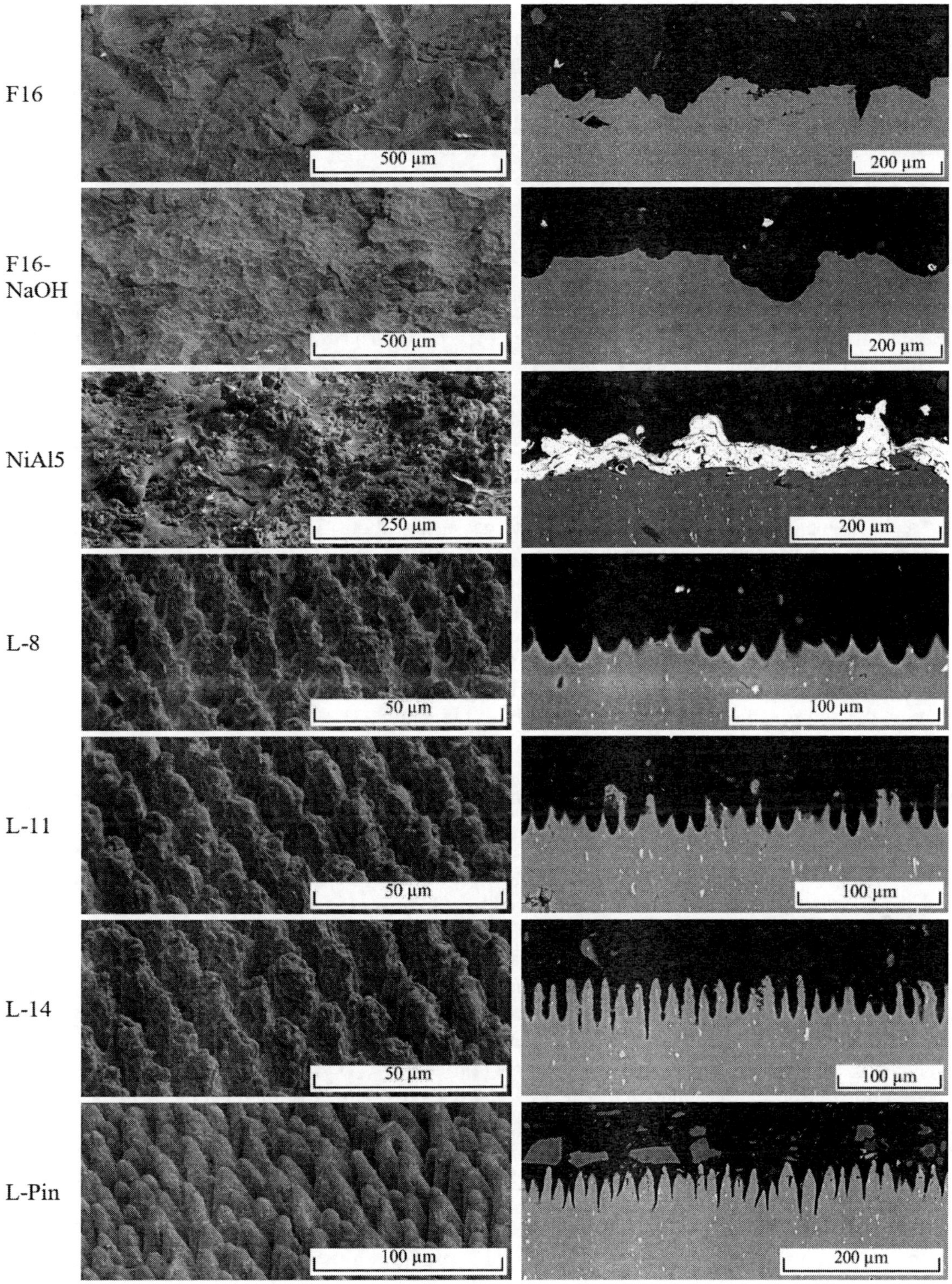

Figure 1. SE images of topography (left) and cross-sectional BSD images (right).

F16 surfaces were generated with WFA F16 corundum (particle size distribution: 1000–1400 μm) using a blasting pressure of 0.2 MPa, a blasting distance of 100 mm, a blasting angle of 75° and a blasting time of 10 s. In addition to the F16 specimens, alkaline etching was carried out on grit blasted surfaces to modify the coarse roughness features with much finer sub-structures. For generating the F16-NaOH specimens, a blasting pressure of 0.3 MPa was applied to increase the vertical extent of the structures. Alkaline etching was conducted with 2 % NaOH solution (343 K / 5 min). Afterwards, the sheets were submerged into 50 % nitric acid solution (ambient temperature / 2 min) to clean the surface.

For generation of the NiAl5 surfaces, a nickel-aluminum 95 wt%-5 wt% coating was deposited on the aluminum. Prior to the coating process, WFA F24 corundum (particle size distribution: 600–850 μm) was applied with a blasting distance of 100 mm, a blasting angle of 75°, a blasting pressure of 0.2 MPa and a treatment time of 10 s to enhance the adhesion of the coating. The coating was applied by electric wire arc spraying, using a VISU ARC 350 arc spray system with a Schub 5 spraying gun (Co. Oerlikon Metco, Wohlen, Switzerland). Thereby, a current of 150 A, a voltage of 30 V, a spraying distance of 130 mm, an air pressure of 3.5 bar, a feed speed of 0.6 m/s and a row spacing of 5 mm were used.

The laser structuring processes were conducted by a Nd:YVO4 nanosecond laser system (specifications: wavelength = 532 nm, pulse duration = 10 ns, max. mean power = 13 W, focus diameter = 15 μm; manufacturer: Co. Spectra Physics®, Santa Clara, USA). Multiple, line-wise scanning of the specimen's surface area with overlapping single pulses was performed for the realization of the surface micro structures. For generating the L-Pin structures, a defined energy input above the material-specific threshold laser fluence was applied to the aluminum surface. Therefore, a laser intensity of 3–6 J/cm² was realized by using a defocused laser spot measuring about 55 μm in diameter. Applying the aforementioned conditions, stochastically arranged micro structures (cone-like protrusions - CLP) were generated on the specimen surface. L-8, L-11 and L-14 surfaces were processed with a focused laser beam, a pulse frequency of 200 kHz and a number of 8, 11 and 14 scans. The resulting material ablations on the aluminum surface measured approximately 14 μm in diameter. By setting the distance between the material ablations slightly below their diameter (13 μm), the material in between was perforated. The resulting surface structure was determined by pin-like profile elements distributed in a grid. Thereby, the number of scans adjusted the height of the pins.

2.3. Specimen geometry and strength testing

Figure 2 shows the geometry of the hollow cylinder specimens. The outer diameter d_o = 28 mm and the inner diameter d_i = 23 mm. The length of the metal cylinder l_m = 40 mm and the length of the polymer cylinder l_p = 60 mm. For the metal cylinder, the free testing length $l_{m,f}$ = 20 mm. For the polymer cylinder, the free testing length $l_{p,f}$ = 30 mm. In case of the laser-structured specimens, $l_{p,f}$ was reduced to 10 mm since the maximum twist angle of the testing machine (90°) would have been exceeded otherwise.

The specimens were manufactured using a heat conduction hot-pressing process. Figure 3 illustrates the hot-pressing tool. During the hot pressing process an isobaric pressure (0.2 MPa) was applied normal to the interfacial zone of the specimen. The copper block was heated until the polymer in the interface melted. Afterwards, the copper block was actively air-cooled. After the joining process, the specimens were reworked by turning to ensure equal dimensions as well as proper centricity for testing. [11, 12] provide a detailed description of the manufacturing process.

The strength testing was conducted with a PTT 250 K1 hydraulic testing machine (Co. Carl Schenck AG, Darmstadt, Germany). The applied testing speeds were determined from the desired strain- and shear-rate of the PA6, whereby the much stiffer aluminum was considered rigid. The shear rate was set to 0.002 1/s (laser-structured: 5 °/min, others: 15 °/min) and the strain rate was set to 0.0002 1/s (laser-structured: 0.12 mm/min, others: 0.36 mm/min). The interlaminar shear strength τ_{max} and the interlaminar tensile strength σ_{max} were calculated from maximum torque T_{max} and maximum force F_{max} before fracture, respectively (Equations (1, 2)). [11] provides a detailed description of the specimen testing procedure.

$$\tau_{max} = \frac{16T_{max}d_o}{\pi(d_o^4 - d_i^4)}, \quad \sigma_{max} = \frac{4F_{max}}{\pi(d_o^2 - d_i^2)} \tag{1, 2}$$

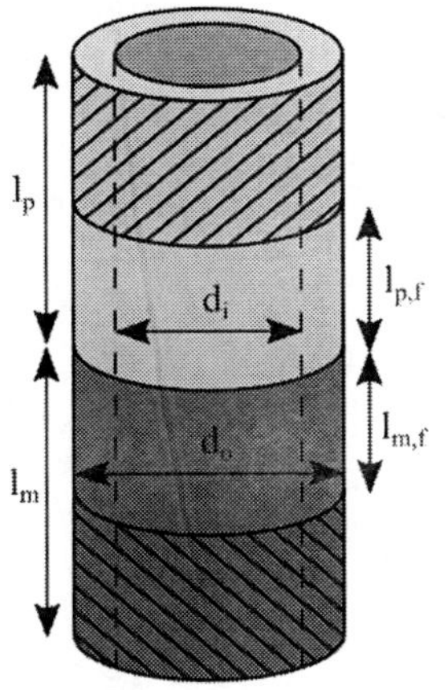

Figure 2. Hollow cylinder specimen [16].

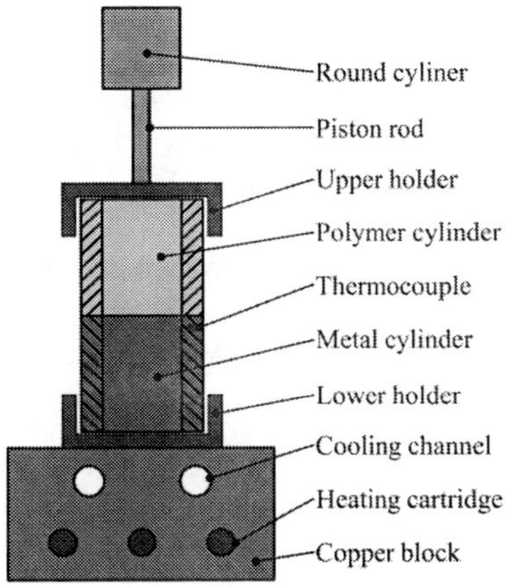

Figure 3. Heat conduction hot pressing tool [16].

2.4. Determination of surface parameters

The surface roughness measurements were carried out with a Hommel-Etamic® T8000 stylus profiler (Co. Jenoptik AG, Jena, Germany). Figure 4 illustrates how the height signal y is recorded as function of distance x. Five measurements with a sampling length of l_r = 2.5 mm and an evaluation length of $l_n = 5l_r$ were recorded and evaluated for each surface structuring process. A stylus tip with 2 µm radius and 60° tip angle was used for capturing a high amount of profile details.

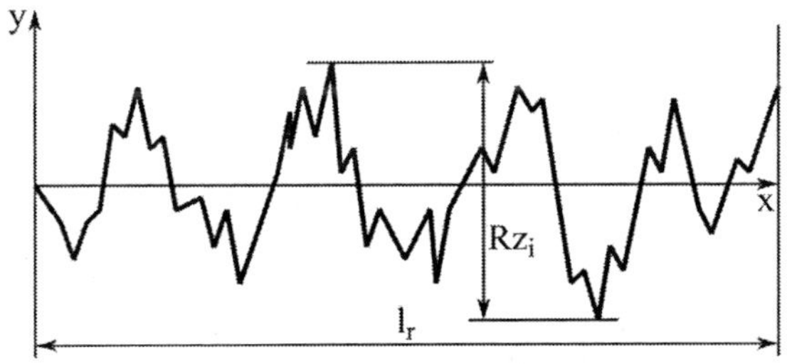

Figure 4. Schematic representation of the surface roughness profile.

Rz (Equation (3)) denotes the average maximum profile height of the roughness features (Figure 4) within 5 times the sampling length l_r.

$$Rz = \frac{1}{5}\sum_{i=1}^{5} Rz_i \tag{3}$$

Ra (Equation (4)) denotes the absolute value of the arithmetical mean height of the roughness features.

$$Ra = \frac{1}{l_n}\int_0^{l_n} |y(x)|dx \tag{4}$$

The root mean square slope $R\Delta q$ (Equation (5)) indicates the root mean square of the local gradient dy/dx along l_n.

$$R\Delta q = \sqrt{\frac{1}{l_n}\int_0^{l_n}\left[\frac{dy(x)}{dx}\right]^2 dx} \quad (5)$$

Figure 5 illustrates how the fractal dimension was derived from the interface line using the box-counting algorithm. Fifteen cross-sectional images were recorded per treatment using a LEO1455VP SE microscope (Co. Carl Zeiss Microscopy GmbH, Jena, Germany). The images were further processed and binarized with GIMP 2 image manipulation software. Embedding resin and metal surface were separated from each other by the fuzzy select function that allows area selection based on colour similarity. Afterwards, a MATLAB® algorithm was used for the identification of the interface line. Finally, the box-counting algorithm was applied. Thereby, the image is divided in squares of size r. For each r_i, a certain number of squares n_i, containing at least one black pixel, exists. Beginning from the smallest possible box size of r_{min} = 1 pixel, the box size is increased stepwise by a power of 2 till the maximum box size r_{max}, covering the complete image, is reached. [16] provides a detailed description of the fractal dimension determination approach in connection with PMH-interfaces.

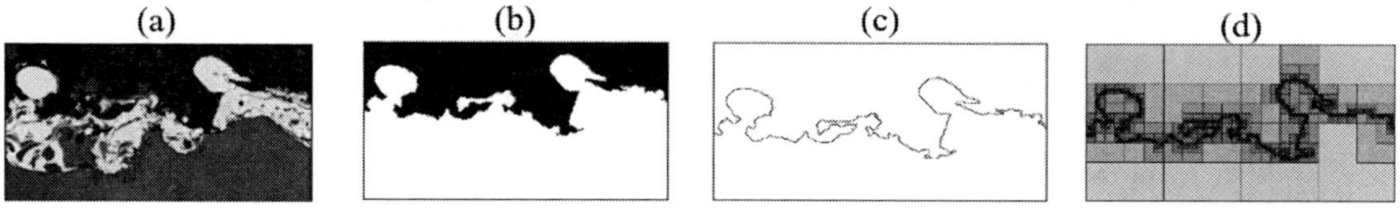

Figure 5. Process of image analysis for determination of fractal dimension (a) Cross-section (b) Binary image (c) Interface line (d) Box-counting algorithm.

The local dimension between individual box sizes d_i is calculated by Equation (6).

$$d_i = \log_2 n_{i+1} - \log_2 n_i \quad (6)$$

The fractal dimension of the interface line D equals the mean value of all local dimensions (Equation (7)).

$$D = \frac{1}{k}\sum_{i=1}^{k} d_i \quad \text{with } k = \log_2 r_{max} \quad (7)$$

3. Results and discussion

Table 2 summarizes the test results as well as surface characteristics for all investigated surface pretreatments (SD denotes the standard deviation). Thereby, the shear strength varied in a broad range from 11.5 MPa to 32.5 MPa and the tensile strength from 2.02 MPa to 26.0 MPa. F16 Treatment offered the second lowest interlaminar strength in both load directions. Additional etching (F16-NaOH treatment) even lowered the strength; presumably due to a leveling of sharp edges and undercuts at the roughness features. NiAl5 treatment provided higher strength than the blasting treatment, but lower strength than every laser treatment. For the L-8, L-11 and L-14 treatments, the strength increased with the number of scans. The highest strength was achieved with the L-Pin treatment.

Table 2. Roughness parameters, fractal dimension and interlaminar strength of the investigated surface structuring methods, mean values ± 1 *SD*.

Treatment	Ra / µm	Rz / µm	RΔq	D	τ_{max} / MPa	σ_{max} / MPa
F16-NaOH	20.5 ± 2.4	124 ± 9	0.773 ± 0.024	1.093 ± 0.032	11.5 ± 2.0	2.02 ± 0.55
F16	22.0 ± 1.9	131 ± 8	0.806 ± 0.020	1.102 ± 0.016	14.2 ± 0.4	4.16 ± 0.69
NiAl5	11.6 ± 0.7	80.4 ± 4.2	0.828 ± 0.031	1.123 ± 0.016	17.0 ± 0.1	5.81 ± 0.03
L-8	4.95 ± 0.15	32.2 ± 0.4	0.919 ± 0.115	1.153 ± 0.012	21.9 ± 0.8	7.92 ± 0.24
L-11	4.92 ± 0.21	31.0 ± 1.1	0.970 ± 0.121	1.195 ± 0.013	26.7 ± 0.8	10.2 ± 1.3
L-14	4.38 ± 0.21	27.5 ± 2.2	0.929 ± 0.164	1.231 ± 0.011	29.6 ± 0.2	17.3 ± 0.8
L-Pin	5.59 ± 0.17	45.6 ± 1.2	0.971 ± 0.025	1.233 ± 0.015	32.5 ± 1.4	26.0 ± 1.6

Figure 6 shows the interlaminar shear and tensile strength in relation to the evaluated roughness parameters. A linear fit was carried out in each case to obtain an additional indication how suitable the respective parameter is to predict the interlaminar strength.

For both roughness height parameters *Rz* and *Ra*, the strength decreases with increasing values for two reasons. Firstly, the surface roughness measurement underestimates the profile height difference especially for the high strength providing laser treated specimens, since the stylus tip is not able to penetrate the tight profile valleys completely. Despite the real profile height of L-8 and L-14 differs around factor two (see Figure 1), the measured surface roughness values (*Ra*, *Rz* and *RΔq*) are almost identical. Secondly, *Ra* and *Rz* do not consider undercuts, sub-structures as well as structure density and aspect ratio of the profile elements. The coarse and widely spaced F16 / F16-NaOH roughness features have high *Ra* and *Rz*, but provide only low strengths. Hence, the coefficients of determination R^2 are low in all cases. Thus, as already stated in state-of-the-art literature, surface roughness height parameters are not suitable for strength prediction.

The root mean square slope *RΔq* shows much higher accordance ($R^2 = 0.9$ for τ_{max}, $R^2 = 0.62$ for σ_{max}) since it is a direct measure for structure density as well as aspect ratio. Both τ_{max} and σ_{max} increase with *RΔq*. Especially for F16, F16-NaOH and NiAl5 treated specimens, a monotonic increase can be observed. However, the loss of detail when recording the surface roughness profile also leads to an underestimation of *RΔq* and thus, the tightly spaced L-8, L-11, L-14 and L-Pin structures show considerable deviation from the fitted function. Moreover, the anisotropic orientation of the L-8, L-11 and L-14 profile elements leads to high standard deviations in *RΔq*.

Fractal dimension *D* shows the highest accordance of all evaluated parameters ($R^2 = 0.98$ for τ_{max}, $R^2 = 0.83$ for σ_{max}). In contrast to the tactile surface roughness measurement, the determination of the interface line from cross-sectional images captures all details as far as the image resolution allows. Thus, undercuts, sub-structures and structure density are considered when calculating *D*.

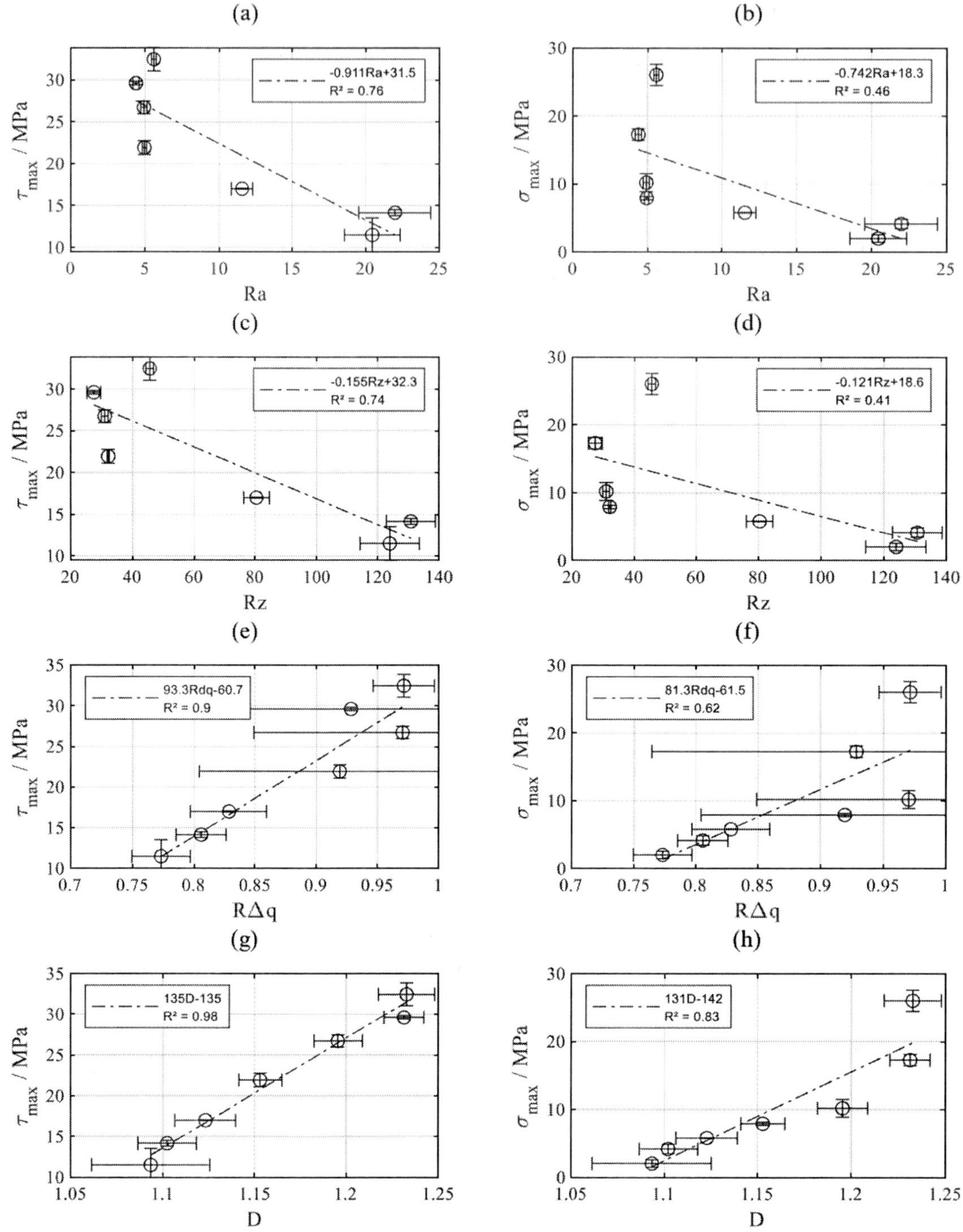

Figure 6. Interlaminar shear- and tensile strength in relation to *Ra*, *Rz*, *R*Δ*q* and *D*, mean values ± 1 *SD*.

IOP Conf. Series: Materials Science and Engineering 1147 (2021) 012021 doi:10.1088/1757-899X/1147/1/012021

4. Conclusions

In the present work, the suitability of tactile measured surface roughness parameters and fractal dimension for surface topography assessment and corresponding interlaminar strength prediction of mechanically interlocked polymer-metal-hybrids has been investigated. The surface topographies were generated by corundum blasting, combined corundum blasting and etching, thermal spraying and laser structuring. For the determination of the interlaminar strength in shear as well as tensile direction, butt-bonded hollow cylinder specimens were used. The joining of specimens was carried out by heat conduction hot pressing.

The results indicate that the interlaminar strength can be predicted from the surface topography by the help of appropriate roughness parameters. Assuming that the adhesion resulting from mechanical interlocking mainly depends on structure density and the aspect ratio of the roughness features as well as the presence of undercuts and sub-structures, surface roughness height parameters alone are not able to give reliable predictions. This is confirmed by the poor correlation between Ra as well as Rz and the interlaminar strength. The surface roughness slope parameter $R\Delta q$ can lead to good accordance since it is directly connected to structure density and aspect ratio of the roughness features. However, the roughness measurement device must be able to record the actual roughness profile adequately. The tactile measurement of tightly spaced, laser-structured topographies yielded a massive underestimation of the actual $R\Delta q$ since the stylus tip was not able to penetrate the profile valleys completely.

Fractal dimension, though, resulted in the best correlation to the interlaminar strength for all investigated specimens. Since the interface line is derived from cross-sectional images, undercuts and small-scale roughness features can be considered. Therefore, fractal dimension is most likely to be recommended for interlaminar strength prediction of mechanically interlocked polymer-metal-hybrids.

Acknowledgments

This study was supported by the Federal Cluster of Excellence EXC 1075 “MERGE Technologies for Multifunctional Lightweight Structures”. The support by Christoph Wollschläger, Torsten Rabes, Michael Zychski, Paul Seidel, Christian Loos, Claudia Albero Rojas and Steffen Clauß (all from the Institute of Materials Science and Engineering) is gratefully acknowledged.

References

[1] Katayama S and Kawahito Y 2008 Laser direct joining of metal and plastic *Scripta Materialia* **59** 1247-1250

[2] Bergmann J P and Stambke M 2012 Potential of Laser-manufactured Polymer-metal hybrid Joints *Physics Procedia* 84-91

[3] Heckert A and Zaeh M F 2015 Laser surface pre-treatment of aluminum for hybrid joints with glass fiber reinforced thermoplastics *Journal of Laser Applications* **27** S29005

[4] Roesner A, Scheik S, Olowinsky A, Gillner A, Reisgen U and Schleser M 2011 Laser Assisted Joining of Plastic Metal Hybrids *Physics Procedia* **12** 370-377

[5] Mitschang P, Velthuis R and Didi M 2013 Induction Spot Welding of Metal/CFRPC Hybrid Joints *Advanced Engineering Materials* **15** 804-813

[6] Saborowski E, Dittes A, Lindner T and Lampke T 2021 Nickel-Aluminum Thermal Spray Coatings as Adhesion Promoter and Susceptor for Inductively Joined Polymer-Metal Hybrids *Polymers* **13** 1320

[7] Wagner G, Balle F and Eifler D 2013 Ultrasonic Welding of Aluminum Alloys to Fiber Reinforced Polymers *Advanced Engineering Materials* **15** 792-803

[8] Steinert P, Dittes A, Schimmelpfennig R, Scharf I, Lampke T and Schubert A 2018 Design of high strength polymer metal interfaces by laser microstructured surfaces *IOP Conference Series: Materials Science and Engineering* **373** 1-9

[9] Liu F C, Liao J and Nakata K 2014 Joining of metal to plastic using friction lap welding *Materials & Design* **54** 236–244

[10] Haberstroh E and Sickert M 2014 Thermal Direct Joining of Hybrid Plastic Metal Components *KMUTNB International Journal of Applied Science and Technology* **7** 29-34

[11] Saborowski E, Kießling R, Dittes A, Paczkowski G, Ihlemann J and Lampke T 2019 Determination of the strength of polymer-metal interfaces under mixed mode loading using butt-bonded hollow cylinders *International Journal of Adhesion and Adhesives* **89** 30–39

[12] Saborowski E, Dittes A, Steinert P, Lindner T, Scharf I, Schubert A and Lampke T 2019 Effect of Metal Surface Topography on the Interlaminar Shear and Tensile Strength of Aluminum/Polyamide 6 Polymer-Metal-Hybrids *Materials* **12** 2963

[13] Pan Y, Wu G, Huang Z, Li M, Ji S and Zhang Z 2017 Effect of surface roughness on interlaminar peel and shear strength of CFRP/Mg laminates *International Journal of Adhesion and Adhesives* **79** 1-7

[14] Schuberth A, Göring M, Lindner T, Töberling G, Puschmann M, Riedel F, Scharf I, Schreiter K, Spange S and Lampke T 2016 Effect of new adhesion promoter and mechanical interlocking on bonding strength in metal-polymer composites *IOP Conference Series: Materials Science and Engineering* **118** 012041

[15] Amend P, Pfindel S and Schmidt M 2013 Thermal Joining of Thermoplastic Metal Hybrids by Means Of Mono- and Polychromatic Radiation *Physics Procedia* **41** 98–105

[16] Saborowski E, Steinert P, Dittes A, Lindner T, Schubert A and Lampke T 2020 Introducing Fractal Dimension for Interlaminar Shear and Tensile Strength Assessment of Mechanically Interlocked Polymer–Metal Interfaces *Materials* **13** 2171

[17] Byskov-Nielsen J, Boll J V, Holm A H, Højsholt R and Balling P 2010 Ultra-high-strength micro-mechanical interlocking by injection molding into laser-structured surfaces *International Journal of Adhesion and Adhesives* **30** 485–488

[18] Rodríguez-Vidal E, Sanz C, Soriano C, Leunda J and Verhaeghe G 2016 Effect of metal micro-structuring on the mechanical behavior of polymer–metal laser T-joints *Journal of Materials Processing Technology* **229** 668-677

[19] Lindner T, Saborowski E, Scholze M, Zillmann B and Lampke T 2018 Thermal Spray Coatings as an Adhesion Promoter in Metal/FRP Joints *Metals* **8** 1-10

[20] Lindner T, Friederichs C, Zillmann B, Hockauf K, Wielage B and Lampke T 2014 Development of an integration zone for joining polymer-metal hybrid structures *Euro Hybrid Materials and Structures 2014*

[21] Kleffel T and Drummer D 2017 Investigating the suitability of roughness parameters to assess the bond strength of polymer-metal hybrid structures with mechanical adhesion *Composites Part B: Engineering* **117** 20-25

[22] Schricker K, Stambke M, Bergmann J P, Bräutigam K Henckell P 2014 Macroscopic Surface Structures for Polymer-metal Hybrid Joints Manufactured by Laser Based Thermal Joining 2014 *Physics Procedia* **56** 782–790

[23] Chen C Q L, Scott W Barker T M 1999 Effect of metal surface topography on mechanical bonding at simulated total hip stem–cement interfaces *Journal of Biomedical Materials Research* **48** 440-446

[24] Amada S and Yamada H 1996 Introduction of fractal dimension to adhesive strength evaluation of plasma-sprayed coatings *Surface and Coatings Technology* **78** 50-55

[25] Saborowski E, Scholze M, Lindner T and Lampke T, 2017 A numerical and experimental comparison of test methods for the shear strength in hybrid metal/thermoplastic-compounds *IOP Conference Series: Materials Science and Engineering* **181** 012031

Hybrid casting – An investigation into the interface of high-pressure die-cast intrinsic aluminum-PEEK-CFRP hybrid composites

A Schmid[1,3*], T Haubold[3], K Koschek[3], A Marx[2], L Pursche[3], A Struß[3], K Thiel[3], M Wiesing[3] and M Busse[1,3]

[1] University of Bremen, Bibliothekstr. 1, 28359 Bremen, Germany
[2] Faserinstitut Bremen e.V., Gebäude IW3, Am Biologischen Garten 2, 28359 Bremen, Germany
[3] Fraunhofer-Institut für Fertigungstechnik und Angewandte Materialforschung IFAM, Wiener Str. 12, 28359 Bremen, Germany

*e-mail: armin.schmid@ifam.fraunhofer.de

Abstract. The joining of plastic-based fiber composites and light metals is increasingly becoming the focus of lightweight construction applications. To date, conventional mechanical and adhesive joining techniques are still the predominant means to produce hybrid composites. The aim of this work was to create a hybrid composite of CFRP and aluminum in a novel way using the economic high-pressure die-casting process without the use of additional joining elements. The casting process implies the direct contact of molten metal (AlSi10MnMg) and thermally resistant polyetheretherketone (PEEK) under short-term temperatures of about 700 °C. The material and process parameters lead to a material transition that spatially separates the CFRP and the aluminum while guaranteeing a firm bond with lap shear strengths up to 22 MPa. This work structurally and chemically characterizes the interface between Al and PEEK in the hybrid composite. The results of the TEM investigations into the boundary region between PEEK and the AlSi10MnMg alloy show that both materials are locally joined without gaps. Furthermore, the chemical EDX, XPS and IR spectroscopy results indicate a tendency for the bond to form through the thermal alteration of the polymer and the associated modification of the bonding possibilities in the direct contact area.

1. Introduction

The optimization of performance and efficiency in the technical field are driving forces for lightweight construction, and the realization of lighter components always underpins the further development of innovation. In addition to the use of alternative materials, the combination of different types of materials within a structure offers great potential for composite and lightweight design [1, 2]. Especially with the increasing use of CFRP, the hybrid joining of light metals and carbon fiber reinforced thermoplastics is a major research focus in the field of joining technology [3, 4].

The joining of different materials has a complex requirement profile. In addition to the compatibility of the mechanical and thermal properties, the electrochemical interaction in the contact area of the joined materials is of particular relevance [5, 6]. Accordingly, there are many potential solutions, with the most established being conventional joining techniques based on mechanical and adhesive bonding. However, these joining processes have specific disadvantages, such as additional weight input from fasteners or surface pretreatments. [7].

Within the framework of this research, hybrid joints that can be implemented without these limitations represent a major research focus. For example, [8] and [9] described welding techniques, which imply the melting of the fiber composite component's plastic to join both materials.

In addition, primary forming processes, such as plastic injection molding and light-metal-based high-pressure die-casting, make it possible to reduce the number of production steps when primary forming and joining processes are carried out simultaneously.

[10] showed that it is possible to generate a direct composite consisting of polyetheretherketone (PEEK) based CFRP and aluminum in the high-pressure die-casting process without involving an induced form closure. Hereby, the bond between CFRP and aluminum, with tensile shear strengths of around 16 MPa, is generated mainly by a 250 µm thick PEEK layer, which is applied to the CFRP in the area of the joining surface before the high-pressure die-casting process. PEEK also enables simultaneous electrochemical separation of aluminum and carbon fibers [10]. Within the scope of the further development of the hybrid connection of PEEK and aluminum using the high-pressure die-casting process, lap shear strengths of up to 22 MPa are achieved. By comparison, [11], joining PEEK and the 6061-T6-Al alloy, and [12], with a composite of CF-PEEK and the AA6082-T6-Al alloy, achieved comparable strengths in the range of 20 MPa without additional joining elements or strength-enhancing measures such as surface treatments or structuring.

Meanwhile, [13] and [14] already showed that a direct connection between PEEK and metal without form or force fit is technically feasible. Oxidic compounds that can form a metal-O-C bond appear to be of particular relevance here [15, 16].

The aim of this work is, therefore, to use microscopic (TEM) as well as spectroscopic means (EDX, XPS, IR spectroscopy) to show how a short-term temperature of around 700 °C within the high-pressure die-casting process affects the PEEK, the formation of the interface, and the bonding of PEEK and aluminum in the hybrid compound.

2. Experimental and materials

2.1. Materials

A carbon fiber reinforced PEEK composite laminate (ThermoPlastic Consolidated Laminate, Tenax-E TPCL PEEK HT-A40, Teijin Carbon, Wuppertal, Germany) with a five-layer fabric [0°/90° +/- 45° 0°/90° +/-45° 0°/90°] and the dimensions 100 mm x 40 mm with a thickness of 1.5 mm was used to produce the hybrid composites. In the area of the surface to be joined (40 mm x 40 mm), the CFRP inserts were coated with a 250 µm thick PEEK film (type Aptiv 1000, Victrex plc, Lancashire, UK), as shown in Figure 1. The coating was applied using a hot-pressing process, performed on a Rucks KV 228 press. The process takes place at a maximum temperature of 350 °C and with a consolidation pressure of 4.2 bar (holding time of 20 minutes).

The work was carried out at Faserinstitut Bremen e.V.

Figure 1. The CFRP laminate coated with PEEK.

An AlSi10MnMg alloy (Silafont 36, Rheinfelden Alloys GmbH & Co. KG, Rheinfelden, Germany) was used for the metal component.

2.2. Composite manufacturing and mechanical testing

Before the casting process, the CFRP components were dried for 24 hours at 120 °C in a circulating air chamber furnace (N250/85 HH, Nabertherm GmbH, Germany), and the joining surfaces were cleaned with acetone directly before pouring.

Using a temperature-resistant adhesive, the CFRP inserts were fixed selectively in the die-casting tool and away from the joining area. Subsequently, the CFRP was partially cast-on with the AlSi10MnMg alloy using a cold-chamber high-pressure die-casting machine (Bühler SC N/66, Bühler AG, Uzwil, Switzerland). Before each casting cycle, the tool was cleaned to remove any residue and coated with the release agent SL 1697 S from Chem-Trend (Chem-Trend GmbH, Maisach Gernlinden, Germany). Relevant process parameters selected were a plunger speed of 2.35 m s^{-1}, a melt temperature of 700 °C, an average mold temperature of 150 °C and a holding pressure of 1000 bar (holding time of 10 s). The PEEK-coated CFRP was placed in the mold so that the joining surface was at the furthest point in relation to the gate system. After the hybrid cast structures (Figure 2) were removed from the tool, they were quenched in a water bath at a temperature of 20-30 °C. The samples were stored at room temperature until heat treatment.

The heat treatment was performed in a chamber furnace (N 300/WAX, Nabertherm GmbH, Lilienthal, Germany). The samples were stored in the furnace for two hours at 190 °C, in accordance with a T5 heat treatment.

The high-pressure die-casting experiments took place at Fraunhofer IFAM.

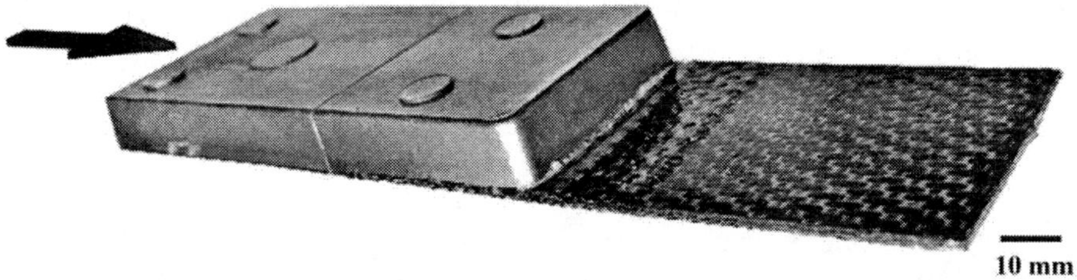

Figure 2. A high-pressure die-cast Al-PEEK-CFRP hybrid composite The arrow shows the entry point of the melt during the mold-filling.

The single-lap test specimens were tested according to DIN EN ISO 527-4 with a tensile testing machine from Zwick (Z250, ZwickRoell GmbH & Co. KG, Ulm, Germany).

The tests were performed at Faserinstitut Bremen e.V.

2.3. Analysis methods

2.3.1. TEM (transmission electron microscopy) und EDX (energy-dispersive X-ray spectroscopy)

For the structural and elemental investigation of the interface between the Al alloy and PEEK, TEM analysis was carried out using a Tecnai F20 S-TWIN microscope (FEI now Thermo Fisher Scientific, Eindhoven, Netherlands) with a field emission gun operated at 200 kV, resulting in a point resolution of 2.4 Å. TEM images were acquired with a 1k × 1k slow-scan CCD camera embedded in a Gatan GIF2001 image filter. In the interface between Al and PEEK, elemental analysis was performed by means of EDX using the EDAX r-TEM-EDX Detector of the TEM operated in scanning mode (STEM). The EDX detector has an energy-resolution of 136eV measured at Mn-Kα. The STEM images were obtained by a Fischione HAADF-STEM Detector.

The TEM samples were prepared via the well-known focused ion beam (FIB) technique in an FEI Helios 600 dualbeam machine, as described in [17]

The TEM and EDX analyses were performed at Fraunhofer IFAM.

2.3.2. XPS (x-ray photoelectron spectroscopy) and solution test of PEEK. For the XPS analysis, the exposed fracture surface of a mechanically tested sample was used. The measurements were performed in regions, which were characterized by the absence of significant PEEK residues as indicated by a visual inspection using a light microscope (adhesive failure mode). The area of analysis was 0.4 x 0.4 mm. The XPS measurements were performed using a Thermo K-Alpha K1102 system and monochromatized Al K-Alpha irradiation (1486.7 eV). The angles of incidence and emission were 57.4 and 0 ° respectively. Spectra were acquired in Constant Analyser Energy mode (CAE) at 150 eV (survey spectra) and 20 eV (high-resolution spectra). The samples were neutralized using a combination of low-energy electrons and argon ions. The binding energy scale of the spectrometer was calibrated using the main peaks of Cu, Ag and Au. Moreover, the binding energy scale was referenced to the C1s binding energy of surface organic compounds at 285.0 eV. The elemental quantification was performed using instrument specific elemental relative sensitivity factors. The depth of information was around 10 nm.

In order to validate irreversible bond formation, the fractured aluminum surface was extracted in a 1:3 boiling solution of 9-fluorenon and diphenyl ether for 5 min, which represents a good solvent for PEEK at minor concentrations. However, strongly adsorped PEEK monolayers are not desorped under these conditions as shown by preliminary experiments on Ti coated Si-wafer (not shown).The XPS analysis and solution experiments were carried out at Fraunhofer IFAM.

2.3.3. IR Spectroscopy (Infrared spectroscopy). The IR measurements were performed on a Hyperion 3000 IR microscope (Bruker Corporation, Massachusetts, USA). The IR spectra were recorded with an ATR objective in emission mode in the range of 600 - 4000 cm^{-1} and a scan number of 64. The data were then evaluated with the Opus software. The samples were measured on the unaffected pure PEEK foil, on the fracture surface of a composite sample after the tensile shear test as well as in cross section of the interface between aluminum and PEEK. The reference spectrum used was that of pure PEEK.

The IR spectroscopy was performed at Fraunhofer IFAM.

3. Results and discussion

3.1. Mechanical testing

Mechanical testing of the single lap shear specimens resulted in an average lap shear strength of 20.55 ± 0.99 MPa (sample size of five), calculated with a bonding area of 1600 mm^2. The maximum value was about 22 MPa. Based on these strengths the interface between Al and PEEK was investigated in detail. The typical fracture surfaces and the characteristic failure behavior of the samples are shown in Figure 3.

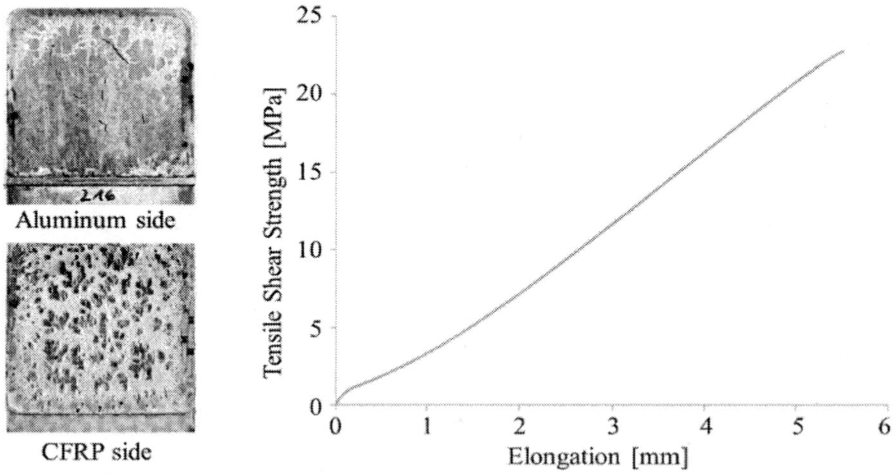

Figure 3. The fracture surfaces after the lab shear test (left) and an exemplary stress-strain curve (right).

3.2. TEM

Figure 4 depicts the prepared FIB lamella, and the results of the TEM analysis are present in Figure 5. The images show that there is a gap-free and material-locking connection between the PEEK and the aluminum-based alloy. The atomic plane distances measured within the scope of the measurement accuracy illustrate the local structural differences. In aluminum, the planes show a distance of about 2.1 Å. In the interface between aluminum to PEEK, however, the atomic lattice seems to be widened and locally has a lattice plane distance of about 2.9 Å. The widened lattice structure could indicate an oxide complex formation.

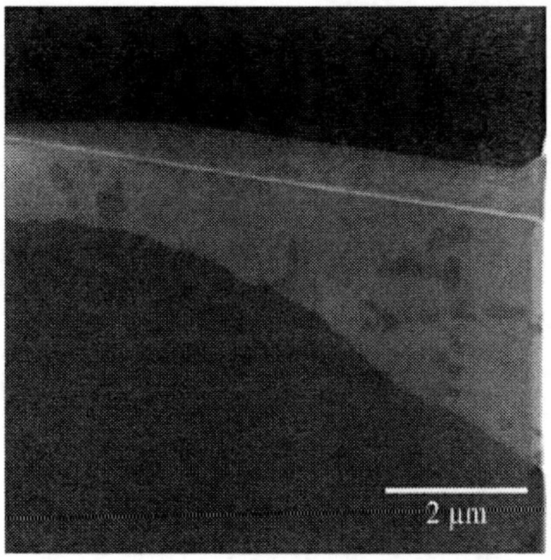

Figure 4. The prepared FIB lamella from the combination of PEEK and AlSi10MnMg (with tendentially contained silicon phases).

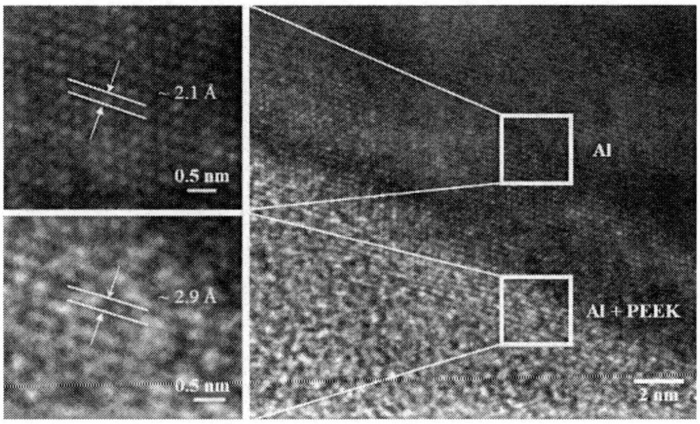

Figure 5. An HRTEM image of the connection between AlSi10MnMg and PEEK. The marked sections illustrate the different atomic lattice structures of both the aluminum alloy and the interface between Al and PEEK.

3.3. EDX

Figure 6 shows the results of the line scan (EDX) of the same sample used in the TEM. Focusing the supposed relevant chemical constituents, it can be seen that in the investigated section of the interface between the AlSi10MnMg alloy and PEEK, the elements magnesium and oxygen are predominantly present, with a local increase in intensity ranging from 8 nm to 11 nm. The superficial precipitation of Mg in Al alloys is known. However, it is not possible to draw specific conclusions on the binding conditions using the EDX analysis procedure. Based on the changes in the atomic lattice structure shown in the TEM analysis, it is assumed that the interface between the AlSi10MnMg alloy and PEEK within the sample could be an oxidic bond. The lattice spacing of 2.9 Å, determined within the scope of the measurement accuracy, is consistent with the results according to [18], which reported an atomic lattice spacing of 2.97 Å in an MgO lattice. In addition, PEEK and metal are clearly separated. This probably indicates adsorption. Based on this, a solution test is carried out below. This allows the detection of irreversibly adsorbed PEEK layers, but also their chemical characterization.

The potential for the formation of metal-polymer complexes in such hybrid composites has been discussed in several publications. [14], for example, showed that a $MgCO_3$ compound is formed when an AZ31 magnesium alloy is thermally joined with PEEK. [13] dealt with the metallization of PEEK with pure aluminum, aluminum bronze and copper. In the investigation, liquid metal droplets were applied to the plastic, which guaranteed a certain comparability to this work. Using TEM and EDX analyses, the potential presence of copper(I)-oxide copper(II)-oxide as well as copper(II)-hydroxide and copper(I)-oxide were demonstrated.

The work of [19] also supports the hypothesis that complex formations between a metal and a polymer are possible in principle. The authors produced a chemical compound of different materials by coating aluminum with polyamide. Specifically, an Al-O-C complex was detected by means of XPS analysis, initiated or formed by the cleavage of the carbonyl group (C=O) of the polymer.

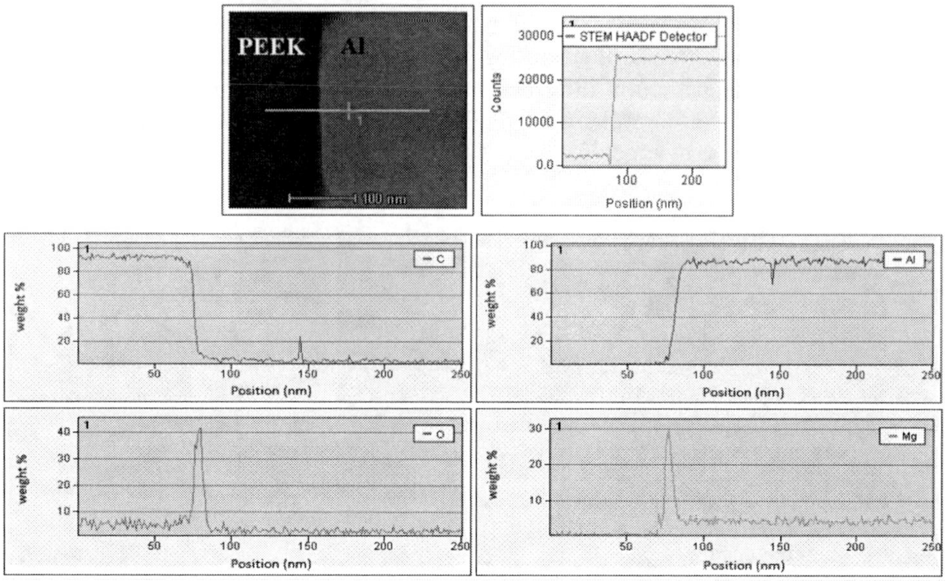

Figure 6. The result of a line scan (upper left) in the transition region between PEEK and the aluminum alloy; the measuring range is visible in the STEM image (upper right). The measured intensities (note the scales) of the respective elements are shown in the individual diagrams (below). With reference to the investigated area, the alloying elements manganese and silicon are excluded from the shown result, because of their noisy like progression of element distribution and the low detected concentrations compared to the main present elements.

3.4. XPS and solution test of PEEK

The area of the aluminum sided fracture surface of the sample examined by XPS is shown and marked in Figure 7. On the left, the considered area after the lap shear test is shown, on the right the same area after the solution test.

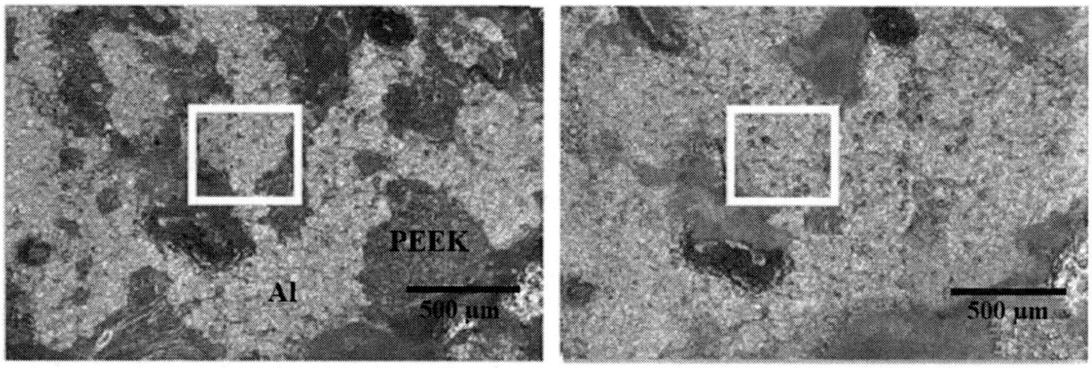

Figure 7. Area of the fracture surface of the hybrid PEEK-aluminum sample examined in the XPS analysis (white framed). On the left, the surface is visible directly after the lap shear test; on the right, after an additionally performed PEEK solution bath.

The results of the analysis show that directly after the lap shear test, the apparently metal like surface still contained an average between 75.8 and 77.2 At.-% carbon and 18.0 and 18.7 At.-% oxygen, as shown in Table 1. The comparatively high percentage of organic matter indicates that PEEK residues still adhered to the metal within the measuring range.

Table 1. The concentration of the elements detected on the fracture surface, shown in At.-%. The positions represent individual measuring points in the measuring range. The measuring sensitivity was about 0.1 At.-%.

	C	O	Al	Mg	F	Na
Pos.1	77.2	18.0	3.4	0.5	0.3	0.6
Pos.2	75.8	18.7	3.8	0.7	0.8	0.2

After the solution bath, the polymer residues still adhering to the fracture surface could be identified analytically. The measured percentage of carbon was between 68.4 and 79.3 At.-% and the percentage of oxygen was between 17.4 and 23.5 At.-%, as shown in Table 2. Analytically, it was still possible to detect significant adhesive residues of PEEK after the solution test. The approach to expose stronger adsorbed polymer layers using adequate solvents, is well established [20; 21]. The parameters and conditions of the solution bath for PEEK were determined empirical. From experiments and experiences with this for metal-to-PEEK bonds tailored method it is also possible to dissolve strong adsorbed polymer chains, which illustrates the efficiency and potential of this approach. The microscopic images (Figure 7) show a notable removale of polymer residues from the surface.

Table 2. The concentration of the elements detected on the fracture surface, shown in At.-%. The positions represent individual measuring points in the measuring range. The measuring sensitivity was about 0.1 At.-%.

	C	O	Al	Mg	F	Na
Pos.1	79.3	17.4	2.4	0.4	0.3	0.1
Pos.2	70.4	22.7	4.8	1.3	0.4	0.1
Pos.3	68.4	23.5	5.3	1.5	0.7	0.1

The results show that PEEK is at least partially strongly adsorbed at the interface. This explains the observed high adhesion. The presence of the macroscopic gel residues on the surface also indicates that crosslinking reactions have taken place, which are known to be caused by oxidative degradation of PEEK by e.g. O_2, like also shown in [22]. Consequently, the gel residues bind to the surface and cannot be removed. Overall, the results indicate a multilayer interface structure:

Al | MgO | PEEKadsorbed | PEEK, PEEKGel | PEEK

It has already been shown [23] that 24-hour heat treatments at temperatures above 319 °C can lead to crosslinking in PEEK. The occurrence of these structurally modified molecules was demonstrated by [23] through analogous chemical solution experiments. A possible quantification of the thermally induced change in PEEK under tribological load as a function of oxygen availability was described in [24], whereby crosslinking processes could be detected within two hours.

In general, the interaction and mechanisms behind the stronger adsorption of single polymer layers is complex and especially for the contact of metal melt and PEEK still unknown. The origin of the adhesion could only be suggested, based on combining the results of the different investigation methods. From microscopic analysis a mechanical interlocking can be excluded. Also an interdiffusion of the different materials does not occur and the EDX results show a direct transition of the components. Further assumed the bonding strengths is mainly based on physisorption, such as van der Waals and polar forces.

The complexity of metal-to-plastic adhesion is also shown in [25; 26]. While using the Good-Girifalco model [25] to describe the irreversible bonding of polystyrene and aluminium, the thermodynamic consideration is not sufficiently explained. [26] postulate that also the time depended absorption kinetics should be taken into account.

3.5. IR Spectroscopy

Finally, IR spectroscopy was performed to describe the possible process-related influences on the structural properties of PEEK. The results in the form of a comparison of the IR spectra of the unaffected PEEK film (Figure 8, top) and the thermally influenced PEEK in the interface to the metallic casting are shown as an example in Figure 8 (bottom) for the fingerprint range of PEEK up to 1800 cm^{-1}.

The spectra show a change in peak maxima relations as a result of the high-pressure die-casting process. The carbonyl peaks at 1653 cm^{-1} and 1307 cm^{-1} were related to the peaks at 1280 cm^{-1}, belonging to the stretching vibrations of the diphenyl ether groups and the phenyl peak at 1490 cm^{-1}.

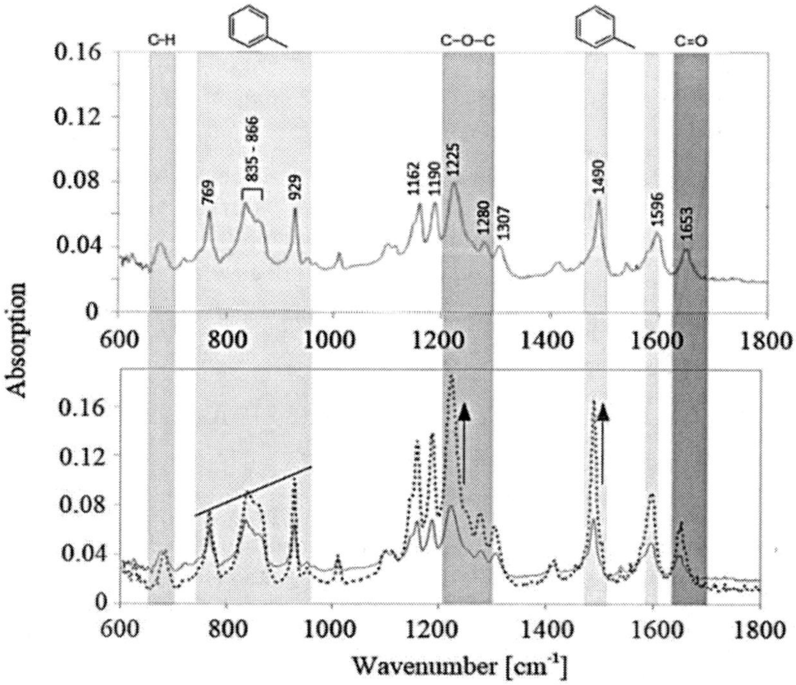

Figure 8. The upper part of the diagram shows bands of the IR spectrum characteristic of non-processed PEEK. The results of the measurement after the high-pressure die-casting process are presented in the lower part of the diagram (dashed lines) and superimposed with the measurement of the upper part of the diagram of non-processed PEEK (continuous line). The arrows show how the peaks of the two spectra rise or fall in relation to each other.

The high-pressure die-casting process entailed a decrease of carbonyl group peaks in relation to the ether and phenyl groups peak maxima. This shift of absorption intensities indicates a decrease in carbonyl groups, which could have been caused by decomposition or by a possible reaction with the aluminum. Basically, a reduction of carbonyl groups in neat PEEK polymer could result an ordered polymer structure and by this a higher degree of crystallization. However, the clear change in the symmetry of the signals between the aromatic C-H vibration, i.e. 769 cm^{-1} and 928 cm^{-1}, in the fingerprint area indicate the opposite [22]. In particular, the decrease of the peak at 769 cm^{-1} proves a reduced crystallinity of PEEK in the interface.

Based on the results, the structural change of the PEEK in the interface region can be attributed to process-related thermal degradation processes. It is assumed that this, in the form of a thermally induced modification of the PEEK, enables the connection between PEEK and aluminum.

4. Conclusions

This work has shown that it is possible to produce a hybrid composite of CFRP, PEEK and the AlSi10MnMg aluminum alloy in an innovative way using the high-pressure die-casting process. Results prove that maximum tensile shear strengths of around 22 MPa can be achieved.

It has been showed by means of TEM and EDX analyses that the PEEK and the aluminum alloy form a gap-free bond in the direct contact areas where no outgassings are present. Microscopic investigations show no significant undercuts at micro- and nanometer level. In the contact area of both materials, a clear presence of the elements magnesium and oxygen could be detected at the nanometer level. These findings strengthen the suggestion that a potential connection of PEEK and AlSi10MnMg is made via MgO.

With regard to the thermal influence under process temperatures up to 700 °C, an XPS analysis showed that the PEEK is present in a structurally modified form in the contact area with aluminum. Even after treatment with solvent, more strongly adsorbed polymer layers could still be analytically detected for PEEK.

Analyses of the IR spectrum of PEEK in the contact area of both materials showed that changes in the polymer structure had occurred, which can be associated with initial degradation processes. There is a likelihood that initial degradation of the PEEK results further bonding possibilities of the PEEK to aluminum.

The results of this work allow a first description of the bond between PEEK and aluminum and the bond strengths obtained. Based on the investigations the form fit and the interdiffusion can be excluded. Especially the EDX and XPS analysis provide the first evidence that the bonding could be mainly based on interactions at a nanoscale level. Further investigations are required in future work. The interface between Al and PEEK should be investigated in more detail, e.g. using XRD analysis to characterize the MgO lattice structures described.

Acknowledgments
We thank the Deutsche Forschungsgemeinschaft (DFG) for the support. We also thank the technical staff of the participating institutes for their support in the realization of this project.

References

[1] Walter S and Schmidt M 2008 Carbon Footprints und Carbon Label – eine echte Hilfe bei der Kaufentscheidung? *uwf* **16,** 175–181
[2] Friedrich H E and Krishnamoorthy S K 2017 *Leichtbau als Treiber von Innovationen,* ed H Friedrich ATZ/MTZ-Fachbuch (Wiesbaden: Springer Vieweg)
[3] Kolesnikov B, Herbeck L and Fink A 2008 CFRP/titanium hybrid material for improving composite bolted joints *Compos. Struct.* **83** 368-380
[4] Lambiase F, Paoletti A, Grossi V and Di Ilio A 2017 Friction assisted joining of aluminum and PVC sheets *J. Manuf. Process* **29** 221-231

[5] Tan X, Zhang J, Shan J, Yang S and Ren J 2015 Characteristics and formation mechanism of porosities in CFRP during laser joining of CFRP and steel *Compos. B. Eng.* **70** 35-43
[6] Li S, Khan H A, Hihara, L H, CongH and Li J 2018 Corrosion behavior of friction stir blind riveted Al/CFRP and Mg/CFRP joints exposed to a marine environment *Corros Sci* **132** 300-309
[7] Lambiase F and Genna S 2018 Experimental analysis of laser assisted joining of Al-Mg aluminium alloy with Polyetheretherketone (PEEK) *Int. J. Adhes. Adhes.* **84** 265-274
[8] Lionetto F, Balle F and Maffezzoli A 2017 Hybrid ultrasonic spot welding of aluminum to carbon fiber reinforced epoxy composites *J. Mater. Process. Technol.* **247** 289-295
[9] Mitschang P, Velthuis R and Didi M 2013 Induction Spot Welding of Metal/CFRPC Hybrid Joints *Adv. Eng. Mater.* **15** 804-813
[10] Schmid A, Arnaut K, Clausen J, Koerdt M, Struss A, Wöstmann,F J and Busse M 2017 Intrinsic Aluminum CFRP Hybrid Composites Produced in High Pressure Die Casting with Polymer Based Decoupling Layer *Key Eng. Mater.* **742** 197–204
[11] Huang Y, Meng X, Wang Y, Xie Y and Zhou L 2018 Joining of aluminum alloy and polymer via friction stir lap welding J. *Mater Process. Technol.* **257** 148-154
[12] Slattery E 2017 *Friction Lap Spot Joining of Aluminium Alloy 6082-T6 to Carbon Fibre-Reinforced Polyetheretherketone* Dissertation School of Engineering Department of Mechanical Manufacturing Engineering University of Dublin
[13] Winkler R 2018 *Haftmechanismen von Metallen (Cu, Al) appliziert durch Draht-Lichtbogenspritzen auf Polymeroberflächen (PEEK)* Dissertation Werkstoffe und Werkstofftechnische Anwendungen Band 75
[14] Wang H, Li N and Liu L 2019 Investigation of the Joining Technology of FRP/AZ31B Magnesium Alloy by Welding and Riveting Hybrid Bonding Method Mater. **12**
[15] Lee L 1994 Molecular Bonding and Adhesion at Polymer-Metal Interphases *J. Adhes.* **46** 15 - 38
[16] Sugama T, Carciello N R and Miuro M 1991 Adhesion of crystalline polyphenyletheretherketone (PEEK) in metal to metal joints Int. J. Adhes Adhes **12** 27-37
[17] Mayer J, Giannuzzi, L, Kamino, T, and Michael, J 2007 TEM Sample Preparation and FIB-Induced Damage *MRS Bulletin* **32** 400-407
[18] Schubert R 2009 *Entwicklung einer Software zur Analyse von magnetooptischen Daten* Abschlussarbeit Universität Osnabrück
[19] Ho P S, Rubloff G, Bartha J W, Hahn P O, LeGoues F K and Silverman B D 1985 Chemical Bonding and Reaction at Metal/Polymer Interfaces *J. Vac. Sci. Technol. A* **3** 739-745
[20] Guiselin O 1992 Irreversible Adsorption of a Concentrated Polymer Solution *EPL* **17** 225-230
[21] Napolitano S, Sferrazza M 2017 *How irreversible adsorption affects interfacial properties of polymers*. Adv Colloid Interface Sci **247** 172-177
[22] Day M, Sally D and Wiles D M 1990 Thermal degradation of poly(aryl-ether-ether-ketone): Experimental evaluation of crosslinking reactions *J. Appl. Poly. Sci.* **40** 1615-1625
[23] Cebe P 1988 Annealing study of poly(etheretherketone) *J. Mater .Sci.* **23** 37S21–3731
[24] Zhang M Q, Lu Z P and Friedrich K 1997 Thermal analysis of the wear debris of polyetheretherketone *Tribol. Int.* **30** 103-111
[25] Napolitano S and Wübbenhorst M 2011 The lifetime of the deviations from bulk behaviour in polymers confined at the nanoscale *Nat Commun* **2** 1-7
[26] Housmans C, Sferrazza, M and Napolitano S 2014 Kinetics of Irreversible Chain Adsorption *Macromolecules* **47** 3390-3393

Influence of EDM generator programs on shape and surface roughness of Ni-Ti sheets

T Berger[1,*], A Martin[1], K Navickaitė[2] and A Schubert[1]

[1] *Chemnitz University of Technology, Professorship Micromanufacturing Technology, Reichenhainer Str. 70, 09126 Chemnitz, Germany*
[2] *Beckmann-Institut für Technologieentwicklung e.V., Annaberger Str. 73, 09111 Chemnitz, Germany*

*e-mail: thomas.berger@mb.tu-chemnitz.de

Abstract. In the last decades caloric cooling technologies received an increasing scientific and industrial interest as promising alternatives to conventional vapor compression technology. Theoretically, efficiency of caloric technologies is more than 40% higher than for the conventional vapor compression cycle. Particularly interesting is elastocaloric cooling, which is an emerging technology, occurring due to the uniaxial mechanical loading of active regenerators. These are made of shape memory alloys, such as Ni-Ti. The regenerators must have a high fatigue resistance to withstand multiple loading cycles and simultaneously possess excellent elastocaloric properties. Therefore, it is crucial to establish a manufacturing process that allows fabricating thin regenerator elements with profound surface quality and as little impact on their functionality as possible. Considering the challenges of conventional machining technologies due to the specific characteristics of Ni-Ti, electrical discharge machining (EDM) demonstrates a high potential to meet the requirements for micro-shaping surfaces for elastocaloric applications. This paper presents fundamental analyses of the influence of EDM generator programs on the resulting removal depth, removal width and surface roughness of Ni-Ti sheets. Based on the perspective application in elastocaloric applications, adequate EDM programs are selected with regard to the targeted precision, surface roughness and minimal surface roughness.

1. Introduction and fundamentals of elastocaloric cooling

Around 8.5% of the globally consumed electricity in 2019 was used for the space cooling, which is also responsible for about 1 Gt of CO_2 emissions in that year [1]. The European Union set a clear target to become a first climate neutral continent by 2050 [2]. In order to achieve this goal, it is necessary to improve cooling and refrigeration technologies, so that environmentally hazardous substances would be completely phased out from the usage and at the same time an energy efficient way for cooling and refrigeration would be provided. Therefore, it is necessary to develop not-in-kind cooling technologies that would be environmentally friendly alternatives to conventional vapor compression (CVC) technology. One of such technologies is elastocaloric cooling, which was listed as one of the promising alternatives to CVC [3]. This is because, theoretical second law efficiency of caloric technologies can reach more than 60%, while for the small-scale appliances using the vapour compression technology it is generally below 20% [4,5]. Moreover, the hazardous gases are eliminated from the cooling cycle, since the active material is a solid matrix, which responses to the applied external stimuli by creating a cooling effect [6,7]. The heat transfer fluid is water or water-based solution of corrosion inhibitors.

Elastocaloric cooling is driven by the elastocaloric effect (eCE), which is a property of a material to produce heat under an external uniaxial stress [8,9]. The magnitude of a caloric effect in a certain material is quantified by the isothermal entropy change and the adiabatic temperature change [8,10,11]. The eCE is found in polymers, such as natural rubber, and shape memory alloys (SMA) such as the near-equiatomic Ni-Ti [8,10]. In order to obtain the elastocaloric effect, stress induced transformation is required. When SMA in the austenitic phase is loaded, at the critical stress the transformation to the martensitic phase begins until the de-twinned phase is reached [10]. When unloaded, the material returns to its original, austenitic state. The transformation from the austenitic to the martensitic phase causes heat generation, while reverse transformation forces the material to absorb heat from the surroundings [10].

The eCE is a reversible process, even though, hysteresis is associated to both transformations. Nevertheless, the eCE can be used for cooling and / or heat pumping applications, when a regenerative thermodynamic cycle is exploited to build up a temperature span between the heat source and the heat sink. In order to achieve this, a regenerator, which is a structure with high surface to volume ratio made of an elastocaloric material (eCM), must undergo multiple tension / compression cycles. Figure 1 shows a sketch of a double corrugated plate which was designed for perspective use as a part of an active elastocaloric regenerator [12,13].

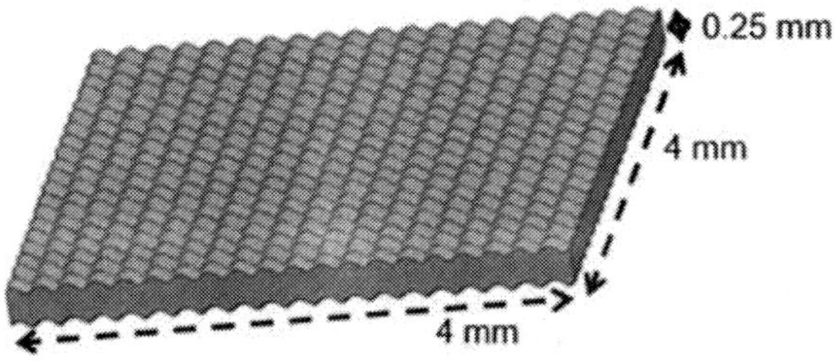

Figure 1. A differential volume of a double corrugated regenerator plate for elastocaloric applications.

Due to the high stresses and strains that must be applied to achieve the phase transformation and exploit the eCE, fatigue resistance of the active materials becomes an issue limiting their functionality. The life time of the constructed regenerators strongly depends on their surface roughness and the selected machining methods [14]. Thus, it is very important to find a suitable machining process and parameters in order to produce fully functional and mechanically stable regenerators.

Since mechanical machining of the thin Ni-Ti sheets is challenging due to the reduced mechanical stability of the sheets, electrical discharge machining (EDM) offers a high potential for structuring the regenerator-plates and controlling their resulting fatigue resistance due to its ablation mechanism. The challenge is to find applicable machining parameters that do not affect the material's functional properties too negatively.

2. Fundamental influences of EDM on the material properties of Ni-Ti

Ni-Ti is an excellent SMA and it is widely applied in biomedical and engineering fields for manufacturing bone implants, actuators, chevrons [4], which all require high surface quality. It was demonstrated that the functionality of Ni-Ti is highly affected by machining methods [15,16] and fabrication routines. The structural fatigue of Ni-Ti is influenced by chemical impurities with oxides and carbides when manufacturing thin films [7,8] or porous Ni-Ti structures and formation of heat-affected zones (HAZ) when using laser cutting or high discharge energy in EDM [16]. These chemical impurities not only affect the temperature at which the direct and reverse phase transition takes place, but also act as crack nucleation sites [4]. These cracks can lead to the failure of the material.

Moreover, the stress distribution is non-uniform in the machined areas compared to the base workpiece material, resulting in different mechanical properties within the active material. For example, the resistance to tensile or compressive stress is reduced at the machined areas [16], which lead to more rapid cold deformation. Once this state is reached, the material is permanently disabled.
According to the current state of the art, EDM offers the possibility for high-precision surface structuring, without significant change of the material's properties in the near of the outer surface. Recent researches focused on micro-EDM of Ni-Ti by investigating electrical and non-electrical machining parameters. It can be derived that an increase in discharge energy increases material removal rate (MRR) but also tool wear rate (TWR) and surface roughness (SR). Increasing capacitance [17–21], pulse current[20,22–24], pulse on time [20,22,23] gap voltage [17,18] and discharge voltage [19,21] were all found to increase discharge energy. Melting temperature and thermal conductivity are the most significant workpiece material related, non-electrical parameters [19,23] since an increase of one or both results in lower removal rate and higher tool wear.

The discharge energy is the parameter with most influence on the HAZ, where mechanical properties and microstructure are affected by the heat of the machining process [25]. This results in mechanical properties that differ from the unaffected basic material. By applying a controlled, low discharge pulse energy and thus targeted reduction of the thickness of HAZ, machining of complex micro-geometries in Ni-Ti with high surface integrity and low surface roughness down to $Ra = 0.1$ µm were realized [26–28]. Differential scanning calorimetry (DSC) results of micro-EDMed samples revealed that by applying discharge pulse energies lower than 10 µJ, it is possible to realize only negligible thermal influences [22,29].

3. Experiments

Even if the state-of-the-art indicates that micro-EDM has the potential to machine Ni-Ti with high precision and high surface quality, the process continues to face challenges related to too high thermal damages on the machined surfaces. When reducing discharge energy, a low MRR and a high TWR are expected, which represent significant hindrances for machining the required regenerator part surfaces. As a result, there is a lack of knowledge about how the machining process must be designed to reduce inefficient discharges and thus reduce the thermal damage. In addition, the special geometry of the regenerators, as shown in Figure 1, requires additional demands on the process regarding the design of the clamping system and the support of dielectric. For this reason, experiments were carried out to determine the basic machinability of a Ni-Ti sheet with available erosion programs. Fundamental analyses of the influence of the chosen EDM generator programs on the resulting shape of the removals, the surface roughness as well as the material removal rate and the tool wear rate were carried out.

3.1. Experimental setup

The experimental analyses were planned for a ZK genius 700 sinking EDM machine from Zimmer & Kreim GmbH (*Brensbach, Germany*). This system allows simultaneous 4-axes machining with a positioning accuracy of ± 1 µm for the linear stages and ± 0,001 ° for the rotational C-axis [30]. The machining experiments were carried out in a bath of EDM oil of the type Ionoplus IME-MH from Oelheld GmbH (*Stuttgart, Germany*).

For the determination of the basic machinability of Ni-Ti, linear one-axis sinking experiments in Z-direction were carried out using electrode rods made of copper with a diameter of 5.0 mm. A Ni-Ti sheet from Memry Corporation, USA, with the dimensions of 39.0 x 50.0 x 0.6 mm³ was used as workpiece. X-ray fluorescence analysis was carried out to characterize the chemical composition of the alloy. The results indicate atomic contents of 50.5 at% Ni and 49.5 at% Ti.

Figure 2 shows a photograph of the machining setup, implemented in the sinking EDM machine. The copper tool electrode is clamped in a collet and fixed on the Z-axis of the EDM-machine using a zero-point clamping system. The Ni-Ti sheet is fixed in a standard workpiece clamping system, which is mounted on the machine table.

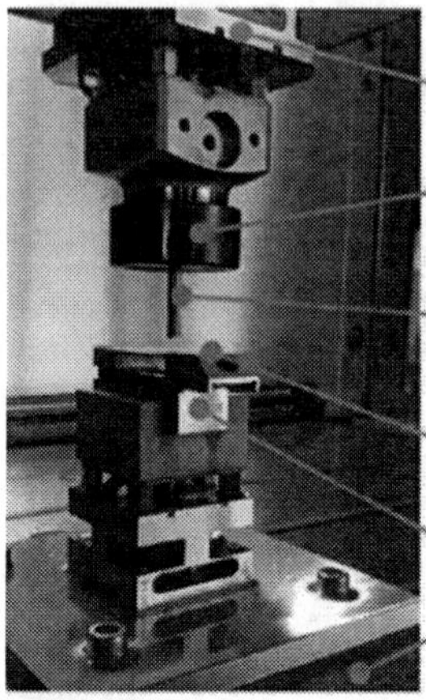

Figure 2. Photograph of the tool electrode and the Ni-Ti workpiece implemented in the sinking EDM machine.

3.2. Design of experiments

To analyze the basic machinability of Ni-Ti, the three generator programs named Cu-Ti 30-30, Cu-Ti 15-15 and Cu-Ti 9-9 were selected. The respective discharge parameters, which are accessible via the machine tool's database, are carted in table 1.

Table 1. EDM processing parameters.

Generator program	VDI class	U_{ign} [V]	U_{oc} [V]	T_{on} [ms]	T_{off} [ms]	C [nF]
Cu-Ti 9-9	9	150	140	8	12	1
Cu-Ti 15-15	15	150	140	20	20	10
Cu-Ti 30-30	30	270	260	50	100	6100

The three available generator programs had been designed for machining TiAl6V4 by the manufacturer of the machine tool. All the chosen programs use an iso-pulse generator and positive tool polarity. They differ in terms of ignition voltage (U_{ign}), open-circuit voltage (U_{oc}), discharge pulse duration (T_{on}) and pause times between the discharge pulses (T_{off}) as well as capacitance (C). This results in different discharge energies. Hence, different targeted surface roughness according to VDI 3400 classification [31] are specified by the last number of the respective program names. In this context, the program Cu-Ti 30-30 provides the highest discharge energy and consequently the highest surface roughness to be expected.

The target depth of each removal was set to 0.2 mm, which almost corresponds to the amplitude of the corrugation presented in Figure 1. In order to ensure a constant initial state, the electrode was dressed after each machining experiment. To assure the exchange of dielectric in the working gap, the immersed flushing of the dielectric bath was assisted by flushing strokes of the Z-axis. The effective removal time between the single flushing strokes was set to 1 s.

4. Results

For the qualitative and quantitative analyses of the erosion results, images of the removals were captured and measured with a Keyence VK-9700 laser scanning microscope (*Osaka, Japan*). The measurement results were evaluated with Mountains Map® analysis software. The diameter D, the removal depth T and the surface roughness values Sa and Sz were determined. For the determination of D, T, Sa and Sz the measurements were repeated 3 times on different measuring areas and the mean value and the standard deviation were calculated. In addition, the length of the electrode wear W_l was derived and the material removal rate MRR was calculated according to Equation 1, where V_Z represents the volume of a cylindrical remocal and t the machining time.

$$MRR = \frac{V_Z}{t} = \frac{\frac{\pi}{4} D^2 T}{t} \tag{1}$$

4.1. Microscopic characterization

Figure 3 shows top view images of the removal results realized with the three chosen generator programs.

a) Cu-Ti 30-30 b) Cu-Ti 15-15 c) Cu-Ti 9-9

Figure 3. Top view images of the erosions realized with the chosen EDM programs.

As can be seen, the machining was successful with all three generator programs. The removal in Figure 3a) appears more uneven than the removals in Figures 3b) and 3c), which appear comparable to one another. Nevertheless, a lower roughness is expected for the removal in Figure 3c). However, both erosions machined with Cu-Ti 15-15 and Cu-Ti 9-9 show erratic deepened structures in the center of the removal area, which are caused by undesired discharges due to poor flushing conditions. This indicates that the flushing strategy needs to be adapted to achieve a sufficiently high precision for the prospective machining of the regenerator shapes. Detailed views of the removal results are shown in Figure 4.

a) Cu-Ti 30-30 b) Cu-Ti 15-15 c) Cu-Ti 9-9

Figure 4. Microscope images of the erosions realized with the chosen generator programs, magnification: 50 x. The red arrows in figure 4 a) highlight cracks as a result from high discharge energy for generator program Cu-Ti 30-30, while lower discharge energies machined with Cu-Ti 15-15 (Figure 4 b) and Cu-Ti 9-9 (Figure 4 c) did not cause cognizable cracks.

There are significant differences in terms of the resulting surface quality. While a very rough surface was created with Cu-Ti 30-30, the surface quality with Cu-Ti 15-15 and Cu-Ti 9-9 is significantly increased. The resulting surface with Cu-Ti 30-30 is shown in Figure 4 a). Due to the high discharge energy, discharge craters with lateral dimensions around 150 µm are recognizable. As highlighted by the red arrows, the surface is characterized by cracks that result from high discharge energy and thermal stress. Thus, Cu-Ti 30-30 is not suitable for machining Ni-Ti for elastocaloric applications, since it is expectable that the fatigue strength will be exceeded too quickly. At lower discharge energies machined with Cu-Ti 15-15 (Figure 4 b) and Cu-Ti 9-9 (Figure 4 c) no cracks were detected. The average lateral dimensions of the single discharge craters were determined to be 15 µm for Cu-Ti 15-15 and 9 µm for Cu-Ti 9-9, respectively. Thus, these generator programs are more applicable for machining Ni-Ti for eCE-regenerators.

4.2. Analyses of shape and surface roughness

The experimental results were evaluated by calculating the aerial roughness values with MountainsMap® analysis software. Figure 5 shows the resulting arithmetical mean height *Sa* and Figure 6 the maximum height *Sz* achieved with the used generator programs as bar charts.

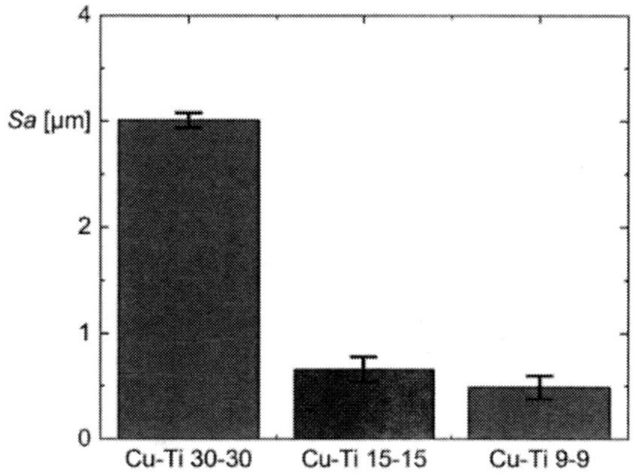

Figure 5. Arithmetical mean height *Sa* achieved with the three EDM generator programs.

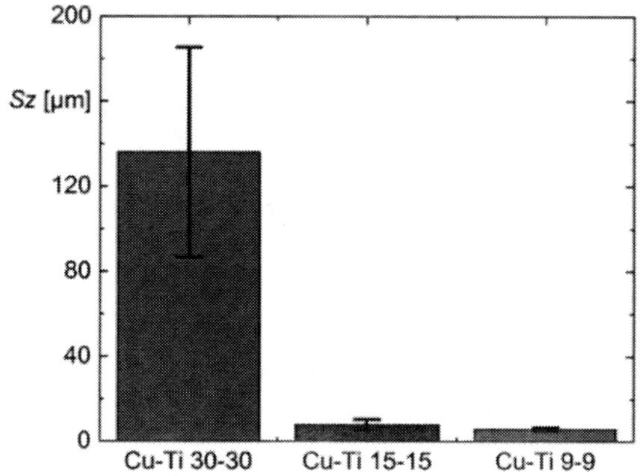

Figure 6. Maximum height *Sz* achieved with the three EDM generator programs.

The highest roughness values of $Sa = 3.01 \pm 0.07$ µm and $Sz = 136.1 \pm 49.4$ µm were measured for Cu-Ti 30-30 as expected from the qualitative inspection. Here, the standard deviation of 49.4 µm for *Sz* is comparatively high due to the large size of the discharge craters. The lowest roughness values of $Sa = 0.49 \pm 0.11$ µm and of $Sz = 7.88 \pm 0.88$ µm were achieved with the generator program Cu-Ti 9-9. However, the targeted roughness values according to VDI 3400 [32], which are shown in Table 2, were not achieved for all eroding-programs except for *Sa* using Cu-Ti 30-30. This is due to the material characteristics of the Ni-Ti workpiece, which differ in terms of melting temperature and thermal conductivity from TiAl6V4, for which these generator programs were designed.

Table 2. Surface classification according to VDI 3400 [31].

	VDI 30	VDI 15	VDI 9
Sa [µm]	3.21	0.56	0.28
Sz [µm]	11.7 – 15.8	2.2 – 3.7	1.3 – 2.2

The resulting diameters *D* were measured on the bottom of the single removals. The results are shown in the bar chart of Figure 7, where the error bars highlight the calculated deviations of single measurement values.

IOP Conf. Series: Materials Science and Engineering 1147 (2021) 012023 doi:10.1088/1757-899X/1147/1/012023

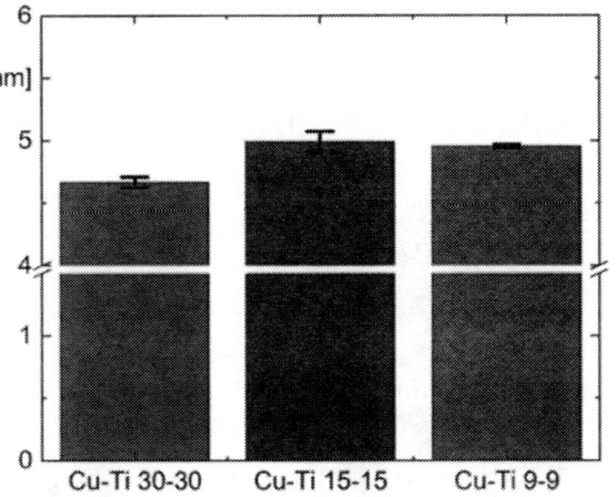

Figure 7. Removal diameter D on the bottom of the removals realized with the three EDM generator programs.

All removal diameters are narrower than the diameter of the used electrode, which exhibits 5.0 mm. This is caused by to the process-related circumferential wear on the tool electrode. The largest diameter of (4.994 ± 0.080) mm was achieved with the generator program Cu-Ti 15-15. This means that the circumferential tool wear was the lowest. The highest circumferential tool wear occurred with Cu-Ti 30-30, where a diameter of (4.667 ± 0.043) mm was measured.

The measured removal depths T are shown in Figure 8 and the corresponding tool wear lengths W_l can be seen in Figure 9. The error bars represent the measured Sz values of figure 6, which illustrates the deviation of the surface peaks and valleys from the measured values.

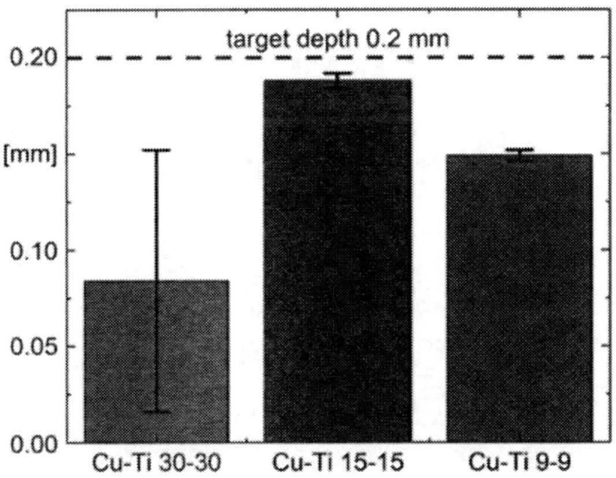

Figure 8. Removal depth T realized with the three EDM generator programs

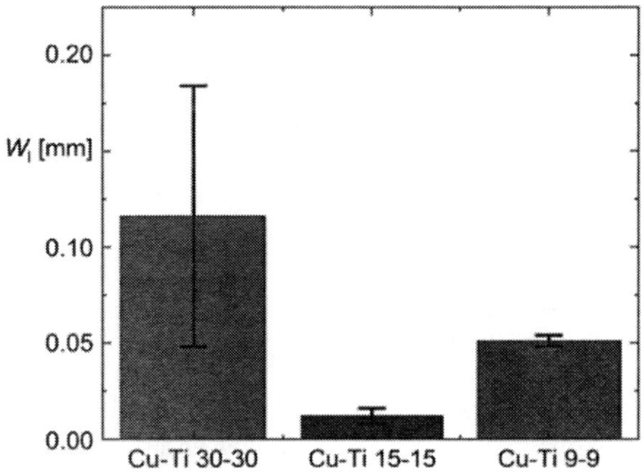

Figure 9. Tool wear length W_l realized with the three EDM generator programs

With the generator program Cu-Ti 15-15 the maximum depth T of 188 µm was realized, which represents the slightest deviation from the targeted depth of 200 µm as highlighted by the dashed horizontal line. Correspondingly, the lowest tool wear length W_l of 12 µm was detected for this program. In contrast, the smallest depth of 84 µm along with the highest tool wear length of 116 µm were measured with the generator program Cu-Ti 30-30. Hence, regarding the removal accuracy and the wear behavior, Cu-Ti 15-15 represents the best choice of the selected generator programs for machining Ni-Ti.

4.3. Analysis of the material removal rate

The material removal rates *MRR* were calculated based on the results of the removal diameter and the removal depth as well as under consideration of the machining time of the three generator programs. The measured machining times were 0.2 min for Cu-Ti 30-30, 45.0 min for Cu-Ti 9-9 and 105.0 min for Cu-Ti 15-15. The resulting *MRR* are shown in Figure 10.

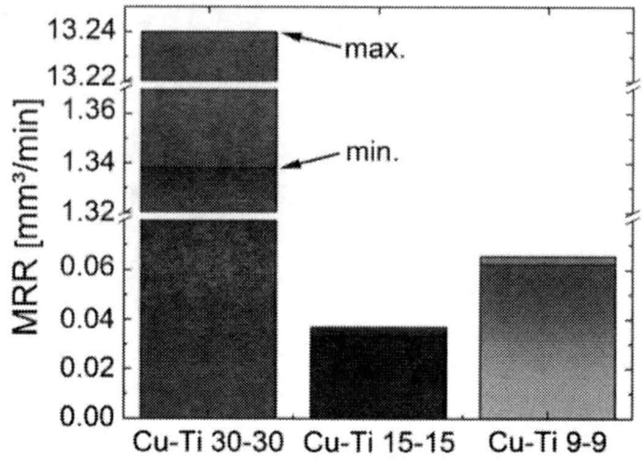

Figure 10. Material removal rate *MRR* realized with the three EDM generator programs.

The maximum and minimum values were calculated according to the deviations described for the removal depth (Figure 8) and diameter (Figure 7). For a useful comparison of the strongly deviating results, the bar chart includes two interruptions within the value ranges (0.08…1.32) mm³/min and (1.37…13.22) mm³/min. These interruptions were chosen, because the average *MRR* of 7.18 mm³/min achieved with Cu-Ti 30-30 is significantly higher than the average *MRR* of 0.06 mm³/min achieved with Cu-Ti 9-9 and the lowest average *MRR* of 0.04 mm³/min achieved with Cu-Ti 15-15. These deviations are caused by the significantly differing discharge energies, which on the one hand results in significantly more material removal volume per time with Cu-Ti 30-30 but on the other hand also causes high tool wear. The high tool wear rate and the large deviations detected for this generator program reveal that Cu-Ti 30-30 is not suitable for precise machining of components for elastocaloric applications. However, the programs Cu-Ti 15-15 and Cu-Ti 9-9 offer the potential for processing such elements.

5. Summary and conclusion

For the topic component manufacturing of elastocaloric applications apart from the actual state of the art, the influence of available sinking EDM generator programs on the resulting removals of Ni-Ti was presented. Experiments were carried out with a commercially available setup. The applicability of three generator programs was examined regarding their fundamental applicability for machining a Ni-Ti sheet. The resulting removal widths, removal depths and surface roughness values *Sa* and *Sz* were analyzed. Furthermore, the tool wear rate and the material removal rate were calculated.

It was shown that the generator programs Cu-Ti 15-15 and Cu-Ti 9-9 offer the potential for prospective machining of Ni-Ti sheets for elastocaloric applications, since they produced low roughness and crack-free surfaces due to significantly lower discharge energies compared to Cu-Ti 30-30. Thus, higher surface quality, reduced tool wear and higher precision was achieved. When focusing on elastocaloric applications, where the main goal is to achieve the lowest possible surface roughness and heat affected zones, Cu-Ti 9-9 offers the highest potential.

There is a significant need for further researches to achieve roughness values that meet the requirements of VDI 3400 on Ni-Ti and to assure the functionality of the machined parts for elastocaloric applications. In addition, erratic structures were detected in the center of the removals, which indicate that the dielectric flushing needs to be optimized.

References

[1] International Energy Agency - IEA 2020 Cooling - More efforts needed *Int. Energy Agency*
[2] European Union 2020 2030 Climate Target Plan
[3] Goetzler W, Zogg R, Young J and Johnson C 2014 *Energy savings potential and RD & D opportunities for non- vapor-compression HVAC technologies* (Burlington, USA)
[4] Kabirifar P, Žerovnik A, Ahčin Ž, Porenta L, Brojan M and Tušek J 2019 Elastocaloric cooling: State-of-the-art and future challenges in designing regenerative elastocaloric devices *Stroj. Vestnik/Journal Mech. Eng.* **65**
[5] Kitanovski A 2020 Energy Applications of Magnetocaloric Materials *Adv. Energy Mater.* **10** 1903741
[6] Smith A, Bahl C R H, Bjørk R, Engelbrecht K, Nielsen K K and Pryds N 2012 Materials challenges for high performance magnetocaloric refrigeration devices *Adv. Energy Mater.* **2**
[7] Kirsch S-M, Welsch F, Michaelis N, Schmidt M, Wieczorek A, Frenzel J, Eggeler G, Schütze A and Seelecke S 2018 NiTi-Based Elastocaloric Cooling on the Macroscale: From Basic Concepts to Realization *Energy Technol.* **6** 1567–1587
[8] Ossmer H, Chluba C, Krevet B, Quandt E, Rohde M and Kohl M 2013 Elastocaloric cooling using shape memory alloy films *J. Phys. Conf. Ser.* **476**
[9] Ossmer H, Lambrecht F, Gültig M, Chluba C, Quandt E and Kohl M 2014 Evolution of temperature profiles in TiNi films for elastocaloric cooling *Acta Mater.* **81**
[10] Kitanovski A, Tušek J, Tomc U, Plaznik U, Ožbolt M and Poredoš A 2015 *Magnetocaloric Energy Conversion* (Cham: Springer International Publishing)
[11] Tušek J, Engelbrecht K, Mañosa L, Vives E and Pryds N 2016 Understanding the thermodynamic properties of the elastocaloric effect through experimentation and modelling *Shape Mem. Superelasticity* **2**
[12] Navickaitė K, Mocerino A, Cattani L, Bozzoli F, Bahl C R H and Engelbrecht K 2019 Heat transfer in double corrugated tubes *The 9th International Conference & Workshop REMOO-2019* (Hong Kong, China)
[13] Navickaitė K, Noel D, Bahl C R H and Engelbrecht K 2017 Passive heat transfer enhancement in 3D corrugated tube *9th World Conf. Expreimental Heat Transf. Fluid Mech. Thermodyn.*
[14] Engelbrecht K, Tušek J, Eriksen D, Lei T, Lee C Y, Tušek J and Pryds N 2017 A regenerative elastocaloric device: Experimental results *J. Phys. D. Appl. Phys.* **50**
[15] Mwangi J W, Nguyen L T, Bui V D, Berger T, Zeidler H and Schubert A 2019 Nitinol manufacturing and micromachining: A review of processes and their suitability in processing medical-grade nitinol *J. Manuf. Process.* **38** 355–369
[16] Mwangi J W 2019 *Micro-Electrical Discharge Machining of Nitinol for medical aplications* (Chemnitz Technical University)
[17] Nirala C K, Reddy B and Saha P 2012 Optimization of Process Parameters in Micro Electro-Discharge Drilling [micro EDM-Drilling]: A Taguchi Approach *Adv. Mater. Res.*
[18] Rasheed M S, Abidi M H, El-Tamimi A M and Al-Ahmari A M 2013 Investigation of Micro-EDM Input Parameters on Various Outputs in Machining Ni-Ti Shape Memory Alloy Using Full Factorial Design *Adv. Mater. Res.*
[19] Saxena H, Kumar H and Shukla S 2014 Overviewof Micro-EDM Process Parameters in Machining *Int. J. Eng. Sci.*
[20] Daneshmand S, Kahrizi E F, Neyestanak A A L and Ghahi M M 2013 Experimental investigations into electro discharge machining of NiTi shape memory alloys using rotational tool *Int. J. Electrochem. Sci.* **8**
[21] Wohlschlögel M, Steegmüller R and Schüßler A 2014 Effect of inclusion size and distribution on the corrosion behavior of medical-device grade nitinol tubing *J. Mater. Eng. Perform.* **23**

IOP Conf. Series: Materials Science and Engineering 1147 (2021) 012023 doi:10.1088/1757-899X/1147/1/012023

[22] Daneshmand S, Kahrizi E F, Lotfineyestanak A A and Monfared V 2014 Optimization of electrical discharge machining parameters for NiTi shape memory alloy by using the Taguchi method *J. Mar. Sci. Technol.* **22**
[23] Gaikwad V and Jatti V S 2018 Optimization of material removal rate during electrical discharge machining of cryo-treated NiTi alloys using Taguchi's method *J. King Saud Univ. - Eng. Sci.* **30**
[24] Natarajan N and Arunachalam R M 2011 Optimization of micro-EDM with multiple performance characteristics using taguchi method and grey relational analysis *J. Sci. Ind. Res. (India).*
[25] Miraoui I, Boujelbene M and Zaied M 2016 High-Power Laser Cutting of Steel Plates:Heat Affected Zone Analysis *Adv. Mater. Sci. Eng.*
[26] Benavides G L, Bieg L F, Saavedra M P and Bryce E A 2002 High aspect ratio meso-scale parts enabled by wire micro-EDM *Microsyst. Technol.* **8**
[27] Liu J F and Guo Y B 2014 Micro Mechanical Properties of EDMed Nitinol Surface Using Nanoindentation *Proc. 4th Int. Conf. Nanomanufacturing*
[28] Mwangi J W, Zeidler H, Berger T and Schubert A 2017 Influence of dielectric fluid and tool electrode choice on micro-EDMed nitinol surface integrity *Proceedings of the 17th International Conference of the European Society for Precision Engineering and Nanotechnology, EUSPEN 2017*
[29] Mwangi J W, Bui V D, Thüsing K, Hahn S, Wagner M F-X and Schubert A 2020 Characterization of the arcing phenomenon in micro-EDM and its effect on key mechanical properties of medical-grade Nitinol *J. Mater. Process. Technol.* **275**
[30] Zimmer & Kreim GmbH 2018 Data Sheet genius 700
[31] NB Boumi AG 2018 Data Sheet surface roughness

Characteristics of cracks formed at the surface of NiTi during a single cycle of pseudoelastic deformation

K E Freiberg[1], R Wonneberger[1,2] and A Undisz[2,*]

[1] Friedrich-Schiller-Universität Jena, Otto Schott Institute of Materials Research, Löbdergraben 32, 07743 Jena, Germany
[2] Technische Universität Chemnitz, Institute of Materials Science and Engineering, Erfenschlager Straße 73, 09125, Chemnitz, Germany

[*]e-mail: andreas.undisz@mb.tu-chemnitz.de

Abstract. Pseudoelastic deformation of NiTi usually results in cracks at the surface. Cracking is promoted by surface oxide layers that form, e.g., during heat treatments required for shape-setting of minimally invasive implants. In connection with the advancing miniaturization of minimally invasive implants, the question arises whether their mechanical integrity may get impaired by such cracks. In the present work, the characteristics of the cracks was investigated in cross section with the help of targeted preparation using focused ion-beams. SEM and TEM on samples after a single cycle of pseudoelastic deformation revealed that cracks extend perpendicular to the loading direction in the surface oxide layer and change to angles between 90° and 45° in the Ni-rich layer below. Pores observed in the surface oxide close to the Ni-rich layer did not prevent the extension of cracks towards the NiTi bulk. When the cracks reach the NiTi, blunting of the crack tip was observed. The crack length essentially corresponds to the thickness of the surface oxide layer and the Ni-rich layer. The findings provide data for estimating crack propagation in according implants in the future.

1. Introduction

Pseudoelastic NiTi is a key material for developing novel implants for minimally invasive application [1, 2]. A current trend is the miniaturization of according implants to reduce the amount of material inserted into the human body and risks associated with surgery [3]. The miniaturization of implants, e.g., stents, brings challenges regarding their mechanical stability and integrity during extended application under alternating loads. In this context, potential damage to the surface of NiTi as a result of a single cycle of pseudoelastic deformation during the implantation procedure is of specific interest [4-6]. It is important to note, that the deformation strain associated with collapsing and deploying such implants during the implantation procedure is usually significantly larger than during the following service life. It was shown that a single cycle of pseudoelastic deformation can lead to damage at the natural surface of NiTi, especially in combination with previously formed surface oxide layers. With respect to the thickness of surface oxide layers, different outcomes were observed, including the formation of cracks or flaking of particles [7]. Such damage to the natural surface of NiTi may not only compromise the mechanical integrity of minimally invasive implants, but it may also expose Ni-rich sublayers with consequences regarding the compatibility of the material in a specific biological environment. Additionally, the cracks at the surface of the material may grow during the often small but alternating loads, eventually impairing the mechanical integrity of the implant by fatigue fracture. Thus, the characteristics of such cracks are of interest. Considering the observation of pores close to the oxide / metal interface in a previous study, their arrangement and size with respect to the heat treatment time was also of interest [4]. A layer of pores parallel to the surface may affect the direction of cracks and help to guide them away from the bulk. Visualization of cracks has so far mainly been carried out using scanning electron microscopy in top view [8, 9]. However, important aspects including crack length, propagation or deflection and details regarding pores remain unclear.

In the present work, the characteristics of cracks at the natural surface of NiTi was observed in cross section using scanning electron microscopy and transmission electron microscopy. An important aspect is if cracks extend into the NiTi bulk already after a single cycle of pseudoelastic deformation.

2. Experimental and materials

2.1. Material condition and heat treatment

The study was carried out using pseudoelastic NiTi wire according to specification SE508 with a diameter of ~ 160 μm (Fort Wayne Metals). The surface of the wire was mechanically polished by the supplier. Wire segments were subject to annealing in a muffle furnace in air atmosphere as detailed in [10]. The annealing temperature was set to 540°C and the annealing to 2 min, 5 min, 10 min, 30 min and 60min. After the heat treatment, the wire surface displayed a coloration as a result of the thin surface oxide layer. Of each condition, a wire segment was strained to the pseudoelastic limit of ~ 6 %.

2.2. Preparation of cross sections and analysis using scanning and transmission electron microscopy

For visualizing details of cracks and pores in the surface oxide layer by scanning electron microscopy (SEM) and transmission electron microscopy (TEM), cross sections were prepared using a focused ion beam system (FIB, FEI Helios NanoLab 600i) [11]. After identifying a region of interest, the surface was covered locally by a Pt-layer with a thickness of ~ 50 nm applying electron-beam deposition inside the FIB. After tilting the sample by 52° perpendicular to the ion-beam, a second layer of protective Pt with a thickness of ~ 1 μm was deposited using the ion-beam. Slope cuts to a depth of 5 μm along a length of 10 μm were cut at 30 kV and 2.4 nA using the ion-beam perpendicular to the wire surface. After slope cutting, cleaning steps were applied by reducing the acceleration voltage step-wise to 5 kV in order to increase the quality of the cross sections by removing material affected by high ion energy.

Immediately after preparation, an overview of the samples along slope cutes was gained using the secondary electron/ in-lens detector of the FIB/SEM system. Although the imaging quality of the FIB/SEM system was limited, a layered structure in the surface region and features resembling pores and cracks became visible. For higher detail, lamellae for TEM were prepared and observed using a Jeol Jem-3010 operating at 300 kV.

For preparation of TEM lamellae, regions of interest were subject to a second slope cut parallel to the first one. Further preparation was carried out according to the common lift-out procedure for preparation of TEM lamellae. Final etching of the lamellae was done in several steps, reducing the acceleration voltage from 30 kV to 2 kV.

3. Results and discussion

3.1. Length and orientation of cracks after a single cycle of pseudoelastic deformation

In the following Figures, exemplary images of slope cuts are displayed for wires after heat treatment for 5 min (Figure 1), 10 min (Figure 2) and 30 min (Figure 3). In each Figure, the wire axis was set horizontally. The thickness of the surface oxide layer increases with annealing time from values of ~ 69 nm after 5 min to ~ 258 nm after 30 min (Table 1). Already during annealing for 5 min, a Ni-rich layer formed below the surface oxide layer as a consequence of the preferential oxidation of Ti, depleting the sublayer by Ti. The formation of a Ni-rich layer is well documented in the literature. The thickness of the Ni-rich layer was in the range of ~ 48 nm after 5 min to ~ 168 nm after 30 min of annealing (Table 1), close to the values determined for the surface oxide layer. For annealing times of 30 min and longer, it was shown, that the Ni-rich layer consists of intermetallic Ni_3Ti that does not exhibit pseudoelasticity as observed for NiTi [4]. For annealing times below 30 min, the phase of the Ni-rich layer has not been identified unambiguously.

Independent of the annealing time, cracks in the surface oxide layer and the Ni-rich layer are visible. The cracks expand from the wire surface into the material and end when reaching the NiTi bulk (Figures 1 to 3). It is interesting to note, that single cracks always damage both, the surface oxide layer and the Ni-rich layer. Cracks exclusively in one of the layers were not observed. Whereas the orientation of the cracks in the surface oxide layer is essentially perpendicular to the surface and thus perpendicular to the applied load, the orientation of cracks often changes when entering the Ni-rich layer towards an angle between 90° and 45° to the loading direction. This indicates towards a brittle behavior of the surface oxide layer and some ductility of the Ni-rich layer. The extension of cracks into the NiTi bulk is apparently hindered by plastic deformation in the vicinity of the crack tip, as visible from the blunting of the crack tip displayed in Figure 3.

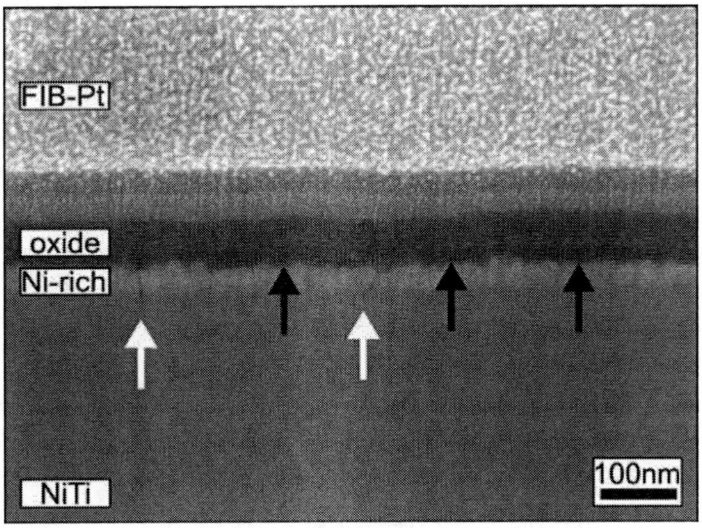

Figure 1. SEM image of a slope cut at the surface of a NiTi wire after heat treatment for 5 min. Between surface oxide layer and NiTi, a Ni-rich layer has formed. Cracks are visible in both, the surface oxide layer and the Ni-rich layer (indicated by white arrows). Pores are visible as dark spots at the bottom of the surface oxide layer (indicated by black arrows).

Table 1. Thickness of the surface oxide layer and the Ni-rich layer formed after annealing for 5 min, 10 min and 30 min, respectively.

Annealing time (min)	Thickness of surface oxide layer (nm)	Thickness of Ni-rich layer (nm)
5	69 ± 8.2	48 ± 8.8
10	115 ± 7.7	89 ± 10.3
30	258 ± 28.8	168 ± 20.7

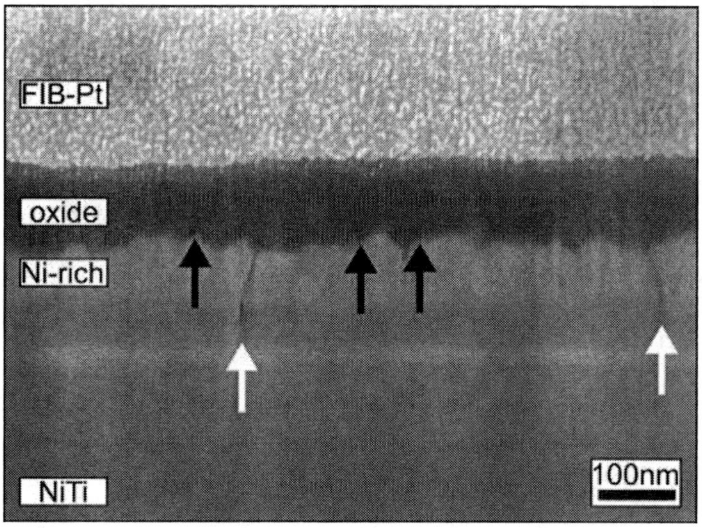

Figure 2. SEM image of a slope cut at the surface of a NiTi wire after heat treatment for 10 min. Cracks are visible in the surface oxide layer and the Ni-rich layer (white arrows). Pores are visible at the bottom of the surface oxide layer (black arrows).

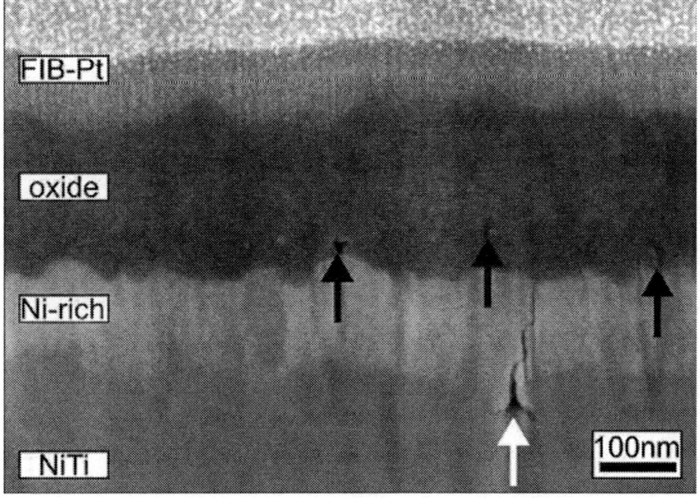

Figure 3. SEM image of a slope cut at the surface of a NiTi wire after heat treatment for 30 min. A crack is visible in the surface oxide layer and the Ni-rich layer (white arrow). The crack stopped when reaching the NiTi bulk and a blunt crack tip is visible. Pores are contained at the bottom of the surface oxide layer (black arrows).

3.2. Effect of annealing time on the occurrence of pores in the surface oxide layer

Independent of the annealing time, pores were identified in the surface oxide layer close to the Ni-rich layer. The formation of the pores is likely a result of the Kirkendall effect. As mentioned earlier, such pores are of specific interest here, since they may help to deflect cracks and guide them away from the bulk. However, the limited resolution of the FIB/SEM system did not allow for identification of further details regarding arrangement and quantity (number/size) of the pores. According to the Kirkendall effect, the quantity of the pores is likely to increase with annealing time, if the thickness of the surface oxide layer increases.

For analyzing the effect of the annealing time on the formation of pores, TEM lamellae were prepared for samples annealed for 2 min to 60 min. Pores are visible in all cases inside the surface oxide layer close to the interface to the Ni-rich layer / NiTi (Figure 4). Although difficult to judge from quasi 2-dimensional samples, the quantity of pores appears to be similar for each condition, whereas the thickness of the surface oxide layer increases notably. This is in contrast to the expected behavior and indicates towards a change of the mechanism of oxide layer growth during annealing. Considering earlier work, the surface oxide layer was reported to grow inward for temperatures < 450°C and outward for higher temperatures [12, 13]. During heating the samples in the present work to 540°C, the surface oxide layer grew inward for ~ 1.5 min before the growth direction changed to outward [10]. The observed arrangement and quantity of the pores would be consistent with the conclusion that the pores essentially form during the initial inward growth of the surface oxide layer. A potential effect of the pores to deflect cracks would thus be independent of the annealing time. For further assessment, a potential impact of the annealing temperature should be considered.

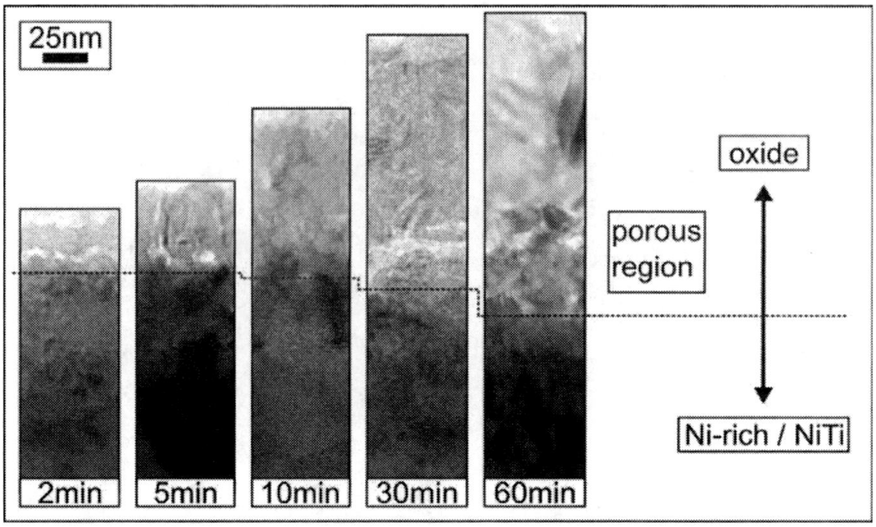

Figure 4. TEM images of cross sections at the surface of NiTi wires after annealing for 2 min to 60 min (left to right). The arrangement and quantity of pores is similar independent of the annealing time, whereas the thickness of the surface oxide layer increases notably.

3.3. Impact of pores inside the surface oxide layer on flaking of oxide particles and propagation of cracks into the bulk

In a previous work it was demonstrated that oxide particles may flake off the NiTi wire during a single cycle of pseudoelastic deformation. In order to visualize details of a flaking site, a TEM lamella was prepared in cross section together with a lamella from a site without flaking (Figures 5 and 6). In the specific case, the previous deformation of the wire was increases to slightly exceed the limits of the pseudoelastic deformation. Whereas the pseudoelastic deformation reverts during unloading, a small amount of plastic deformation remained, facilitating the identification of cracks after unloading of the sample.

The characteristics of a crack is shown in Figure 5 for a site without flaking. The behavior is as observed previously using FIB/SEM. The crack extends essentially perpendicular to the loading direction into the material. Crack deflection at the pores at the bottom of the surface oxide layer is not visible. For the lamella attained from the flaking site, some oxide is visible at the top of the Ni-rich layer (Figure 6). In this case, the layer of pores has apparently acted as weak region, guiding a crack parallel to the surface. However, cracks are visible in the Ni-rich layer that remained attached to the wire in all observed cases.

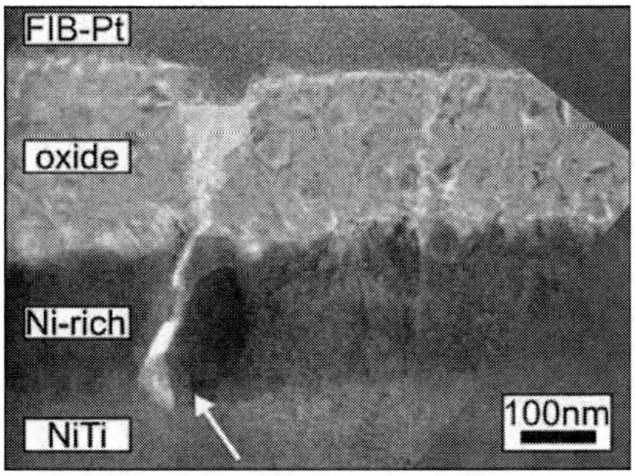

Figure 5. TEM image of a NiTi wire after 30 min of annealing in cross section. The behavior is as observed previously using FIB/SEM.

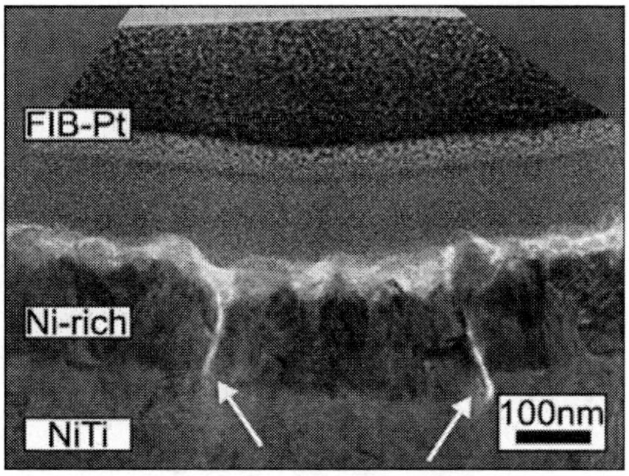

Figure 6. TEM image of a NiTi wire after 30 min of annealing in cross section at a flaking site. The layer of pores has acted as weak region, helping to guide a crack parallel to the surface. Cracks in the Ni-rich layer are visible.

4. Conclusions

Details of cracks in the natural surface of NiTi after a single cycle of pseudoelastic deformation were visualized in cross section using SEM and TEM. The following aspects are compiled regarding the characteristics of according cracks:

- cracks in the surface oxide layer extend perpendicular to the loading direction, consistent with the brittle behavior of the surface oxide layer
- cracks in the Ni-rich layer below the surface oxide layer exhibit an angle between 90° and 45° to the loading direction, indicating some ductility of the layer
- cracks stop when reaching the NiTi bulk and blunting of the crack tip is observed
- pores form independent of the annealing time at the bottom of the surface oxide layer and contribute to flaking of oxide particles, but do not prevent the formation of cracks in the Ni-rich layer

An estimation of the crack length becomes possible considering the thickness of the surface oxide layer and the thickness of the Ni-rich layer. Both layers crack during the first cycle of pseudoelastic deformation. Accordingly, the crack length corresponds to the combined thickness of both layers and is expected to remain < 200nm for conditions representative for application. Thus, the cracks do not immediately limit the mechanical integrity, even of delicate NiTi minimally invasive implants with feature sizes below 100 µm. The potential growth of such cracks during small but alternating loads remains to be assessed in the future. Regarding the compatibility of the material in a specific biological environment, it is interesting to note that flaking of oxide particles does not necessarily expose the Ni-rich layer below the oxide, since the separation occurs along the pores at the bottom but inside the oxide. Accordingly, some oxide remains on the material. However, since oxide particles with dimensions of a few µm and exhibiting sharp edges may cause adverse reactions in a biological host, their formation should be avoided as discussed in [7].

Acknowledgments
This work was supported by the German Research Foundation (grant numbers UN 341/3-1 and Inst 275/ 391-1).

References

[1] Duerig T W 2002 The use of superelasticity in modern medicine *MRS Bull* **27** 101-104
[2] Pelton A R, Duerig T W, Berg B, Hodgson D, Mertmann M, Mitchell M, Proft J, Wu M and Yang J 2005 Nitinol medical devices *Adv Mater Process* **163** 63-65
[3] Zende S, Freiberg K E, Dorner F, Feth N A and Undisz A 2019 Corrosion Resistance of Nitinol Wires After Deformation *Shape Mem Superelast* **5** 346-351
[4] Freiberg K E, Bremer-Streck S, Kiehntopf M, Rettenmayr M and Undisz A 2014 Effect of thermomechanical pre-treatment on short- and long-term Ni release from biomedical NiTi *Acta Biomater* **10** 2290-2295
[5] Racek J, Sittner P, Heller L, Pilch J, Petrenec M and Sedlak P 2014 Corrosion of NiTi Wires with Cracked Oxide Layer *J Mater Eng Perform* **23** 2659-2668
[6] Hirmanova K, Pilch J, Racek J, Heller L, Sittner P, Recman L, Petrenec M and Sedlak P 2014 Physical Simulation of the Random Failure of Implanted Braided NiTi Stents *J Mater Eng Perform* **23** 2650-2658
[7] Undisz A, Schrempel F, Wesch W and Rettenmayr M 2009 In situ observation of surface oxide layers on medical grade Ni-Ti alloy during straining *J Biomed Mater Res A* **88A** 1000-1009
[8] Hessing C, Frenzel J, Pohl M and Shabalovskaya S 2008 Effect of martensitic transformation on the performance of coated NiTi surfaces *Mat Sci Eng a-Struct* **486** 461-469
[9] Shabalovskaya S, Undisz A, Rettenmayr M, Rondelli G 2009 Functionality of Bare Nitinol Surfaces *Funct Mater Lett* **2** 67-71
[10] Undisz A, Fink M and Rettenmayr M 2008 Response of austenite finish temperature and phase transformation characteristics of thin medical-grade Ni-Ti wire to short-time annealing *Scripta Mater* **59** 979-982
[11] Freiberg K E, Hanke R, Rettenmayr M and Undisz A 2016 Surface Preserving Targeted Preparation using Focused Ion Beam Demonstrated by the Example of Oxide layers on Ni-Ti Alloys *Prakt Metallogr-Pr M* **53** 193-205
[12] Undisz A, Schrempel F, Wesch W and Rettenmayr M 2012 Mechanism of oxide layer growth during annealing of NiTi *J Biomed Mater Res A* **100A** 1743-1750
[13] Undisz A, Hanke R, Freiberg K E, Hoffmann V and Rettenmayr M 2014 The effect of heating rate on the surface chemistry of NiTi *Acta Biomater* **10** 4919-4923

Vacuum brazing and heat treatment of NiTi shape memory alloys

W Tillmann, A Eilers and T Henning*

Institute of Materials Engineering, TU Dortmund University, 44227 Dortmund, Leonhard-Euler-Straße 2, Germany

*e-mail: tim.henning@tu-dortmund.de

Abstract. The pseudoelasticity of NiTi shape memory alloys is a unique material property which can be characterized by a complete recovery of a previously impressed component shape by a change of the thermal or mechanical load conditions after deforming. In contrast to the elastic deformation of ordinary materials like steels, twentyfold higher elastic strain rates up to 10 % are possible due to a temperature or a stress induced diffusion-free transformation of the crystal lattice between the austenite and martensite phases. Therefore, these superelastic alloys are frequently used as actuators, implants or stents so that there is an extraordinary high requirement of reliability and biocompatibility. In terms of joining, vacuum brazing might be a particularly suitable method to produce joined components which preserve a maximum of pseudoelasticity. Within the present research, it was shown that the vacuum brazing process at 1180 °C using pure niobium is well integrable into a solution annealing and a shape annealing heat treatment in a single furnace run. This led to a distinct tension plateau at around 285 MPa with an almost R-phase-free conversion of NiTi. Furthermore, it was proven that the share of the superelastic and proeutectic NiTiNb-phase was significantly increased with the dwell time.

1. Introduction

Shape memory alloys are metals that are able to "remember" their original shape. The mechanism of the memory effects is a change of the crystalline structure depending on temperature or stress level. The high temperature phase is austenite while the low temperature phase is martensite. The latter can occur in a twinned or detwinned structure. There are three shape memory effects: one-way memory effect, two-way memory effect and pseudoelasticity.

One-way effect: The crystal structure at room temperature is the twinned martensite. When mechanical stress is applied the martensite gets detwinned so that a plastic deformation of the component occurs. A subsequent heat treatment (above austenite-finish temperature) leads to a change of the crystalline structure to the high temperature phase austenite. This is accompanied by a deformation to the original shape of the component. During the subsequent cooling, the twinned martensite is formed without further deformation.

Two-way effect: In this case a change of temperature leads to a deformation of the component. Each of the crystalline structures (martensite and austenite) has a different shape. To obtain this effect, the component has to "learn" the two different shapes using a shape-set heat treatment.

Pseudoelasticity: The crystal structure at ambient temperature is the high temperature phase austenite. During the application of mechanical stress, the austenite transforms into martensite. After a stress relief the martensite transforms into austenite and the original shape is obtained. This behavior leads to a tension plateau in the stress strain curve which results in high reversible strains up to approx. 10 % [1-4].

Beside the austenitic and martensitic phase, a so-called R-phase can occur. This phase is martensitic and forms due to precipitations or cold deformation. The presence of the R-phase increases the transition temperature due to a previous transformation to austenite [5-7]. The most common shape memory alloys are NiTi materials which are also biocompatible. To obtain the shape memory effects the ratio of Ni and Ti is essential. Small deviations of the chemical composition can change the material properties such as transformation temperatures and the shape memory effect. The applications of the shape memory alloys are especially the medical and actuator technology [3, 4, 8-10].

Joining of shape memory alloys is still today extremely challenging since the main requirements for the joint is to preserve the maximum of the shape memory effects as well as to assure a high biocompatibility [3, 4, 8-10]. In addition, there is a highly limited range of suitable parameter with regard to the alloy composition and in particular to the heat treatment. Adhesive bonding, clamping and crimping are used if there is just a low demand to the strength and the joint is not part of the actual functionality [8, 9, 11, 12]. The following statements refer to NiTi shape memory alloys, but are generally also valid for other ones. Today, laser welding is state of the art and frequently used to obtain high strength joints of shape memory alloys [13-26]. This is mainly due to a very high energy density which is furthermore extremely accurate to the local position so that the total heat input as well as the size of the heat affected zone (HAZ) are well controllable. In the HAZ, there is just a slight grain growth while the weld metal has a coarse-grained and cast-like structure [13, 17, 18]. As a consequence of this, other welding technologies like WIG or plasma welding are only used sporadically because the size of the HAZ and the weld pool are usually larger which affects the shape memory affects significantly [15, 25, 26]. However, laser welding of shape memory alloys is extensively researched but the joint geometry is technically limited to wire-to-wire or to butt joints. Furthermore, the size of a laser welded joint is between 0.5-4.0 mm and as a consequence of this, the joints are not able to obtain the maximum possible expansion rate of the base material [14, 16, 20]. In detail, this results from the precipitation of some martensitic phases in the weld metal during rapid solidification within the regular austenitic microstructure. During loading these martensitic phases are detwinned and thus, dislocations will be formed which avoid a complete martensite to austenite retransformation on load reduction below the tension plateau. In addition to this, brittle phases such as Ti_2Ni and Ni_3Ti are frequently present in the weld metal [14]. Both aspects lead to an irreversible elongation even if there is just a load applied to the component which is normally in the elastic range for shape memory alloys [20]. Nevertheless, laser welding is well suited to produce high strength joints of NiTi shape memory alloys when considering these explanations for the design of the component.

From another point of view, vacuum brazing might be able to preserve a significant higher level of pseudoelasticity than laser welding since the thickness of the joint is just below 0.1 mm in general. In brazing it is well known, that the strength of a joint is highly enhanced if the thickness of the braze metal is near the zero gap [27, 28]. Due to this, it is imaginable that this effect on the strength is possibly transferable to the pseudoelasticity of shape memory alloys and in particular if the braze metal itself consists of superelastic phases. Vacuum brazing offers the advantage to connect large areas or even several joints at the same time and unite the joining and the heat treatment in a single furnace run.

However, before discussing the vacuum brazing in detail, the heat treatment on the shape memory alloys will be described extensively hereafter. Due to the high oxygen affinity of titanium, NiTi shape memory alloys are usually produced by induction or arc melting in vacuum systems. After this, an optional homogenization annealing at approximately 1000 °C can take place before hot forming (open-die forging, hot rolling) between 600-900 °C and subsequent cold forming (cold rolling, wire drawing) which is carried out to generate a specified amount of dislocations. In this condition, the material features poor mechanical properties and almost no shape memory effects [29, 30]. A fine-grain microstructure is achieved by multiple cold forming with particular specific degrees of deformation and intermediate annealing. A final shape-set heat treatment under load between 350-550 °C enables the shape memory properties to be set in a defined manner. In detail, the remaining dislocations and Ni_4Ti_3-phases were reduced specifically by the occurring recovery process. Thus, the remaining amount of Ni_4Ti_3-precipitations is a function of the heat-treatment and as a consequence of this the austenite-finish temperature and the strength level of the tension plateau are adjustable for each individual shape memory alloy [29, 31]. The Ni_4Ti_3-phase is thermodynamically stable between 300-500 °C so that it is practicable to use a shape-set temperature of 510-530 °C [29]. As explained before, this martensitic R-phase will primarily transfer to austenite on very small load application so that there is at first a non-linear response. Anyway, this finally results in a very small pseudoplastic elongation what can easily be considered in technical applications. It should be noted that a heat treatment above the maximum shape-set temperature of 550 °C is barely investigated with respect to pseudoelasticity and only applied by very few scientists in terms of vacuum brazing [32-36]. Due to the necessity of a total base material heat treatment during vacuum brazing, a solution annealing process between 830-1180 °C of the base material is automatically carried out with the brazing. As a result, previously precipitated Ni_4Ti_3-phases which impair pseudoelasticity are brought into solution in the nickel mixed crystal and suppressed in precipitation by rapid pressure gas quenching during cooling [37-39]. Compared to laser welding, a vacuum furnace process is a fairly slow process so that there will be a noticeable grain growth of the originally ultra-fine-grained shape memory alloy. This can also cause an irreversible expansion of the components [30]. It is vital to know for the present research work, that the grain growth is primarily increasing with temperature and time but is slowed down by the total amount of Ni_4Ti_3-phases as well as by dislocations [1, 30].

Vacuum brazing of shape memory alloys as well as the effect of solution annealing on the pseudeoelastic behavior is still today not investigated adequately by far. There is some literature for which pure copper was used in terms of infrared brazing NiTi [36]. In detail, the strategy was to substitute the nickel of the shape memory alloy by copper due to the formation of $Ti_{50}Ni_{50-x}Cu_x$ ($x \leq$ 30 wt.-%). It was shown that the presence of the Ti(Cu,Ni)-phase, which is considered superelastic, increased with the dwell time. However, the most relevant research on brazing NiTi shape memory alloys is based on using niobium as a brazing filler metal. As well as NiTi, niobium exhibits biocompatibility which is essential for the use in the medical technology. Grummon et al. and Wang et al. investigated the brazing of NiTi joints using niobium as brazing filler metal. In **figure 1** the quasi-binary NiTi-Nb phase diagram is shown.

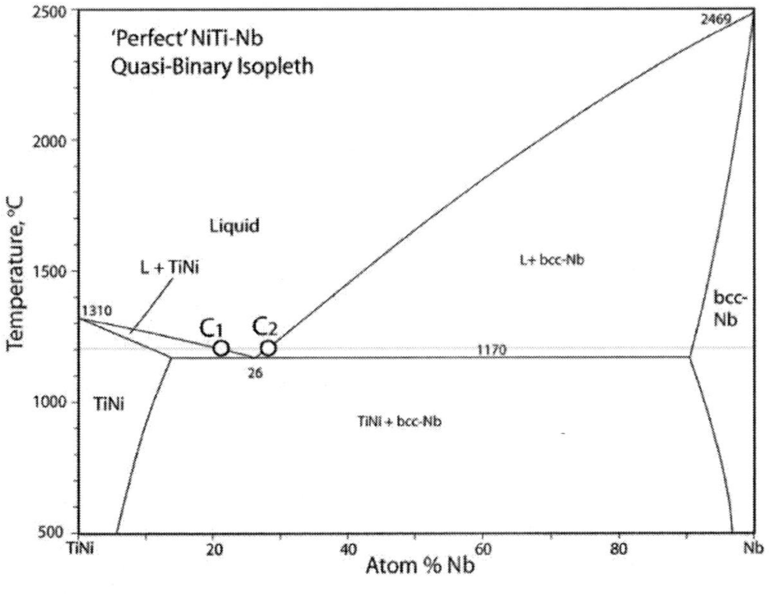

Figure 1. Quasi-binary NiTi-Nb phase diagram [32].

At niobium content of 26 at.-% at 1170 °C, there is a eutectic point. The vacuum brazing process of Grummon et al. were conducted at a temperature of 1180 °C with different dwell times (1 s to 6 min) using a pure niobium foil (thickness: 50 μm). A subsequent heat treatment was conducted at 350 °C for 90 min. The microstructure of the joints consisted of pro eutectic NiTi dendrites as well as a eutectic NiTi-Nb phase (**figure 2**). As one can see for the horizontal line at 1180 °C, there is a solid-liquid equilibrium for the phases of concentration C_1 and C_2. As a consequence of their different niobium content there is a bidirectional diffusion leading to the liquid quasieutecic phase.

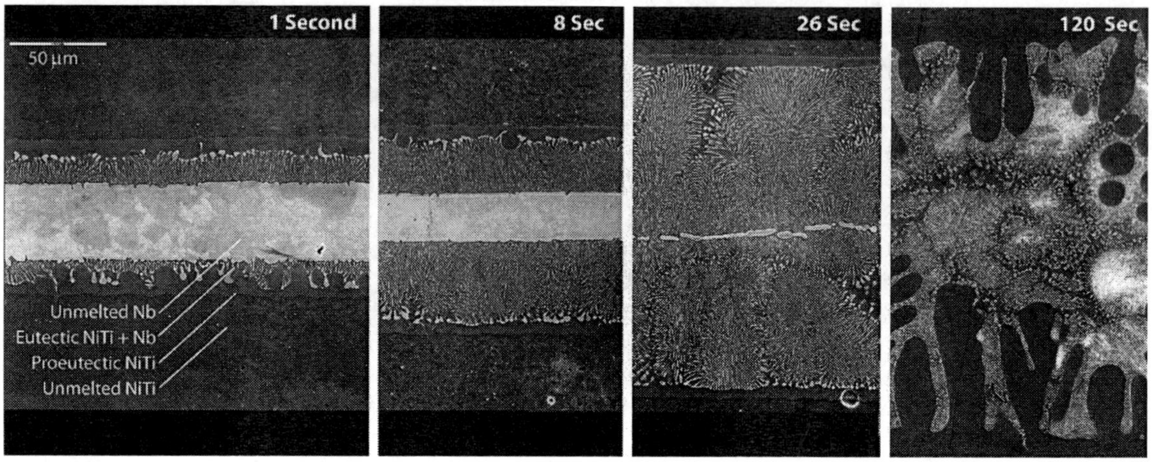

Figure 2. SEM images of NiTi brazing joints with Nb as filler metal alloy. Different dwell times at a brazing temperature of 1180 °C [32].

There was a tensile strength of the joints of about 800 MPa and a tension plateau in the stress strain curve was proven after brazing. Wang et al. used a NiTi-Nb powder (ratio 3:2) for brazing with a brazing temperature of 1180 °C (4 – 8 min) and a subsequent heat treatment at 530 °C (30 min). Comparable to the results of Grummon et al. the microstructure consisted of a proeutectic and a eutectic phase [32-34, 40].

The main objective of the project is to evaluate the effects of a vacuum heat treatment and vacuum brazing process on the microstructure and pseudoelasticity of NiTi and the joint properties as well with special regard to the pseudoelasticity. Since there are very few and hardly any studies on these interactions during vacuum brazing of NiTi, the present research is characterized to give an evidence of feasibility. As a matter of course, the first task is to investigate the effect of a solution annealing process in a vacuum furnace on the base material properties. In foresight to limit the expected increase in grain size during this process, it might be reasonable to use a cold-formed NiTi alloy which was never done before. As a result of the initial recovery processes during heating up, an inhibition of grain growth and thus, a significantly better pseudoelasticity may be achievable. Hence, the effect of solution annealing between 800-1180 °C will be investigated for both, a regular shape-set condition as well as a cold-formed condition. In this context, a comprehensive vacuum furnace process will be designed which includes solution annealing, brazing, quenching, and shape-set. In addition, NiTi shape memory alloys will be brazed at 1180 °C for a dwell time of one second to 34 minutes by using a pure niobium foil of thickness 50 μm.

2. Experimental and materials

The heat treatment and brazing procedures were carried out in a horizontal vacuum furnace type EU80/1H (Schmetz Inc.) in a molybdenum heating chamber. NiTi-wires of diameter 1.6 mm were used for the investigations on the effect of solution annealing on the base material properties. As explained in the introduction, one set of the wires were in a regular straight annealed condition (50.5 at.-% Ni and 49.5 at.-% Ti, Imasys Inc.) whereas a second set was in a cold-formed condition with a degree of deformation of φ=40,7 % (51.5 at.-% Ni and 48.5 at.-% Ti, Ingpuls Inc.) which is in contrast to the regular wire not pseudoelastic but ultra-fine grained. In order to prevent reactions of NiTi with the molybdenum, the wires were placed inside of an Al_2O_3 tube. After the evacuation to a high vacuum level better than $6.2*10^{-6}$ mbar, the wires very slowly heated up to 200 °C with 10 K/min and hold for 10 minutes. This was to assure an equal temperature distribution within the furnace as well as to prevent a formation of Ni_4Ti_3 which is thermodynamically possible from 250 °C [38]. Furthermore, there is no effect of this temperature to the wires. Afterwards, a very fast heating rate of 100 K/Min was set to minimize the total heat input into the base material which was intended to limit the grain growth which is considered to decrease pseudoelasticity. Solution annealing temperatures of 800 °C and 1180 °C were investigated for a fix dwell time of five minutes. Argon was used as quenching gas with a partial pressure of 1.5 bars to avoid an extensively reaction of titanium with nitrogen what is used in general. Furthermore, the rapid cooling to room temperature was also necessary to suppress the precipitation of Ni_4Ti_3-phases which impair pseudoelasticity and shifts the transformation temperature. Hence, a convective cooling gas flow was used by a heat exchanger with an integrated fan. Afterwards the wires were directly shape-set annealed for five minutes at 510 °C. An exception of this was given for some regular wires which were solution annealed to 800 °C and then shape-set annealed at 400 °C for five minutes so that a formation of the R-phase is intended which will be used as a reference to show up this effect. Tensile testing was carried on an Instron Inc. testing machine type E10BMTM1020 at a test speed of 0.02 mm/s at room temperature by using an inductive displacement sensor. It is well known that the microstructural transformation from austenite to martensite is exothermic. Because the stress-strain behavior is very temperature dependent a convective ambient air flow was used to assure the comparability of the results [41]. Three wires of length 150 mm were examined for each parameter and 25 mm were used for clamping on each side.

Due to the fact that applicability of vacuum brazing is clearly intended to join larger areas, two NiTi samples of size 10x10x2 mm were used (49.5 at.-% Ni and 50.5 at.-% Ti, Euroflex Inc.). Inbetween a pure niobium foil of thickness 50 µm was placed which has a melting point of 2477 °C. As explained in the introduction, an eutectic phase in the quasi-binary NiTi-Nb system is present by 26 at-% Nb at 1170 °C. Therefore, brazing was conducted at 1180 °C by using a heating rate of 10 K/min. In order to investigate the effect of the dwell time on the microstructural composition of the braze metal, a dwell time of one second to 34 minutes was used. Vacuum cooling was set after brazing and no additional shape-set heat treatment was carried out. For tensile testing two NiTi plates of 30x20x2 mm were brazed with a dwell time of 6 min on the narrow front side. Afterwards a non-standardized flat tensile test geometry was cut out by using wire-cut EDM (Electrical Discharge Machining). The brazed samples were cross-sectioned and polished finally with a one micrometer diamond suspension. A scanning electron field emission microscope type JSM7001F (Jeol Inc.) as well as an integrated EDS-detector (Oxford Inc.) were used for the microstructural analysis of the cross-sections.

3.Results and discussion

3.1. Effect of the high temperature heat treatment on the NiTi base material properties

The results of tensile testing were shown in **figure 3** for which the original state of the regular straight annealed condition is shown in a line whereas the cold-formed condition is shown in a dashed line. For reasons of clarity, only one curve is shown.

As expected, a well superelastic stress-strain behavior was proven for the common delivery condition which has a clearly tension plateau of around 500 MPa at which the austenite to martensite transformation takes place. However, there were two level of slope visible prior to the beginning of the plateau which might be a consequence of the clamping device. A maximum tensile strength of 1800 MPa was determined for the cold-formed state without pseudoelastic behavior. Through a vacuum shape-set annealing of theses wires at 510 °C (curve c)), a tension plateau at around 270 MPa was achieved and thus, they had a pseudoelastic behavior now. However, there was a non-linear stress-strain relation at the beginning of load application what is characteristic for the presence of the R-phase. Based on this it can be assumed that most of the dislocations were relieved by forming Ni_4Ti_3. Due to the vacuum shape-set annealing, the tension plateau of the regular wires was reduced to approximately 350 MPa and a lower maximum elastic strain was observed. Furthermore, there were still two level of slope visible at the beginning of the curve. It could be determined that a shape-set annealing temperature of merely 400 °C after solution annealing at 800 °C is not useful because the stress-strain curve indicates the presence of the R-phase (curve d). In contrast to this, a linear stress-strain behavior was found for all the solution annealed and at 510 °C shape-set annealed wires of the originally regular condition (curves e-f). In addition, it could be shown that the level of the tension plateau was increased by 35 MPa to 285 MPa, if the solution annealing temperature was increased from 800 °C to 1180 °C which means brazing temperature. A similar behavior was found for the originally cold-formed condition so that the level of the tension plateau was 305 MPa for a solution temperature of 1180 °C whereas it was just 200 MPa for the solution annealed wires at 800 °C. To summarize the results of tensile testing, it was found that the designed vacuum process is suitable to assure a pseudoelastic behavior of NiTi base materials after the brazing cycle. Furthermore, it could be proven that the approach to use cold-formed shape memory alloys for brazing might be very promising since there was a pseudoelastic behavior afterwards. Actually, the presence of the R-phase was indicated for these samples but this can be hopefully prevented in future by increasing the shape-set temperature or the dwell time as well. The results achieved are of course not comparable to the ultra-fine grained regular condition of NiTi shape memory alloys but still show a high degree of the tension plateau and very high strain rates. The effect of the heat treatment on the grain size, on the formation of phases, and on the pseudoelasticity will be examined more detailed in future.

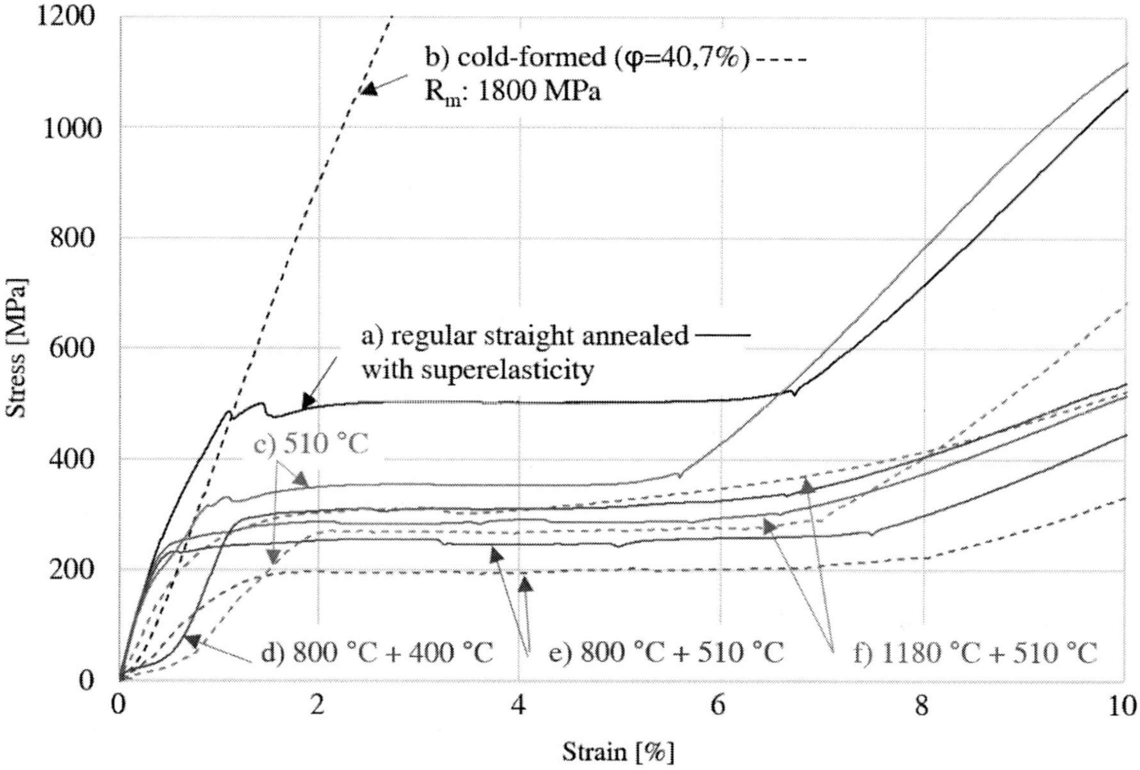

Figure 3. Effect of vacuum heat treatment on the stress-strain behavior of NiTi-wires, a) regular as-delivered condition, b) cold-formed as-delivered condition, c) shape-set annealed, d-f) solution annealed and shape-set annealed. A dwell time of five minutes was used for all given temperatures in the figure.

IOP Conf. Series: Materials Science and Engineering 1147 (2021) 012025 doi:10.1088/1757-899X/1147/1/012025

3.2 Microstructural analysis of niobium brazed NiTi

In **figure 4** the SEM image of a cross section of a brazed joint with brazing parameters of 1180 °C and 6 min dwell time is shown. In the center of the brazing zone a NiTi-Nb eutectic is formed with a proportion of approx. 23 % niobium (see element analysis in figure 4, position 3). Thus, the brazing seam has a much larger thickness (100 – 150 µm) than the used niobium foil (50 µm). The composition of the eutectic (23 % Nb) is consistent with those in other research projects (20 – 26 % Nb) [33, 42-44]. Between the base material and the eutectic, proeutectic dendrites are present. This phase consists of approx. 7 % niobium (figure 4, position 1) which matches the solubility of niobium in NiTi at the brazing temperature of 1180 °C (figure 1).

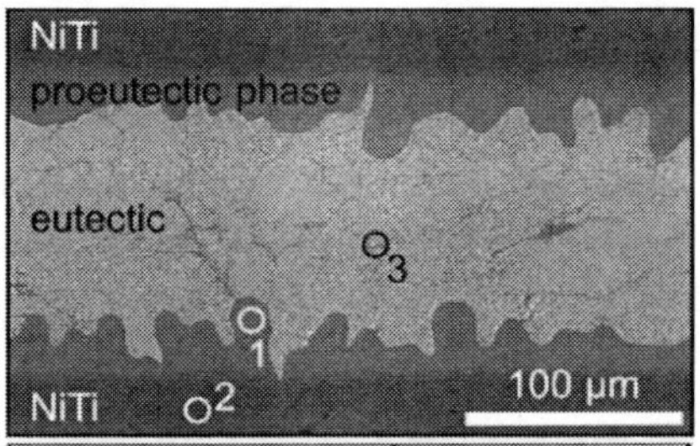

position	element [At.-%]		
	Ni	Ti	Nb
1 (proeutectic dendrites)	45,1	47,9	7
2 (base material)	49,5	50,5	-
3 (eutectic)	37,3	39,4	23,3

Figure 4. SEM image with element analysis of a NiTi brazed joint with pure niobium (50µm) as brazing filler metal. Brazing parameter: 1180 °C, 6 min.

The effect of the dwell time at a brazing temperature of 1180 °C on the microstructure is shown in **figure 5**. Both phases (eutectic and proeutectic) change with increasing dwell time. Due to the longer period of time at a higher temperature, more diffusion processes occur. This leads to a finer grained eutectic phase with increasing holding time (compare figures 5a, b and c). The whole niobium foil has reacted with the base material to the eutectic after a dwell time of only 1 s. Furthermore, an increasing dwell time leads to a growth of proeutectic dendrites. The average thickness of this phase increases from approx. 10 µm (dwell time 1 s) to 15 µm (1 min) to 25 µm (34 min).

For tensile testing one specimen was brazed at 1180 °C for 6 min. The tensile strength of the specimen is 1022 MPa. The fractured surface of this specimen is shown in figures 5d and e. The failure occurred in the eutectic and partly in the proeutectic phase. Though, for validation of the tensile strength and the failure mechanism more tensile tests are necessary.

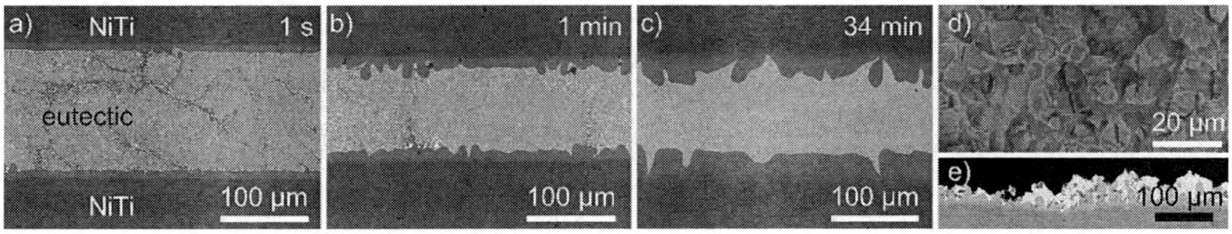

Figure 5. a-d) SEM images of NiTi joints with different holding times (brazing temperature 1180 °C), d) SEM Images of the fractured surface of a tensile specimen, e) cross section of the fractured surface.

4. Conclusions

It could be proven that solid solution annealing between 800-1180 °C with a subsequent shape-set heat treatment of 510 °C in a common vacuum furnace preserved a high level of pseudoelasticity for NiTi-wires. A regular straight annealed condition as well as a cold-formed condition (φ=40,7 %) were used in the original state. It was found that there was a largely R-phase free conversion for the regular condition after heat treatment leading to tension plateau of 285 MPa for the wires annealed to 1180 °C. This was 35 MPa higher than for the wires which were annealed at only 800 °C. In contrast to this, there was a non-linear stress-strain behavior present for the heat-treated wires which were in the cold-formed state before. This suggests the presence of the R-Phase which indicates an incomplete dissolvement of dislocations. Nevertheless, it could be proven for the first time that these wires offer a high degree of pseudoelasticity. The level of the tension plateau was 305 MPa for a solution annealing temperature of 1180 °C whereas it was just 200 MPa for the solution annealed wires at 800 °C.

Furthermore, it was shown that vacuum brazing of NiTi using pure niobium led to high quality and sound joints. In accordance to the literature, the microstructure consisted of a quasieutectic NiTi-Nb phase and proeutectic phase which was increased with the dwell time. A tensile strength of 1022 MPa was achieved for a brazed sample with a dwell time of six minutes. The fracture was within the eutectic phase so that it will be interesting to investigate the effect of higher dwell times on the tensile strength and the pseudoelasticity in future in detail.

In future, the project will investigate the effect of the heat treatment process on NiTi materials in detail and systematically for the regular pseudoelastic condition as well as for the cold-formed condition. In this regard, the grain growth, the microstructure and the pseudoelastic behavior are of main interest. As a consequence of these findings, limiting process variables for brazing with niobium are derived. In addition to brazing with Nb, NbZr1, Cu, and AuCu65 are promising brazing alloys to achieve a partial pseudoelastic behavior within a NiTi/NiTi-joint.

Acknowledgments
The authors thank Michael Hartung for his support within this paper.

References

[1] Hebda D A and White S R 1995 Effect of training conditions and extended thermal cycling on nitinol two-way shape memory behaviour *Smart Mater. Struct.* **4** 298-304
[2] Otsuka K and Ren X 2005 Physical metallurgy of Ti–Ni-based shape memory alloys *Prog. Mater. Sci.* **50** 511-678
[3] Bäker M 2014 *Funktionswerkstoffe: Physikalische Grundlagen und Prinzipien* (Wiesbaden: Springer) pp 29-31
[4] Langbein S and Czechowicz A 2013 *Konstruktionspraxis Formgedächtnistechnik Potentiale – Auslegung – Beispiele* (Wiesbaden: Springer) pp 3-29
[5] Miyazaki S and Otsuka K 1986 Deformation and transition behavior associated with the R-Phase in Ti-Ni Alloys *Metall. Mater. Trans.* A **17** 53-63
[6] Michutta J, Somsen C, Yawny A, Dlouhy A and Eggeler G 2006 Elementary martensitic transformation processes in Ni-rich NiTi single crystals with Ni4Ti3 precipitates *Acta Mater.* **54** 3525-3542
[7] Kroeger A, Dziaszyk S, Frenzel J, Somsen C, Yawny A., Dlouhy, A. and Eggeler G. 2008 Direct transmission electron microscopy observations of martensitic transformations in Ni-rich NiTi single crystals during in situ cooling and straining *Mater. Sci. Eng.* A **481-482**, 452-456
[8] Chau E T F 2007 *A Comparative study of joining methods for a SMART aerospace application* Doctorate thesis Cranfield University.
[9] Akselsen O M 2010 Joining of shape memory alloys *Intech Open Science* 83-209
[10] Bahraminasab M and Bin Sahari B 2013 NiTi Shape Memory Alloys, Promising materials in orthopedic Applications *Intech Open Science* 261-278
[11] Mehta K and Gupta, K 2018 *Fabrication and processing of shape memory alloys* (Springer) pp 39-60
[12] Matsumura H, Tanoue N, Yanagida H, Atsuta M, Koike M and Yoneyama T. 2003 Adhesive bonding of super-elastic titanium–nickel alloy castings with a phosphate metal conditioner and an acrylic adhesive. *J. Oral Rehabil.* **30** 653–658

[13] Haas T 1996 *Laser-Strahlschweißen von NiTi-Formgedächtnislegierungen* Dissertation Universität Karlsruhe
[14] Gugel H 2010 *Laserschweißen artgleicher und artfremder Materialkombinationen mit Nickel-Titan Formgedächtnislegierungen* Dissertation Ruhr-Universität Bochum
[15] Kimura K and Tobushi H 1996 TIG Welding and shape memory effect of TiNi shape memory alloy *J. Intel. Mat. Syst. Str.* **7** 646–655
[16] Tam B 2010 *Micro-Welding of Nitinol shape memory alloy* Masterthesis Waterloo
[17] Song Y G, Li W S, Li and Zheng Y F 2008 The influence of laser welding parameters on the microstructure and mechanical property of the as-jointed NiTi alloy wires *Mater. Lett.* **62** 2325–2328
[18] Li M G, Sun D Q, Wang W Q, Han Y W and Gu X Y 2010 Effects of laser welding parameters on microstructure and mechanical properties of laser welded TiNi shape memory alloy (SMA) wires *Lasers in Engineering* **20** 167–177
[19] Raju R, Salamat-Zadeh F and Brriesci, G. 2016 Laser micro welding of Nitinol for cardiovascular applications *2nd World Congress on Electrical Engineering and Computer Systems and Science (EECSS'16) Budapest, 16. – 17. August* ICBES 108-1-108-8
[20] Oliveira J P 2016 *Laser welding of shape memory alloys* Dissertação Universidade Nova De Lisboa
[21] Falvo A, Furgiuele F M and Maletta C 2005 Laser welding of a NiTi alloy: Mechanical and shape memory behaviour *Mater. Sci. and Eng.* A **412** 235–240
[22] Tuissi A, Besseghini S, Ranucci T, Squatrito F and Pozzi M 1999 Effect of Nd-YAG laser welding on the functional properties of the Ni–49.6 at.% Ti. *Mater. Sci. Eng.* A **273–275** 813–817
[23] Pequegnat A 2014 *Novel laser based NiTi shape memory alloy processing protocol for medical device applications* Doctoral thesis University of Waterloo
[24] Falvo A, Furgiuele F M and Maletta C 2008 Functional behaviour of a NiTi-welded joint: Two-way shape memory effect *Mater. Sci. Eng.* A **481–482** 647–650
[25] Eijk C v d, Fostervoll H., Sallom Z K and Akselsen O M 2003 Plasma welding of NiTi to NiTi, stainless steel and Hastelloy C276. *ASM Materials Solutions Conference, Pittsburgh, 13-15 October*
[26] Vondrus P, Kolarik L, Kolarikova M 2012 Plasma arc welding of NiTi and 304 Steel. *23rd International DAAAM Symposium* **23** 1039-1042
[27] Hedin E 2018 Influence of gap clearance on brazing joint strength for Ni and Fe based filler metals *Proceedings from the 7th international brazing and soldering conference, New Orleans, USA* 155-160
[28] Way M, Willingham C and Goodall R 2020 Brazing filler metals *Int. Mater. Rev.* **65:5** 257-285
[29] Mehnert A 2018 *Wärmebehandlung von superelastischen Nickel-Titan-Drähten die zur Formveränderung führt* Dissertation Ludwig-Maximilians-Universität München
[30] Burow J 2010 *Herstellung, Eigenschaften und Mikrostruktur von ultrafeinkörnigen NiTi-Formgedächtnislegierungen* Dissertation Ruhr-Universität Bochum
[31] Pelton A, DiCello J and Miyazaki S 2000 Optimisation of processing and properties of medical grade Nitinol wire *MITAT* **9(2)** 107-118
[32] Grummon D S, Low K B and Foltz J 2007 A new method for brazing Nitinol based on the quasibinary TiNi-Nb System. *48th AIAA/ASME/ASCE/AHS/ASC Structures, Structural Dynamics, and Materials Conference, Honolulu, Hawaii, 23.-26. April*
[33] Grummon D S, Shaw J A and Foltz J 2006 Fabrication of cellular shape memory alloy materials by reactive eutectic brazing using niobium *Mater. Sci. Eng.* A **438–440** 1113-1118
[34] Wang L, Wang C and Dunand, D C 2015 Microstructure and strength of NiTi-Nb eutectic braze joining NiTi wires *Metall. Mater. Trans.* A **46A** 1433-1436
[35] Shaw J A, Grummon D S and Foltz J. 2007 Superelastic NiTi honeycombs: Fabrication and experiments *Smart Mater. Struct.* **16** 170-178

[36] Yang T Y, Shiue R K and Wu SK 2004 Infrared brazing of Ti50Ni50 shape memory alloy using pure Cu and Ti-15Cu-15Ni foils *Intermetallics* **12** 1285-1292
[37] Bitzer M 2014 *Pulvermetallurgische Funktionsbauteile aus NiTi- und NiTi-X Legierungspulvern* Dissertation Forschungszentrum Jülich
[38] Kaack M 2002 *Elastische Eigenschaften von NiTi-Formgedächtnis-Legierungen* Dissertation Ruhr-Universität Bochum
[39] Undisz A 2009 *Optimierung von strukturellen und funktionellen Eigenschaften von NiTi mit Pseudoelastizität/Formgedächtnis für den medizinischen Einsatz* Dissertation Friedrich-Schiller-Universität Jena
[40] Wang L, Wang C, Zhang L C, Chen L, Lu W and Zhang D 2016 Phase transformation and deformation behavior of NiTi-Nb eutectic joined NiTi wires *Sci. Rep.* **6** 1-11
[41] Johe R 2009 Tensile testing of Nitinol tubes and wires with higher strain rates *J. Mater. Eng. Perform.* **18** 836-842
[42] Bewerse C, Brinson L C and Dunand D C 2015 Microstructure and mechanical properties of as-cast quasibinary NiTi-Nb eutectic alloy *Mater. Sci. Eng.* A **627** 360-368
[43] Piao M, Miyazaki S, Otsuka K and Nishida N 1992 Effects of Nb addition on the microstructure of Ti-Ni alloys *Mater. Trans.* **33** 337-345
[44] Wang Y, Cai X Q, Yang Z W, Wang D P, Liu X G and Liu Y C 2017 Effects of Nb content in Ti-Ni-Nb brazing alloys on the microstructure and mechanical properties of Ti-22Al-25Nb alloy brazed joints *J. Mater. Sci. Technol.* **33** 682-689

IOP Conf. Series: Materials Science and Engineering 1147 (2021) 012026 doi:10.1088/1757-899X/1147/1/012026

Simulation of thermally sprayed coating properties considering the splat boundaries

K Bobzin, W Wietheger, H Heinemann* and F Wolf

Surface Engineering Institute (IOT), RWTH Aachen University
Kackertstraße 15, 52072 Aachen, Germany

*e-mail: heinemann@iot.rwth-aachen.de

Abstract. The properties of thermally sprayed coatings depend heavily on the solidification process during the impact on the substrate. The simulation of the coating build-up is of essential importance in order to understand the processes behind it. The main objective of this study is the development of a coating build-up model to predict the properties e.g. thermal conductivity of plasma sprayed coatings. Therefore, a CFD simulation based on a VOF approach coupled with the momentum source method is applied. 3D-simulations of the impact of multiple alumina particles are performed to observe the coating build-up. Subsequently, the result is transferred to another model to calculate the coatings thermal conductivity. Since the resolution of the mesh does not allow a proper representation of the air gap at the splat boundaries between two solidified particles, a new function is developed to cover such phenomena. This function enables a broader representation of the splat boundaries and an adjustment of the coating properties in these areas. To validate this function, multiple simulations with different particle temperatures and velocities are calculated. Likewise, different assumptions for the thermal conductivity values in the gap are investigated. The results of the calculated properties correspond with experimental values.

1. Introduction

The properties of thermally sprayed coatings are determined during the solidification process of the liquid particles, but the exact details of this process are challenging to observe [1]. Simulations are an important tool to not only predicted the results of otherwise time and cost intensive experiments, but also a way to analyse certain physical phenomena that would be impossible to observe in experiments. The calculation time is the bottleneck resource of most simulations and thus poses a problem in the development of simulations. Especially in such highly dynamic simulation of the coating build-up of thermally sprayed coatings the required simulation time is high, due to fast particles, requiring an extremely small time step, and small features of the coating itself, requiring a small element size.

In 2008 the authors developed a model to simulate the impact of thermally sprayed particles on a smooth and rough surface [2] and compared the results with experimentally sprayed coatings [3]. This model was extended to a state in which it was capable of simulating the impact of multiple particles, resulting in a coating thickness of about 70 μm [4]. J. Prehm et. al. developed a function to simulate the solidification process by increasing the viscosity of the particles once the temperature drops below the solidification temperature [5]. A similar approach was pursued by the authors of this study in a work in 2019, in which the solidification formulation of the simulation software Ansys Fluent was replaced with a momentum source function, to immobilize the particles [6]. Both of these functions replaced a complex solidification model with simple approximations, thus lowering the required calculation time. This has led to a significant time saving in the simulation and now enables coating build-up simulations of not only individual particles but of entire coatings in a reasonable time. Other work concentrated more on the processes during simulated solidification. H. Fukanuma et. al. studied the effect of trapped gas in simulations of thermally sprayed coatings [7] and H. B. Parizi et. al. studied the solidification process of particles dropped onto non-smooth surfaces [8]. Both discovered that the contact area of the particle and the surface is a crucial parameter for the solidification process. However, in most simulations the tiny gap of air that gets trapped between particles and the substrate and between particles itself is ignored as the simulations cannot capture such small features. This makes the simulations less realistic.

Figure 1 shows a SEM-image of an alumina coating in the upper left corner. The ceramic coating exhibits features such as porosities, air gaps between the different splats and cracks within the splats. In the lower left and upper right corner these coating features are schematically represented and a close-up is used to demonstrate where no coating material is present or where there are gaps, and therefore affect the coating properties. In the lower right corner of figure 1 the schematic representation is overlaid by an arbitrary mesh with a relatively small element size of $L_{mesh} < 1$ μm, which would lead to unacceptable long calculation times. However, it is also evident that even with this small element size not all features of the coatings can be captured, at the same time a reduction of the element size is no longer justifiable due to the high computational costs.

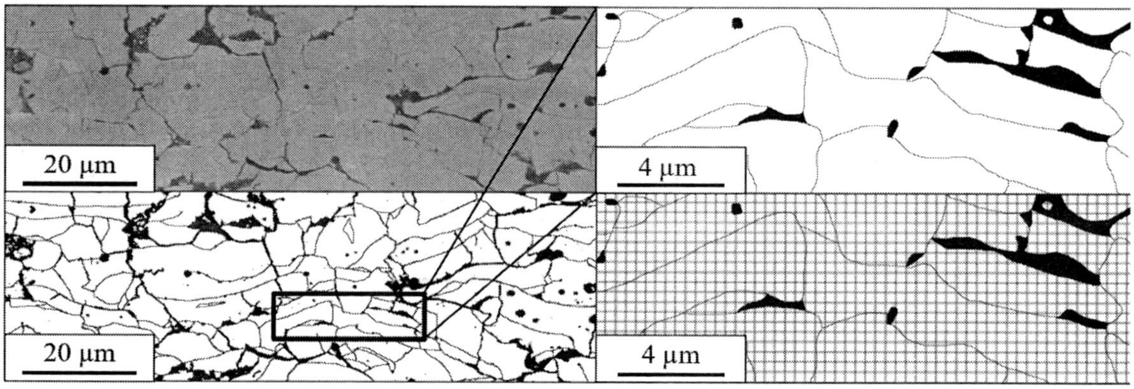

Figure 1. SEM-image of an plasma sprayed alumina coating (top left), schematic representation of the coatings with visualized air gaps (bottom left), close-up of the schematic representation (top right), mesh overlay on the close-up demonstrating the necessity of small element sizes to cover all coating features (bottom right).

In order to resolve this contradiction - the ability to perform fast simulations and simultaneously predict correct coating properties - this publication presents a method to successfully simulate the air gap between two splats on a coarser mesh for thermally sprayed coatings. The coating build-up of alumina coatings were simulated and the coatings' thermal conductivity was calculated using this approach.

2. Experimental and materials

A 3-D particle coating build-up model was created in which the developed functions are implemented. The model is based on a previous publication of the authors [6]. In this model the impact of a thermally sprayed particle onto a flat substrate was simulated. A momentum source function was used to simplify the calculation of the solidification process. This model was extended for this study to simulate the impact of multiple particles in one single calculation domain, thus creating a representing volume element (RVE) of the coating. Furthermore, export functions were implemented to simplify the monitoring of the particle values, like temperature and velocity. The most important step is the inclusion of a gap function, which alters the coating properties in the vicinity of the interface between different splats thereby emulating an air gap. The precise details of this functions will be explained later on. The created RVE was then transferred to another model, in which the effective thermal conductivity was calculated by applying a heat flow at the upper boundary.

The coating build-up model uses the Volume of Fluid (VoF) approach to model the two phases air and alumina. The simulation model consists of 3 boundaries and the interior domain as shown in figure 2. The domains dimensions are 225 µm x 225 µm x 75 µm. The calculation mesh consists of 330,000 cells with an edge length of L_{mesh} = 2.25 µm. The upper side of the model represents an inlet where the molten particles are generated and air is injected. The properties of the generated particles can be found in table 1. The lower boundary, the substrate, performs as a wall, where the particles impact and solidify. The outer sides of the domain are an outlet, where excessive volume of the particles and the air can leave the domain.

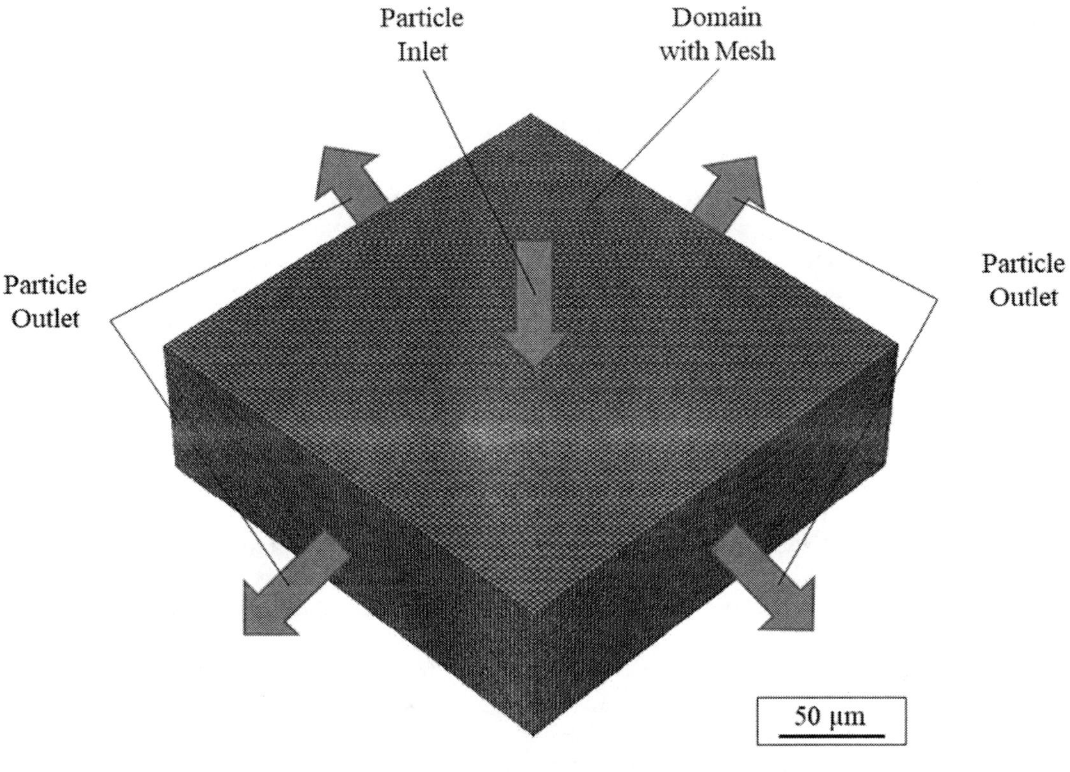

Figure 2. Schematic representation of the calculation domain, with one inlet on top, 4 outlets on the sides and the solid wall on the bottom.

The following underlying assumptions are made:

- The viscous model was set as laminar
- The energy model was enabled and all other models were disabled
- The interior domain is filled with air at the start of the simulation
- The locations of the particles at the inlet boundary randomly distributed according to a gauss distribution
- The temperatures and velocities at the inlet are equal for the particulate and the air phase
- The temperature at the lower substrate is set to T = 300 K

Table 1. Material properties and process parameters.

parameter [unit]	value
domain size [μm×μm]	225 × 225
particle diameter [μm]	70
particle velocity [m/s]	180/200/220
particle/gas temperature [K]	2,473/2,673/2,873
melting temperature [K]	2,345 [9]
melting energy [J/g]	1,060 [9]
thermal conductivity alumina [W/(m·K)]	30 [10]
thermal conductivity air [W/(m·K)]	0.0242
viscosity alumina [kg/(m·s)]	0.055 [11]
density alumina [kg/m³]	3,950 [10]
heat capacity alumina [J/(kg·K)]	900 [11]

The simulations were implemented in Ansys Fluent 18.2 and computed on 8 Intel HNS2600BPB Platinum 8160 2.1 GHz CPUs with 1,024 MB of RAM each. The computation time was approximately 24 hours. To avoid refining the mesh and increasing the calculation time, a function is created to emulate a realistic gap between two solidified particles while simultaneously keeping the calculation time low. For this new function, a user defined scalar (UDS) is created, which consecutively enumerates the particles starting from one. If two particles are next to each other, fluent calculates the median value of the user defined scalar values in the cells between the two particles. When particle 1 and particle 2 solidify directly next to each other, the cell in the middle between the two will have a UDS value of 1.5. As a consequence, if a cell has an UDS value, that is not an integer, it has to be part of the gap area.

With the presented functions it is now possible to successful isolate the areas between two particles and to assign a different thermal conductivity to the area between these particles. The formulation for the thermal conductivity values chosen in this work is shown in equation (1). The function then adjusts every gap between two particles into the formulation (1) for a gap between particle 1 and particle 2. Such that even for a gap between e.g. particle 1 and 3 the equation (1) can be used.

$$\lambda = \lambda_{max} \cdot \left[\cos((1 - i_{count}) \cdot 2 \cdot \pi) \cdot (0.5 - \frac{x_{\%}}{2}) + (0.5 + \frac{x_{\%}}{2})\right] \quad (1)$$

λ: thermal conductivity [W/(m·K)]
λ_{max}: maximal thermal conductivity [W/(m·K)]
i_{count}: loop variable [-]
$x_{\%}$: percentage of minimal thermal conductivity in the gap [-]

Figure 3 graphically shows the difference between equation (1) and a physically correct thermal conductivity. While the solid line shows an exemplary course for the thermal conductivity for a realistic coating with a small air gap in between, the dotted line shows the course of the adjusted thermal conductivity as used in this model. In the first case the thermal conductivity drops steeply and remains for a very short interval at the value for air of λ_{Air} = 0.0242 W/(m·K). The smoother dotted line shows the course of the developed function, which can then be captured even by a coarser mesh.

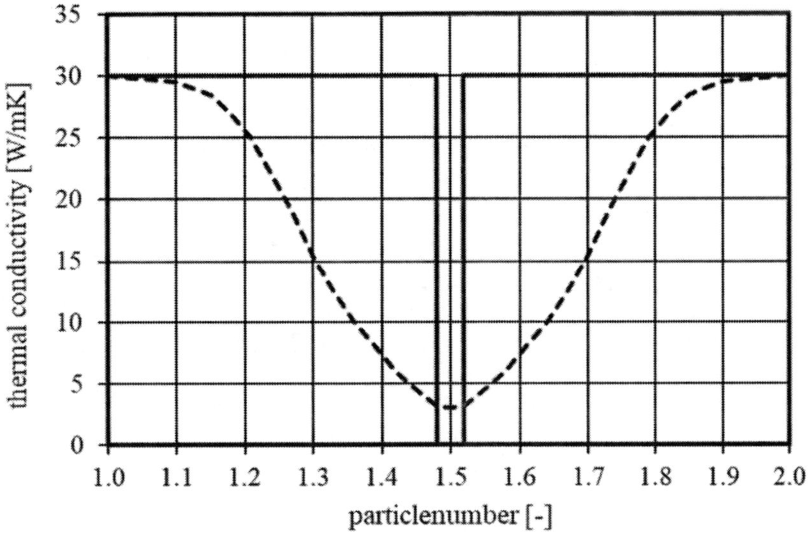

Figure 3. Visual representation of the realistic (solid line) and the approximated (dashed line) thermal conductivity, over the value of the particle counting variable.

As a first guess, an approximation to calculate the most fitting value for the reduced thermal conductivity in the gap used by the model $x_{\%}$ was carried out. For this approximation the gap is assumed to be represented by a serial connection of two thermal resistances for the gap itself and the alumina around $R_{serial} = R_{gap} + R_{alumina}$. The thermal resistance of the model R_M needs to equal the thermal resistor of a realistic gap R_{serial}, which is expressed in equation (2).

$$\frac{L_M}{\lambda_M \cdot A_M} = R_M = R_{serial} = R_{gap} + R_{alumina} = \frac{L_{gap}}{\lambda_{gap} \cdot A_{gap}} + \frac{L_M - L_{gap}}{\lambda_{max} \cdot A_{gap}} \tag{2}$$

R:	thermal resistance [K/W]
L:	length of the heat resistor [µm]
A:	area of the heat resistor [µm²]

As the areas of both resistors are equal $A_M = A_{gap}$, this results in equation (3).

$$\left(\frac{L_{gap}}{\lambda_{gap}} + \frac{L_M - L_{gap}}{\lambda_{max}}\right)^{-1} = \frac{\lambda_M}{x_M} \tag{3}$$

From SEM images of plasma sprayed coatings alumina coatings it was estimated that the length of the gaps between the particles are roughly $L_{gap} = 0.01$ µm. This value represents an initial guess and shouldn't be emphasized too much, as it could be adjusted later on. The thermal conductivity of the real gap was assumed to be $\lambda_{gap} = \lambda_{Air} = 0.0242$ W/(m·K). Furthermore, the gap width in the model was observed to be about 7 mesh cells wide, so $L_M = 7 \cdot 2.25$ µm = 15.75 µm was assumed. At last the expression of equation (1) was used for the thermal conductivity of the model λ_M, resulting in

$$\left(\frac{0.01\,\mu m}{0.0242\,\frac{W}{mK}} + \frac{15.75\,\mu m - 0.01\,\mu m}{30\,\frac{W}{mK}}\right)^{-1} = \frac{30\,\frac{W}{mK}}{15\,\mu m} \cdot \left[\cos\left((1-i) \cdot 2 \cdot \pi\right) \cdot (0.5 - \frac{x_{\%}}{2}) + (0.5 + \frac{x_{\%}}{2})\right]$$

This equation was integrated over i, resulting in

$$\int_1^2 \left(\frac{0.01\mu m}{0.0242\frac{W}{mK}} + \frac{15.75\mu m\text{-}0.01\mu m}{30\frac{W}{mK}} \right)^{-1} = \int_1^2 \frac{30\frac{W}{mK}}{15.75\mu m} \cdot \left[\cos\left((1\text{-}i)\cdot 2\cdot\pi\right)\cdot(0.5 - \frac{x_{\%}}{2}) + (0.5 + \frac{x_{\%}}{2})\right]$$

$$<=> 1.066 = 1.905 \cdot (0.5 + 0.5 \cdot x_{\%})$$

$$<=> 0.119 = x_{\%}$$

Thus it was calculated that a suitable approximation for the reduction of the thermal conductivity in the gap was $x_{\%} = 11.9$ %. As the underlying calculations are only an estimation in the first simulations a value $x_{\%} = 10$ % was assumed in a first step as it simplifies the monitoring of the function.

After the calculation of the coating build-up, the resulting coating structure was transferred into a thermal conductivity calculation model. In this model a heat flow of $\dot{q} = 2{,}000{,}000$ W/m² was applied onto the upper surface of the simulated RVE. This heat flow results in temperature difference, which can then be used to calculate an effective thermal conductivity λ_{eff} of the simulated coating structure. A schematic representation of the thermal conductivity calculation model is shown in figure 4.

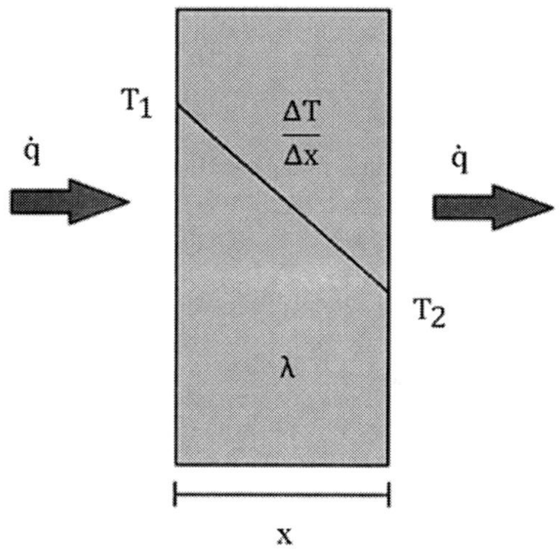

Figure 4. Schematic representation of the thermal conductivity calculation model.

The effective thermal conductivity λ_{eff} is then calculated according to equation (4) with the value of the applied heat flow $\dot{q}$, the thickness s of the created coating and the temperature difference $\Delta T = (T_1 - T_2)$, which can be monitored through the created export function.

$$\lambda_{eff} = \frac{\dot{q}}{s \cdot \Delta T} \tag{4}$$

3. Results and discussion

Figure 5 and 6 show a completed simulation of the particle impact model. The particle inlet temperature for this run was 2,673 K and the particle velocity was 200 m/s. While figure 5 shows the numbered particles, in figure 6 the adjusted thermal conductivity resulting from the implemented gap function is shown. It is clearly visible that at the centre of each particle the thermal conductivity remains at the value of alumina of 30 W/(m·K). At the edge and in the transition region to different particles this value

IOP Conf. Series: Materials Science and Engineering 1147 (2021) 012026 doi:10.1088/1757-899X/1147/1/012026

drops to 3 W/(m·K), so exactly the desired 10 % of the original thermal conductivity. Therefore, it can be concluded that the basic operability of the developed function is given.

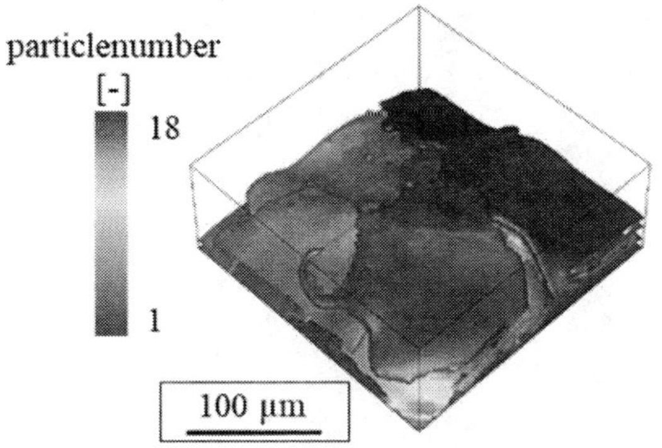

Figure 5. Numbered particles of the completed simulation of thermally sprayed particles.

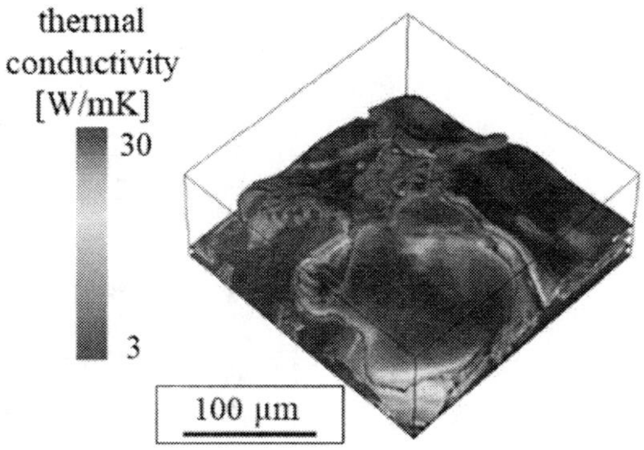

Figure 6. Thermal conductivity of the completed simulation of thermally sprayed particles.

While examining the completed simulations, it was noticed, that the border zone of the domain was not sufficiently filled with particles, as seen in figure 7. This can occur due to the Gaussian distribution of the particles onto the calculation domain. In order to circumvent that issue, every part except the interior 150 µm × 150 µm domain was cut and disregarded, which is shown in figure 8. The remaining part now represents a reduced RVE for the simulated coatings.

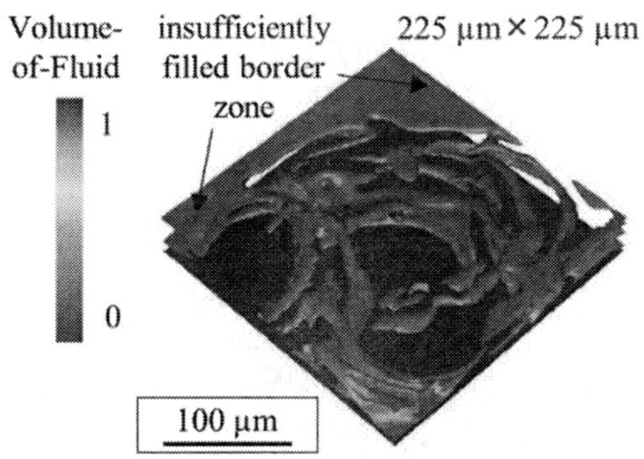

Figure 7. Porosity ($VoF_{Alumina}$ < 0.5 is displayed in green) of a completed simulation.

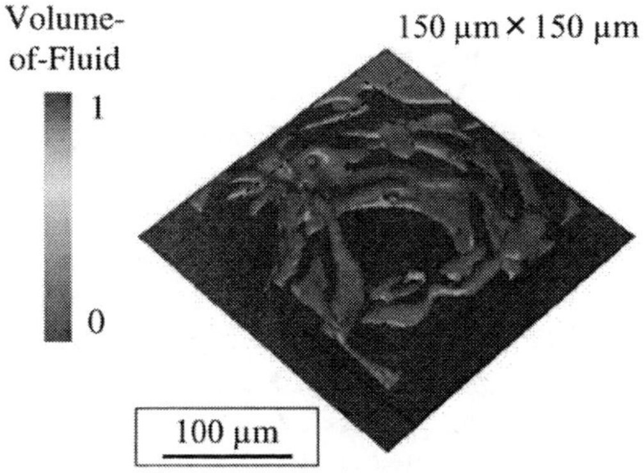

Figure 8. Apparent porosity after cutting the zones at the model boundaries to the reduced RVE.

Following these first runs a series of simulations, each with different values for percentage of the thermal conductivity in the gap $x_{\%}$, were calculated to investigate its influence on the effective thermal conductivity of the coatings. The results of these adjustments are shown in Figure 9. It can be clearly observed, that the gap value has a direct influence on the effective thermal conductivity of the coating. The calculated thermal conductivity varies between 5.8 W/(m·K) and 11.4 W/(m·K) depending on the chosen gap value. As expected, an increasing conductivity in the gap $x_{\%}$ leads to an increased effective thermal conductivity of the coatings. The implemented function and the gap value work as intended.

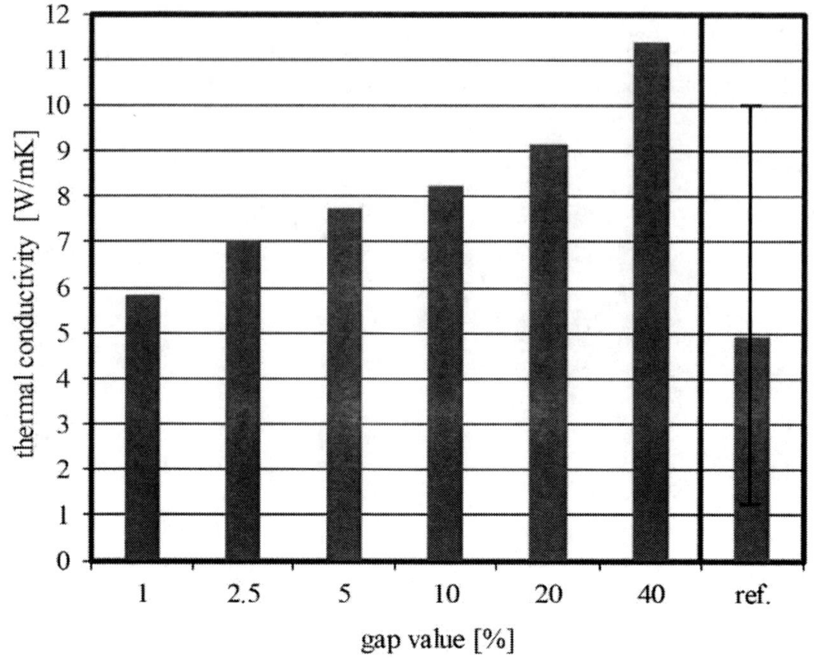

Figure 9. Visual representation of the resulting thermal conductivity with different values for the thermal conductivity of the gap and the reference value, calculated from table 2; the error bar indicates the minimum and maximum reported values in literature.

Additionally, to the calculated values a reference value for plasma sprayed alumina coatings is given in figure 9. The reference value is a mean value of various thermal conductivities found in literature as shown in table 2. This value is even below the effective thermal conductivity for a value of x% = 10 %. Two possible explanation could lead to this overestimation of the thermal conductivity by the model. First of all, the presented model is unable to simulate cracks inside solidified particles. As seen in figure 1, apart from the gaps between the splats, there exist multiple cracks in thermally sprayed ceramic coatings. These cracks are additional obstacles to the heat flow and thus they would reduce the effective thermal conductivity. Secondly, this model ignores any occurring phase transformations and treats all apparent Al_2O_3 as α-Al_2O_3. It is known in literature, that plasma sprayed coatings with higher content of α-Al_2O_3 exhibits a higher thermal conductivity than coatings with higher contents of γ-Al_2O_3 [12]. However, plasma sprayed coatings mostly consist of γ-Al_2O_3 [13], therefore it seems reasonable that the model overestimates the effective thermal conductivity compared to experimental values. For the reasons mentioned above, the previously estimated gap value was therefore retained at $x_{\%} = 10$ % and not adjusted to fit the literature values.

Table 2. Effective thermal conductivities λ [W/(m·K)] of thermally sprayed alumina coatings in literature.

1.1-1.5 [14]
2.1-2.3 [15]
2.45-9.99 [16]
3-7 [17]
5.9 [18]
3.64 [19]
10 [20]
Mean value: 4.89

By varying the particle temperature and velocity, the influence of these particle inflight properties on the effective coating properties was investigated. As mentioned, during this procedure the gap value was kept constant at $x_{\%} = 10$ %. The resulting thermal conductivity and the porosity of the coatings are displayed in table 3. In general, the thermal conductivity varies only in a limited range between 7.87 W/(m·K) and 8.49 W/(m·K). Furthermore, no clear trend is observable. Two reasons could provide a potential explanation. On the one hand, again the missing modelling of cracks and phase transition could be the reason for these small deviations. On the other hand, the limited variety of the chosen inflight parameters could lead to the observed limited differences in thermal conductivity. As the porosity varies also very little and is not affected by the first mentioned causes, the second explanation seems therefore more probable.

Table 3. Results of the coating models thermal conductivity with different values for the particle injection temperature and velocity.

particle temperature [K]	particle velocity [m/s]	gap-value $x_{\%}$ [%]	domain-size [μm×μm]	thermal-conductivity λ_{eff} [W/(m·K)]	porosity [%]
2,473	180	10	150×150	8.24	8.1
2,673	180	10	150×150	8.05	8.2
2,873	180	10	150×150	8.13	8.1
2,473	200	10	150×150	8.49	8.4
2,673	200	10	150×150	8.29	8.1
2,873	200	10	150×150	7.87	7.9
2,473	220	10	150×150	8.33	8.2
2,673	220	10	150×150	8.03	8.0
2,873	220	10	150×150	7.94	8.0

4. Conclusions and outlook

To successfully simulate the coating build-up of thermally sprayed coatings, including the air gaps between two solidified particles, the necessary mesh size would be extremely small. An approach to implement this gap by a function while keeping the calculation time low was presented in this work. A particle impact model followed by the calculation of the effective thermal conductivity were created and a function, which detects the area between two particles and adjusts the thermal conductivity to a percentile of the particles' material thermal conductivity was implemented. The following results were observed:

- A function to identify the individual particles was successfully implemented
- The consecutive gap function was able to reduce the thermal conductivity at the splat boundaries
- The presented model for the coating build-up is able to simulate an RVE for a ceramic coating, by controlling the gap value $x_{\%}$, the coatings thermal conductivity could be adjusted
- The simulated coatings exhibited a higher effective thermal conductivity than the reported literature values for an estimated gap value $x_{\%}$; the missing representation of crack formation and of phase transitions are the believed reasons for this difference
- Variations of the particle in-flight properties did not lead to significant changes of the effective thermal conductivity; the limited range of the input parameters is the most probable cause

Future work needs to concentrate on a broader variation of the particles' in-flight properties to simulate a possible difference due to changes in the process parameters of the plasma spraying process. Furthermore, distributions of particle velocities and temperatures coming from an existing plasma generator model will be used in this model. Finally, the implementation of models for crack propagation and phase transition are crucial to predict correct coating properties in the future.

Acknowledgments
The presented investigations were carried out at RWTH Aachen University within the framework of the Collaborative Research Centre SFB1120-236616214 "Bauteilpräzision durch Beherrschung von Schmelze und Erstarrung in Produktionsprozessen" and funded by the Deutsche Forschungsgemeinschaft e.V. (DFG, German Research Foundation). The sponsorship and support is gratefully acknowledged. Simulations were performed with computing resources granted by RWTH Aachen University under project rwth0570.

References
[1] Goutier S, Vardelle M and Fauchais P 2011 Last developments in diagnostics to follow splats formation during plasma spraying *J. Phys.: Conf. Ser.* **275** 12003
[2] Bobzin K, Bagcivan N, Parkot D and Petkovic I 2008 Simulation of PYSZ-particles impact and solidification in atmospheric plasma spraying coating process *Thermal spray crossing borders: ITSC 2008 - International Thermal Spray Conference & Exposition, Maastricht, The Netherlands, June 2 - 4, 2008* ed E Lugscheider (Düsseldorf: DVS-Verl.) pp 560–4
[3] Bobzin K, Bagcivan N, Parkot D and Petković I 2010 Simulation of PYSZ particle impact and solidification in atmospheric plasma spraying coating process *Surf. Coat. Technol.* **204** 1211–5
[4] Beck T *et al* 2010 Modeling of Coating Process, Phase Changes, and Damage of Plasma Sprayed Thermal Barrier Coatings on Ni-Base Superalloys *Adv. Eng. Mater.* **12** 110–26
[5] Prehm J, Xin L, Möhwald K and Bach F-W 2011 *Coupled coating formation simulation in thermal spray processes using CFD and FEM*
[6] Bobzin K, Öte M, Knoch M A, Alkhasli I and Dokhanchi S R 2019 Modelling of particle impact using modified momentum source method in thermal spraying *IOP Conf. Ser.: Mater. Sci. Eng.* **480** 12003
[7] Fukanuma H, Huang R, Tanaka Y and Uesugi Y 2009 Mathematical Modeling and Numerical Simulation of Splat Cooling in Plasma Spray Coatings *J Therm Spray Tech* **18** 965–74
[8] Parizi H B, Rosenzweig L, Mostaghimi J, Chandra S, Coyle T, Salimi H, Pershin L, McDonald A and Moreau C 2007 Numerical Simulation of Droplet Impact on Patterned Surfaces *J Therm Spray Tech* **16** 713–21
[9] Burcat A, Ruscic B and Chemistry 2005 *Third millenium ideal gas and condensed phase thermochemical database for combustion (with update from active thermochemical tables)*
[10] Ruys A J 2019 *Alumina Ceramics: Biomedical and Clinical Applications (Woodhead Publishing series in biomaterials)* (Duxford, United Kingdom: Woodhead Publishing)
[11] Bates J L, McNeilly C E and Rasmussen J J 1971 Properties of Molten Ceramics *Ceramics in Severe Environments: Proceedings of the Sixth University Conference on Ceramic Science North Carolina State University at Raleigh December 7-9, 1970 (Materials Science Research* vol 5*)* ed W W Kriegel and H Palmour (Boston, MA: Springer) pp 11–26
[12] Heintze G N and Uematsu S 1991 Plasma sprayed γ-and α-Al2O3 coatings **28** 41–68
[13] McPherson R 1989 A review of microstructure and properties of plasma sprayed ceramic coatings *Surf. Coat. Technol.* **39-40** 173–81
[14] Ravichandran K S, An K, Dutton R E and Semiatin S L 1999 Thermal Conductivity of Plasma-Sprayed Monolithic and Multilayer Coatings of Alumina and Yttria-Stabilized Zirconia *J American Ceramic Society* **82** 673–82
[15] Buzovkina T B, Sokolova T V, Obukhov R I, Uspenskaya R I and Degen M G 1972 Effect of Structural Parameters and of Temperature on the Effective Thermal Conductivity of Plasma-Deposited Coatings of Aluminum Oxide *High Temperature* **10** 345–8
[16] Fiedler H C 1983 The Effect of Structure on the Thermal Conductivity of Plasma Sprayed Alumina *MRS Proc.* **30**
[17] Pawłowski L, Martiin C and Fauchais P 1983 The Application of Infrared Thermography in Testing the Coatings and Optimizing the Plasma Spraying Process *10th International*

Thermal Spraying Conference: Lectures of the same called international conference in Essen on 2nd to 6th May 1983 (DVS-Berichte vol 80*)* (Düsseldorf: Dt. Verl. für Schweißtechnik) pp 31–5

[18] Bjorneklett A, Nilsen L M, Kristiansen H and Storfossene T 1992 Coating System for Electrical Applications *Thermal spray: international advances in coatings technology: Proceedings of the International Thermal Spray Conference, 28 May - 5 June 1992, Orlando, Florida USA* ed C C Berndt (Materials Park, Ohio: ASM International) pp 829–34

[19] Nakahira H, Tani K, Miyajima K. and Harada Y 1992 Anisotropy of Thermally Sprayed Coatings *Thermal spray: international advances in coatings technology: Proceedings of the International Thermal Spray Conference, 28 May - 5 June 1992, Orlando, Florida USA* ed C C Berndt (Materials Park, Ohio: ASM International) pp 1011–7

[20] Taylor R 1998 Thermal conductivity determinations of thermal barrier coatings *Materials Science and Engineering: A* **245** 160–7

IOP Conf. Series: Materials Science and Engineering 1147 (2021) 012027 doi:10.1088/1757-899X/1147/1/012027

Parameter study on the electrical resistivity of plasma sprayed TiO_x/Cr_2O_3 coatings

K Bobzin, W Wietheger, H Heinemann and A Schacht*

Surface Engineering Institute (IOT), RWTH Aachen University
Kackertstraße 15, 52072, Aachen, Germany

*e-mail: schacht@iot.rwth-aachen.de

Abstract. Thermally sprayed coatings based on TiO_x/Cr_2O_3 gain increased attention for the application as resistive heating elements. The higher resistivity of ceramic compared to metallic coatings improves the geometric flexibility of the coating as no long or complex conductor paths are required to achieve sufficient resistances. The main objective of this work is to understand the influence of the process parameters on plasma sprayed TiO_x/Cr_2O_3 coatings. Therefore, a design of experiment method is conducted to investigate the effects on the coatings' resistivity. Initially, the feedstock of TiO_x with a weight proportion of 20 % Cr_2O_3 is characterized. A three cathode-plasma spray gun with an Ar/H_2 plasma gas mixture is used to apply the coatings. The current, the amount of H_2 in the plasma gas mixture as well as the stand-off distance of the spray gun are varied according to a circumscribed central composite (CCC) design. These three factors are found to be significant in terms of influencing the coatings' resistivity and a corresponding regression model is established. Furthermore, a significant phase change of the powder due to the spraying process is detected by means of XRD.

1. Introduction

Thermally sprayed electrically conductive coatings are gaining in importance especially for the use as surface heating elements. Possible areas of application include in particular production processes such as injection molding and die-casting. Ceramic coatings intrinsically exhibit a higher electrical resistivity than metallic coatings. This property improves the geometric flexibility since no long or complex conductor paths are required to achieve the necessary resistance values. As a result, planar heating elements can be applied in the conventional coating thickness range for thermal spraying. Furthermore, the application of these planar heating elements leads to homogeneous temperature distribution on the surface [1]. Titanium suboxides have proved to be suitable materials [2], as even a slight oxygen deficiency decreases the electrical resistivity of the initially insulating TiO_2 by several orders of magnitude [3]. TiO_x may form a homologous series of Magnéli phases with the formula Ti_nO_{2n-1} [4]. These phases exhibit an increased stability at elevated temperatures when Cr_2O_3 is added [2]. TiO_x and Cr_2O_3 are able to form another homologous series of Andersson phases with the formula $Cr_2Ti_{n-2}O_{2n-1}$ [4]. The influence of thermal spraying processes on the electrical resistivity of TiO_x was investigated by Floristán et al. [5] and Colmenares-Angulo et al. [6]. Floristán et al. observed a correlation of the electrical conductivity of TiO_x coatings with residual stress-induced cracking. They further state that the H_2 content in the plasma gas mixture results in a reduction of titanium dioxide, while air cooling and high spray distances lead to increased oxidation. Colmenares-Angulo et al. report an anisotropic behaviour of the electrical properties of plasma sprayed TiO_x coatings. Similar investigations with a mixture of TiO_x and Cr_2O_3 in the feedstock were conducted by Trache et al. [7]. However, the electrical

IOP Conf. Series: Materials Science and Engineering 1147 (2021) 012027 doi:10.1088/1757-899X/1147/1/012027

resistivity of these coatings was determined perpendicular to the coatings' surface instead of in-plane. In case of a planar heating elements, the electrical current is applied in-plane, thus these properties are relevant.

This study takes a systematic approach to reveal correlations between the plasma spraying parameters and the resulting electrical resistivity of coatings based on titanium suboxides within a regression model. Therefore, the electrical resistance of the deposited coatings was measured by four-terminal sensing. Together with the coatings' thickness, the electrical resistivity was calculated. In addition, the phase composition of the powder feedstock as well as the deposited coatings was analyzed by means of X-ray diffraction.

2. Experimental and materials

In this study, coatings of TiO_x/Cr_2O_3 were applied on Al_2O_3 coated mild steel C45W (1.1730) substrates (WERTZ GmbH, Aachen, Germany). The TiO_x/Cr_2O_3 feedstock (Ceram, Albbruck-Birndorf, Germany) as well as the Al_2O_3 feedstock Amdry™ 6062 (Oerlikon Metco, Pfäffikon, Switzerland) were jointly fused and crushed. The nominal grain size distribution alongside the weight composition according to the manufacturer certificates are given in Table 1. Both powders were applied using the three-cathode plasma torch TriplexPro™-210 (Oerlikon Metco). The Al_2O_3 coating's sole purpose is to electrically insulate the TiO_x/Cr_2O_3 heater coating from the steel substrate.

Table 1. Nominal grain size distribution and weight composition according to the manufacturer certificates of used feedstocks.

feedstock	grain size distribution [µm]	Al_2O_3 [%]	TiO_2 [%]	Cr_2O_3 [%]	SiO_2 [%]	Fe_2O_3 [%]	CaO [%]	MgO [%]	Na_2O [%]
Amdry™ 6062	-45 +22	Bal.	0.02	-	0.02	0.01	0.03	0.02	0.14
TiO_x/Cr_2O_3	-25 +5	0.08	Bal.	19.95	0.04	0.08	0.03	0.06	-

The current, the gas flow of H_2 within the plasma gas mixture as well as the stand-off distance between torch nozzle and substrate were varied within five levels each. The parameter variations were conducted according to a circumscribed central composite (CCC) design [8] and are listed in Table 2. Specimens S15-1 through S15-6 conform to the center point of the applied CCC design, which was set following earlier findings [1]. The CCC design was chosen to cover a wide range of these three spray parameters in a systematic approach.

The phase composition of the TiO_x/Cr_2O_3 feedstock as well as the TiO_x/Cr_2O_3 coatings were determined by means of x-ray diffraction (XRD 3000, GE Inspection Technologies, Hürth, Germany) with a Cu-K_α x-ray tube source. All patterns were recorded using a current of $I = 40$ mA, voltage of $U = 40$ kV, step width of $s = 0.05\,°$ and scanning time per step of $t = 25$ s. Post-processing was abandoned to ensure distortion-free results.

Table 2. Overview of spray parameters and their spray order within the conducted design of experiment.

plasma gas flow rate Ar [SLPM]	60	feeding rate [g/min]	23.6
nozzle diameter [mm]	9	robot velocity [mm/s]	1
carrier gas flow rate Ar [SLPM]	7.5	meander width [mm]	5
injection nozzle diameter [mm]	2	number of passes [-]	10

specimen	spray order	current [A]	gas flow H_2 [SLPM]	stand-off distance [mm]
S1	7	400	3,6	105
S2	14	500	3,6	105
S3	17	400	8,4	105
S4	3	500	8,4	105
S5	2	400	3,6	135
S6	10	500	3,6	135
S7	13	400	8,4	135
S8	5	500	8,4	135
S9	1	366	6	120
S10	9	534	6	120
S11	6	450	2	120
S12	16	450	10	120
S13	20	450	6	95
S14	12	450	6	145
S15-1	19	450	6	120
S15-2	4	450	6	120
S15-3	8	450	6	120
S15-4	11	450	6	120
S15-5	15	450	6	120
S15-6	18	450	6	120

The thickness of the non-magnetic Al_2O_3 and TiO_x/Cr_2O_3 coatings was determined by the magnetic induction principle according to ISO 2178 using the Dualscope® MP 40 (Helmut Fischer GmbH, Sindelfingen, Germany) with a measuring probe of the type EGABW1.3. For each specimen, ten individual spots were measured to calculate a mean thickness value. As the specimens were sprayed with ten passes each and a constant powder feed rate, the resulting coating thickness is an indicator for the deposition efficiency. The electrical resistance R was measured by means of four-terminal sensing (TM-508A, Isothermal Technology, Southport, United Kingdom) parallel to the surface. The distance between the measuring probes was $l = 75$ mm. Using the coating's thickness d as well as its width of $w = 40$ mm the electrical resistivity ρ was calculated according to equation (1). Since the resistivity is an intrinsic property, it allows for an appropriate comparison between the different coatings.

electrical resistivity:
$$\rho = R \cdot \frac{d \cdot w}{l} \quad (1)$$

R: electrical resistance
l: measuring length
d: coating thickness
w: coating width

To visualize the correlation between the three spray parameters current, flow rate of H_2 as well as stand-off distance and the coatings' resistivity the software MATLAB R2017b (MathWorks, Natick, Massachusetts, USA) and its evaluation tools for design of experiments was employed. The CCC design was evaluated with the response surface method (RSM) to comply with the quadratic equation (2), whereby Y resembles the coatings' resistivity, β_i resemble estimates and X_i resemble the factors current, flow rate of H_2 and stand-off distance. Multiplications of X_i equal interactions of these factors. Since the numeric values of the factors in this study significantly differ from each other, all values were normalized between 0 and 1 for their respective minimum and maximum values in Table 2 at the start of the evaluation. For visualization, however, the normalized values were reverted again. The quality of the estimates is described with an adjusted R^2. The fitting of the estimates depends on the sequence, in which they are added to the fitting equation. Thus, multiple sequences, which started the fit with different estimates, were tested. An included analysis of variance determined the significance of influence of each individual factor and its interactions. The threshold value for the significance level was set to $p = 0.05$.

$$Y = \beta_0+\beta_1X_1+\beta_2X_2+\beta_3X_3+\beta_{12}X_1X_2+\beta_{13}X_1X_3+\beta_{23}X_2X_3+\beta_{11}X_1^2+\beta_{22}X_2^2+\beta_{33}X_3^2 \quad (2)$$

Y: response X_i: factor – linear effect X_i^2: factor – quadratic effect
β_i: estimates X_iX_j: factor – interaction

3. Results and discussion

A summarized overview of the measured values in this study is available online at [9]. Exemplary micrographs of S5 and S8 with an electrical resistivity of $\rho_5 = 1.592$ Ωmm and $\rho_8 = 0.312$ Ωmm are depicted in Figure 1. The TiO_x/Cr_2O_3 coating of S5 shows a more porous structure than S8. While both coatings were deposited at the same stand-off distance of 135 mm, the difference originates from the lower energy input into S5 by the plasma due to a lower current and proportion of H_2 as secondary plasma gas.

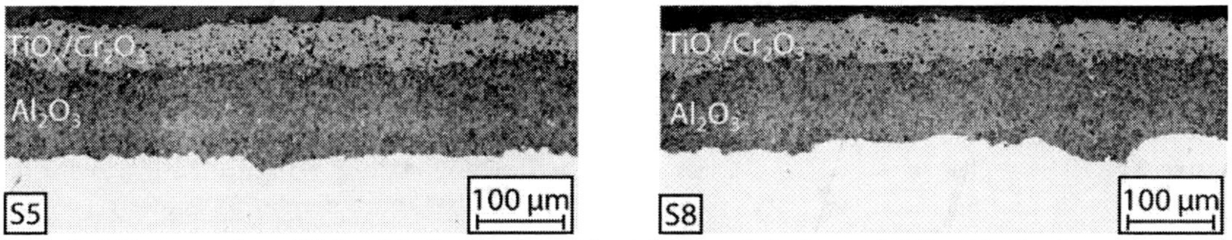

Figure 1. Micrographs of S5 with the highest electrical resistivity and S8 with the lowest electrical resistivity.

The XRD patterns of the TiO_x/Cr_2O_3 feedstock and the coatings with the lowest and highest resistivity, S8 and S5 respectively, are depicted in Figure 2, while the used JCPDS numbers are summarized in Table 3. Within the feedstock, clear signs of the existence of the Magnéli phase Ti_4O_7 and $CrTi_2O_5$ are detected besides a variety of additional Magnéli and Andersson phases and small amounts of Cr_2O_3. However, reduction and oxidation of these substoichiometric phases during plasma spraying result in a multitude of different oxygen levels and thus, a varying range of values of n in the homologous series of Andersson and Magnéli phases. It is observed that Ti_4O_7 is not present in the coating. The different degrees of reduction and oxidation may occur due to a combination of the various values of factors like actual particle diameter, grain size, melt degree of the particles, energy distribution and particle trajectory within the plasma flame. As the XRD patterns within these homologous series are close to each other and partially overlapping as reported by Somiya et al. [10], those phases are given in areas to give a clearer view of the XRD patterns.

Table 3. Summary of the used JCPDS numbers.

Phase	JCPDS number
Ti_4O_7	01-072-1724
Ti_nO_{2n-1}	00-050-0787, 00-051-0641, 00-050-0788, 00-050-0789, 00-050-0790, 00-050-0791
$CrTi_2O_5$	01-079-0302
$Cr_2Ti_{n-2}O_{2n-1}$	00-035-0098, 00-035-0099, 00-035-0100, 00-030-0420
TiO_2	01-076-1938
Cr_2O_3	01-084-0313

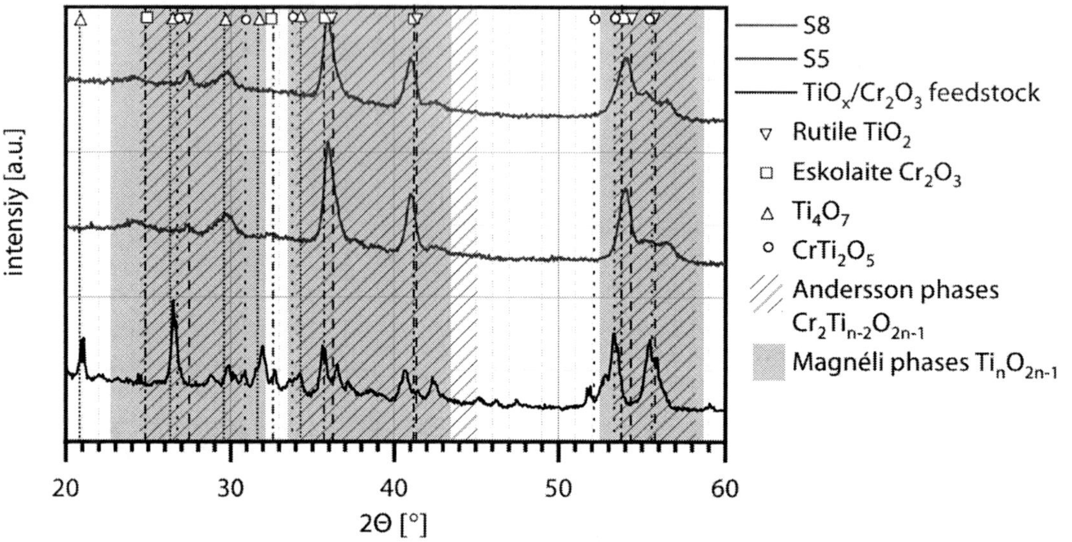

Figure 2. XRD patterns of TiO_x/Cr_2O_3 feedstock as well as the coating S8 with the lowest resistivity and S5 with the highest resistivity. Andersson and Magnéli phases are displayed as areas as the peaks shift slightly in relation to values of n.

The coatings show minor proportions of rutile and eskolaite, while the amount of rutile is almost negligible in the feedstock. Among the coatings, no major difference is noticeable. The peak heights of various overlapping Andersson phases and Magnéli phases have slightly different values, however, no specific phase stands out. This corresponds with earlier findings by the authors shown in Figure 3, where heating cycles were conducted with TiO_x/Cr_2O_3 coatings [1]. This is particularly remarkable, since the phase composition of the TiO_x/Cr_2O_3 powder in the current study differs from the one observed in the previous study, when a different batch of the same powder was used. Since the TiO_x/Cr_2O_3 powder consists of suboxides, which form during its production, the proportion of the different suboxides may vary according to slight changes or maybe disturbances in the process. This may result in the observed difference of phase composition within different powder batches. The phase composition of the coatings remained the same regardless of the actual phase composition of the TiO_x/Cr_2O_3 feedstock. This leads to the assumption that the reduction and oxidation of the feedstock reach a rather stable equilibrium during plasma spraying.

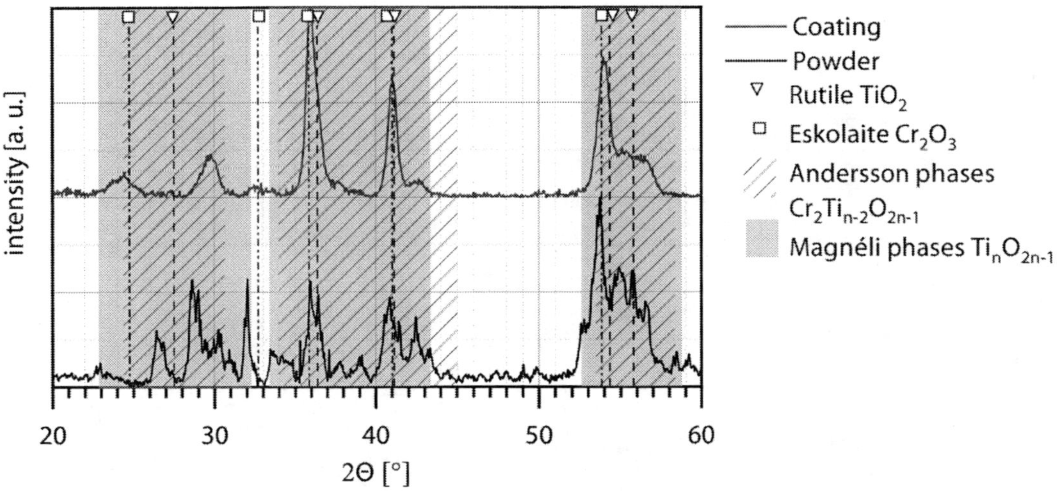

Figure 3. XRD patterns of TiO_x/Cr_2O_3 powder and plasma sprayed coating according to [1].

Table 4 lists the fitted values for equation (2) as well as the respective significance levels for each estimate. Since all p values are lower than the chosen threshold of $p = 0.05$, all three varied parameters X_i, their interactions X_iX_j and the quadratic terms X_i^2 exhibit a significant influence on the resistivity of the plasma sprayed TiO_x/Cr_2O_3 coatings. The stand-off distance certainly exhibits the least significant influence out of these three parameters. The adjusted R squared of the fitted equation results to $R^2_{adj} = 0.95008$, showing a rather good conformity of the chosen model with the experimental results.

Table 4. Fitted estimates β_i and their respective significance levels p.

	estimate β_i	significance level p
intercept	0.930	3.040E-05
X_1 (current)	-1.413	2.088E-04
X_2 (gas flow H_2)	-2.062	8.570E-06
X_3 (stand-off distance)	0.971	3.016E-03
X_1X_2	1.652	3.670E-05
X_1X_3	-1.269	3.298E-04
X_2X_3	-1.058	1.160E-03
X_1^2	0.736	1.965E-03
X_2^2	1.177	5.570E-05
X_3^2	0.396	4.949E-02

In Figure 4, the contour plot of the resistivity is shown in dependence of the current and the gas flow of H_2. The stand-off distance is kept constant at 120 mm as its influence on the resistivity is the smallest of the three. A rather wide valley with an almost constant resistivity forms in the response surface. The resistivity increases at the edges of the investigated parameter field. This is especially noticeable at a low level of current and gas flow of H_2. In this case, the particles experience less energy input from the plasma, which may result in a higher porosity of the coating. A higher porosity yields a higher resistivity. High particle temperatures at high levels of current and gas flow of H_2 may result in increased re-oxidation and hence increase the resistivity of the coating as well. The reduction and oxidation of the coatings within the observed room of spray parameter is not detectable by XRD. Nonetheless, the electrical properties of titanium suboxide are highly susceptible to changes of the substoichiometry. Liu and West made a similar observation in their study on electrical properties of sintered titanium suboxides

[11]. The model predicts a minimum resistivity of $\rho_{min} = 0.312$ Ωmm for a current of I = 503 A, an H_2 gas flow of $\dot{V} = 6.2$ SLPM and a stand-off distance of s = 121 mm. This set of parameters is close to specimen S10, which resulted in a resistivity of $\rho_{S10} = 0.390$ Ωmm. This result is in the same order, which shows consistency with the model's prediction.

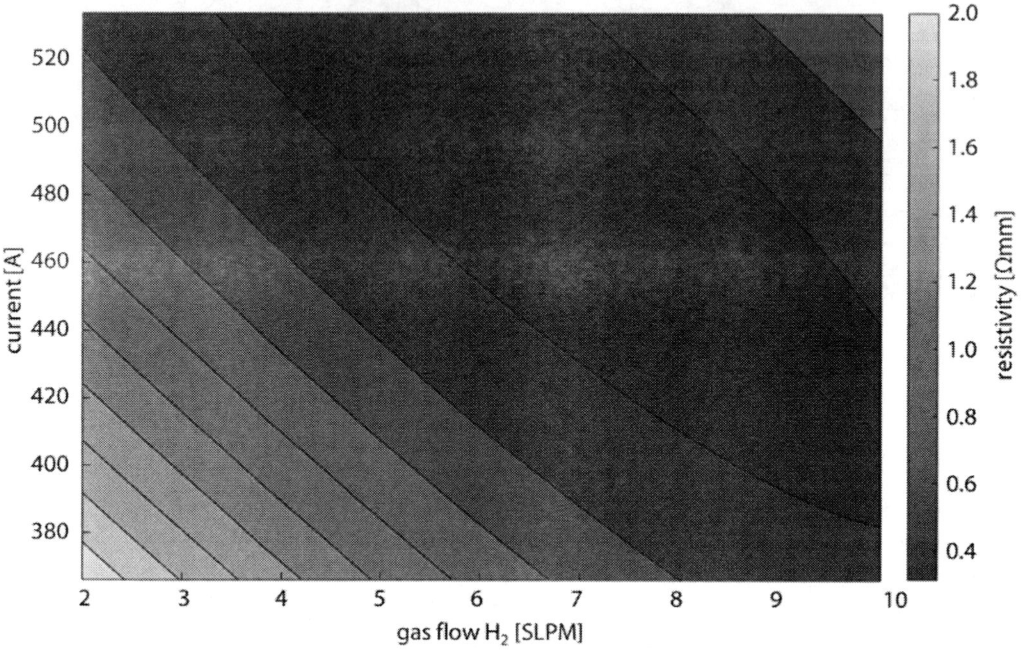

Figure 4. Contour plot of the resistivity depending on the current and the gas flow H_2 with a constant stand-off distance of 120 mm.

The electrical resistivity of the TiO_x/Cr_2O_3 coatings, measured in this study, range from $\rho_{min} = 0.312$ Ωmm to $\rho_{max} = 1.592$ Ωmm. The detected range is comparable to the minimal electrical resistivity of $\rho = 0.5$ Ωmm for TiO_x coatings reported by Floristán et. al [5]. The reported minimum electrical resistivity for a plasma sprayed coating made of TiO_2 with a weight proportion of 20 % Cr_2O_3 is $\rho = 2{,}000$ Ωmm [7], resulting in a difference of over three orders of magnitude. Within that study, however, the resistivity was measured through-thickness of the coating, while the herein presented resistivity was measured in-plane. This leads to the assumption that plasma sprayed TiO_x/Cr_2O_3 coatings exhibit anisotropic behaviour of the electrical resistivity, which may result from less splat boundaries in-plane than through-thickness. A similar anisotropy of about four orders of magnitude was found by Colmenares-Angulo et al. for plasma sprayed TiO_2 coatings [6]. They state that the difference results from differing stoichiometric distribution among splat regions.

The individual effects of each parameter are depicted in Figure 5. Furthermore, the correlation between the coatings' resistivity and the net power of the plasma is added for comparison. An increase of the gas flow of H_2 leads to a decrease of the resistivity. This correlation is described by a decreased porosity and the reducing effect of H_2 on the particles, which results in a slightly increased oxygen deficiency in the suboxide phases. An increase of the current results in a decreased resistivity as well. This can be explained by the increase of available energy to accelerate the chemical reduction reaction. The overlapping effect shows in the graph of the net power, too. The stand-off distance, however, raises the electrical resistivity as it is increased. Mixing of the plasma jet with the surrounding atmosphere increases proportionally to the stand-off distance. The particles therefore experience a stronger oxidation and less reduction with the H_2 within the plasma gas mixture. Additionally, the kinetic energy of the particles is less, which could result in a less dense coating, and thus raise its specific resistivity.

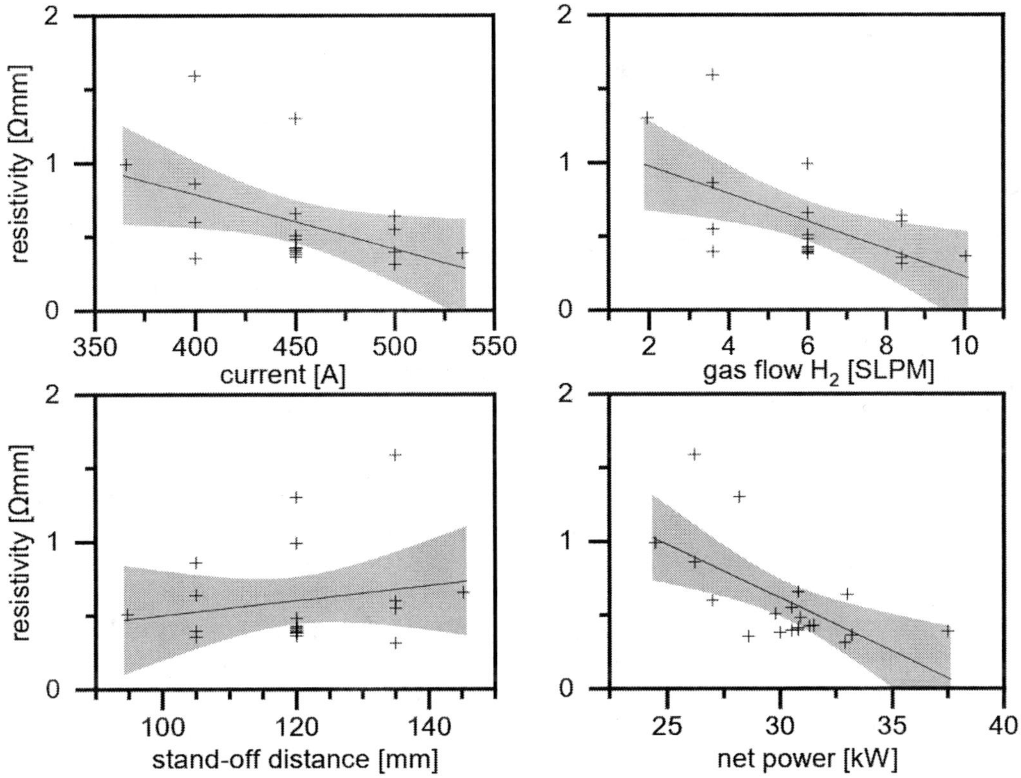

Figure 5. Prediction slice plots at the center point. The red line shows the linear correlation, while the band resembles the 95 % confidence band of the resistivity for each individual parameter as well as the occurring net power of the plasma. The marks resemble the measured values. The remaining parameters were set to the values given in Table 2.

Within the variation of the stand-off distance the coating thickness of the TiO_x/Cr_2O_3 coatings is negatively proportional as depicted on the left side of Figure 6. Hence, the deposition efficiency is decreasing with an increasing stand-off distance. The depicted scattering of the thickness results from additional influences on the deposition efficiency from the varied current and H_2 gas flow.

On the right side of Figure 6, the correlation between the electrical resistivity and the coating thickness of TiO_x/Cr_2O_3 is shown. At coating thicknesses lower than 50 µm, an increased scattering of the resistivity is observed. This may result from a larger proportion of coating roughness to coating thickness. Large coating roughness in a thin coating may form narrow places, which in turn raise the electrical resistivity of the coating.

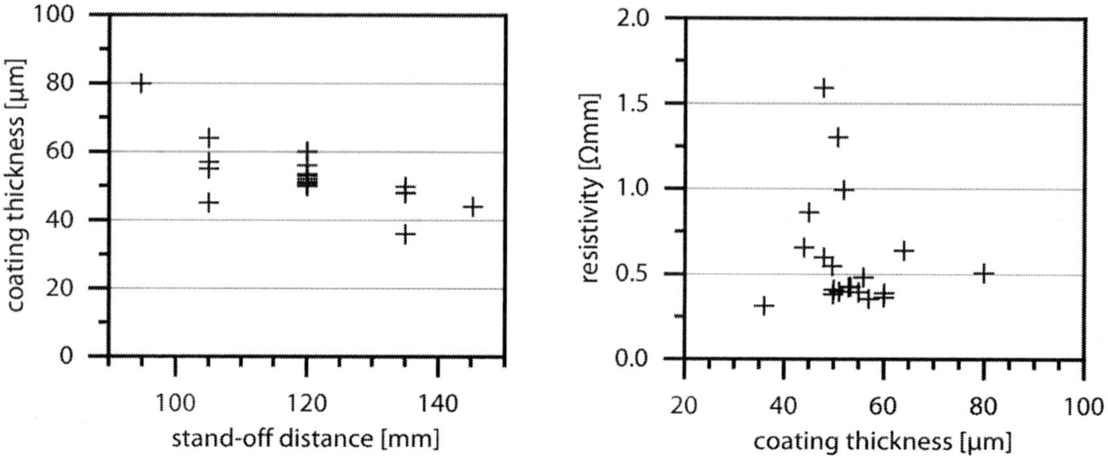

Figure 6. Coating thickness of the TiO_x/Cr_2O_3 coatings depending on the stand-off distance (left) and its electrical resistivity over the coatings thickness (right).

4. Conclusions

A regression model for the influence of the current, the H_2 proportion within the plasma gas mixture and the stand-off distance on the in-plane electrical resistivity of plasma sprayed TiO_x/Cr_2O_3 coatings was successfully established. While all three factors are significant, the effect resulting from the stand-off distance is least significant. According to the established model, the minimal electrical resistivity of TiO_x/Cr_2O_3 coatings is $\rho_{min} = 0.312$ Ωmm. While the plasma spraying process alters the phase composition of the TiO_x/Cr_2O_3 feedstock considerably, comparison with previous studies showed that the resulting phase composition within the TiO_x/Cr_2O_3 coating is hardly affected.

Further investigations have to be conducted to verify the established regression model. Therefore, particle states as particle temperature and velocity within the plasma jet will be measured to associate these with the resulting microstructure and hence electrical resistivity of the coatings. The exact correlation between the phase compositions of TiO_x/Cr_2O_3 feedstocks with the resulting phase composition of TiO_x/Cr_2O_3 coatings must be examined in more detail. SEM and EDX analysis of the feedstocks as well as the coatings will yield significant insight. The clarification will help in understanding and controlling the possible tolerances for future application.

Acknowledgments

The presented investigations were carried out at RWTH Aachen University within the framework of the Collaborative Research Centre SFB1120-236616214 "Precision Melt Engineering" and funded by the Deutsche Forschungsgemeinschaft e.V. (DFG, German Research Foundation). The sponsorship and support are gratefully acknowledged.

References

[1] Bobzin K, Wietheger W, Knoch M A and Schacht A 2020 Heating behaviour of plasma sprayed TiO_x/Cr_2O_3 coatings for injection moulding *Surface and Coatings Technology* **399** 126199

[2] Scheitz S, Toma F-L, Berger L-M, Puschmann R, Sauchuck V and Kusnezoff M 2011 Thermally sprayed multilayer ceramic heating elements *Thermal Spray Bulletin* **11** 88–92

[3] Hayfield P C S 1981 Electrode material, electrode and electrochemical cell *US4422917*

[4] Andersson S, Sundholm A and Magnéli A 1959 A Homologous Series of Mixed Titanium Chromium Oxides $Ti_{n-2}Cr_2O_{2n-1}$ Isomorphous with the Series Ti_nO_{2n-1} and V_nO_{2n-1} *Acta Chemica Scandinavica* **13** 989–97

[5] Floristán M, Fontarnau R, Killinger A and Gadow R 2010 Development of electrically conductive plasma sprayed coatings on glass ceramic substrates *Surface and Coatings Technology* **205** 1021–8
[6] Colmenares-Angulo J R, Cannillo V, Lusvarghi L, Sola A and Sampath S 2009 Role of process type and process conditions on phase content and physical properties of thermal sprayed TiO_2 coatings *J Mater Sci* **44** 2276–87
[7] Trache R, Berger L-M, Toma F-L, Saaro S, Lima R and Marple B 2011 Electrical Resistivity of Thermally Sprayed Cr_2O_3-TiO_2 *Coatings Conference proceedings: Int. Thermal Spray Conf., 27.-29. Sept. 2011, Hamburg*
[8] Montgomery D C 2013 Design and analysis of experiments (Hoboken NJ: John Wiley & Sons Inc)
[9] Bobzin K, Wietheger W, Heinemann H and Schacht A 2020 Supplementary data for: Parameter Study on the Eelctrical Resistivity of Plasma Sprayed TiO_x/Cr_2O_3 Coatings http://hdl.handle.net/21.11102/32b4f333-d6df-45c9-a539-e90a1162fc13
[10] Somiya S, Hirano S and Kamiya S 1978 Phase Realtions of the Cr_2O_3-TiO_2 System *J. Solid State Chem.* **25** 273–84
[11] Liu Y and West A R 2013 Semiconductor-Insulator Transition in Undoped Rutile, TiO_2, Ceramics *J. Am. Ceram. Soc.* **96** 218–22

Impact resistance of WC-Co reinforced HVAF-sprayed FeCrB-coatings

K Bobzin, W Wietheger, H Heinemann, J Sommer and M Schulz*

Surface Engineering Institute, RWTH Aachen University
Kackertstrasse 15, 52072 Aachen, Germany

*e-mail: schulz@iot.rwth-aachen.de

Abstract. This study describes the development of cost-effective FeCrB-coatings and the evaluation of these novel coatings against impact loads. The cost reduction is achieved by using a new and economical iron-based feedstock material with a small particel size distribution, which can be processed at high feeding rates with the HVAF process. The reduced particle size distribution enhances the application of near-net-shape coatings, minimizing the need for expensive, time-consuming grinding postprocesses. Thicker coatings usually result in higher thermal insulation, whereas thin near-net-shape coatings can reduce the thermal insulation. Dense FeCrB and FeCrB/WC-Co coatings were applied by HVAF with feeding rates of $\dot{m}$ = 200 g/min and a particle size distribution of -20 + 3 µm. Through curvature measurements during the application it is shown that by integrating WC-Co into FeCrB the tensile residual stress state of the coating was minimized, which in turn increased the fracture toughness of the coating. The evaluation of the impact resistance was investigated by an impact test. The results show that the crater volume was halved by adding WC-Co in the FeCrB-coating. However, it is shown that the plastic deformability of the coating is minimized and a stronger cracking behavior can be observed.

1. Introduction

Iron-based coatings are in the current focus of research, since there is an increasing demand for economical and ecological coatings for the protection against corrosion and wear. In order to provide an adequate protection, established coatings such as Cr_3C_2/NiCr are often applied using high velocity oxygen fuel (HVOF) or high velocity air fuel spraying (HVAF). Due to the higher costs, such coatings are only an economic solution for applications where very high loads are present. A much more cost effective process is wire arc spraying (WAS). The high cost efficiency results from the high application rates of $\dot{m} \leq 500$ g/min [1] and the usage of cost-effective feedstock materials. A typical iron-based feedstock material, that is applied by means of WAS are FeCrBSiMnC cored wires. However, due to the high roughness in as-sprayed condition, expensive and time-consuming grinding is often necessary. Furthermore, the high porosities of η = 3-20 % [2] of WAS coatings, require high coating thicknesses of 800-1,200 µm [3, 4] to ensure sufficient corrosion protection in some applications. Sealers can improve the corrosion properties, yet they need again an additional post-production step. Sufficient corrosion protection could also be achieved with coatings with lower coating thicknesses and lower porosity using HVOF or HVAF, where the porosity is $\eta \leq 2\%$ [2]. If hard materials, e.g. WC, are sufficiently attached in a surrounding matrix, the wear properties of these composite coatings can be improved. For example, Terajimi et al. showed that the wear can be significantly reduced by increasing the WC/Co content in a FeCrMoCB-WC/Co coating against an Al_2O_3 ball in a reciprocating wear test [5]. Similar results were observed by Bolelli et al. with FeCrAl-WC coatings [6]. They compared the wear behavior of the iron-based coatings with established HVOF-sprayed coatings made of WC/CoCr. While the wear coefficient was comparable, the iron-based coatings showed a significantly more brittle behavior. They explain this with a lower cohesion and higher oxide content of the HVOF-sprayed coating. In [7] FeCrNiSiBC+WC coatings were applied by HVAF. This process is a modification of the established HVOF system, where instead of oxygen, air is used as oxidizing agent. In this study from Bolleli et al. it is shown that by the integration of WC particles in the FeCrNiSiBC coating the wear coefficient can be reduced by an order of magnitude. However, they also show that an exact process control is necessary, since otherwise un- or partially molten particles can weaken the cohesion of the coating and micro cracks along the splat boundaries occur during the wear test.

In own previous studies the basic wear and corrosion properties of FeCrB/WC-Co-coatings were investigated [8–10]. In [8] crack-free FeCrB/WC-Co coatings were produced with a powder mass flow rate of $\dot{m}$ = 40 g/min and a particle size distribution of -45 +11 µm. By reducing the particle size distribution to -32 +11 µm homogeneously molten particles were realized [9]. Through the finer distribution and the higher possible gas flow rates compared to HVOF, crack-free coatings with an increased feed rate of $\dot{m}$ = 200 g/min could be achieved. The higher mass flow rates of the iron-based coatings are benefical for a further reduction of the production costs compared to the state of the art. In [10] a FeCrB/WC-Co-powder with a particle size distribution of -20 +3 µm and the same chemical composition as described in [8, 9] was used to achieve a near-net-shape coating. The wear in the pin-on-disk test was comparable to the results obtained in [9], but the corrosion resistance was significantly improved.

Periodic impacts, such as those caused by vibrations in machines, are a special type of load for coatings. Therefore, the influence of WC-Co in the FeCrB/WC-Co coating on the impact behavior is investigated in this study. Knotek et al. have introduced an impact machine to evaluate the impact behavior of thin CVD and PVD coatings [11]. Based on this, Bouzakis et al. [12] made the first basic investigations with this new investigation method. However, thermal sprayed coatings have not been investigated very much so far. First investigations on thermally sprayed Cr_3C_2/NiCr coatings were made in 2017 [13]. Results on FeCrB/WC-Co are not yet available and are discussed in this study.

2. Experimental and materials

As substrate flat samples of 1.0038 (EN 10027, S235JR) were used. Prior to the coating process, all samples were grit blasted with corundum (F16) at a pressure of p = 0.4 MPa to roughen the surface. For the application of the coatings, commercially available powders from Above Material Technology Co, Ltd, Beijing, China, were used. These powders are distributed in Germany by the company GTV Luckenbach under the trade names 80.44.0S-0 and 80.44.0S-3. The WC/Co content is 15 percent by weight in the WC/Co-containing coating. The chemical composition of the powders is given in Table 1. While the high Cr content shall provide adequate corrosion protection, the contents of hard phase-forming elements C and B shall ensure adequate wear protection [3, 4].

Table 1. Particle size distribution, manufacturer information of the chemical composition [wt.-%] of the feedstock powder used for the application of the coatings and a SEM image of the used powder.

Coating	Particle size distribution	Fe	Cr	B	C	W	Co
F20	-20 +3 µm	Bal	22-26	4.3-5	0.4-0.7	-	-
W20	-20 +3 µm	Bal	20-23	3.5-4	1-1.2	10-12	1-1.5

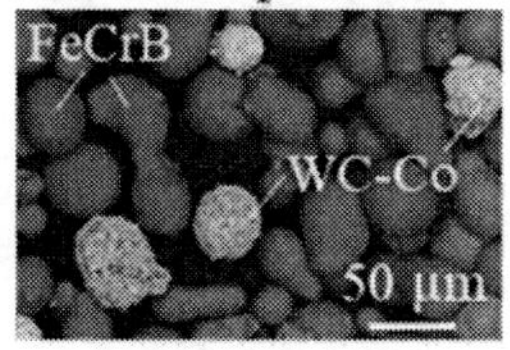

Subsequently, the samples were coated with the parameters given in Table 2. All coatings of this study were applied with the HVAF system AK-07 (Kermetico Inc., Benicia, USA). The parameter set for F20 was developed based on experience with other Fe-based powders. However, no dense coating could be achieved with this parameter set and the WC-Co-containing powder. Therefore, the spray distance was increased and the pressure of the compressed air was reduced to increase the heat input to the particles. Through stronger flattening of the hotter particles a dense coating could be applied in case of W20, too.

Table 2. Spraying parameters of the investigated samples.

Sample name	Propane [MPa]	Compressed air [MPa]	Surface speed [mm/s]	Powder feed rate [g/min]	Stand-off-distance [mm]	Cooling air [MPa]
F20	0.56	0.72	4,500	200	180	0.15
W20	0.56	0.63	4,500	200	350	0.15

In a preliminary investigation it was shown that the coating thickness per pass has a considerable influence on the crack formation in the coating [8]. It is assumed that the thermal load per pass is higher and thus the thermally induced residual stresses are increased. In order to reduce these residual stresses, the samples were fixed on a cylindrical sample holder. The sample holder rotates, while the HVAF gun movement is restricted to a linear movement parallel to the rotation axis of the sample holder. Due to the high surface velocity of 4,500 mm/s the thermal load per pass is reduced. The exact design can be taken from [9].

The evaluation of the microstructure was performed with the light microscope Axiophot (Zeiss, Oberkochen, Germany). The phase composition was determined by X-ray diffraction, XRD. The Seifert XRD 3000 X-ray diffractometer (GE Sensing Inspection Technologies GmbH, Hürth, Germany) was used for this purpose. For the measurements a Cu-anode was utilized. The angle of incidence was $\omega = 10\,°$ and the step size was set to $\Delta 2\theta = 0.05\,°$. The determination of the Vickers micro hardness HV was performed according to DIN EN ISO 6507 with the hardness tester Micromet 1 (Bühler Ltd, Illinois, USA). To determine the hardness, a load of $F = 0.98$ N and a holding time of $t_{mh} = 15$ s were used at 15 measuring points in each coating. The fracture toughness was calculated by use of indention method [14]. For the determination, cracks were created in the cross sections of the coatings using the hardness measurement system Wilson Hardness 452 SVD device (Bühler Ltd, Illinois, USA) with a Vickers indenter. A high load of load $F_P = 98$ N was used to produce sufficient cracks. The maximum crack length c_{max} was determined with the light microscope. The Young's modulus E was determined with the Fischerscope HM2000 (Helmut Fischer GmbH, Sindelfingen-Maichingen, Germany). The fracture toughness was calculated according to equation 1 [14].

$$\text{Fracture toughness } K_{IC}\text{:} \quad K_{IC} = 0.015 \left(\frac{E}{HV}\right)^{\frac{1}{2}} \cdot \left(\frac{F_P}{C_{max}^{1.5}}\right) \qquad (1)$$

As shown in [6], the residual stress state of the coating has a significant influence on its wear resistance, which is why the residual stresses of the coatings were measured in this study. During the coating process, residual stresses are generated primarily by impact and shrinkage of the particles. The impact can induce compressive residual stresses, whereas the hindered shrinkage of the particles induces tensile residual stresses. In addition, thermal residual stresses occur after the coating process due to the different thermal expansion coefficients of the coating and the substrate. The residual stresses lead to a curvature of the coating. This curvature is determined by means of ICP8 sensor (ReliaCoat Technologies, LLC, Setauket, USA) and three precision lasers. Through the measured curvature the residual stresses in the coating are calculated. From the literature it is known that tensile residual stresses can enhance crack propagation, therefore the results of this investigation are compared with the fracture toughness [15].

The evaluation of the impact resistance was performed using the impact tester Apollo NXG (impact-bz Ltd, London, UK). In this test, a WC-Co ball with a diameter Ø = 5 mm impacts cyclically a sample with a defined force $F_N = 1{,}000$ N. The frequency was set to $f = 50$ Hz and $n = 500{,}000$ impacts were performed. The crater volume was determined by means of a confocal laser microscope VKX 210 (Keyence Corporation, Osaka, Japan). The wear mechanisms were investigated by scanning electron (SEM) and light microscopy. The SEM investigations were performed with the scanning electron microscope Phenom XL (Thermo Fisher Scientific, Waltham, USA). For an evaluation of the crater volume, both samples were ground identical. In analogy to the previous investigations, a 10 µm diamond pad was used as the last step [8–10]. To influence the residual stresses in both coatings equally, a fully automatic grinding routine with the grinding machine Saphier 550 (ATM, Golling, Austria) was used, see Table 3.

Table 3. Grinding routine used for the investigated samples.

	Step1	Step 2	Step 3	Step 4	Step 5
Grinding Pad [*ANSI Grit]	180 SiC*	400 SiC*	800 SiC*	1,200 SiC*	10 µm Diamond
Velocity [rpm]	100	100	100	100	100
Force [N]	30	30	30	30	30
Time [min]	6	2	2	2	6

In order to prove how far the grinding process influences the impact resistance, coatings in the as sprayed condition were also examined. All experiments were repeated three times on each samples, to ensure that the results are statstically reliable.

3. Results and discussion

Figure 1 shows the morphology of the investigated coatings. For both coatings a coating thickness of about 150 µm was obtained. Figure 1a shows the cross section of the coating F20 without WC-Co inclusions. No splat boundaries can be identified, which suggests a low oxidation of the particles in flight. Only in the vicinity of the surface small pores can be identified, due to the missing peening-effect of subsequent impacting particles. Figure 1b shows the cross section of the WC-Co containing FeCrB coating W20. Also in this coating, it is hardly possible to identify the splat boundaries. In contrast to the WC-Co free coating, no pores can be identified even in the vicinity of the surface. This can be explained by an additional peening-effect due to the WC particles. Both coatings suggest a good bonding to the substrate, since little to no pores at the interface to the substrate are visible.

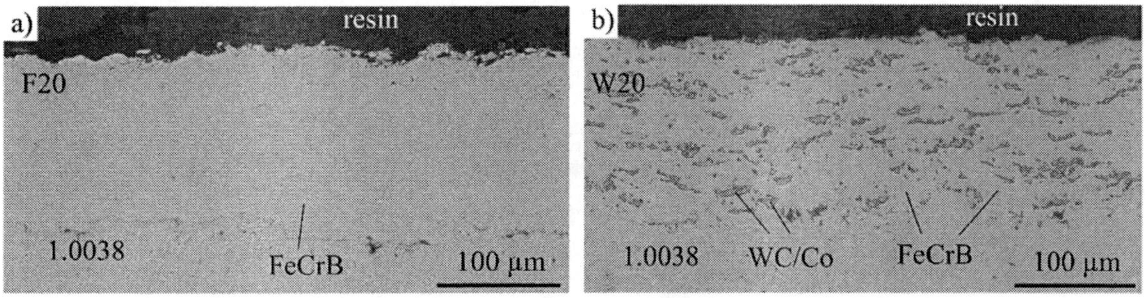

Figure 1. Cross section of the examined coatings: a) F20, b) W20.

Figure 2 shows the diffractogram of the coatings. Unwanted phases such as oxides could not be detected by XRD. As expected, the detected main peaks can be assigned to Fe and WC in the case of WC-Co containing coating. Furthermore, the presence of borides in the coating is evident. The broadening of the peaks and the reduction of the peak intensity also suggests nanocrystalline phases [16].

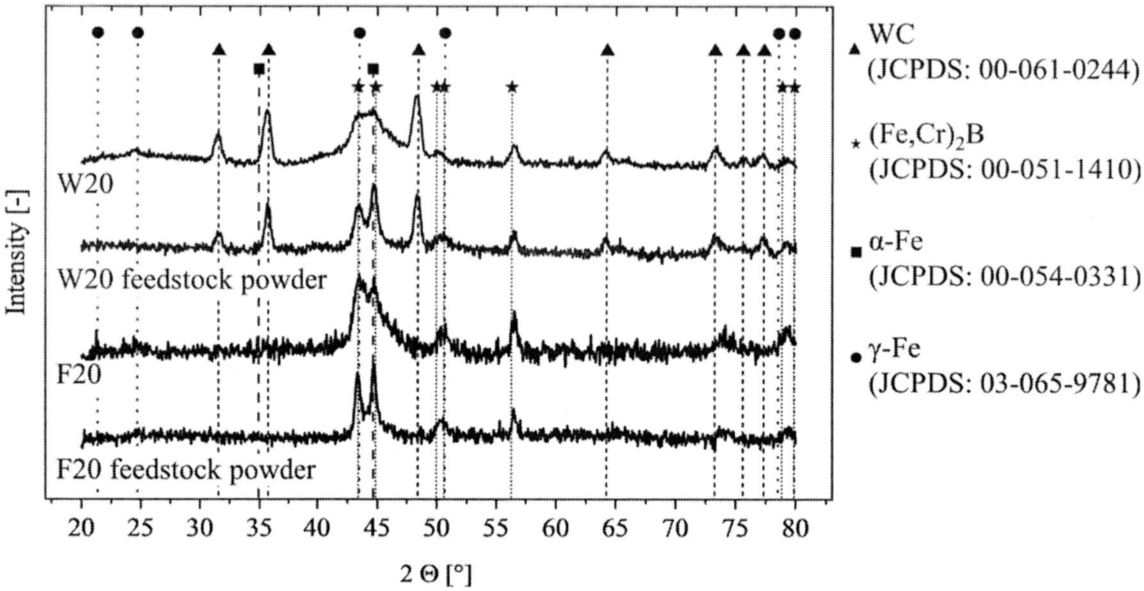

Figure 2. X-Ray diffractogram of the examined coatings. Only the dominant peaks of the phases are shown.

Figure 3 shows the results of the micro hardness measurements. Both coatings have a comparable micro hardness of around 900 HV0.1. The quartiles shown in this diagram indicate the limit where 75 % of all hardness measurements are below the 3rd quartile or above the 1st quartile. The medians are in the middle of these quartiles, showing that this measured micro hardness values have a normal distribution. The small width between the quartiles and the position of the medians show a homogeneous hardness distribution in the coating. However, a significant influence on the hardness by the integration of WC-Co in FeCrB could not be observed. At the chosen load, the hardness seems to be more influenced by the FeCr matrix than by the WC-Co inclusions. Only at individual points in the vicinity of WC-Co-particles a significantly higher hardness value can be measured in the W20 coating, which in turn is shown in the higher maximum value of the coating.

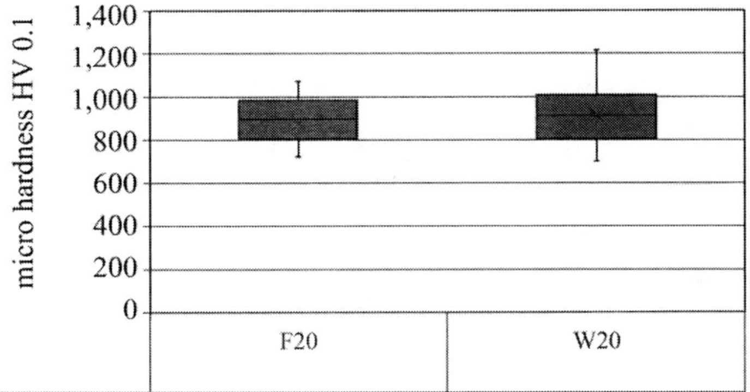

	F20	W20
Minimum	724	702
1. Quartile	807	804
Median	896	913
Mean Value	877	913
3. Quartile	985	1,008
Maximum	1,073	1,215

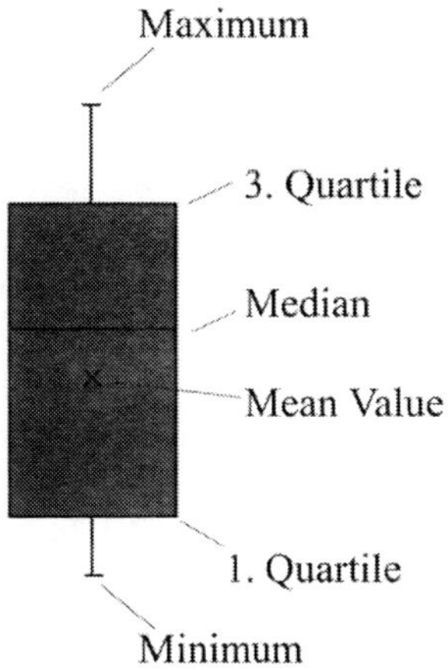

Figure 3. Vickers micro hardness HV0.1 of the examined coatings.

Figure 4 shows the residual stresses which are introduced by the coating process. Since the used sample holder could not be used with the ICP8 sensor, the velocity at which the spray gun passes over the sample was reduced to $v_{rel} = 2250$ mm/s and the mass rate was set to $\dot{m} = 100$ g/min, to achieve a comparable coating thickness per pass of 12.5 µm/pass. The remaining parameters are shown in Table 2. The coatings were also applied with a coating thickness of 150 µm. Both coatings exhibit tensile residual stresses. With regard to the small particle size and the low melting temperature of the FeCrB particles [8], a completely or partially molten state can be assumed. Due to the rapid cooling, the flattened particles remain in a strong tensile residual stress state. A comparison between the coating W20 and F20 shows that in the WC-Co-containing coating W20 significantly lower tensile residual stresses exist. The WC-Co particles impact on the sample in a solid state. The impact of the heavy WC-particles introduces compressive residual stresses that counteract the tensile residual stresses. Since the influence of the quenching stresses is greater than the compressive residual stresses induced by the peening effect, the resulting deposition stresses are on tension.

The thermally induced residual stresses resulting from the shrinkage of the substrate and the coating after the deposition process are slightly higher in the case of W20. Due to the hotter process parameters, originating from the lower volume flow of compressed air, the particles were presumably heated to higher temperatures. Consequently, the coating and its substrate reached higher temperature as well, Figure 4. However, the thermally induced residual stresses are an order of magnitude lower than the deposition residual stresses, which is why the resulting residual stresses are predominantly determined by the deposition residual stresses. It can be assumed that the crack propagation behavior and consequently the impact behavior is influenced by the residual stress state of the coating. In order to show the influence of the residual stresses on the crack propagation behavior, fracture toughness tests were performed.

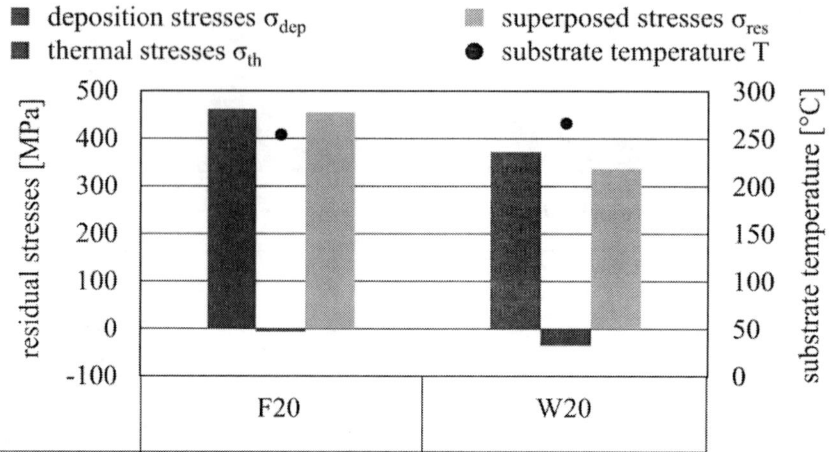

	F20	W20
σ_{dep} [MPa]	462	372
σ_{th} [MPa]	-7	-35
σ_{res} [MPa]	455	337
● T [°C]	254	266

Figure 4. Residual stress states and substrate temperatures of the examined coatings systems measured by ICP8 Sensor.

Figure 5 shows the results of the fracture toughness tests. The coating W20 shows a slightly higher fracture toughness. The higher variation of the WC-Co-containing coating can be explained by the more inhomogeneous microstructure of the coating. In accordance with the results of the hardness measurement, a higher maximum value is also shown for the fracture toughness of the coating W20. A reason for this is that the indenter penetrates less in the vicinity of WC-Co inclusions, resulting in shorter crack lengths. Another reason that has to be considered is that the higher tensile residual stresses in the F20 coating promote the crack propagation in the coating and therefore slightly lower values were measured in the coating F20.

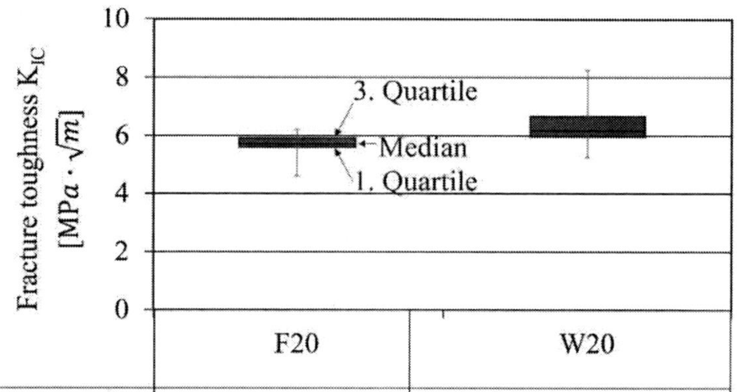

	F20	W20
Minimum [$MPa \cdot \sqrt{m}$]	4.6	5.2
1. Quartile [$MPa \cdot \sqrt{m}$]	5.6	6.0
1.Median [$MPa \cdot \sqrt{m}$]	5.7	6.2
3. Quartile [$MPa \cdot \sqrt{m}$]	5.9	6.6
Maximum [$MPa \cdot \sqrt{m}$]	6.2	8.3

Figure 5. Fracture toughness of the examined coatings.

To analyse the influence of the previously determined coating properties on the impact behaviour, the crater volume of the coatings was determined as a quantifiable variable of the impact behaviour. Figure 6 shows the crater volumes of the F20 and W20 coatings after the impact tests. It is shown that the crater volume was significantly reduced by integrating WC-Co particles in FeCrB. It can be assumed that the plastic deformability was strongly reduced by the incoporated WC-Co particles. However, it should be noted that the real volume can be a little higher due to spallings at the interface. The crater volume of the as sprayed coatings cannot be determined exactly due to the roughness peaks, but the ratio between the crater volume in polished and as sprayed conditions is comparable between F20 and W20. HVOF sprayed Cr_3C_2/NiCr coatings were already used as a reference for wear and corrosion resistance in a previous study [8]. The impact behavior of a Cr_3C_2/NiCr coating was investigated in a previous study [13]. The results show that with the same test parameters and device a comparable crater volume could be achieved with the coating W20.

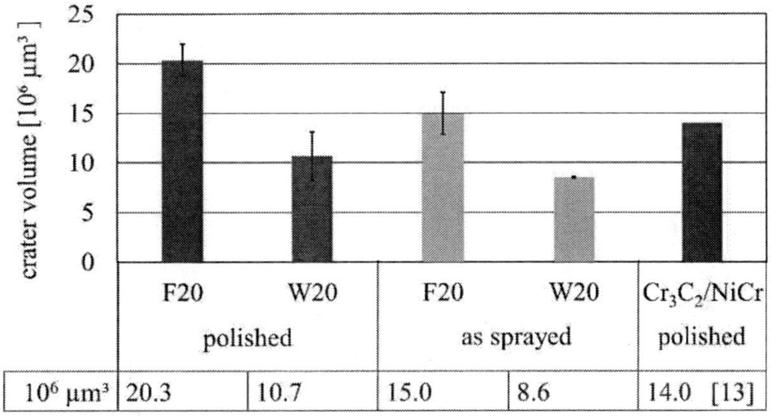

Figure 6. Crater volume of the examined coatings, the error bar indicates the standard deviation.

Figure 7 shows detailed images of the impact craters of the polished samples. In sample F20, circular cracks can be detected in the peripheral zone. These cracks are caused by high stresses at the edge of the impact and indicate a cohesive failure. While these cracks are relatively small in case of the coating F20, they are very large in case of W20. In W20 large craters can be identified, which also indicate an adhesive failure, what is also seen in the cross section of this coating after the impact test, Figure 7e. In the impact ground cracks can only be detected for W20. The corresponding cross sections show that the cracks are in the vicinity of the surface and have little depth. The coating F20 shows no cracks at the impact ground. It can be assumed that the energy on impact at F20 was reduced by plastic deformation, whereas in the case of W20 the energy led to cracking due to the lower deformability. The hypothesis is supported by the cross sections of the coatings. The stronger deformation at F20 can be recognized by the distinct curvature of the coating. Cracks can only be identified at the edges. The curvature is much less pronounced for W20 and a much stronger crack network can be identified.

Both coatings show a radial crack pattern. In case of F20 many smaller radial cracks with a maximum length of $l_c \approx 300$ µm can be identified. The coating W20 shows fewer but significantly longer radial cracks with a maximum length of $l_c \approx 950$ µm. A more detailed analysis of the cracks, Figure 7c and 7f, shows that such cracks propagate mainly along weak points like partially molten particles, pores and splat boundaries. This observation is consistent with the observations of Bolelli et al. of the crack propagation of an HVAF-sprayed FeCrNiSiBC coating in the ball-on-disk test [7]. The coating W20 shows an adhesive failure at the bottom of the impact crater. In addition, a significantly more pronounced crack network can be identified. The detail images show that cracks propagate between the WC-Co-inclusions. Here, especially the cohesion between the Co-matrix and the iron-based coating seems to be insufficient.

IOP Conf. Series: Materials Science and Engineering 1147 (2021) 012028 doi:10.1088/1757-899X/1147/1/012028

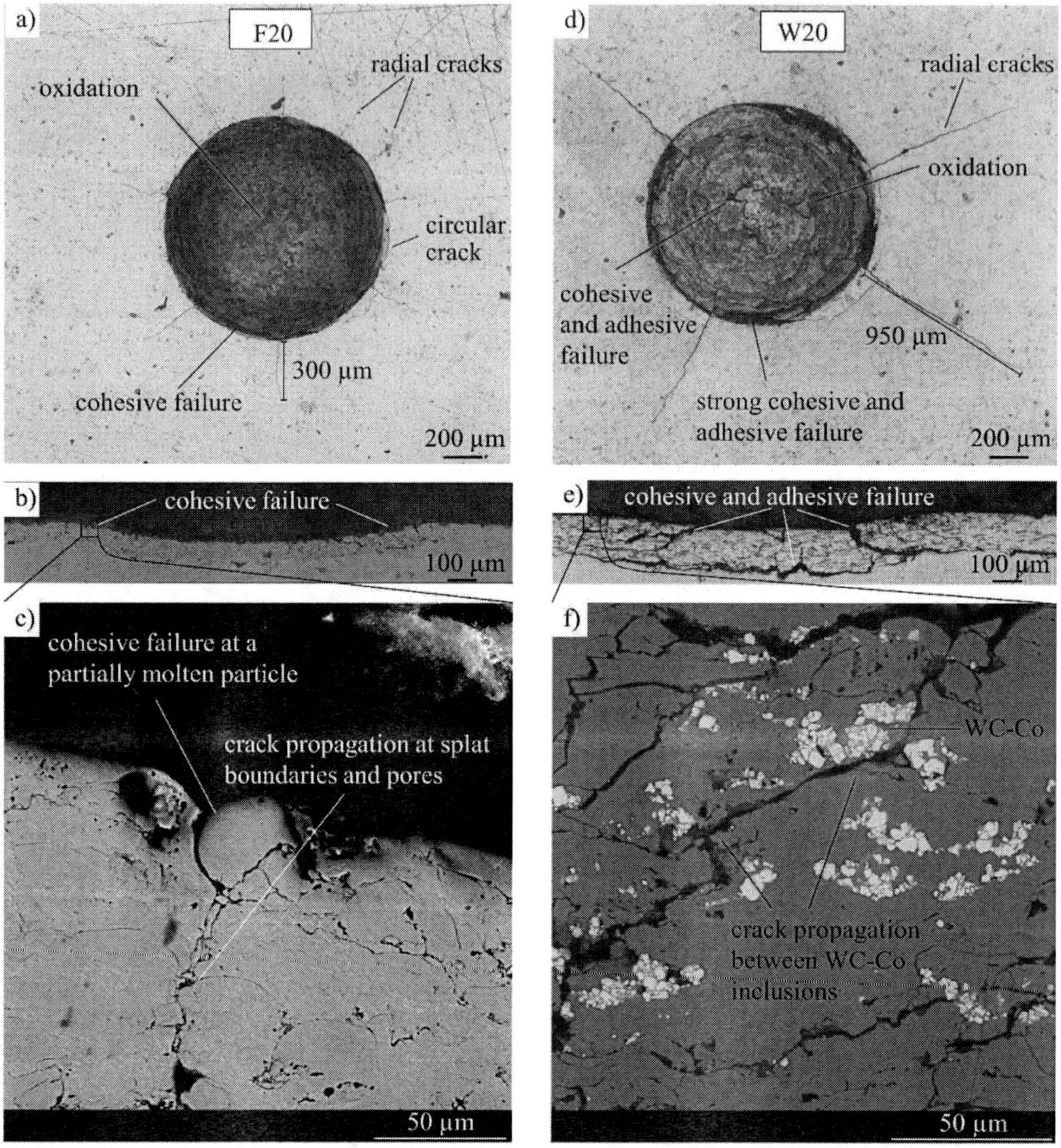

Figure 7. Wear marks of F20 (a-c) and W20 (d-f): a,d: overview of the impact crater; b,e: cross section of the impact crater; detail view of the introduced cracks.

The results of the fracture toughness test show that the indentation depth was reduced in W20 and a stronger crack propagation was most probably prevented by a reduced tensile residual stress state. However, with higher indentation forces of the impact tester, a significantly more pronounced crack network could be detected in case of W20. This is due to the fact that cracks are concentrated between the WC/Co particles. Due to the cyclical impacts, such cracks are ripped off more strongly, which results in longer cracks compared to F20. Therefore, the main influence on crack formation in the coating seems to be the distribution of the WC-Co particles and the indention force. Reduced tensile residual stresses appear to be advantageous, which is shown by an increased fracture toughness. The results show that the integration of WC-Co can significantly improve the dimensional stability under impact stresses, but that the WC-Co content must be chosen with regard to the load, to avoid cracks and delaminations.

4. Conclusions

The aim of this study was to investigate the influence of WC-Co-inclusions in FeCrB coatings. For this purpose, HVAF-sprayed FeCrB and FeCrB/WC-Co coatings were applied with a powder feed rate of $\dot{m} = 200$ g/min and a coating thickness of $s = 150$ µm. Impact tests with a load of $F = 1{,}000$ N and a number of cycles of $n = 500{,}000$ were performed. The residual stresses, fracture toughness, micro hardness, phase composition and microstructure of the coating were investigated to identify the influences on the impact behavior. The results can be summarized as follows:

- The crater volume can be reduced significantly by integrating WC-Co into the FeCrB coatings. However, the WC-Co inclusions cause the cracks to agglomerate between the WC-Co particles. The splat boundaries between the WC-Co particles and the FeCr matrix are an additional weak point in the coating where cracks can develop and propagate.

- The low tensile residual stress state of the WC-Co-containing coating W20 are the reason for the higher fracture toughness. However, the high impact loads in the impact test outweigh the fact that the coating can be deformed less plastically due to the integration of the WC-Co particles and tends to crack more easily.

- Cracks propagate primarily at weak points in the coatings, such as pores and splat boundaries of partially molten particles.

In previous studies it could be shown that the integration of WC-Co can be useful for certain wear loads [8–10]. In the case of impact loads, dimensional stability can be significantly improved, but crack propagation is also increased. To minimize crack propagation, the amount of WC-Co-particles must be chosen with regard to the load.

Acknowledgments
This work based on the DFG project BO 1979/45-1 "Ressource efficient production of novel Fe-Based coatings for large-area applications by AC-HVAF coating process". The authors gratefully acknowledge the financial support of the German Research Foundation (DFG).

References

[1] Fauchais P and Vardelle A 2012 Thermal sprayed coatings used against corrosion and corrosive wear *dvanced plasma spray applications* 3–38
[2] Espallargas N 2015 *Future development of thermal spray coatings: Types, designs, manufacture and applications (Woodhead Publishing series in metals and surface engineering* number 65*)* (Cambridge, UK: Woodhead Publishing is an imprint of Elsevier)
[3] Schwenk A HVOF-K2-Spritzen von Eisenbasislegierungen für Anwendungen in der Druckindustrie *Proceedings of 8.HVOF-Kolloquium* **2009**
[4] Osmanda A, Battenbough A and Staines A 2012 Fe-Cr and Fe-Cr-Ni based HVOF Coating Systems Customised for Wear and Corrosion Protection *Protection Thermal Spray Bulletin* **64**
[5] Terajima T, Takeuchi F, Nakata K, Adachi S, Nakashima K and Igarashi T 2010 Composite coating containing WC/12Co cermet and Fe-based metallic glass deposited by high-velocity oxygen fuel spraying *Journal of Alloys and Compounds* **504** S288-S291
[6] Bolelli G, Börner T, Bozza F, Cannillo V, Cirillo G and Lusvarghi L Cermet coatings with Fe-based matrix as alternative to WC–CoCr: Mechanical and tribological behaviours *Surface and Coatings Technology* **206** 4079–94

[7] Bolelli G, Börner T, Milanti A, Lusvarghi L, Laurila J, Koivuluoto H, Niemi K and Vuoristo P 2014 Tribological behavior of HVOF- and HVAF-sprayed composite coatings based on Fe-Alloy+WC–12% Co *Surface and Coatings Technology* **248** 104–12
[8] Bobzin K, Öte M, Knoch M A and Sommer J 2018 Development of HVAF-sprayed novel Fe-based coatings for large area applications *Thermal Spray Bulletin* **70** 38–45
[9] Bobzin K, Öte M, Knoch M A and Sommer J Influence of powder feed rate on corrosion and wear properties of Fe-based HVAF coatings *IOP Conf. Ser.: Mater. Sci. Eng.* **373** 12012
[10] Bobzin K, Öte M, Knoch M A and Sommer J 2019 Influence of Powder Size on the Corrosion and Wear Behavior of HVAF-Sprayed Fe-Based Coatings *J Therm Spray Tech* **28** 63–75
[11] Knotek O, Bosserhoff B, Schrey A, Leyendecker T, Lemmer O and Esser S 1992 A new technique for testing the impact load of thin films: the coating impact test *Surface and Coatings Technology* **54-55** 102–7
[12] Bouzakis K-D, Asimakopoulos A, Michailidis N, Kompogiannis S, Maliaris G, Giannopoulos G, Pavlidou E and Erkens G 2004 The inclined impact test, an efficient method to characterize coatings' cohesion and adhesion properties *Thin Solid Films* **469-470** 254–62
[13] Bobzin K, Zhao L, Öte M, Königstein T and Steeger M 2018 Impact wear of an HVOF-sprayed Cr 3 C 2 -NiCr coating *International Journal of Refractory Metals and Hard Materials* **70** 191–6
[14] Wachtman J B 1996 *Mechanical properties of ceramics (A Wiley-Interscience publication)* (New York: Wiley)
[15] Nelson D 1982 *Effects of Residual Stress on Fatigue Crack Propagation // Residual stress effects in fatigue: A symposium (ASTM special technical publication* vol 776*)* (Philadelphia, Pa. (1916 Race St., Philadelphia 19103): ASTM International; American Society for Testing and Materials)
[16] Spieß L, Teichert G, Schwarzer R, Behnken H and Genzel C 2009 *Moderne Röntgenbeugung* (Wiesbaden: Vieweg+Teubner)

High temperature treatment effects on the microstructure and properties of a plasma sprayed 25 mol% Al_2O_3- 25 mol% Cr_2O_3- 50 mol% TiO_2 coating

M Grimm*[1], S Conze[2], L-M Berger[2], R Drehmann[1] and T Lampke[1]

[1] Materials and Surface Engineering Group, Institute of Materials Science and Engineering, Chemnitz University of Technology, D-09107 Chemnitz, Germany
[2] Fraunhofer IKTS, Fraunhofer Institute for Ceramic Technologies and Systems, D-01277 Dresden, Germany

*e-mail: maximilian.grimm@mb.tu-chemnitz.de

Abstract. In this study, the influence of heat treatments up to 1200°C in vacuum on coatings prepared by atmospheric plasma spraying (APS) from a blend of 25 mol% Al_2O_3, 25 mol% Cr_2O_3 and 50 mol% TiO_2 on microstructure, porosity, hardness and sliding wear resistance was investigated. The well-known transformation from α-Al_2O_3 to γ-Al_2O_3 occurred as a result of the spray process, as well as small amounts of titanium were found in Cr_2O_3 splats due to an interaction between the feedstock particles. No significant changes were found after a heat treatment at 400°C or 800°C. Heat treatment at 1200 °C led to complete coating delamination. Besides the retransformation of γ-Al_2O_3 to α-Al_2O_3, interactions between Cr_2O_3 and TiO_2 splats and the formation of $CrTi_2O_5$ was observed. The formation of $Al_2Ti_7O_{15}$ was observed in result of a reaction at the outer regions of the Al_2O_3 splats. At the same time, a reduction in porosity and healing of microcracks were observed as a result of sintering processes, which leads to an increase of the coating hardness to 1170 HV0.2.

1. Introduction

Coatings sprayed from the single oxides of the Al_2O_3-Cr_2O_3-TiO_2 system and some commercially available binary compositions show multifunctional properties and are widely used for many technical applications. Ternary compositions are promising for further improvement of their performance and the extension of the application range [1-3]. As shown in previous studies, such coatings can effectively be produced by atmospheric plasma spraying (APS) using ternary blends of the three oxides. When spraying these powder blends, the following main observations were made:

Due to the specific behavior of the components (differences in the difficulty of melting factor (DMF) [4] and the thermal diffusivity), the spraying process can act selectively [2]. Thus, considerable differences between the chemical composition of the feedstock powder and the coating can occur. However, this can be largely avoided by the spray parameter selection and a suitable process control [2]. The coating properties are influenced by both the elemental distribution and phase composition. Different interactions between the different particles were observed despite the very short process times. In all previous studies, small amounts of Ti were detected in Cr_2O_3 lamellae by energy dispersive X-ray spectroscopy (EDS) point measurements [1-3]. Results for Al_2O_3-based ternary coatings revealed that Cr and Ti atoms can be dissolved in γ-Al_2O_3 affecting the properties [3].

Ceramic coatings are often applied at high temperatures [5,6]. For this reason, the effects of high temperature treatments on the microstructure and the resulting properties of the coating are of interest. Since as-sprayed coatings are in a thermodynamic non-equilibrium state many different effects can be expected in result of a thermal post-treatment or in high temperature service conditions.

The disadvantageous phase transformation from α-Al_2O_3 in the feedstock material to γ-Al_2O_3 in the coating, occurring due to the high cooling rates and preferential formation of γ-Al_2O_3 nuclei above a certain degree of undercooling [7], is typical for thermal spraying. The reverse transformation is described to be complete at about 1180 °C (with sufficient heat-treatment time) [8-11]. Such phase transformations in plasma sprayed coatings have also been reported for $(Al,Cr)_2O_3$ solid solutions [12]. A study on Al_2O_3-based coatings produced by water-stabilized plasma spraying shows that typical additives such as TiO_2 and Cr_2O_3 influence the transformation temperature of α-Al_2O_3 in the heat treatment [13]. The reverse transformation is associated with the formation of cracks in the coatings due to the density differences between γ-Al_2O_3 and α-Al_2O_3 [9-11].

In result of the spray process, oxygen-deficient TiO_x feedstock powders can be partially oxidized to stoichiometric TiO_2 in the coating [2]. It can be assumed that a heat treatment can have a further effect depending on the atmosphere. It is well known that in atmospheric conditions above 1000 °C Cr_2O_3 tends to oxidize to hazardous volatile Cr(VI)-oxide [14]. For plasma sprayed coatings of the binary system Cr_2O_3-TiO_2, significant changes in the microstructure and the formation of the E-phase ($Cr_2Ti_2O_7$) were observed as a result of high temperature sliding wear tests at 800 °C [15].

As a result of a heat treatment of plasma sprayed coatings, a densification of the coating due to sintering processes is often described [12,15].

In this study, APS coatings from a powder blend consisting of 50 mol-% titania and 25 mol-% each of alumina and chromia were prepared. Heat treatments of these coatings were carried out in order to investigate their effects on the microstructure, hardness and sliding wear resistance of the coatings.

2. Experimental and materials

The feedstock composition was obtained by blending fused and crushed Al_2O_3, Cr_2O_3 and TiO_2 powders, which are described in Table 1 and were already characterized in previous studies [1,3]. The blend consisted of 50 mol% TiO_2 and 25 mol% each of Al_2O_3 and Cr_2O_3, corresponding to 24.6 wt-% Al_2O_3, 36.7 wt-% Cr_2O_3, 38.6 wt-% TiO_2 or 12.5 at-% each of Al, Cr, Ti and 62.5 at-% oxygen.

Table 1. General information on the feedstock powders, particle size according to the supplier information, granulometric data from laser light diffraction analysis.

			Granulometric data in µm		
Material	Supplier	Particle Size	d10	d50	d90
Al_2O_3	Saint Gobain Coating Solutions, Avignon France	-45 + 15 µm	21	34	56
Cr_2O_3	GTV, Luckenbach, Germany	-45 + 15 µm	19	34	54
TiO_2	Ceram Ingenieurkeramik, Albbruck-Birndorf, Germany	-45 + 20 µm	22	39	61

The substrates made of low-carbon steel (S235JR, 1.0038) were grit blasted (3 bar, 20 mm distance, 70° angle) with alumina (EK-F 24) and cleaned in an ultrasonic ethanol bath before depositing the coating. The APS process was performed with an F6 torch (GTV, Luckenbach, Germany) mounted on a six-axis robot using a stationary fixture for the samples with parameters given in Table 2. The parameter set selected resulted to dense coatings with a thickness of ~100 µm and a chemical composition similar to that of the feedstock powder.

Table 2. Parameters of APS process.

Current	Plasma Gas Flow Rate Argon	Hydrogen	Spraying Distance	Traverse Speed	No. of Passes	Powder Feed Rate
600 A	41 l/min	11 l /min	110 mm	0.4 m/s	4	25 g/min

The heat treatment of the coated samples was carried out at 400, 800 and 1200 °C in a vacuum furnace at a pressure of 10^{-4} bar. Vacuum was selected to protect the substrate from oxidation and Cr_2O_3 from oxidation to volatile CrO_3. Low heating and cooling rates of 200 K/h were applied. The holding time at the treatment temperatures was 6 hours.

The preparation of coating cross sections was made according to the standard metallographic procedure. In order to investigate the microstructure and local compositions, the coatings were studied by scanning electron microscopy (SEM) (Leo 1455VP, Zeiss, Oberkochen, Germany) equipped with an EDS detector (GENESIS, EDAX, Mahwah, NJ, USA) using an acceleration voltage of 25 kV. To determine the porosity, five SEM images taken at a magnification of 500x, were evaluated for each coating using an image analysis method (Olympus, Shinjuku, Japan). By using the backscattered electron detector (BSD), the distribution and homogeneity of the elements were visualized by different gray levels. The local chemical composition of individual splats was studied for each coating by five EDS point measurements. The measuring points were located at positions where an influence of surrounding areas on the measuring process could be largely excluded.

In order to investigate the changes in microstructure that occur during the spray process and as a result of the heat treatment, the phase compositions of the powder blend, the as-sprayed coating and the coating heat-treated at 1200°C were studied by X-ray diffraction (XRD) (D8 Discover diffractometer, Bruker AXS, Billerica, MA, USA) using Co Kα radiation with a tube voltage of 40 kV and a tube current of 40 mA. The diffraction patterns were measured for a 2θ range from 10° to 130°, with a step size of 0.01° and a dwell time of 1.5 s/step.

The Vickers microhardness of the coatings was measured by 10 indentations with a test load of 2.94N on the cross sections using a Wilson Tukon 1102 device (Buehler, Uzwil, Switzerland). For measurement of the dry, unidirectional sliding wear resistance, coatings which were ground ($R_a < 1$ μm) and tested at room temperature with a ball-on-disk tribometer (Tetra, Ilmenau, Germany). For the as-sprayed coating and the coatings heat-treated at 400 °C and 800 °C, three runs were carried out on the rotating, coated sample according to the test parameters given in Table 3. The volume of the resulting wear tracks was determined using an optical 3D profilometer MikroCAS (LMI, Teltow, Germany).

Table 3. Sliding wear test parameters.

Force	Radius	Speed	Cycles	Wear Distance	Counter Body Material	Diameter
10 N	5 mm	0.05 m/s	3183	100	Al_2O_3	6 mm

3. Results and discussion

Despite the low coating thickness and the low heating and cooling rates selected, at 1200 °C the coating delaminated from the substrate. However, the coating itself remained not damaged. Thus, the characterization was carried out with the exception of the sliding wear test.

Figure 1 displays the SEM images of the coatings. Some specific features of thermal spray coatings, such as the splat-wise build up, pores and some microcracks as well as some not or only partially melted particles can be observed. The three components of the coating are clearly identifiable in the images due to their different gray levels in the BSE contrast. The light gray areas correspond to Cr_2O_3, the dark gray areas belong to Al_2O_3 and the medium gray areas consist of TiO_x or TiO_2. Due to the high DMF [2,4] and low thermal diffusivity non-melted particles are mostly Al_2O_3. Non-melted Al_2O_3 particles as well as Al_2O_3 and Cr_2O_3 splats are predominantly in direct contact with titania splats (Fig. 1a). The as-sprayed coating and coatings heat-treated at 400°C and 800°C show very similar microstructures (Fig. 1a-c). The coating heat-treated at 1200°C shows a significantly lower amount of microcracks (Fig. 1d). Original porosity decreased due to sintering, but new porosity is formed in result of solid state diffusion processes. In the coating heat-treated at 1200°C, the contrast between Cr_2O_3 and TiO_2 splats is significantly lower. Overall, the proportion of light gray areas associated with Cr_2O_3 appears to decrease in favor of gray TiO_x areas. In addition, the outer areas of Al_2O_3 splats started to react. The former interface of the Al_2O_3 lamellae with the surrounding material shows a different grayscale. These observations indicate that more intensive diffusion processes take place at 1200 °C, significantly changing the microstructure and phase composition of the coating.

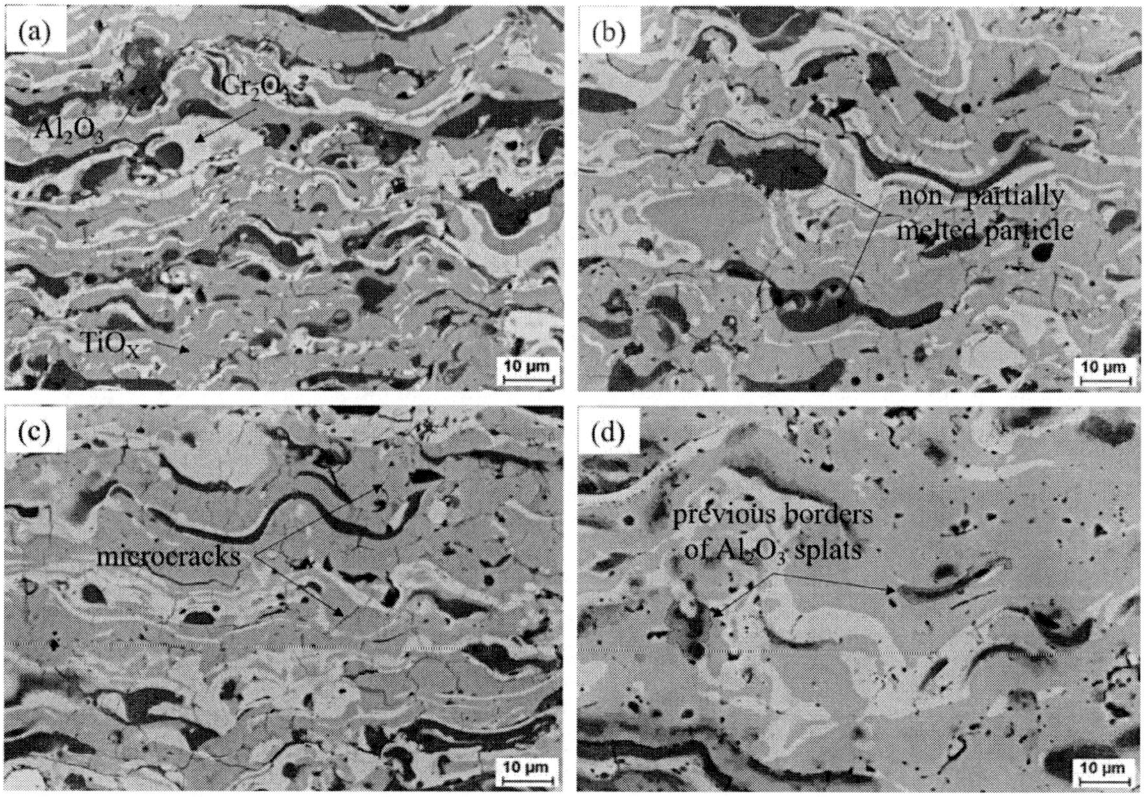

Figure 1. SEM image of (a) as-sprayed coating and coatings heat-treated at (b) 400°C, (c) 800°C and (d) 1200°C.

The local chemical composition determined by EDS point measurements of the different areas in the SEM images are shown in Figure 2. Neither Ti nor Cr were detected in the dark areas corresponding to Al_2O_3. As also shown in previous studies [1-3], low amounts of Ti (1-2 at.-%) can be reliably detected in the Cr_2O_3 splats as a result of the spray process. The heat treatment did not affect the composition of both Cr_2O_3 and Al_2O_3 areas, regardless of the annealing temperature. Interestingly, changes in the chemical composition of the medium gray TiO_x/TiO_2 areas were found. After heat treatment at 1200°C, 5-12 at.-%. Cr and traces of Al were found in these areas. The diffusion directions during heat treatment seemed to be the opposite compared to the spray process. As a result of the heat treatment at 1200°C Cr_2O_3 lamellae disappear due to interaction with the surrounding TiO_x areas.

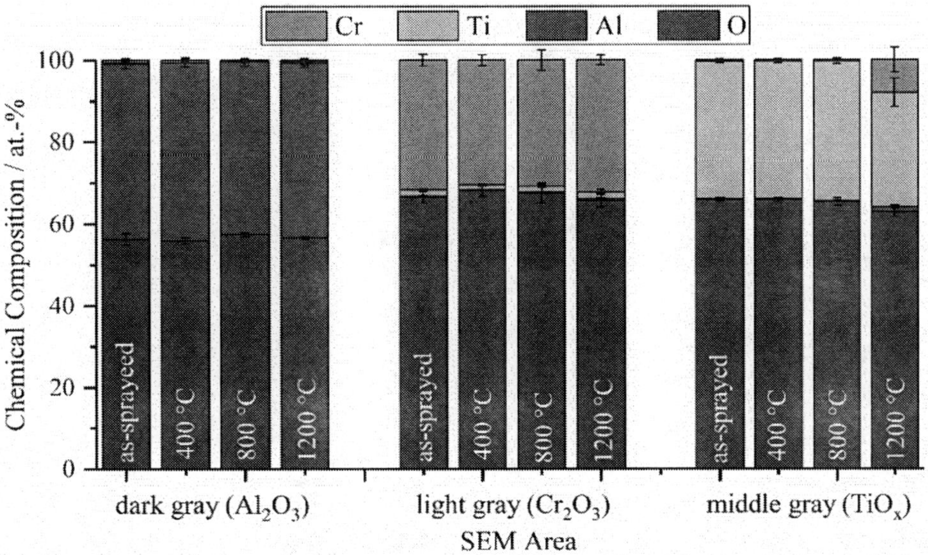

Figure 2. Chemical composition of splats with different grayscale (Fig.1) determined by EDS point measurements.

The XRD patterns shown in Figure 3 clearly reveal a change of the phase composition of the coating compared to the feedstock powder. The powder blend consisted mainly of α-Al_2O_3, Cr_2O_3 and a variety of TiO_x phases. Gravimetric measurements in another study have indicated that x in this TiO_x powder is about 1.9 [16]. The XRD pattern of the as-sprayed coating indicates the expected occurrence of the typical phase transformation from α-Al_2O_3 to γ-Al_2O_3. This is connected with the appearance of new peaks at $2\theta = 53.8°$ and $2\theta = 79.6°$ associated with γ-Al_2O_3 and the decrease of the peak intensities of α-Al_2O_3. A further change in the phase composition is noticeable by analyzing the TiO_2 or TiO_x peaks. In the as-sprayed coating, additional peaks associated with the rutile structure of TiO_2 at $2\theta = 48.3°$ and $2\theta = 64.0°$, a considerable broadening of the left shoulder of Cr_2O_3 at $2\theta = 42.2°$ as well as the simultaneous loss of TiO_x peak intensities are observed. This shows that oxidation of the non-stoichiometric titanium oxide phases during the APS process leads to the formation of stoichiometric rutile. The evaluation of the XRD pattern of the coating heat-treated at 1200°C is challenging, as a result of the intensive interaction between the components. Due to the formation of solid solutions, some peaks are shifted compared to their standard position. In addition, the local composition is sometimes varying. However, it is noticeable that the intensity of the α-Al_2O_3-related peaks increases significantly. This indicates a retransformation of γ-Al_2O_3 into α-Al_2O_3 above 800 °C, as already described in other studies [5-8]. The decrease of the intensity of Cr_2O_3 and TiO_2 is connected with the formation of new phases. The additional peaks are identified as $Al_2Ti_7O_{15}$ and $CrTi_2O_5$ and described in the literature [17,18]. It is assumed that vacuum conditions during heat treatment, support the formation for these phases. Both phases contain Ti^{3+} and Ti^{4+} atoms. Accordingly, they are sensitive to oxidation at higher temperatures in atmospheric conditions. Taking into account the results of the EDS point measurements, it can be assumed that Cr_2O_3 lamellae react with the surrounding $TiO_{2/x}$ splats and form $CrTi_2O_5$. Furthermore, it is suspected that $Al_2Ti_7O_{15}$ is formed at the outer regions of the Al_2O_3 splats. Moreover, the transformation from α-Al_2O_3 to γ-Al_2O_3 can increase the reactivity in solid state reactions above 800 °C.

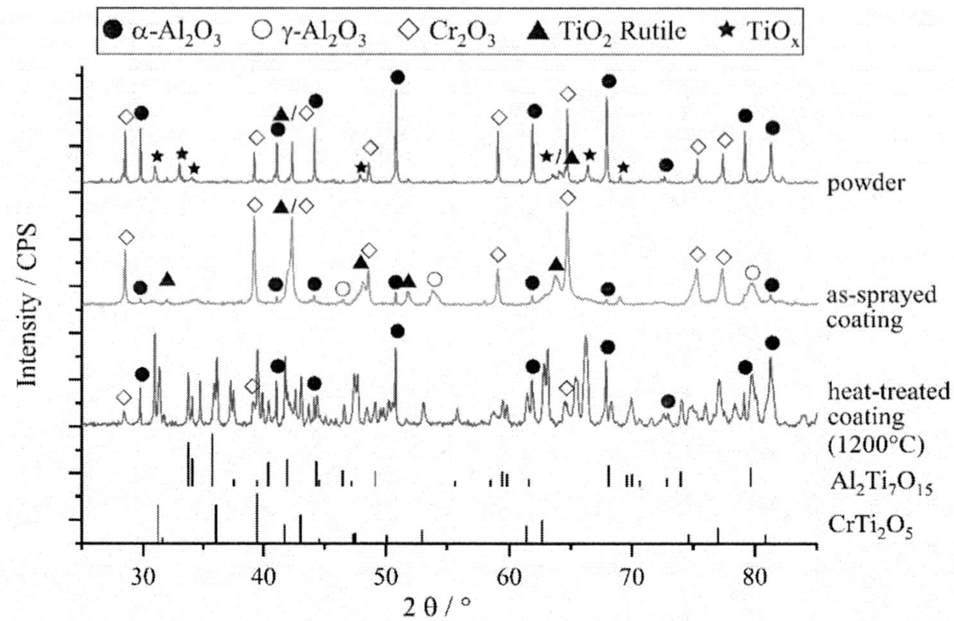

Figure 3. XRD patterns of the powder blend, the as-sprayed and the coating after heat-treatment at 1200 °C.

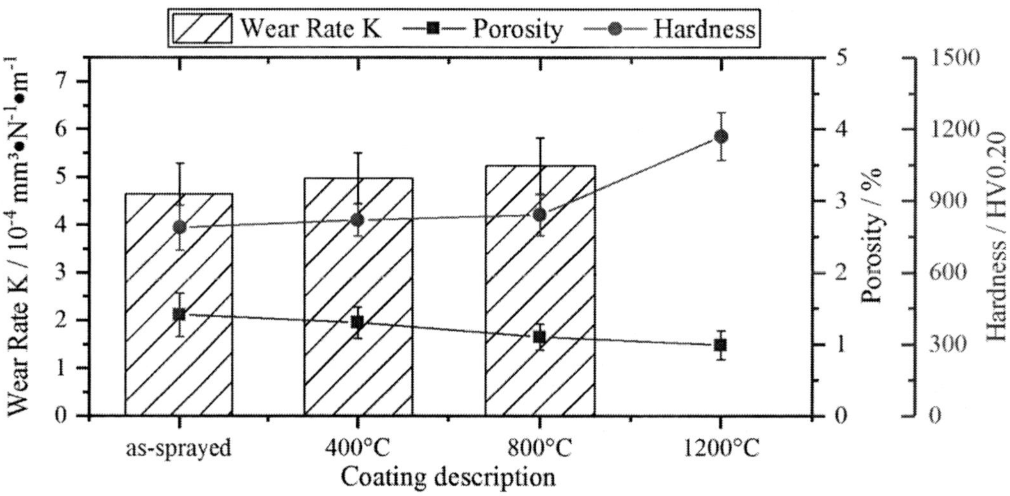

Figure 4. Ball-on-disc wear rate, porosity and hardness for the as-sprayed and heat-treated coatings.

Figure 4 shows the results of the porosity, hardness and sliding wear resistance measurements. No significant changes in the coating properties are found due to a heat treatment up to 800°C. This in good agreement with the corresponding microstructures of the coatings. Indeed, slightly reduced porosities are measured, which also lead to a slight increase in hardness and can be presumably associated with sintering processes. Due to the increased number of microcracks which appear as a result of additional stresses caused by the heat treatment, decrease of porosity does not improve the sliding wear resistance. In summary, porosity, hardness as well as the coating wear rate K after heat treatment at 400°C and 800°C are at a similar level compared to the as-sprayed coating. These changes are less significant than found for plasma sprayed coatings of the binary system Cr_2O_3-TiO_2, as a result of high temperature sliding wear tests at 800 °C [15]. The delaminated coating heat-treated at 1200 °C has the lowest porosity of 0.99 ± 0.2 % probably due to healing of microcracks. Due to the extensive microstructural changes, this coating has a hardness of 1170 (±101) HV0.2, which is approximately 50% higher than for the as-sprayed coating. Further studies are required to determine the reason for the hardness increase. It has to be clarified, if this can be attributed entirely to the healing of microcracks and phase transformation from γ- to α-Al_2O_3 or is also influenced by the newly formed $CrTi_2O_5$ and $Al_2Ti_7O_{15}$ phases. The influence of the vacuum on phase formation has to be studied as well.

4. Conclusions

The effects of heat treatments up to 1200°C in vacuum on the microstructure and properties of APS coatings using a powder blend consisting of 50 mol-% TiO_2 and 25 mol-% each of Al_2O_3 and Cr_2O_3 were investigated. The following conclusions can be drawn:

- The microstructural analyses and investigations of hardness and sliding wear resistance reveal that significant changes occur after heat treatment at 1200°C.
- As already described in other studies, a reverse transformation of γ-Al_2O_3 to α-Al_2O_3 takes place between 800°C and 1200°C.
- As a result of the thermal spray process, small amounts of Ti were detected in Cr_2O_3 splats by EDS point measurements. After heat treatment at 1200°C, the amount of Ti in remaining Cr_2O_3 splats remains unchanged.
- SEM images, EDS point measurements and XRD patterns suggest that heat treatment of the coating at 1200°C causes Cr_2O_3 lamellae to dissolve by reacting with surrounding TiO_2 splats to form $CrTi_2O_5$ in vacuum.
- In addition, new peaks probably associated with $Al_2Ti_7O_{15}$ are also present in the XRD pattern of the coating heat-treated at 1200°C in vacuum. The evaluation of the SEM studies suggests that this phase is located at the outer areas of the Al_2O_3 splats.
- The reactions between the individual splats above 800 °C and the possible influence of the furnace atmosphere on the formation of new phases ($Al_2Ti_7O_{15}$, $CrTi_2O_5$) needs further investigation.
- Through sintering processes and the mentioned phase formation in the coating heat-treated at 1200 °C, the number of microcracks is considerably reduced and the porosity of the coating is decreased.
- These microstructural changes lead to a significant increase in the hardness of the coating, but also to coating delamination. For a successful heat treatment of the coating, the bonding of the coating must be improved by selection of a suitable substrate material and application of a bond coat.

Acknowledgments
The authors would like to thank the Sächsische Aufbaubank for funding the project under contracts 100310631 / 100310633 by the European Structural Fonds EFRE and by the Free State of Saxony.

References

[1] Grimm M, Conze S, Berger L-M, Paczkowski G, Lindner T and Lampke T 2020 Microstructure and Sliding Wear Resistance of Plasma Sprayed Al_2O_3-Cr_2O_3-TiO_2 Ternary Coatings from Blends of Single Oxides *Coatings* **10**(1) 42
[2] Grimm M, Conze S, Berger L-M, Paczkowski G, Drehmann R and Lampke T 2021 Changes in the coating composition due to APS process conditions for Al_2O_3-Cr_2O_3-TiO_2 ternary powder blends *J. Therm. Spray Technol.* **30** 168-180
[3] Conze S, Grimm M, Berger L-M, Thiele S, Drehmann R and Lampke T 2021 Influence of Simultaneous Cr_2O_3 and TiO_2 Additions on the Microstructure and Properties of APS Alumina Coatings *Surf. Coat. Tech.* **405** 126702

[4] McPherson R 1981 The Relationship Between the Mechanism of Formation, Microstructure and Properties of Plasma-Sprayed Coatings *Thin Solid Films* **83** 297-310
[5] Fauchais P L, Heberlein J V R, and Boulos M I 2014 *Thermal Spray Fundamentals: from Powder to Part* (New York: Springer)
[6] Pawlowski L 2008 *The Science and Engineering of Thermal Spray Coatings* 2nd ed. (Hoboken: Wiley)
[7] McPherson R 1973 Formation of Metastable Phases in Flame- and Plasma-Prepared Alumina *Journal of Materials Science* **8** (6) 851–858
[8] Dhakar B, Namdeo A, Chatterjee S and Sabiruddin K 2018 Heat treatment of plasma sprayed alumina-chromia composite coatings Surface Engineering **34**:10 737-746
[9] Heintze G N and Uematsu 1992 Preparation and structures of plasma sprayed γ- and α-Al_2O_3 coatings *Surf. Coat. Tech.* **50** 213-22
[10] Thompson V S. and Whittemore O J 1968 Structural Changes on Reheating Plasma-Sprayed Alumina *Am. Ceram. Soc. Bull.* **47** 637-641
[11] Chráska P, Dubsky J Neufuss K and Pisacka J 1997 Alumina-base plasma-sprayed materials part I: Phase stability of alumina and alumina-chromia *J Therm Spray Technol.* **6** 321-326
[12] Marple B R, Voyer J and Béchard P 2001 Sol infiltration and heat treatment of alumina-chromia plasma-sprayed coatings *J. Eur. Ceram* **21** 861-868
[13] Ilavsky J, Berndt C C, Herman H, Chraska P and Dubsky J 1997 Alumina-Base Plasma-Sprayed Materials – Part II: Phase Transformations in Aluminas *J. Therm. Spray Technol.* **6** 439-444
[14] Kiilakoski J, Trache R, Björklund S, Joshi S and Vuoristo P 2019 Process Parameter Impact on Suspension-HVOF-Sprayed Cr_2O_3 Coatings *J. Therm. Spray. Technol.* **28**(8) 1933-1944
[15] Berger L-M, Woydt M, Saaro S, Stahr C C and Thiele S 2010 *High-Temperature Sliding Wear of APS-Sprayed Coatings of the Cr_2O_3-TiO_2 System* Proc.14th Nordic Symposium on Tribology, Nordtrib, Luleå and Storforsen, Sweden, June 8-11, pp 9
[16] Richter A, Berger L-M, Sohn Y J, Conze S, Sempf K, and Vaßen R 2019 Impact of Al_2O_3-40wt% TiO2 Feedstock Powder Characteristics on the Sprayability, Microstructure and Mechanical Properties of Plasma Sprayed Coatings *J. Eur. Ceram. Soc.* **39**(16) 5391-5402
[17] Remy F, Monnereau O and Casalot A 1987 Titane à valence mixte, un nouvel oxide ternaire $Al_2Ti_7O_{15}$ *J. Solid State Chem.* **76** 167-177
[18] Müller-Buschbaum H and Bluhm K 1988 Weitere magnetische Untersuchungen an $Ti_{3-x}M_xO_5$ Phasen (M=Al^{3+}, Fe^{2+}, Mn^{2+}, Mg^{2+}) mit einem Beitrag über $CrTi_2O_5$ *Z. anorg. Allg. Chem.* **588** 28-34

Influence of the wire diameter and particle fraction of Fe-Al_2O_3/ZrO_2 flux-cored wires processed by wire arc spraying on the abrasive wear properties of the coating

K Gustke*, R Drehmann and T Lampke

Chemnitz University of Technology, Chair of Materials and Surface Engineering

*e-mail: kevin.gustke@mb.tu-chemnitz.de

Abstract. The embedding of ceramic particles of a certain size and concentration in a ductile matrix is a strategy for producing wear-resistant coatings. In wire arc spraying, flux-cored wires with a metallic sheath and particle-filled core can be used for this purpose. The influence on wear properties of wire diameter and particle fraction has hardly been investigated so far. Due to the high kinetics of arc and gas flow, there are considerable challenges with regard to reproducibility, homogeneity and material efficiency. As a basis for a significantly improved understanding of the process, the relationships between the wire diameter, the particle fraction used and the wear resistance of the produced coatings have been investigated. The objects of investigation were flux-cored wires with an iron sheath and a particle filling of Al_2O_3/ZrO_2. By using the ball-on-disc test and rubber wheel test, the wear resistance of the respective coatings was investigated.

1. Introduction

Thermally sprayed wear protection coatings are used in a variety of machine elements to increase the resistance to severe abrasive wear. The main objective is the use of cost-effective base materials enhanced by appropriate surface functionalization. To increase wear resistance, hard material particles in the desired size and concentration can be embedded in a ductile matrix [1, 2]. The matrix material consists mostly of an Fe, Ni or Co alloy. The reinforcing particles usually are metal carbides like WC or TiC [3]. Dallaire and Levert investigated the influence of TiB_2 particles with different volume ratios in wear-resistant austenitic stainless steel coatings produced by arc spraying of flux-cored wires. The flux-cored wires consisted of a sheath of 304 stainless steel, TiB_2 particles, 316L particles and inorganic additives. The wear resistance was investigated using the dry rubber wheel abrasion test. The coatings with a volume ratio of 18 vol.% TiB_2 showed the lowest wear volume loss, which was about 6.5 times lower compared to an austenitic stainless steel coating [1]. Fang et al. studied the coating microstructure and wear properties of TiB_2 reinforced stainless steel 304L and nickel-chromium alloy coatings produced by arc spraying of flux-cored wires with different particle fractions. Dry sliding wear tests were performed using a block-on-ring tribometer at room temperature according to the GB 12444.2-90 standard. The coatings with the finer particle fraction (1 - 10 µm) exhibited clearly lower volume loss compared to the coatings with the coarser particle fraction (10 - 100 µm). Due to the smaller volume, the finer particles are more likely to melt and distribute themselves more homogeneously in the coating [4]. As a result of the high and volatile prices as well as the restricted availability, there are strong efforts from both political and economic perspectives to develop suitable substitute materials for conventional reinforcement particles such as WC, Cr_2C_3 und TiC [5]. An appropriate cost-effective alternative is the use of iron-based matrices with oxide hard particles. Metallic oxides such as alumina

(Al_2O_3) and zirconia (ZrO_2) provide high hardness up to 2100 HV and their compositions (subsequently referred as Al_2O_3/ZrO_2) provide high fracture toughness of up to 7.8 $MPa \cdot m^{1/2}$ [6] exceeding even those of fused tungsten carbides [5, 7–10]. Hwang et al. investigated the influence of Al_2O_3/ZrO_2 (60 wt.% Al_2O_3, 40 wt.% ZrO_2) particles with different volume ratios (10 - 30 wt.%) in wear resistant stainless steel coatings produced by plasma spraying. The wear resistance was examined using the pin-on-disc test. The coating with a 20 wt.% of Al_2O_3/ZrO_2 particles showed the lowest volume loss under various load conditions of only 3 - 5 % compared to the coating without particle reinforcement. The dominant wear characteristics were abrasive wear in the form of scratches and particle breakouts. The breakouts were initiated by particle/matrix interface cracks. The crack initiation is favoured by a high hardness gradient between the matrix and the particles [11]. Another critical aspect is the poor wettability of oxide ceramics like Al_2O_3/ZrO_2 with iron-based melts due to the low chemical interaction. This leads to insufficient bonding of the ceramic particles in the metallic matrix. Brust et al. improved the bonding between the particles of Al_2O_3/ZrO_2 and the metallic matrix by a CVD coating of the particle surface with TiN [2, 5]. The use of the wire arc spraying process in combination with flux-cored wires represents an attractive technique for the production of particle-reinforced metallic coatings [1, 12, 13]. The advantages of the process are the very low process costs, as only an atomizing gas such as nitrogen or compressed air is required [4, 14]. In addition, the process technology and process control enable an automated application [14]. When using flux-cored wires, however, special features must be taken into account. The composition of the wires varies due to the seaming used, the dimension of the sheath material and the proportion of particle filling. The ratio between the sheath material and the particle filling has a significant influence on the process as well as on the properties of the resulting wear protection coating. Due to the significant smaller cross-section area of the sheath material compared to solid wires, the resulting molten metal is considerably smaller. Due to the high thermal energy of the electric arc - especially on the anodic side - a significant volume of the comparatively low-melting metallic sheath material is not only melted but converted into gaseous phase and evaporates. The melted sheath material is sprayed onto the prepared substrate surface by the process gas. At the same time, the particles get extracted out of the flux-cored wire by the negative pressure of the process gas flowing past. A part of the oxide ceramic particles gets in contact with the electric arc and therefore melts partly or fully. Another part remains solid and penetrates the molten metal during the flight phase or by impacting at the substrate surface and is thus integrated into the generated coating [15]. A considerable disadvantage of using the wire arc spraying process is the loss of reinforcing particles during the coating application. Due to the non-perpendicular angle of impact of the particles to the substrate surface as a result of the wide expansion of the spraying jet in the wire arc spraying process, individual particles are not bound into the matrix but are reflected directly from the substrate surface [16]. By increasing the wire diameter, the ratio of sheath material to particle reinforcement deceases. A better understanding of the effects of different wire diameters, sheath-to-filling ratios and particle size fractions on the microstructure and the wear resistance of the generated coatings is the target of the presented research.

2. Experimental setup and materials

The investigated flux-cored wires were composed of an iron sheath and reinforcing particles of Al_2O_3/ZrO_2. The investigated wire variations differed in the diameter, the sheath geometry (width and thickness of the metal tape before seaming) and the contained particle filling fraction (fine (F) and coarse (C) fraction) specified by the manufacturer. The particle fraction is indicated by using d90 as well as d10, which means that 90 % respectively 10 % of all particles are smaller than this threshold. Thus, 80 % of the particles are in between these two threshold values. For example, considering the coarse particle fraction -160/+63, 90 % of the particles should be smaller than 160 µm and 10 % of the particles should be smaller than 63 µm. A corresponding overview of the different wires and the used nomenclature is given in table 1.

Table 1. Properties of the flux-cored wires used (* = manufacturer's specification).

name	wire diameter (mm)	sheath width and thickness (mm x mm)	particle fraction d90/d10* (µm)
FeF_16	1.6	10 x 0.22	-90/+45
FeC_16	1.6	10 x 0.22	-160/+63
FeF_24	2.4	14 x 0.30	-90/+45
FeC_24	2.4	14 x 0.30	-160/+63

The aluminium substrate surface was roughened by means of grit blasting. For this purpose, Alodur EK-F-24 corundum with a particle size in the range of 600 µm to 850 µm was used. The grit blasting was carried out at a pressure of ca. 2 bar with a blasting angle of ca. 70° and a distance of ca. 150 mm. The samples were subsequently cleaned in an ethanol bath using ultrasound and then dried. The coating process was performed with a wire arc spraying system consisting of the VisuArc350 inverter and the Schub5 torch (both Oerlikon Metco, Pfäffikon, Switzerland). The torch was mounted on a five-axis robot. The process parameters, which were selected on the basis of preliminary investigations with the aim of achieving the lowest possible coating porosity, are given at table 2.

Table 2. Wire arc spraying process parameters of flux-cored wires.

name	current (A)	voltage (V)	relative traverse speed (m/s)	spraying distance (mm)	carrier gas pressure (MPa)	spray path offset (mm)	carrier gas
FeF_16	150	35	1	150	0.6	5	Air
FeC_16	150	35	1	150	0.6	5	Air
FeF_24	150	45	1	150	0.6	5	Air
FeC_24	150	45	1	150	0.6	5	Air

Due to the increased dimensions of the iron sheath of the wires with 2.4 mm diameter the voltage was increased from 35 V to 45 V. To achieve a comparable coating thickness, the number of transitions varied between nine and eleven for the different wires. After four transitions, a cooling period of five minutes was implemented to reduce residual thermal stresses. The target coating thickness was 500 µm to 600 µm.

A detailed characterisation of the microstructure has been conducted for the flux-cored wires and the coatings. Metallographic cross-sections were produced using standard metallographic procedures. The analyses were carried out with the scanning electron microscope (SEM) LEO 1455VP (Zeiss, Jena, Germany). A backscattered electron detector was used to visualise the material contrast. The chemical composition was identified with the integrated energy-dispersive X-ray spectroscopy (EDS) system EDS GENESIS (EDS, Mahwah, NJ, USA). The phase composition of the wire sheath, the particle filling and the thermal sprayed coatings was studied by X-ray diffraction (XRD) with the D8 Discover diffractometer (Bruker AXS, Billerica, MA, USA) equipped with a LynxEye XE-T detector using Co Kα radiation with a tube voltage of 40 kV and a tube current of 40 mA. The diffraction patterns were measured for a 2θ range from 20° to 130°, with a step size of 0.02°. Using the Olympus GX51 inverted microscope (Olympus, Hamburg, Germany), the porosity and particle ratio of the coatings were determined using contrast-based quantitative digital image analysis within the Olympus Stream software. The determination of the area fraction of the pores or particles is in line with the usual procedure from the state of the art and serves as an approximation for the existing actual volume fraction. The particle size distribution of the ceramic particles of the flux-cored wires was measured through laser diffraction analysis by a Cilas 920 particle size analyzer (Cilas, Orléans, France). The measuring principle is based on the diffraction of the laser depending on the particle size. The measuring range is 0.3 µm to 400 µm.

The wear behaviour of the coatings was determined under combined sliding and abrasive wear conditions. The surfaces of the coated samples were ground to Ra 0.3 µm in advance. For the evaluation of the sliding wear behaviour in the ball-on-disc test, the Tetra basalt tester (Tetra, Ilmenau Germany) was used. The wear test parameters were based on the ASTM G99 standard. The abrasive wear behaviour was carried out by the rubber wheel test. Test method E according to ASTM G65 standard was selected. The applied abrasive was dried quartz sand with a grain size of -300/+180 µm (d90/d10). Three samples of each coating were investigated for both tests. The applied wear parameters are listed in table 3.

Table 3. Wear test parameters.

Ball-on-Disc Test		**Rubber Wheel Test**	
Normal force	20 N	Normal force	130 N
Wear track radius	6 mm	Wheel diameter	229 mm
Rotation	96 RPM	Rotation	200 RPM
Speed	60 mm/s	Speed	2400 mm/s
Wear track	596 m	Wear track	718 m
Counterbody	Al_2O_3 (ø 6 mm)	Counterbody	Rubberised steel wheel
		Abrasive medium	Quartz sand

The evaluation of the loss of volume as well as the depth and width of the wear tracks was performed using the tactile measurement device Hommel-Etamic T8000 (Jenoptik, Villingen-Schwenningen, Germany) and the 3D profilometer MicroCAD compact (LMI, Teltow, Germany). The tactile measurement of the wear tracks was carried out perpendicular to the movement of the counterbody.

3. Results and Discussion

3.1. Flux-cored wire characterisation

Figure 1 shows the cross sections of two wires with a diameter of 2.4 mm and different particle fractions. It can clearly be seen that the coarse fraction contains considerably larger particles.

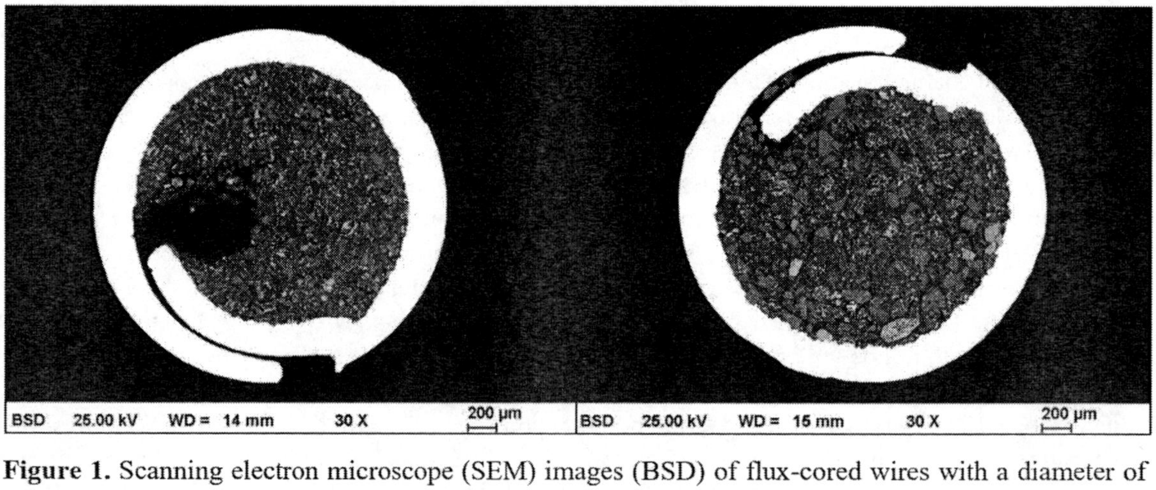

Figure 1. Scanning electron microscope (SEM) images (BSD) of flux-cored wires with a diameter of 2.4 mm (left: fine particle fraction; right: coarse particle fraction).

For an exact determination of the particle distribution, the particle fillings were taken from the wires with a diameter of 2.4 mm. Contrary to the manufacturer's specifications, the particle fractions used are significantly larger. The fine fraction has a particle range (d90/d10) of -83/+8 μm and the coarse fraction of -225/+24 μm. Both fractions have a very high proportion of very fine particles, which is beneficial to wire stability and thus to wire feeding. As shown in figure 2, the particles have a very edged and irregular structure.

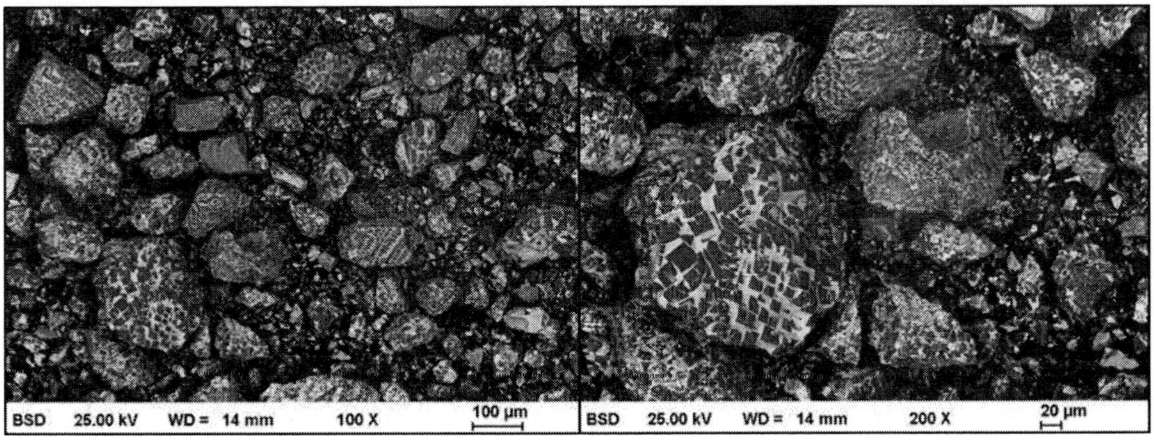

Figure 2. SEM images (BSD) Al_2O_3/ZrO_2 particles (left: overview; right: in detail).

By using the backscattered electron detector (BSD), the composition of the particles becomes clearer. The heterogeneous composition of the particles was investigated in detail by EDS measurements on the cross sectioned particle shown in figure 3. These showed that the grey areas consist of Al_2O_3 and the white areas of a lamellar structure of Al_2O_3 and ZrO_2. The chemical composition of the lamellar structure is 60 at. % O, 30 at. % Al and 10 at. % Zr. However, due to the difficulty of measuring light elements like oxygen by EDS measurements, the oxygen content should be understood as a rough orientation.

Figure 3. Detailed view of a sectioned Al_2O_3/ZrO_2 particle with red arrows indicating Al_2O_3 particles and eutectic areas of Al_2O_3 and ZrO_2.

3.2. Phase formation and chemical composition

The XRD patterns of the flux-cored wire sheath, particle filling and sprayed coatings of the 2.4 mm wires are presented in figure 4. The flux-cored wire sheath consists solely of ferrite with a body-centred cubic structure. The particle filling inside the flux-cored wire is composed of alumina (Al_2O_3), zirconia (ZrO_2) with monoclinic structure and aluminium zirconium oxide ($Al_{0.52}Zr_{0.48}O_{1.72}$) with tetragonal structure. The sprayed coatings contain iron as well as aluminium oxide, zirconium oxide and aluminium zirconium oxide like the flux-cored wire. During the spraying process the iron sheath undergoes oxidation which causes additional iron oxide (FeO) with cubic structure in the coatings. Compared to the XRD measurement of the particle filling only small proportions of Al_2O_3 and ZrO_2 were detected within the coatings although they are clearly visible within the cross-sections.

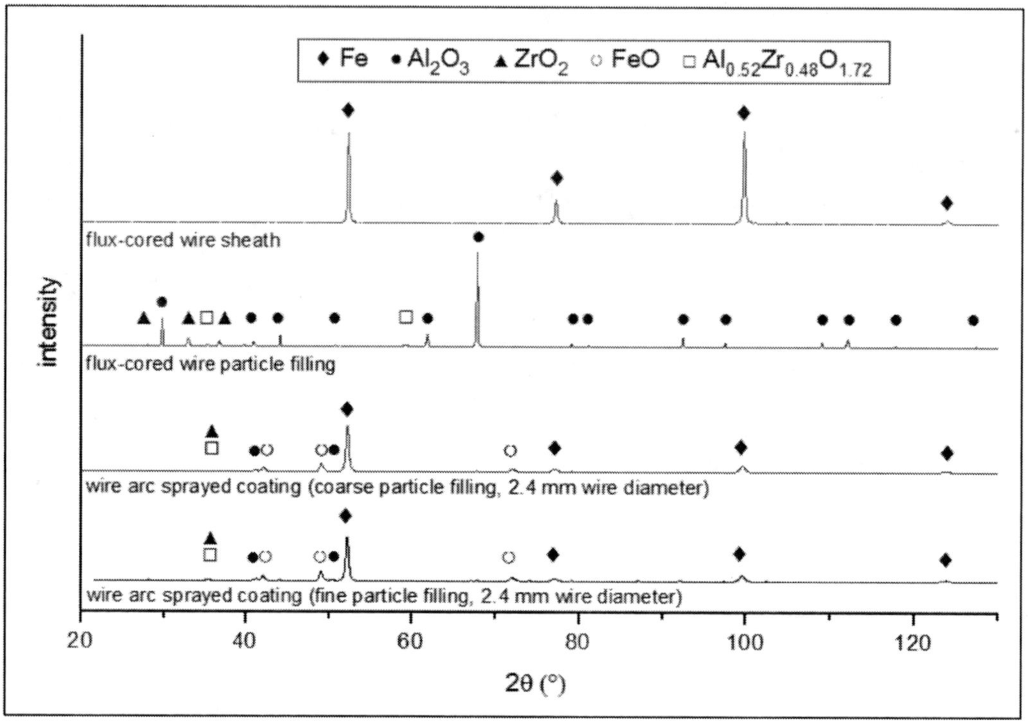

Figure 4. XRD patterns of the flux-cored wire sheath and particle filling as well as the sprayed coatings.

The chemical composition of the coatings was examined by area measurements and are presented in table 4. The investigations show a higher iron content and slightly lower content of aluminium, zirconium and oxygen for the coatings with the finer particle fraction regardless of the wire diameter. The FeF_24_coating presents the lowest proportion of aluminium and zirconium, which is even lower compared to the FeF_16_coating. This could be due to the increased ratio of sheath material to particle filling (2:3) compared to the smaller wire (1:1). In addition, the use of the thicker wires leads to a widening of the spraying jet, whereby more particles hit the substrate surface at a lower angle and after a longer flight phase. Both of these factors reduce the chance of the smaller ceramic particles being integrated into the coating.

Table 4. Chemical composition of wire arc sprayed coatings in at. %, measured by energy-dispersive X-ray spectroscopy (EDS).

Sample	Fe	Al	Zr	O
FeF_16_coating	41.0	20.1	1.5	37.3
FeC_16_coating	38.4	21.5	1.8	38.3
FeF_24_coating	41.6	18.2	1.3	38.9
FeC_24_coating	35.8	22.1	1.9	40.2

The FeC_24_coating contains the highest composition of aluminium, zirconium and oxygen leading to the assumption of the highest particle ratio of all four coatings. The particles of the coarse particle filling fraction are significantly larger and therefore heavier, which means that they are less strongly distracted by the process gas and therefore increased their chance of impacting the substrate surface perpendicularly. This could favour the inclusion of the larger ceramic particles in the coating. The increase of the elements aluminium and zirconium in the chemical composition of the coatings could be an indicator for a higher particle ratio of the coatings with the coarse particle filling. As mentioned before, the detection of light elements like oxygen by EDS measurements is not as accurate.

3.3. Coating microstructure

The microstructure of the thermal sprayed coatings of the flux-cored wires was investigated by SEM to visualize the material contrast of the different elements. Figure 5 shows representative microstructural images. The coatings exhibit a porosity of ca. 3 % to 5 % which is typical for wire arc sprayed coatings. The bright areas are made out of iron, the light grey out of oxidized iron (FeO), the dark grey sections are made out of the Al_2O_3/ZrO_2 particles and the black particles are made out of Al_2O_3. It can be seen very clearly that the particles did not retain their original form. Within the coating, they have a very irregular shape. Sometimes they form a lamellar structure, as it is characteristic for thermal spray coatings. This indicates that the particles are partially or completely melted during the process. Due to the great variation in particle sizes, the particles in the coating are also of very different sizes.

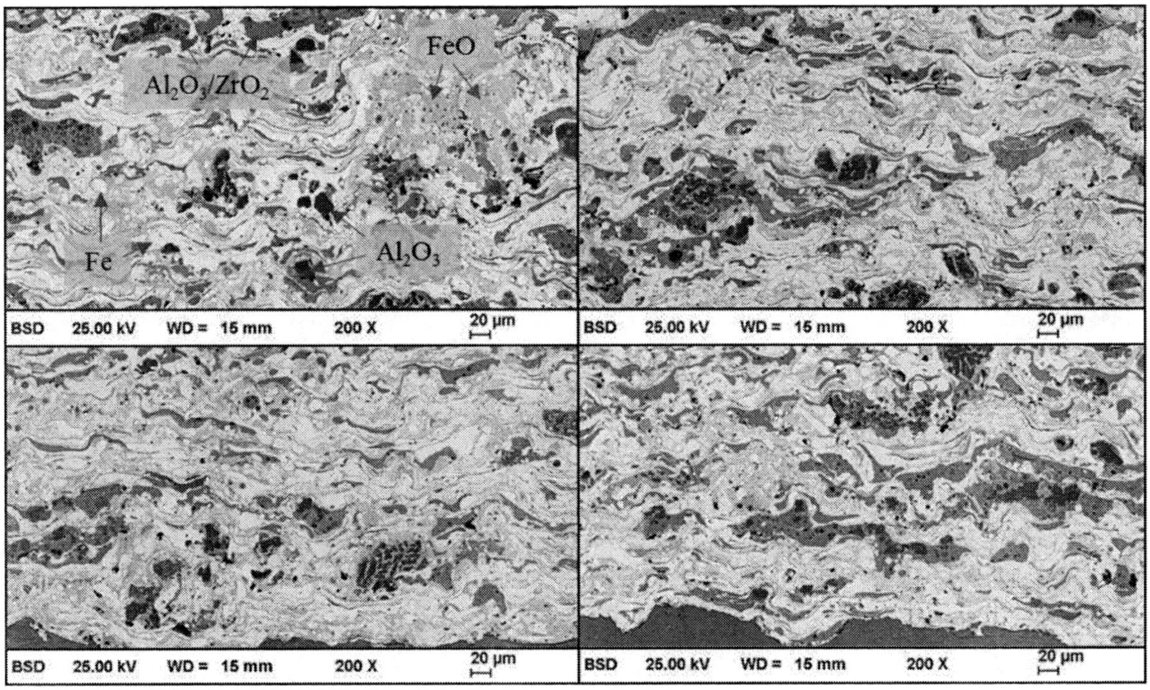

Figure 5. SEM images (BSD) of thermal sprayed coatings (upper left: FeF_16_coating; upper right: FeC_16_coating; lower left: FeF_24_coating; lower right: FeC_24_coating).

Over the entire coating the particle distribution is comparatively homogeneous. When looking at individual areas, however, significant differences can be observed. Like shown in figure 6 there are areas with a large number of particles and areas where hardly any particles are embedded.

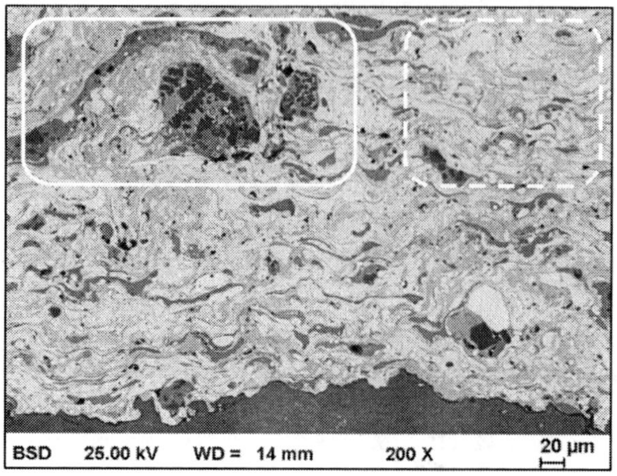

Figure 6. SEM images (BSD) of thermal sprayed FeF_24_coating showing locally inhomogeneity of the particle concentration with high particle concentration in the upper left image corner and low particle concentration in the upper right image corner.

On closer inspection, the particles also have different microstructures. As already shown in figure 3, the particles usually consist of a fine lamellar structure formed by ZrO_2 and Al_2O_3. Some particles additionally integrate Al_2O_3 particles into the lamellar structure. In accordance with the results from Brust et al., cracks are visible in the interface between the particles and the iron matrix like shown in figure 7. These indicate an insufficient adhesion between the two materials.

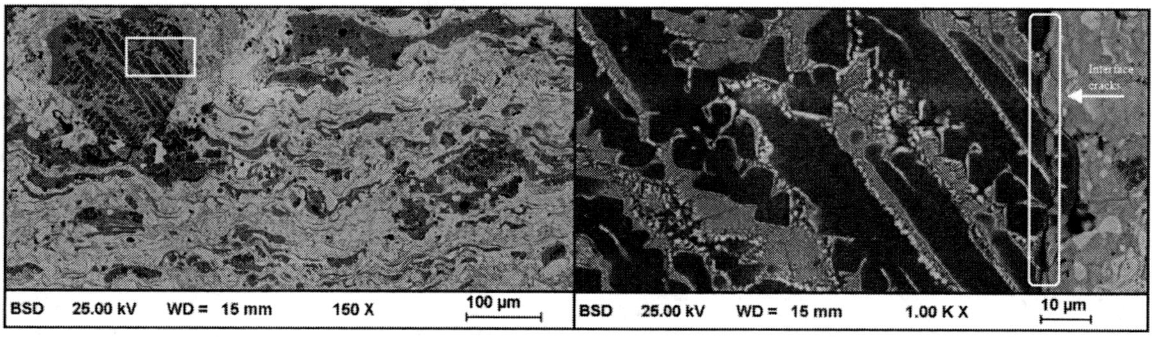

Figure 7. SEM images (BSD) of thermal sprayed FeC_16_coating (left: coating overview; right: enlarged view of the white rectangular area with arrow indicating the area of particle/matrix interface cracks).

The particle ratios of the coatings as well as the loss of the particles during the spraying process are given in table 5. The values are in a similar range for all four coatings. Although the particle-filled volume of the thicker wires is significantly higher than that of the thin wires, their coatings do not contain a significantly increased particle ratio. The FeF_24_coating shows the highest loss of particles during the spraying process. This could be due to the increased ratio of sheath material to particle filling as well as the widening of the spraying jet of the thicker wires like mentioned before. Although the use of thicker wires regarding the fine particle filling results in a similar particle ratio in the coating, the loss of particles is higher. For this reason, thick wires are only a more economical alternative to a limited extent.

Table 5. Particle ratio of the wire arc sprayed coatings using flux-cored wires and loss of the particles during the spraying process.

Sample	Particle ratio of the coating	Loss of particles during the spraying process
FeF_16_coating	24.7 ± 2.1 %	~ 51 %
FeC_16_coating	24.2 ± 4.8 %	~ 52 %
FeF_24_coating	24.5 ± 1.7 %	~ 60 %
FeC_24_coating	28.8 ± 2.8 %	~ 53 %

3.4. Microhardness and wear behaviour

The microhardness of the iron sheath and the Al_2O_3/ZrO_2 particles as well as that of the coating consisting of the iron matrix and the Al_2O_3/ZrO_2 particles embedded in the matrix for the FeC_24 flux-cored wire and coating was investigated. Furthermore, the surface hardness of all coatings was determined. The measured values are summarised in table 6. The hardness of the iron matrix as well as the iron sheath is significantly lower compared to the embedded reinforcement particles. The hardness of the iron matrix increased due to the spraying process and the oxidation of the iron sheath. The higher standard deviation indicates the heterogeneous matrix of Fe and FeO. The hardness of the Al_2O_3/ZrO_2 particles before and after spraying does not differ significantly. No significant differences in surface hardness occurred between the different coatings.

Table 6. Microhardness HV0.5 of the FeC_24 flux-cored wire and coating as well as HV1 of the coating surfaces.

Sample	Matrix	Particles	Surface
FeF_16_coating	-	-	322 ± 22
FeC_16_coating	-	-	347 ± 25
FeF_24_coating	-	-	335 ± 38
FeC_24_coating	315 ± 62	1607 ± 305	338 ± 21
FeC_24 flux-cored wire	236 ± 6	1686 ± 208	-

The investigations carried out showed no significant differences between the coatings in terms of their chemical composition, their phases, their particle ratio and their coating hardness as a function of the wire diameter. This leaves mainly the particle size fraction used as a differentiation criterion between the coatings wear properties.

The results of the ball-on-disc and rubber-wheel test are summarized in figure 8. The wear behaviour under sliding wear conditions in the ball-on-disc-test is slightly influenced by the wire diameter. The coatings produced with the 1.6 mm wires exhibit a lower wear depth and wear volume. This could be due to the slightly higher proportion of Fe respectively FeO which could also act as a lubricant. They also show a significantly lower scattering of the measured values. For both wire diameters the fine particle fraction leads to a lower wear volume loss. This is in coincidence with the investigations of Fang et al. [4]. An explanation is the lower local overload of individual particles using a smaller particle fraction due to a wider distribution of the loads in the contact area to a larger amount of particles [17].

The abrasive wear behaviour of the coating using the rubber wheel test shows also an influence of the particle fraction. The wear depth of the coatings is smaller for the coarser particle fraction regardless of the wire diameter. However, the increased particle surface area, which improves the binding of the particle to the matrix, is a possible explanation. This reduces the probability of the particle breaking out. Another reason is the lower mean free path between the particles. Similar results are presented by Tillmann et al. They discovered a better wear resistance of coatings with larger reinforcing particles under abrasive wear using the rubber wheel test [17]. The integration of larger particles considering a similar particle ratio leads to a smaller mean free path between the particles. The wear volume is only influenced by the particle fraction for the 1.6 mm wires. The coatings with the coarser particle fraction show a lower wear volume. The wear volume of the 2.4 mm wire coatings is at a similar level.

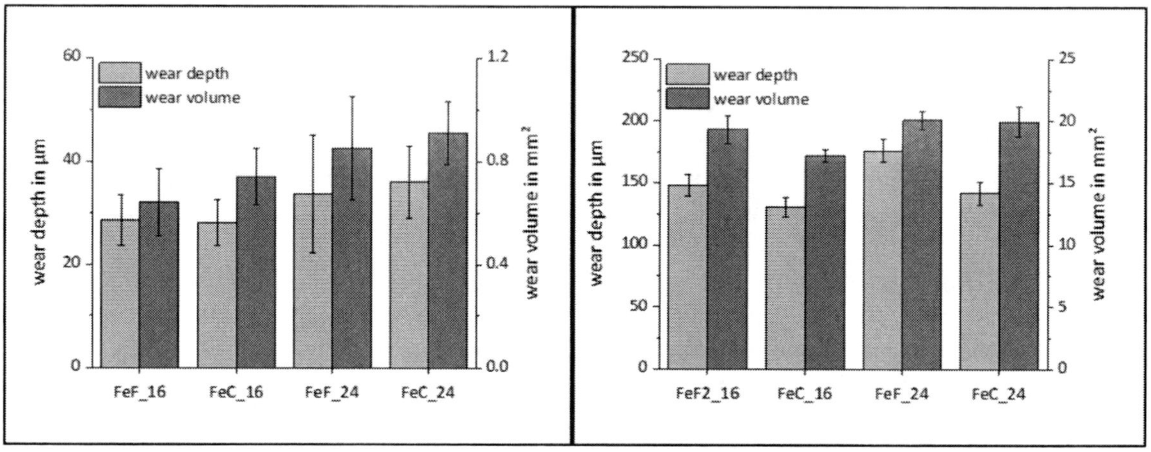

Figure 8. Results of wear investigations of wire arc sprayed Fe-Al_2O_3/ZrO_2 coatings (left: ball-on-disc test; right: rubber wheel test).

4. Conclusions

The influence of the wire diameter and particle fraction of flux-cored wires on the wear properties of the sprayed coatings was examined. The wires were processed using the wire arc spraying technology. The use of flux-cored wires with a diameter of 2.4 mm was successfully demonstrated. However, the significant increase in wire diameter is reflected neither in a significantly higher application rate nor in a considerably increased particle density.

The microstructures of the coatings show 24.2 to 28.8 % of reinforcing particles in the matrix. Investigations of the coating microstructures showed a strong deviation of the reinforcing particle sizes. The particle distribution is comparatively homogeneous over the coating, with locally varying particle concentrations. The coating of the 2.4 mm flux-cored wire with the coarse particle fraction had the highest particle ratio of 28.8 %.

Investigations of the phases within the coatings and the flux-cored wires by XRD show the oxidation of the iron sheath by the process. Further investigations of the chemical composition by EDS showed a higher concentration of Al, Zr and O in the coatings with the coarse particle fraction. The particles of the coarse fraction are significantly larger and therefore heavier, which means that they are less strongly deflected by the process gas. This favours the inclusion of the particles in the coating.

The hardness of the coatings is very heterogeneous due to the composition of particles and matrix material. The iron matrix reaches average values around 315 $HV_{0.5}$. The reinforcing particles, on the other hand, have average hardness values around 1650 $HV_{0.5}$. The hardness values of the coatings surfaces are in a similar range for all four coating systems without significant differences.

This is also reflected in the results of the wear tests. The investigation of sliding friction wear by means of the ball-on-disc test showed a slightly higher volume loss and a higher wear depth for the coatings produced with the 2.4 mm wire. The use of the fine particle fraction resulted in a slightly lower wear volume for both wire diameters.

When investigating the abrasive wear behaviour with the rubber wheel test, the wire diameter had less effect. However, the use of the coarser particle fraction results in a lower depth of wear. The wear volume is also lower in the coatings of the 1.6 mm wires using the coarser particle fraction. No difference in wear volume can be detected in the coatings of the 2.4 mm wires.

Acknowledgments

The IGF project 19.866 BR / DVS No. 02.106 of the research association “Forschungsvereinigung Schweißen und verwandte Verfahren e. V. des DVS, Aachener Straße 172, 40223 Düsseldorf“ was, on the basis of a resolution of the German Bundestag, promoted by the Federal Ministry of Economic Affairs and Energy via AiF within the framework of the programme for the promotion of joint industrial research. The Autors thank AiF for financially supporting the project. Further gratitude goes to the members of the project-accompanying committee and of DVS who are supporting the project with enthusiasm.

References

[1] Dallaire S and Levert H 1997 Development of cored wires for improving the abrasion wear resistance of austenitic stainless steel Journal of Thermal Spray Technology 6 456–62

[2] Brust S, Röttger A, Kimm J, Usta E and Theisen W 2017 Manufacturing of Hard Composite Materials on Fe-Base with Oxide Particles Key Engineering Materials 742 106–12

[3] Berns H 1998 Hartlegierungen und Hartverbundwerkstoffe: Gefüge, Eigenschaften, Bearbeitung, Anwendung (Berlin, Heidelberg, s.l.: Springer Berlin Heidelberg)

[4] Fang J-J, Li Z-X and Shi Y-W 2008 Microstructure and properties of TiB_2-containing coatings prepared by arc spraying Applied Surface Science 254 3849–58

[5] Brust S, Röttger A and Theisen W New wear-resistant materials for mining applications International Conference on Stone and Concrete Machining pp 272–80

[6] Pastor J Y, Poza P, LLorca J, Peña J I, Merino R I and Orera V M 2001 Mechanical properties of directionally solidified Al_2O_3–$ZrO_2(Y_2O_3)$ eutectics Materials Science and Engineering: A 308 241–9

[7] Riedel R 2008 Handbook of ceramic hard materials (Chichester: Wiley)

[8] Lube T, Pascual J, Chalvet F and Portu G de 2007 Effective fracture toughness in Al_2O_3–Al_2O_3/ZrO_2 laminates Journal of the European Ceramic Society 27 1449–53

[9] Carter C B and Norton M G 2007 Ceramic materials: Science and engineering (New York, NY: Springer)

[10] Wu D, San J, Niu F, Zhao D, Huang Y and Ma G 2020 Directed laser deposition of Al_2O_3–ZrO_2 melt-grown composite ceramics with multiple composition ratios Journal of Materials Science 55 6794–809

[11] Hwang B, Ahn J and Lee S 2002 Correlation of microstructure and wear resistance of ferrous coatings fabricated by atmospheric plasma spraying Metallurgical and Materials Transactions A 33 2933–45

[12] Fauchais P L, Heberlein J VR and Boulos M I 2014 Thermal Spray Fundamentals (Boston, MA: Springer US)

[13] Zhu Y L, Ma S N, Yang C H and Fu R F 1999 Investigation on microstructure and tribological properties of cored wire arc sprayed Al/$A_{l2}O_3$ coating Act Metallurgica Sinica 988–94

[14] Lugscheider E and Bach F-W (eds) 2002 Handbuch der thermischen Spritztechnik: Technologien - Werkstoffe - Fertigung (Fachbuchreihe Schweißtechnik vol 139) (Düsseldorf: Verl. für Schweißen und Verwandte Verfahren DVS-Verl.)

[15] Tillmann W and Abdulgader M 2013 Wire Composition: Its Effect on Metal Disintegration and Particle Formation in Twin-Wire Arc-Spraying Process Journal of Thermal Spray Technology 22 352–62
[16] Winkler R, Paczkowski G and Lampke T 2020 Optimisation of the hard material fractions of flux-cored wires with large diameters for arc wire coating Thermal Spray Bulletin 90–6
[17] Tillmann W, Hagen L and Schröder P 2017 Investigation on the Tribological Behavior of Arc-Sprayed and Hammer-Peened Coatings Using Tungsten Carbide Cored Wires Journal of Thermal Spray Technology 26 229–42

Electrochemical testing of thermal spray coatings using gel electrolytes

P Kutschmann*, T Lindner, M Grimm and T Lampke

Chemnitz University of Technology, Institute of Materials Science and Engineering, Materials and Surface Engineering Group, D-09107 Chemnitz, Germany

*e-mail: pia.kutschmann@mb.tu-chemnitz.de

Abstract. The use of electrochemical methods allows fast and inexpensive corrosion measurements of bulk materials with high significance. In the case of thermal spray coatings, electrolyte penetration into open pores up to the substrate material can cause undesired mixed potentials. Furthermore, the implementation of complex geometries or rough surfaces remains a problem. Preconditioning of the surface or the use of the electrochemical cell is required to eliminate leakage. Therefore, reliably measuring corrosion is still a challenging task. This undermines fast monitoring of corrosion performance as a part of the production process. Gelling agents are investigated to increase the viscosity of many electrolytes. A procedure has been developed to determine the concentration level and the mixing conditions. Passivation and pitting-corrosion testing are performed on thermal spray AISI 316L coatings. The electrochemical potential curves as well as the corroded surface layers were studied in comparison to a liquid electrolyte. The suitability of the test on rough surfaces in the sprayed condition was investigated. The results prove the novel approach as an alternative to established electrochemical test methods with extended application range.

1. Introduction

Thermal spray coatings are commonly used to protect surfaces under harsh environmental conditions. The most important applications are providing protection against corrosion and tribological loads. The most challenging requirement for high quality coatings is the corrosion resistance. Complex and cost-intensive preliminary studies are often conducted by standardized test conditions, which are usually developed for bulk material, e.g. salt-spray test, Corrodkote, or potentiodynamic polarization measurement. The open porosity of the coating structure inhibits a selective assessment of the coating properties. In general, the corrosive media penetrate the microporous coating structure and reach the substrate material. This leads to the fact, that the area of the corrosion attack is unknown. Hence, mixed potentials and superimposed corrosion processes are the result of the interaction between coating and substrate.

A quick localized validation test for the quality of coating during production processes is not available so far. Therefore, screening tools for monitoring the quality of coatings during the production process would be beneficial. The most common influencing factors include deviation in quality of feedstock material or fluctuations of process parameters. Electrochemical methods can provide quantifiable results to classify material-specific properties. By electrochemical testing, the relationship of the current density in regards to the potential can be identified. However, the high complexity of the measuring arrangement and the low viscosity of the testing agent have thus far prevented widespread use of this process. To overcome these issues, different adaptions are required.

Representative and limited potential areas are sufficient to gather reliable information. Therefore, the duration time of validation tests can be reduced significantly. Increasing the viscosity can successfully prevent penetration by the corrosive media. Therefore, gels can be a reliable alternative. One established test is the KorroPad indication test, which focuses on the visualization of corrosion processes with indicator-doped gel pads [1–4]. KorroPad is designed for passivating steels to detect their pitting-corrosion behavior [3]. Few studies have been conducted in recent years which focus on high viscosity gels as agents for corrosion monitoring with electrochemical-instrumented experiments [5–10]. The intended applications are the validation of corrosion properties of sculptures and monuments with non-destructive tests [5,9] and the rebuilding of specific atmospheric conditions like moisturizing films on zinc coatings [7] or artificial mud [8]. Most of the studies use agar-gelled NaCl electrolytes [1–10], with little investigation into the properties of these gels [11,12]. Because of their interaction, the combination of thickeners and electrolytes is still a big challenge. In the case of thermal spray coatings, the gel electrolytes show beneficial properties due to the transfer of the corrosion attack to the surface of the coating.

In a first study, the feasibility of gel electrolytes in potentiodynamic polarization measurements of thermal spray stainless-steel coatings on low carbon-steel substrates are determined. The focus is on gel preparation of NaCl and H_2SO_4 electrolytes to prevent infiltration into the specific microstructural coating features and to determine the level of corrosion attack compared to low viscous liquid electrolytes.

2. Experimental and materials

Corrosion-resistant AISI 316L stainless-steel coatings were prepared by atmospheric plasma spraying (APS) and wire arc spraying using either air or nitrogen for atomization. As powder material, the gas-atomized 80.46.1 powder (GTV Verschleißschutz GmbH, Luckenbach, Germany) with a particle size distribution of -53+22 µm was selected and processed with the GTV APS F6 gun. Arc sprayed AISI 316L coatings were deposited by using the wire arc spray unit VISU ARC 350 (Oerlikon Metco, Wohlen, Switzerland) with the Ø 1.6 mm thick 85T wire (TAFA Inc. A Praxair Surface Technologies Company, Concord, NH, USA). The composition of the spray materials and the process parameters are listed in table 1 and table 2.

Table 1. Nominal composition of the AISI 316L feedstock materials.

	Cr	Ni	Mo	Mn	Si	C	P	S	Fe
AISI 316L	mass fraction in %								
Powder	17	12.5	2.5	-	-	0.02	-	1	bal.
Wire	17	12	2.5	2	1	0.08	0.045	0.03	bal.

A low carbon steel EN S235 with the dimension of Ø 40 mm x 10 mm was used as the substrate material. Grit blasting was conducted before the single-side coating process using corundum EK F 24 (-800 +600 µm), air pressure of 2.5 bar, a nozzle-surface distance of 250 mm, and an angle of 70°. Afterward, the substrates were cleaned in an ethanol ultrasonic bath. Prior to the electrochemical corrosion test, the coatings were ground and polished up to a diamond paste of 9 µm.

Table 2. APS and wire arc spray parameters.

Process	Ar (slpm)	H_2 (slpm)	Current (A)	Voltage (V)	Atomizing gas pressure (bar)	Powder flow (g/min)	Spraying distance (mm)	Offset (mm)	Surface velocity (m/s)
APS	50	10	600	-	-	2 x 38	140	5	1
Wire arc spraying - Air	-	-	180	27	4	-	130	6	1.2
Wire arc spraying - N_2	-	-	140	42	7	-	100	5	1

Two types of electrolytes were considered for the electrochemical corrosion tests choosing 0.5 M NaCl to investigate the pitting corrosion behavior and 0.05 M H_2SO_4 to determine the passivation properties of the AISI 316L coatings. The electrolytes were prepared by gelling the liquid electrolyte with polyacrylic acid (Carbopol 71G NF Polymer, Lubrizol, Advanced Materials Europe, Brussels, Belgium). The electrolyte concentration was not adjusted due to the small additions of Carbopol. The amount of Carbopol was adjusted to increase the viscosity up to 450–500 mPa·s suitable to prevent penetration of the open porosity of the thermal spray coatings. Therefore, a mass fraction of 2.5 % Carbopol was added into the NaCl electrolyte and a mass fraction of 6.5 % into the H_2SO_4 electrolyte. The viscosity was measured by tuning fork vibration method using the viscometer (Vibro Viscometer SV-10, A&D Company, Limited, Tokyo, Japan) at temperatures of 25 °C. The NaCl electrolyte was neutralized using NaOH to reach a pH value of 7. The pH value of H_2SO_4 electrolyte was 1.2. Using a motorized stirrer for the gel electrolyte preparation turned out to be the best method to prepare homogeneous gels. Furthermore, the practicality of the gel electrolytes was assessed by determining the temperature-viscosity dependence of the applied electrolyte compositions.

The feasibility of gel electrolytes for corrosion testing of thermal spray coatings was determined by measuring potentiodynamic polarization curves in a 3-electrode cell arrangement. The cell consisted of a silver chloride reference electrode (Ag/AgCl saturated KCl electrode), a Pt plate as a counter electrode, and the sample acted as a working electrode with a testing area of Ø 1 cm. The measurement was controlled with a potentiostat PS6 (Sensortechnik Meinsberger GmbH, Waldheim, Germany). For all tests, the volume of the electrolytes was around 200 ml. After measuring the open circuit potential for 15 minutes, the potential scan was conducted over a range of 950 mV, starting at -250 mV relative to the open circuit potential, and a scan rate of 0.1 mV/s. The result of the polarization measurements were used to derive the corrosion potentials E_{Corr} and corrosion current densities i_{Corr} using the Tafel slopes. A comparison between a liquid and gel electrolyte using the same electrochemical cell and identical parameters was made and the corrosion attack was assessed.

Metallographic investigations were conducted to investigate the corrosion attack. The optical microscope GX51 (Olympus, Shinjuku, Japan) equipped with a SC 50 camera (Olympus, Shinjuku, Japan) was used for analysis of the cross-sections.

3. Results and discussion

3.1. Coating characterization

Figure 1 shows the AISI 316L coatings in the sprayed condition, revealing the altered structural properties with the different thermal spray processes. The APS coating contains lamellar and oxidized splats, non-molten particles, and pores (figure 1a). The porosity content of (6.7 ± 1.6) % and a coating thickness of (287 ± 14) µm were measured. Figure 1b and figure 1c show the typical lamellar micro-structure of wire arc sprayed coatings. Using compressed air as the atomizing gas causes significant oxidation of the spray particles, while nitrogen reduces this effect. With a porosity of (3.1 ± 0.6) % and (1.9 ± 0.3) %, the arc sprayed coatings are denser than the APS coating. The coating thickness is (470 ± 40) µm and (540 ± 21) µm, respectively.

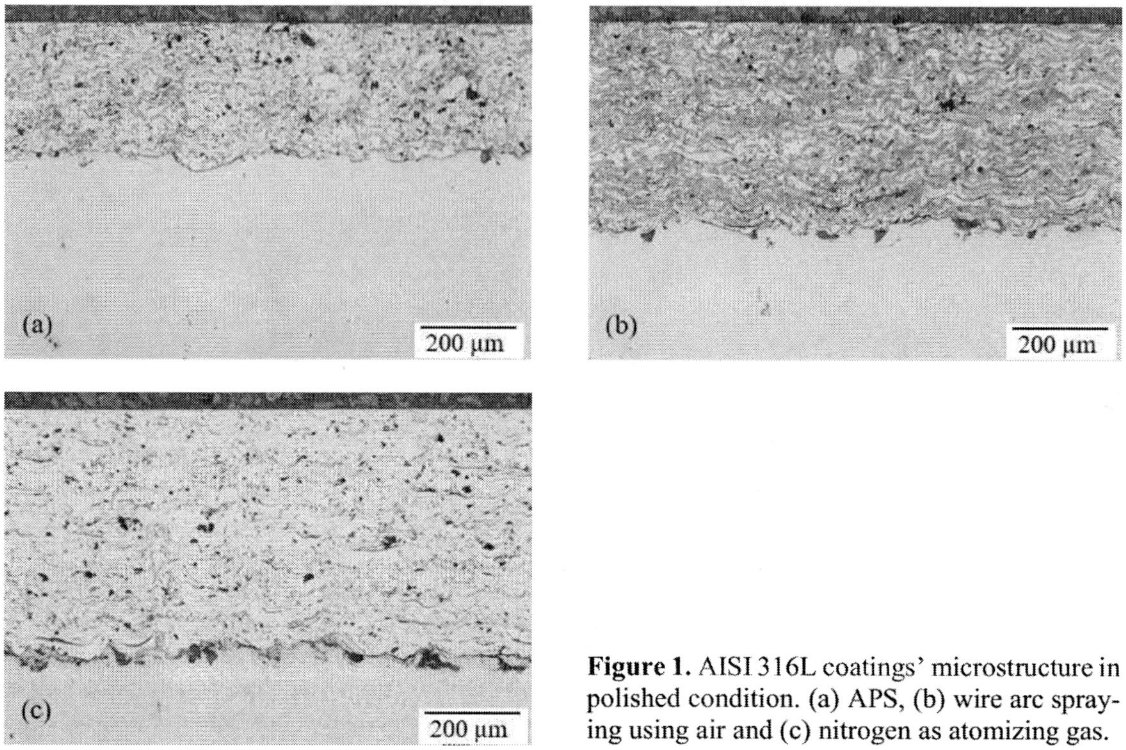

Figure 1. AISI 316L coatings' microstructure in polished condition. (a) APS, (b) wire arc spraying using air and (c) nitrogen as atomizing gas.

3.2. Gel electrolyte characterization

To prepare gel electrolytes, which prevent the penetration in the micropores of thermal spray coatings, the gelling of the 0.5 M NaCl and 0.05 M H_2SO_4 electrolyte was examined as it related to the addition of the thickener Carbopol. The thickening results in the ability of the polyacrylic acid to form a networked microgel structure in aqueous solutions [13]. Figure 2a represents an exponential increase of the viscosity with the amount of Carbopol addition and shows different gelling behavior, which depends on the electrolyte. In general, electrolytes tend to reduce the viscosity of Carbopol gels and additionally show a pH dependency of the gelling properties. Maximum viscosity is achieved by neutralization [14,15]. Hence, a neutralization with NaOH to a pH value of 7 during the preparation of NaCl electrolyte was necessary to increase the viscosity. The viscosity was defined between 450–500 mPa·s for the present study. Repetitions of the viscosity measurements after several days revealed that the gel electrolytes continue to be stable at room temperature, which is beneficial for ease of storage.

Another experiment focused on the temperature stability of the gel electrolytes (figure 2b). The NaCl gel electrolyte with a mass fraction of 2.5 % Carbopol addition shows a steady viscosity level up to 65 °C. In contrast, a steep decline can be observed for H_2SO_4 gel electrolyte with a mass fraction of 6.5 % Carbopol. Figure 2c and figure 2d show images of both gel electrolytes. Due to the gelling agent, the electrolytes become opaque. During mixing, mechanical entrapment and chemical generation caused the formation of bubbles in the H_2SO_4 electrolyte, which could not be prevented in spite of trying several different preparation methods. Kim et al. investigated the microstructure properties of the Carbopol gel network in detail and faced similar problems [16].

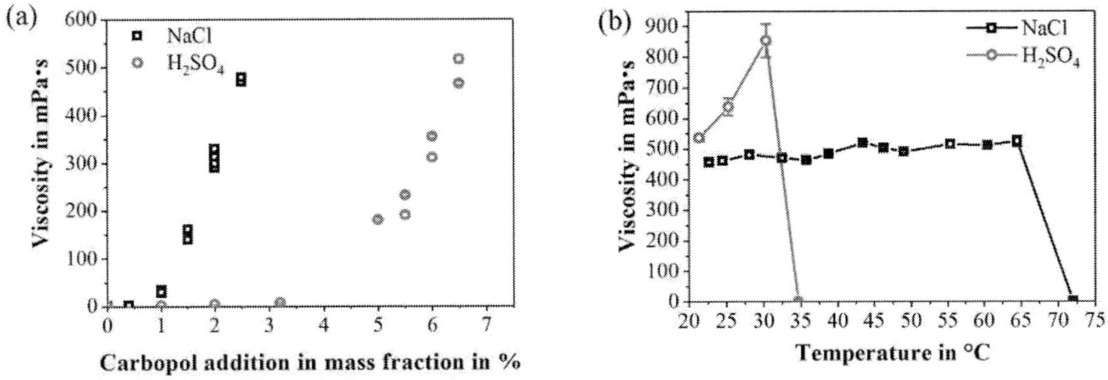

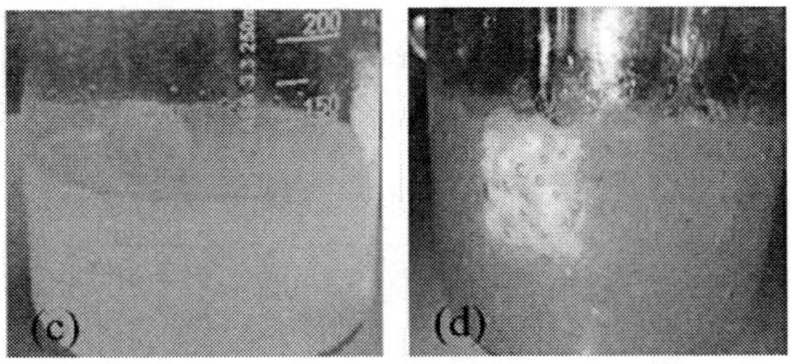

Figure 2. Viscosity of 0.5 M NaCl and 0.05 M H_2SO_4 gel electrolytes; (a) Carbopol addition-viscosity dependence, (b) temperature-viscosity behavior; images of (c) 0.5 M NaCl and (d) 0.05 M H_2SO_4 gel electrolyte.

3.3. Electrochemical-instrumented corrosion tests using gel electrolytes

Figure 3a shows the polarization curves of the APS and wire arc sprayed AISI 316L coatings as well as the low carbon steel substrate using 0.5 M NaCl gel electrolyte in comparison with the corresponding liquid electrolyte. As expected from the galvanic series and the Cr content, the low carbon steel substrate shows the lowest corrosion potential. The wire arc sprayed coating with nitrogen atomizing gas possesses the highest corrosion potential E_{Corr}. With the increase in oxide content, the corrosion potential decreases for APS and air-atomized wire arc sprayed coatings. Figure 3b summarizes the behavior of the gel electrolyte during the potential scan. When the corrosion begins, the gel electrolyte starts to gel further and the corrosion products remain at the surface causing a smoother ascent on the anodic slope (figure 3a). In table 3, the corrosion potentials E_{Corr} and corrosion current densities i_{Corr} were derived from the polarization curves using the Tafel slope. As a result, comparable values could be determined independently of the use of liquid or gel electrolytes. This ensures a direct comparability of the results. However, no significant influence on the characteristics of the current density-potential curve can be observed from the formation of mixed potentials by substrate contact when using liquid electrolytes.

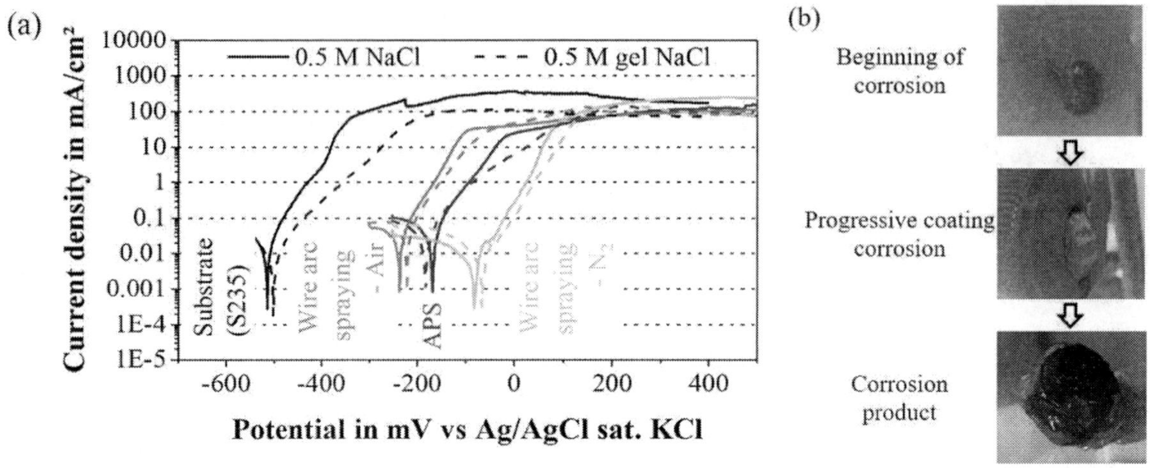

Figure 3. (a) Polarization curves of AISI 316L coatings tested in 0.5 M NaCl electrolyte and gel electrolyte. (b) Gel electrolyte behavior during polarization scan.

Table 3. Summary of the results of the potentiodynamic polarization tests.

	0.5 M NaCl electrolyte	Substrate	APS	Wire arc spraying - Air	Wire arc spraying - N_2
E_{Corr} (mV)	liquid	-515	-168	-237	-99
	gel	-502	-183	-221	-86
i_{Corr} (mA/cm²)	liquid	0.009	0.028	0.023	0.005
	gel	0.007	0.023	0.034	0.007

Figures 4 and 5 compare the corrosion phenomena related to the used electrolyte. In contrast to liquid electrolytes, gels cause a homogeneous corrosion attack. In addition, crevice corrosion at the edge of the test area can be prevented. Due to the open porosity, the liquid electrolyte corrodes the steel substrate and several isolated rust spots appear at the coating's surface. This is in accordance with [5,6,9] who studied agar-based NaCl electrolytes.

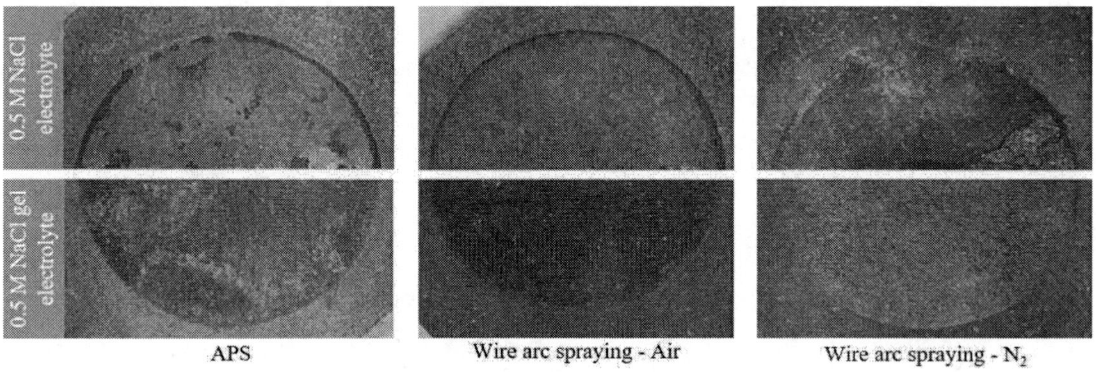

Figure 4. Comparison of the corrosion attack at the AIS 316L coating surface using 0.5 M NaCl electrolyte (upper row) and gel electrolyte (lower row).

Figure 5 shows the cross-sections of the different coatings for NaCl electrolyte in both liquid and gel state. The duration of the experiments at high current densities caused a severe corrosion attack of the coatings. However, the gel electrolyte prevented the penetration of the coatings up to the substrate (figure 5b, d, and f in comparison with figure 5a, c and e).

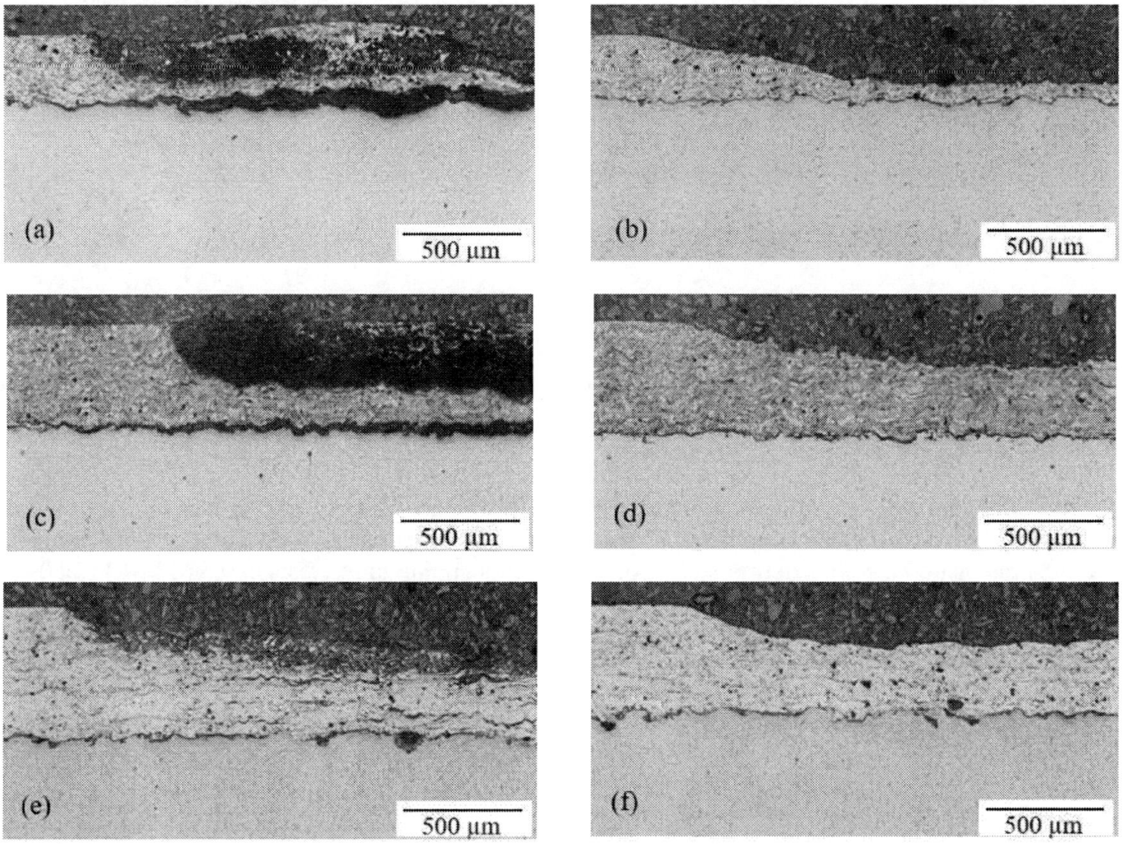

Figure 5. AISI 316L coatings' microstructure after electrochemical corrosion tests in 0.5 M NaCl electrolyte and gel electrolyte, respectively, APS (a & b), wire arc spraying (c & d), and wire arc spraying using nitrogen as atomizing gas (e & f).

Figure 6 shows the first results of the experiments with H_2SO_4 electrolytes. In a direct comparison between liquid and gel electrolytes, similar corrosion potentials can be detected. However, the active and passive sections of the curves deviate. The main issue is the bubble containing gel electrolyte (figure 2d), which leads to insufficient coverage of the corrosion area (figure 6, small image). In addition, the contact area of the sample and the electrolyte varies, particularly because of the infiltration of the coating by the liquid electrolyte, leading to different current densities.

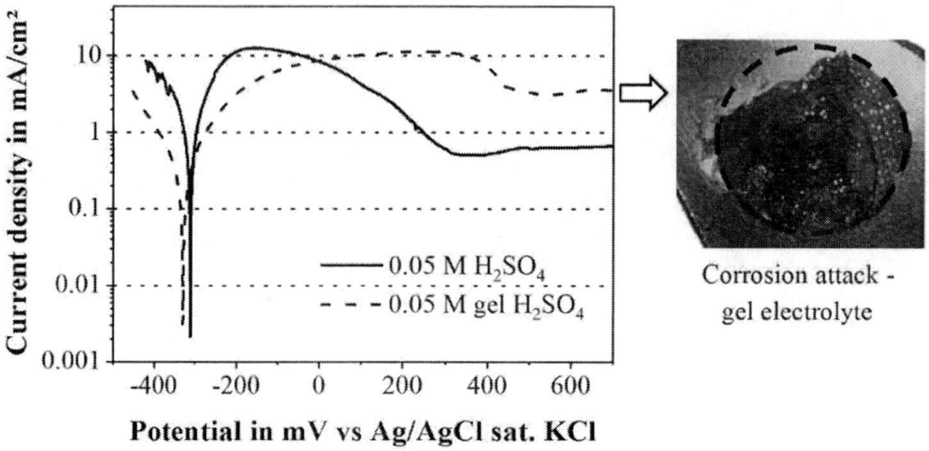

Figure 6. Polarization curves of AISI 316L wire arc sprayed coatings tested in 0.05 M H_2SO_4 electrolyte and gel electrolyte.

4. Conclusions

In this study, the influence of electrolyte viscosity in terms of potentiodynamic polarization measurements of thermal spray coatings was investigated. AISI 316L coatings were produced using APS and wire arc spraying processes with different atomizing gases. The corrosion behavior of the coatings and the substrate material were determined by potentiodynamic polarization tests in 0.5 M NaCl and 0.05 M H_2SO_4 electrolytes. With polyacrylic acid (Carbopol), the viscosity of the electrolytes was successfully increased to 450–500 mPa·s. The used polyacrylic acid, which gels at room temperature, shows disadvantages regarding the production of homogeneous gel electrolytes. However, the suitability of gel electrolytes was proven by comparative measurements with liquid electrolytes. Independent of the coating structure, a comparable trend of the measurement curves could be seen. This proves that mixed potentials exert only a minor influence on the measurement. Cross-section images in the area of corrosion measurement show that infiltration of the coating structure can be prevented by using gel electrolytes. The electrochemical-instrumented tests are capable of revealing variations in the microstructure of coatings such as oxide content, alloying elements, and porosity. Using gel electrolytes in combination with electrochemical-instrumented methods can be beneficial for quality control in the production of thermal spray coatings. Construction of a modified electrochemical test cell would support the transfer of electrochemical testing from theory into practice as well as permits the usage of manifold electrolyte composition. Nevertheless, more experimental test data is necessary to classify the test method for thermal spray coatings.

Acknowledgments

The authors gratefully acknowledge the Arbeitsgemeinschaft industrieller Forschungsvereinigung "Otto von Guericke" e.V. (AiF) in the framework of this work (AiF-No. ZF4139111SU9, ZF4752603SU9 and ZF4820401SU9). The manuscript documents the preliminary study of the ongoing project. The authors thank Elke Benedix and Naman Goyal for their support during the electrochemical testing as well as preparing and characterizing gel electrolytes, respectively.

References

[1] Burkert A, Klapper H S and Lehmann J 2012 Novel strategies for assessing the pitting corrosion resistance of stainless steel surfaces *Materials and Corrosion* **64** 8 675-682

[2] Rosemann P, Müller Th, Babutzka M and Heyn A 2015 Influence of microstructure and surface treatment on the corrosion resistance of martensitic stainless steels 1.4116, 1.4034 and 1.4021 *Materials and Corrosion* **66** 1 45-53

[3] Kauss N, Heyn A, Halle T and Rosemann P 2019 Detection of sensitisation on aged lean duplex stainless steel with different electrochemical methods *Electrochimica Acta* **317** 17-24
[4] Rosemann P and Kauss N 2018 Visualisierung werkstoffbedingter Lochkorrosionsanfälligkeit durch die KorroPad-Prüfung *WOMAG* **7-8** doi: 10.7395/2018/Rosemann1
[5] Ramírez Barat B and Cano E 2019 Agar versus agarose gelled electrolyte for in situ corrosions studies on metallic cultural heritage *ChemElectroChem* **6** 2553–2559
[6] Monrrabal G, Ramírez Barat B and Cano E 2018 Non-destructive electrochemical testing for stainless-steel components with complex geometry using innovative gel electrolytes *Metals* **8** doi: 10.3390/met8070500
[7] Babutzka M and Heyn A 2017 Dynamic tafel factor adaption for the evaluation of instantaneous corrosion rates on zinc by using gel-type electrolytes *IOP Conf. Ser.: Mater. Sci. Eng.* **181** 012021
[8] Spark A J, Cole I, Law D, Marney D and Ward L 2016 Investigation of agar as soil analogue for corrosion studies *Materials and Corrosion* **67** 1 7-12
[9] Monrrabal G, Guzmán S, Hamilton I E, Bautista A and Velasco F 2016 Design of electrolytes for electrochemical studies on metal surfaces with complex geometry *Electrochimica Acta* **220** 20–28
[10] Subbiah K, Velu A, Kwon S J, Lee H S, Rethinam N and Park DJ 2018 A novel in-situ corrosion monitoring electrode for reinforced concrete structures. *Electrochimica Acta* **259** 1129-1144
[11] Umoren S A, Li Y and Wang F H 2010 Synergistic effect of iodide ion and polyacrylic acid on corrosion inhibition of iron in H_2SO_4 investigated by electrochemical techniques *Corrosion Science* **52** 2422-2429
[12] Lee W K, Lim Y Y, Leow A T C, et al. 2017 Factors affecting yield and gelling properties of agar *J Appl Phycol* **29** 1527–1540
[13] Carnalo J O and Naser M S 1992 The use of dilute viscometry to characterize the network properties of carbopol microgels *Collois Polym Sci* **279** 183-193
[14] Technical Data Sheet TDS-730 2010 Viscosity of Carbopol® polymers in aequous systems Lubrizol Advanced Materials Inc., Edition August 13, 2010
[15] Giuseppe E di, Corbi F, Funiciello F, Massmeyer A, et al 2015 Characterization of Carbopol® hydrogel rheology for experimental tectonics and geodynamics *Tectonophysis* **642** 29-45
[16] Kim J Y, Song J Y, Lee E J and Park S K 2003 Rheological properties and microstructures of Carbopol gel network system *Colloid Polym Sci* **281** 614-623

IOP Conf. Series: Materials Science and Engineering 1147 (2021) 012032 doi:10.1088/1757-899X/1147/1/012032

High-temperature wear behaviour of borided Inconel 718 HVOF coatings

M Löbel*, T Lindner, N Hanisch and T Lampke

Materials and Surface Engineering Group, Institute of Materials Science and Engineering, Chemnitz University of Technology, D-09107 Chemnitz, Germany

*e-mail: martin.loebel@mb.tu-chemnitz.de

Abstract. Increasing demands on component performance and efficiency require continuous development efforts in existing material systems and concepts. In addition to the specific material properties, economic aspects have to be taken into account. Thermochemical treatments of iron- and nickel-base alloys allow for a distinct improvement of hardness and wear resistance. The process of boriding enables the highest hardness values and the formation of thermally stable precipitates. Especially nickel-base alloys are suitable for high temperature applications. An economic application of these alloys can be achieved by applying coating technologies and limiting the material usage to the surface. High-velocity oxygen fuel thermal spraying of the nickel-base alloy Inconel 718 and subsequent powder-pack boriding is conducted. Furthermore, the influence of a solution annealing step prior to the boriding process is investigated with the motivation to achieve a homogenisation of the coating. A successful diffusion enrichment and the formation of a precipitation layer could be achieved. The investigation of the resulting properties revealed a distinct increase of hardness and an improvement of wear resistance tested under reciprocating conditions in a wide temperature range.

1. Introduction

Nickel-base alloys are widely applied due to their mechanical properties especially at high temperatures. Furthermore, they exhibit a high corrosion resistance also under harsh environmental conditions e.g. in contact with seawater or at elevated temperatures. However, the wear resistance is rather limited [1–4].

Thermochemical treatments offer a possibility to enhance the property profile. Thermochemical processes can be divided into two main groups. Interstitial hardening can be achieved by nitrogen and carbon enrichment [5, 6]. For higher demands and the application at elevated temperatures precipitation hardening processes are used. The highest hardness values and the formation of thermally stable precipitates can be achieved by boriding. The most common process variant is powder-pack boriding. Besides nickel-base alloys this process is used for various steels and further recently developed materials [7–12]. However, thermochemical treatments are mainly limited to cast alloys. Due to the relatively high cost of nickel-base alloys, an economic application can be achieved by a functional division and the application of steel substrates.

Coating processes like laser cladding and thermal spraying are established for a wide range of technical applications, especially for wear and corrosion protection. Laser cladding of the nickel-base alloy Inconel 625 caused dilution of Fe resulting in a degradation of corrosion properties depending on the coating parameters [13]. Thermal spray processes enable the economic deposition of coatings within a wide range of coating thicknesses. Due to the relatively low thermal input and the absence of metallurgical bonding, the chemical composition of the feedstock material can be retained. For coating processes conducted under normal atmosphere typical structural defects e.g. pores and oxides are formed. The extent of these defects can be reduced by the application of high kinetic processes. For the deposition of metallic coatings and cemented carbides the industrially most widely used process is high-velocity oxygen fuel (HVOF) thermal spraying [14, 15].

The process combination of boriding thermally sprayed coatings of various iron- and nickel-based coatings has been investigated in previous studies. However, the resulting property profile was only characterised at ambient temperature [16, 17].

In the current investigations a process combination of depositing the nickel-base alloy Inconel 718 by HVOF thermal spraying and thermochemical treatment by powder-pack boriding is considered. With the objective to improve internal bonding and reduce heterogeneities one group of the thermally sprayed coatings was solution annealed prior to the thermochemical treatment. An adapted boriding routine was applied to avoid the formation of silicon rich phases. The wear behaviour was investigated in a wide temperature range up to 900 °C under reciprocating conditions in dependence of the treatment condition.

2. Experimental

Coatings of the nickel-base alloy Inconel 718 were produced by applying the high kinetic thermal spray process HVOF. The commercially available powder NI-202-3 (Praxair Inc., Danbury, CT, USA) was used as a feedstock material. A particle size range of -45 +16 µm was specified by the manufacturer. The stainless steel EN 1.4404 (AISI 316L) was used as a substrate material. Pre-treatment was conducted by grit blasting using the blasting medium Alodur EK F 24 with a particle size of -850 +600 µm. A blasting pressure of 2.5 bar, a distance of 200 mm under an angle of 70° was applied. Subsequently, ultrasonic cleaning in ethanol has been carried out. For the coating process the liquid-fuelled HVOF system K2 was used (GTV Verschleißschutz GmbH, Luckenbach, Germany). The spraying parameters are summarised in Table 1.

Table 1. HVOF thermal spray parameters for the deposition of Inconel 718.

O_2	kerosene	λ	nozzle	powder feed rate	spraying distance	relative traverse speed	spray path offset
900 l/min	26 l/h	1.0	150/14	2 x 50 g/min	350 mm	1.0 m/s	5 mm

A final coating thickness of approximately 270 µm was measured after 14 single-layers. For the thermochemical treatment one part has been considered in as-sprayed condition, whereas for the other group a heat treatment has been conducted prior to boriding with the objective to reduce heterogeneities and improve internal bonding. For the heat treatment the vacuum furnace Torvac 12 Mark IV (10^{-4} mbar) has been used. The samples were heated to a temperature of 1100 °C with a rate of 10 K/min. Holding the temperature for a duration of 1 h was followed by unregulated furnace cooling. Boriding has been conducted in a powder-pack process by the company BorTec GmbH & Co. KG with the parameters summarised in Table 2.

Table 2. Boriding parameters.

boriding agent	temperature	duration	atmosphere
Ekabor® Ni	900 °C	5 h	Argon

Cross-sections were prepared by standard metallographic procedures. Light microscopic investigations have been performed with an Olympus GX51 (Olympus, Shinjuku, Japan). For the visualisation of material contrast the scanning electron microscope (SEM) LEO 1455VP (Zeiss, Oberkochen, Germany) equipped with a backscattered electron (BSE) detector has been applied. The chemical composition of the feedstock powder and coatings has been measured by X-ray fluorescence spectroscopy (XRF) using a Fischerscope X-Ray XAN (Fischer, Sindelfingen, Germany) operated with an acceleration voltage of 30 kV. The phase formation was determined by X-ray diffraction (XRD) with a D8-discover diffractometer (Bruker AXS, Billerica, MA, USA). Co Kα radiation and a diffraction angle range of 20° to 130° were used.

Elemental depth profiles were determined by glow discharge spectroscopy (GDOS) with a GDA 750 (Spectruma Analytik GmbH, Hof, Germany) using an anode diameter of 2.5 mm, 800 V, 25 mA and 3 hPa argon pressure.

Selective hardness measurements of the borided surface layer and the coating have been conducted by nanoindentation using an UNAT tester (ASMEC GmbH, Radeberg, Germany), equipped with a Berkovich B16 indenter (tip radius: 0.394 µm). A load of 10 mN has been applied for a holding time of 5 s. The Vickers hardness was deduced from the measured indentation depth using the InspectorX testing software version 3 (ASMEC GmbH, Radeberg, Germany). At least ten single measurements have been considered for the calculation of the average Vickers microhardness and the standard deviation.

The wear behaviour was investigated under reciprocating conditions in a wide temperature range using an SRV tribometer (Optimol Instruments GmbH, Munich, Germany) with the parameters summarised in Table 3.

Table 3. Wear test parameters.

force	frequency	duration	amplitude	counter body	temperature
26 N	40 Hz	900 s	0.5 mm	Al_2O_3 (ø 10 mm)	25 °C; 500 °C; 650 °C; 800 °C; 900 °C

Three measurements have been conducted to determine the average value for every parameter set. The resulting wear tracks were investigated with the optical profilometer MikroCAD 3-D (LMI Technologies Inc., Burnaby, Canada).

3. Results and Discussion

3.1. Feedstock and coating characterisation

Prior to the coating process the feedstock powder was investigated in detail. The surface and cross-sections were investigated in SEM. Representative images are shown in Figure 1.

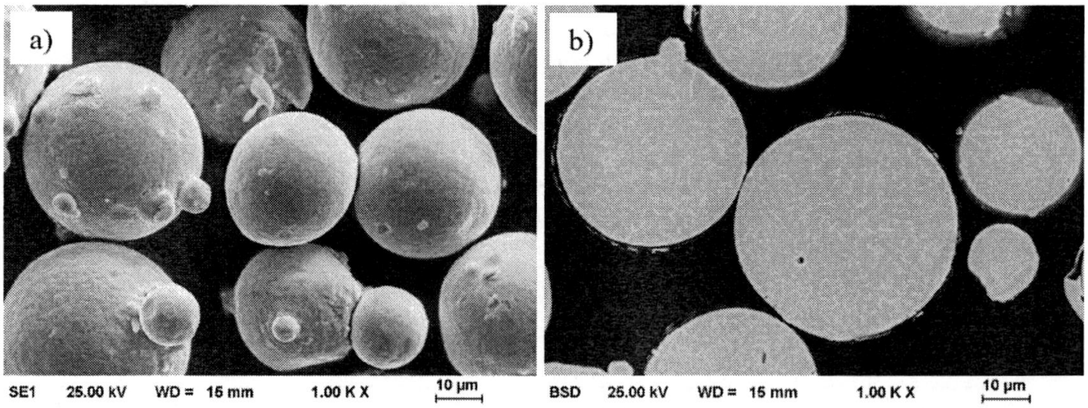

Figure 1. SEM images of Inconel 718 powder: a) surface (SE) and b) cross-section (BSE).

The gas atomized powder has a characteristic spherical shape. A dendritic structure can be observed in the cross-section. Due to the low porosity the powder is well suited for the HVOF process.

The microstructure of the coating in as-sprayed state was investigated in cross-sections by light microscopy. Furthermore, the influence of the additional heat treatment step was investigated. Cross-sections of the coatings in dependence of the treatment condition are shown in Figure 2.

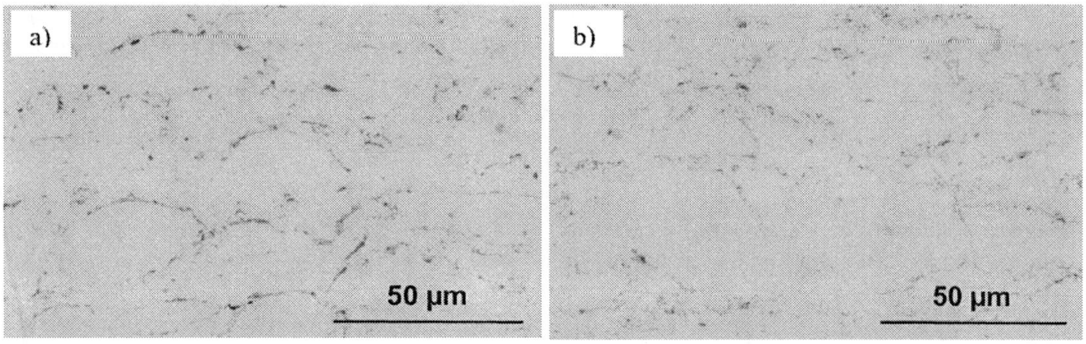

Figure 2. Light microscopy of Inconel 718 HVOF coatings: a) as-sprayed state and b) solution annealed state.

A typical structure occurs for the coating in as-sprayed state. Single spray particles can be observed, which are flattened due to the high kinetic energy of the HVOF coating process. These particles are well bonded. At the boundaries oxides appear due to the interaction with the atmosphere in the coating process. Only a minor content of porosity is formed.

Solution annealing causes an altered coating structure. Particle boundaries are dissolved and the oxide lamellae coagulate, resulting in the reduction of coating defects and a more homogeneous state.

The chemical composition was measured for the feedstock powder and the coatings in as-sprayed as well as solution annealed state by XRF. The results are summarised in Table 4.

Table 4. Chemical composition of the feedstock powder and HVOF coatings in as-sprayed (AS) and solution annealed (SA) state, measured by XRF in wt.%.

	Ni	Cr	Fe	Nb	Mo	Ti
powder	52.5	19.2	18.5	5.3	3.5	0.9
coating (AS)	52.5	19.7	19.0	4.7	2.9	1.2
coating (SA)	53.2	16.9	19.3	4.8	3.1	2.6

Processing of the powder by HVOF thermal spraying and subsequent solution annealing causes only minor changes of the chemical composition. The chemical composition is within the specification of the manufacturer. For the coating in as-sprayed state the molybdenum and niobium content are slightly reduced in comparison to the feedstock powder. The chromium concentration is decreased after the solution annealing treatment. The heat input in the coating and heat treatment process result in diffusion, oxidation or evaporation of individual components and minor changes of the chemical composition.

In general, the changes in the composition are small and the composition of the alloy Inconel 718 has been largely preserved.

3.2. Boriding treatment

The microstructure of the boron-enriched surface layer was investigated in cross-sections using an SEM equipped with a BSE detector to show material contrast. Representative images of the coatings borided in as-sprayed and solution annealed state are shown in Figure 3.

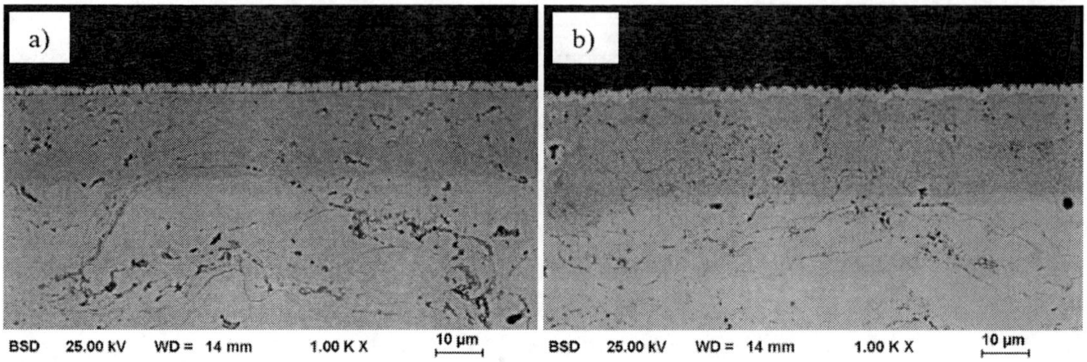

Figure 3. Cross-sections of borided Inconel 718 HVOF coatings: a) borided in as-sprayed state and b) borided after solution annealing.

For the coating borided in as-sprayed state, a uniform dark layer with a thickness of approximately 20 µm can be observed at the surface. The low intensity of backscattered electrons indicates a high concentration of elements with a low atomic number. Thus, an enrichment with boron at the surface can be assumed. The enriched layer has a mostly homogeneous thickness and is well bonded to the underlying material. However, the structure of the thermally sprayed coating comprised of single particles and structural defects can still be observed. In the underlying material the dendritic structure of the initial feedstock powder is retained.

For the coating borided in solution annealed state also a diffusion enriched surface layer is formed. The typical structure of thermally sprayed coatings is partially dissolved after the additional annealing treatment. In contrast to the coating borided in as-sprayed state no dendritic structure occurs, showing that the heat treatment caused homogenisation.

For the investigation of the elemental distribution, depth profiles have been determined by GDOS measurements. The depth profiles of boron and the main alloying elements nickel and chromium are shown in Figure 4.

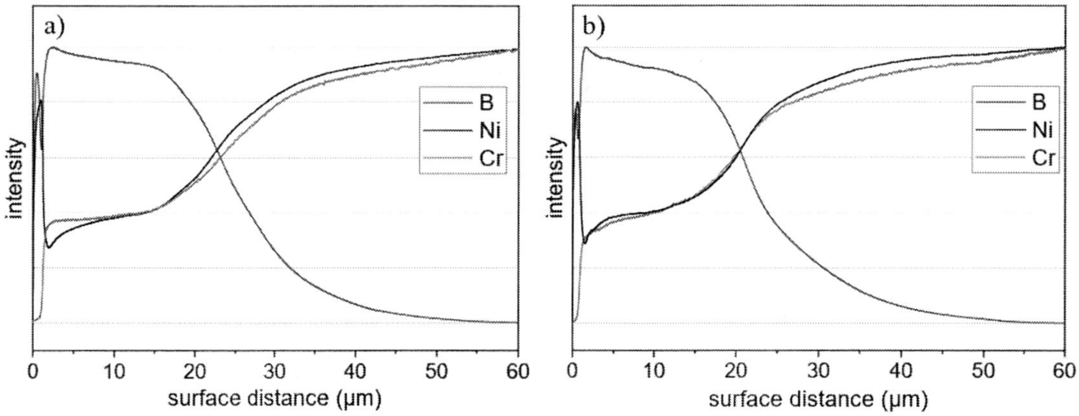

Figure 4. Elemental depth profiles determined by GDOS for borided Inconel 718 HVOF coatings: a) as-sprayed coating and b) solution annealed coating.

For the coatings borided in as-sprayed state a high concentration of boron was determined at the surface. Until a surface distance of approximately 20 µm a high level of boron concentration is retained. At higher depths a sharp decline occurs. With decreasing boron concentration, the concentration of the main alloying elements nickel and chromium increases. However, a continuous transition from the boron-enriched layer to the thermally sprayed nickel-base coating occurs.

For the coatings borided in solution annealed state similar elemental depth profiles were recorded. A high concentration of boron was determined at the surface, which is retained until a surface distance of approximately 20 μm.

For the evaluation of the phase formation in dependence of the conducted treatment, phase analyses by XRD have been conducted. The resulting diffractograms are shown in Figure 5.

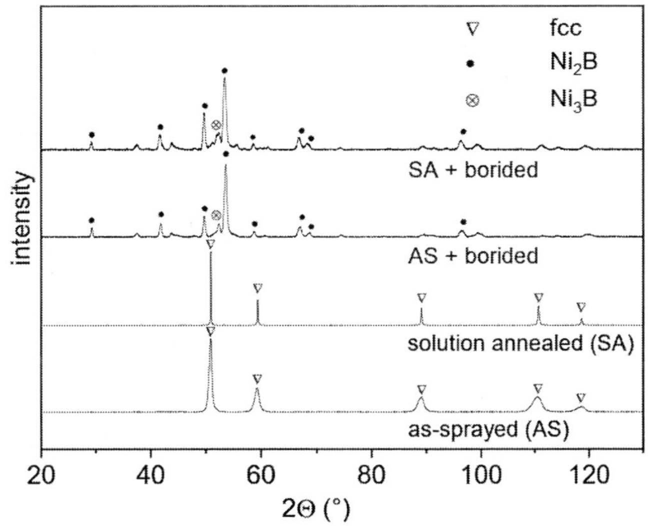

Figure 5. Diffractograms of Inconel 718 HVOF coatings in dependence of the treatment condition.

For the coating in as-sprayed state solely diffraction peaks of a phase with fcc structure occur. Solution annealing results in no distinct changes of the diffractogram, showing that no phase transformation occurred. However, in comparison to the as-sprayed state narrower diffraction peaks can be observed. This behaviour is caused by grain coarsening as a result of the heat treatment.

Boriding of the coatings results in a distinct change of the diffractograms. No diffraction peaks of the fcc phase can be observed after diffusion enrichment with boron. For the coating borided in as-sprayed state high intensity diffraction peaks of the tetragonal precipitates Ni_2B appear. Furthermore, diffraction peaks with low intensity are present, which can be assigned to the orthorhombic Ni_3B phase. Partially overlapping of the diffraction peaks with other phases occurs. The formation of the phase Ni_4B_3, which was proven in previous studies for the borided bulk Inconel 718, could not be detected [18]. Additional minor diffraction peaks are not assignable, showing that minor contents of additional phases were formed. According to the phase analyses by XRD the formation of silicides was prevented by the application of a process routine adapted for nickel-base alloys. A successful formation of precipitates in the boron-enriched surface layer could be achieved.

The additional solution annealing step prior to the boriding did not cause distinct changes of the respective diffractogram. Hence, the phase formation was not influenced.

To investigate the influence of the treatment condition on the resulting properties, hardness measurements have been conducted. The results are summarised in Table 5.

Table 5. Microhardness HV 0.001 of Inconel 718 HVOF coatings borided in as-sprayed and solution annealed state.

	as-sprayed coating	solution annealed coating
B-rich layer	2110 ± 120	2010 ± 80
unaffected material	500 ± 40	440 ± 20

Boriding causes a distinct increase of hardness in comparison to the unaffected material due to the successful formation of a precipitation layer. For the coating borided in as-sprayed state a high hardness of 2110 HV 0.001 was determined. A high standard deviation occurs due to the presence of structural defects in the precipitation layer. For the coating, solution annealed prior to the boriding treatment, the hardness is slightly reduced. Also the standard deviation is reduced indicating the formation of a more homogeneous state and the reduction of structural defects, which was proven by microscopic investigations. Solution annealing reduces the hardness measured in the unaffected material, which can be caused by grain coarsening and the reduction of residual stress.

3.3. Wear behaviour

The results of the wear investigations under reciprocating conditions in dependence of the test temperature and treatment condition are summarised in Figure 6.

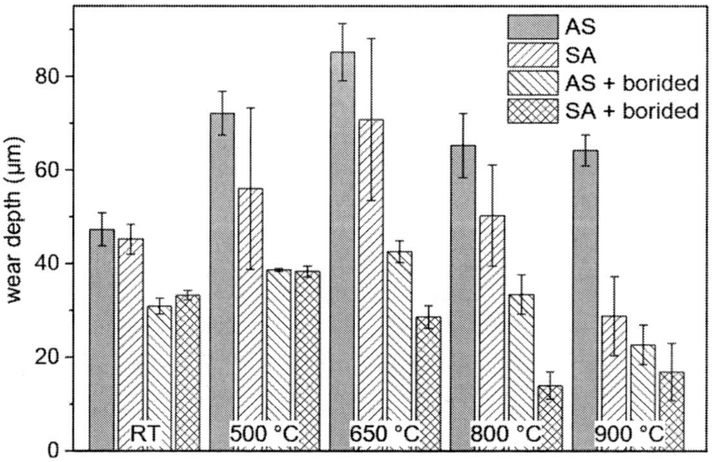

Figure 6. Results of reciprocating wear investigations for Inconel 718 HVOF coatings in a temperature range of 25 °C to 900 °C in dependence of the treatment condition.

The investigations at room temperature reveal a similar wear resistance for the as-sprayed and solution annealed coating. Boriding causes a reduction of the wear depth and hence an increased wear resistance. However, the wear depth exceeds the thickness of the diffusion enriched surface layer, which was determined by SEM and GDOS measurements, showing that failure occurred.

With an increase of temperature, the wear depth distinctly increases for the coating in as-sprayed and solution annealed state. This effect can be caused by thermal softening. For temperatures of ≥ 800 °C the wear depth is reduced. This effect is most pronounced for the coatings in solution annealed state, resulting in an improvement of wear resistance in comparison to the as-sprayed state. Possible reasons are the enhanced internal bonding and the reduction of heterogeneities.

The coatings in borided state show a less distinct dependence on test temperature. This behaviour indicates that thermally stable precipitates were formed. For temperatures of ≥ 800 °C the wear depth is reduced. This effect is most pronounced for the coatings in solution annealed and borided state.

For the investigation of the underlying wear mechanisms the wear tracks were investigated by SEM using a BSE detector to visualise material contrast. Representative images in dependence of the treatment condition and test temperature are shown in Figure 7.

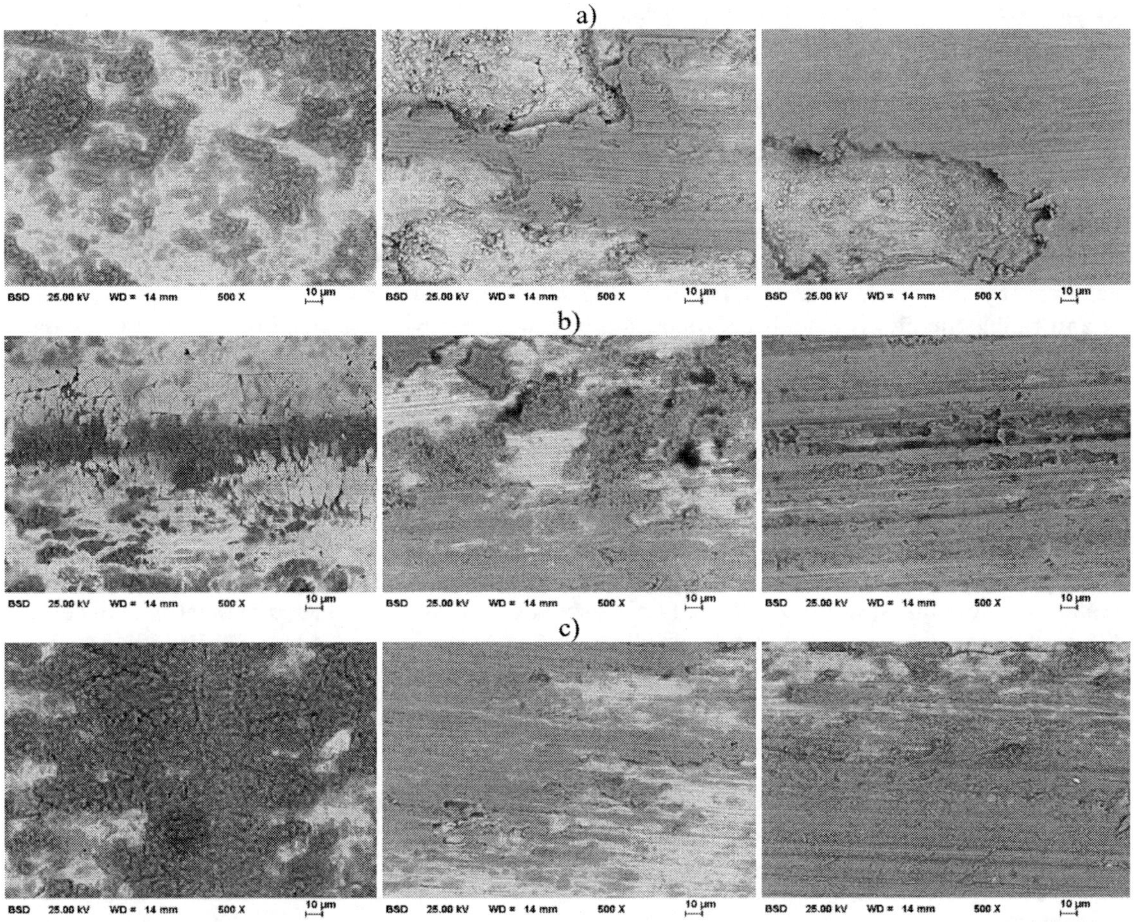

Figure 7. Surface (SEM/BSE) of the HVOF Inconel 718 coatings after reciprocating wear investigations at room temperature (left), 650 °C (middle) and 900 °C (right): a) solution annealed state, b) borided state and c) solution annealed and borided state.

The wear tracks of the coatings in solution annealed state exhibit a distinct material contrast. Dark areas represent a low average atomic number. These areas can be assigned to oxides, whereas the bright appearing areas can be assigned to the metallic coating. The surface is partially covered with oxides showing a rough surface. Hence, a permanent formation and removal of oxides can be assumed. At increased temperature of 650 °C, where the maximum wear depth was determined, the wear track exhibits a changed appearance. The surface is still partially covered with oxides. However, a relatively smooth oxide layer with grooves in wear direction is formed. With a further increase of test temperature (900 °C) the content of the surface covered with oxides increases, enabling a protection of the underlying coating. Partially breakouts and failure of the oxide layer occurs.

For the coating borided in as-sprayed state, the investigations at room temperature result in a distinctly changed appearance. Partially oxides are formed. Furthermore, cracks can be observed, showing that reciprocating load causes failure of the surface layer. For the test conducted at a temperature of 650 °C the wear track is partially covered with oxides, showing a relatively smooth surface. Higher test temperatures result in the formation of a widely closed oxide film. Grooves in wear direction appear, indicating abrasive wear of the oxide film. However, protection of the underlying material could be achieved.

For the coatings borided after initial solution annealing of the coating, distinct oxide formation was observed after testing at room temperature. However, a rough surface appears showing that permanent formation and removal of oxides occurs. At elevated temperature wide areas are covered with a smooth oxide layer, protecting the underlying material.

4. Summary and Conclusions

The process combination of powder-pack boriding thermally sprayed Inconel 718 coatings was successfully realised. A homogenisation of the coating could be achieved by an additional solution annealing step. The chemical composition could be retained by the chosen process route. Powder-pack boriding resulted in the formation of a homogeneous precipitation layer mainly comprised of the tetragonal phase Ni_2B. No delamination occurred and the formation of silicon-rich phases could be prevented by the application of an adapted boriding routine. A significant increase in hardness could be achieved by the thermochemical treatment. Detailed investigations on the wear behaviour under reciprocating conditions have been conducted in a wide temperature range up to 900 °C. A distinct improvement of wear resistance in the investigated temperature range was determined. Furthermore, the temperature dependence was reduced for the borided coating, showing that thermally stable precipitates were formed. However, the wear depth exceeded the thickness of the precipitation layer, hence an adaption of the boriding process is required to enable a higher thickness of the precipitation layer. The combination of solution annealing and subsequent boriding resulted in a further increase of wear resistance. Detailed investigations of the wear tracks revealed the formation of stable oxide films at elevated temperatures, enabling the protection of the underlying material.

The results show that the investigated process combination is a conceivable option for the replacement of bulk nickel-base alloys. Further studies have to be conducted to determine the effect of the investigated process combination on the resulting corrosion properties. A distinct reduction of process cost and duration can possibly be achieved by the integration of the solution annealing treatment in the thermochemical process.

Acknowledgements

The authors gratefully acknowledge the Arbeitsgemeinschaft Industrieller Forschungsvereinigungen "Otto von Guericke" e.V. (AiF) for support of this work (AiF-No. ZF4131908SU9) with funds from the German Federal Ministry for Economic Affairs and Energy.

References

[1] Davis J R 1997 *ASM Specialty Handbook: Heat-Resistant Materials* ASM International: Materials Park, OH, USA

[2] Ghica V G, Buzatu M, Iacob G and Petrescu M I 2014 Electron Microscopy Researches on the Microstructure of Inconel 718 Alloy *Revista de Chimie* **1** 105-109

[3] Osoba L O, Oladoye A M and Ogbonna V E 2019 Corrosion evaluation of superalloys Haynes 282 and Inconel 718 in hydrochloric acid *J. Alloys Compd.* **804** 376–384, https://doi.org/10.1016/j.jallcom.2019.06.196

[4] Doleker K M, Odabas O, Ozgurluk Y, Askerov H and Karaoglanli A C 2019 Effect of high temperature oxidation on Inconel 718 and Inconel 718/YSZ/Gd2Zr2O7 *Mater. Res. Express* **6** 1-9, https://doi.org/10.1088/2053-1591/ab26d8

[5] Bottoli F, Jellesen M S, Christiansen T L, Winther G and Somers M A J 2018 High temperature solution-nitriding and low-temperature nitriding of AISI 316: Effect on pitting potential and crevice corrosion performance *Appl. Surf. Sci.* **431** 24-31, http://doi.org/10.1016/j.apsusc.2017.06.094

[6] Lindner T, Löbel M, Saborowski E, Rymer L-M and Lampke T 2020 Wear and Corrosion Behaviour of Supersaturated Surface Layers in the High-Entropy Alloy Systems CrMnFeCoNi and CrFeCoNi *Crystals* **10(2)** 110, https://doi.org/10.3390/cryst10020110

[7] Kulka M 2019 *Current Trends in Boriding Techniques* Springer Nature Switzerland AG: Cham, Switzerland

[8] Günen A 2019 Properties and High Temperature Dry Sliding Wear Behavior of Boronized Inconel 718 *Metall. Mater. Trans.* A **51** 1-27, https://doi.org/10.1007/s11661-019-05577-3

[9] Davis J R 2002 *Surface Hardening of Steels: Understanding the Basics* ASM International: Materials Park, OH, USA

[10] Lindner T, Löbel M, Sattler B and Lampke T 2019 Surface hardening of FCC phase high-entropy alloy system by powder-pack boriding *Surf. Coat. Tech.* **371** 389-394, https://doi.org/10.1016/j.surfcoat.2018.10.017

[11] Ozdemir O, Omar M A, Usta M, Zeytin S, Bindal C and Ucisik A H 2009 An investigation on boriding kinetics of AISI 316 stainless steel *Vacuum* **83** 175-179, https://doi.org/10.1016/j.vacuum.2008.03.026

[12] Löbel M, Lindner T, Hunger R, Berger R and Lampke T 2020 Precipitation Hardening of the HVOF Sprayed Single-Phase High-Entropy Alloy CrFeCoNi *Coatings* **10(7)** 701, https://doi.org/10.3390/coatings10070701

[13] Abioye TA, McCartney D G and Clare A T 2015 Laser cladding of Inconel 625 wire for corrosion protection *J. Mater. Process* **217** 232-240, https://doi.org/10.1016/j.jmatprotec.2014.10.024

[14] Tucker R C 2013 *ASM Handbook, Volume 5A: Thermal Spray Technology* ASM International: Materials Park, OH, USA

[15] Oksa M, Turunen E, Suhonen T, Varis T and Hannula S-P 2011 Optimization and Characterization of High Velocity Oxy-fuel Sprayed Coatings: Techniques, Materials, and Applications *Coatings* **1(1)** 17–52, https://doi.org/10.3390/coatings1010017

[16] Lindner T, Löbel M, Hunger H, Berger R and Lampke T 2020 Boriding of HVOF-sprayed Inconel 625 coatings *Surf. Coat. Tech.* **404**, https://doi.org/10.1016/j.surfcoat.2020.126456

[17] Lin L and Han K 1998 Optimization of surface properties by flame spray coating and boriding *Surf. Coat. Tech.* **106** 100-105, https://doi.org/10.1016/S0257-8972(98)00501-5

[18] Petrova R S, Suwattananont N and Samardzic V 2007 The Effect of Boronizing on Metallic Alloys for Automotive Applications. *Materials Science & Technology - Automotive and Ground Vehicles symposium* **17-3** 340-345

IOP Conf. Series: Materials Science and Engineering 1147 (2021) 012033 doi:10.1088/1757-899X/1147/1/012033

Influence of diameter of the cored wires on abrasive wear resistance of arc sprayed coatings

H Pokhmurska[1], **M Student**[2,*], **G Paczkowski**[3], **M Holovchuk**[2], **H Chumalo**[2], **V Hvozdetskyi**[2] **and S Schuberth**[3,+]

[1] Lviv Polytechnic National University Department of Welding Manufacture, Diagnostics and Restoration of Metal Structures
[2] Karpenko Physico-Mechanical Institute of the NAS of Ukraine
[3] Materials and Surface Engineering Group, Faculty of Mechanical Engineering, Chemnitz University of Technology, Erfenschlager Straße 73, D-09125 Chemnitz, Germany
[+] present address: Cewotec gGmbH, Jagdschänkenstraße 104, D-09116 Chemnitz, Germany

[*]e-mail: student.phmi@gmail.com

Abstract. To obtain wear-resistant coatings, the cored wires (CW) with a diameter of 1.6...1.8 mm as electrode materials are mainly used. Using the large-diameter CW enable significantly increase the filling of their shells with powder charges and, as a consequence, to essentially raise the content of alloying elements in the coating structure. The model cored wires of two basic alloying systems Fe – Cr – C and Fe – Cr – B with a diameter of 1.6 and 2.4 mm used for spraying. Three types of the cored wires (namely 50CrNi2B5MnSi, 250Cr21WVMnSi, 50Cr6Mn2MoSi) used in this study. It is found that more oxide phase (40…100 %) was in coatings formed with CW with a diameter of 2.4 mm compared with coatings formed with CW with a diameter of 1.6 mm with lower filling. It is established that the abrasive and gas-abrasive wear resistance of coatings with 250Cr21WVMnSi CW decreased, and with 50CrNi2B5MnSi CW increased due to the increase in the thickness of the lamellae in their structure with the use of CW of larger diameter. It was found that the higher the coefficient of microheterogeneity and chromium content in the lamellae, the higher the corrosion resistance of coatings.

1. Introduction

Electric arc spraying of coatings is widely used in industry for protection against corrosion, restoration of the size of machine parts, as well as against abrasive and gas-abrasive wear, both at room and at elevated temperatures [1-9]. For spraying wear-resistant coatings, cored wires (CW) with a diameter of 1.6...1.8 mm are mainly used as electrode materials. Such CW cannot always provide a high filling factor with the charge, and therefore cannot provide the required content of alloying elements in the coatings. At the same time, CW coatings operated under abrasive wear conditions, especially in corrosive environments, require at least 12 wt.% chromium in their composition [10-14]. The use of CW of large diameters makes it possible to significantly increase their filling factor with the charge, and, therefore, to significantly increase the content of alloying elements in the coating structure. However, there are no studies in the world literature that would show the effect of the diameter of the CW and the productivity of spraying coatings on the structure and abrasive wear resistance. This hinders the development of the method of electric arc spraying of coatings and does not allow the creation of new CW with a high content of alloying elements.

2. Experimental and materials

Electric arc coatings (EAC) were obtained using original equipment, developed and manufactured at the Karpenko Physico-Mechanical Institute of the NAS of Ukraine. For spraying, model CW of basic alloying systems Fe – Cr – C and Fe – Cr – B with a diameter of 1.6 and 2.4 mm were used (table 1, figure 1).

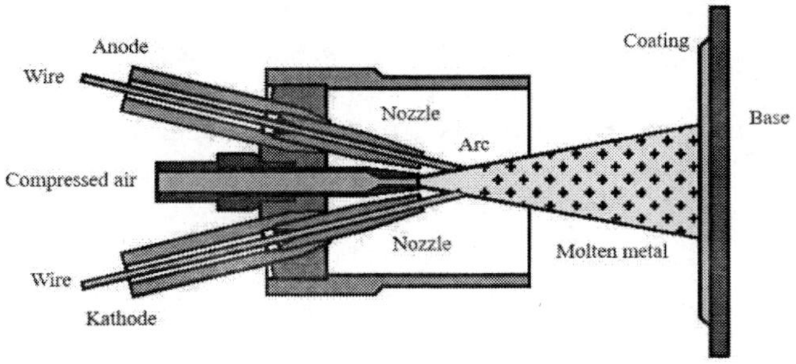

Figure 1. A typical scheme of formation of electric arc coatings.

Table 1. Chemical composition of the used model cored wires.

CW type (CW designation)	Ø CW (mm)	Content of elements (wt. %)									
		Fe	C	Si	Mn	Cr	Ni	B	V	Mo	W
50CrNi2B5MnSi (*A*)	1.6	90.6	0.55	0.19	1.09	0.20	1.38	5.95	-	-	-
	2.4	91.1	0.45	0.30	1.20	0.22	1.75	4.96	-	-	-
250Cr21WVMnSi (*B*)	1.6	74.4	2.80	0.30	0.39	21.10	-	-	0.60	-	0.40
	2.4	74.0	2.51	0.62	0.38	21.40	-	-	0.52	-	0.55
50Cr6Mn2MoSi (*C*)	1.6	91.7	0.41	0.40	1.56	5.40	-	-	-	0.50	-
	2.4	90.6	0.47	0.40	1.40	6.60	-	-	-	0.51	-

To obtain the same chemical composition of cored wires, iron powder was additionally added to the CW charge of a larger diameter. The cross-section of both investigated flux-cored wires is shown in figure 2. Metallographic studies were performed on PO-21 and Carl Zeiss Stemi 2000 optical microscopes and on LEO-840 scanning electron microscopes (with an attachment for microelement analysis) and ZEISS EVO-40XVP (with an INCA Energy X-ray microanalysis system). The phase composition of the coatings was studied on a DRON-3.0 diffractometer in Cu Kα radiation with tube focusing according to the Bragg-Brentano scheme.

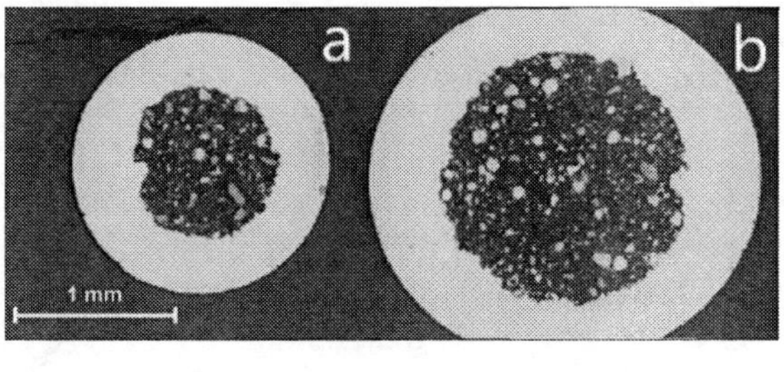

Figure 2. Cross section of flux-cored wires with a diameter of: (*a*) 1.6 and (*b*) 2.4 mm.

The elastic-plastic properties of the developed coatings were investigated by the method of dynamic indentation on the device "Micron-gamma" with automatic registration of load diagrams *P* - *h*, where *P* is the load on the indenter (Berkovich's diamond pyramid), *h* is the depth of its penetration into the coating.

The abrasive wear of the coatings was investigated under the conditions of a rigidly fixed abrasive using an abrasive corundum disk with a grain size of 20 μm in diameter and 150 and 8 mm wide, the rotation frequency of which was 2.7 s^{-1}, and the load in the contact zone - P = (14.7 ± 0.25) N. Gas-abrasive wear of the coatings was investigated at a temperature of 200 °C for 50 hours. Abrasive wear resistance was assessed by the weight loss of the samples after testing at an abrasive speed of 36 m/s. Quartz sand with a grain size of up to 200 μm was used as an abrasive.

EAC is characterized by a heterogeneous structure, the degree of heterogeneity of which was proposed to be estimated by the value of the microheterogeneity coefficient (K_{MH}) according to the formula (*1*):

$$K_{MH} = (\sum_{i=1}^{n} |c_{aver} - c_i|) / n \times c_{aver} \qquad (1)$$

where n is the number of rectangular sections with an area of $35 \cdot 10^{-4}$ mm^2, highlighted in the photo of the EAC microstructure (figure 3), on which the content of each c_i element within a separate lamella and the average content of the coating element are determined by spectral analysis.

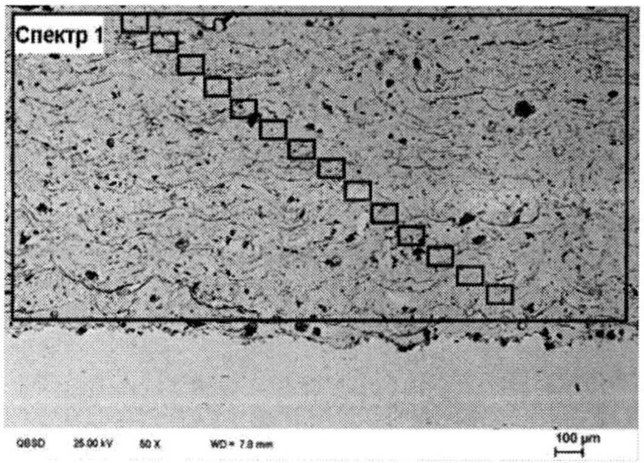

Figure 3. Scheme of carrying out Micro X-ray spectral analysis to determine the local content of alloying elements in the coating.

The corrosion characteristics of the coated samples were determined in potentiodynamic mode on an AutoTafel potentiostat (ACM Instruments) in 3% aqueous NaCl solution. The reference electrode was a silver chloride electrode of the EVL-1M1 type, the auxiliary one was platinum, with a potential change of 1 mV/s. The corrosion rate was determined by extrapolating the linear sections of the polarization curves to the corrosion potential. For electrochemical studies, the working part of the samples with an area of 0.5 cm^2 was isolated, applying an insulating waterproof lacquer coating around it.

3. Results and discussion

3.1. Influence of CW diameter on the structure of coatings from model CW of different diameters

During electric arc spraying of coatings between the ends of the wires there is an arc in which the wires melted, and their melt was dispersed by air jet into droplets, which, hitting the sprayed surface, deformed in the form of "saucers" (lamellae) and layer by layer formed the coating. To analyze the fractional and chemical composition of the drops, they were caught in a snow target. Metallographic and spectral analysis revealed that among the droplets there are completely oxide droplets, droplets with an outer layer of oxides and droplets without oxides in general. It was found that with increasing CW diameter and arc current, the volume of molten metal at the ends of the CW increased and therefore the air jet dispersed it into droplets of larger diameter (figure 4). As a result, the thickness of the lamellae (as a structural feature of the coatings) increased in the coatings of all alloying systems (figure 5).

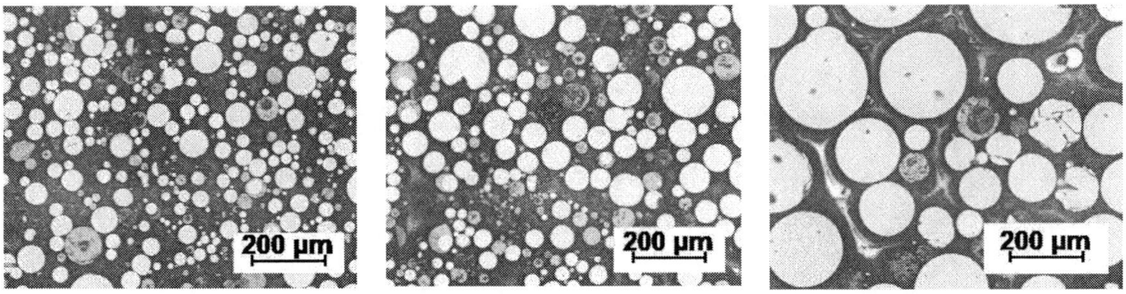

a *b* *c*

Figure 4. Morphology of droplets obtained by spraying of the *A* flux-cored wire into a snow target at an arc current that determines the spraying performance: (*a*) 125, (*b*) 150 and (*c*) 180. This droplet morphology is typical for all studied CW.

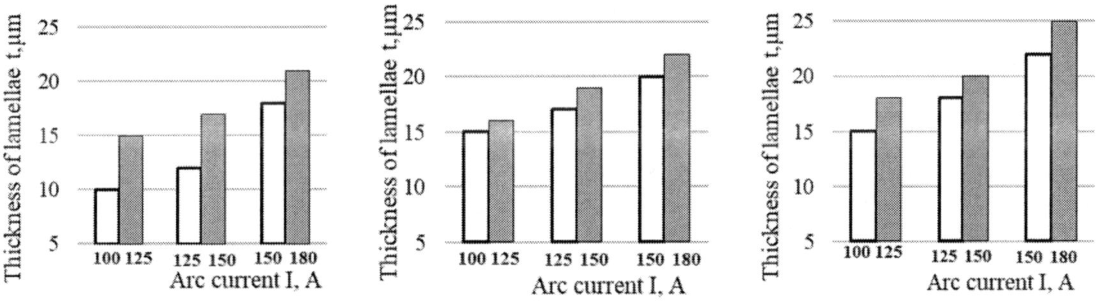

a *b* *c*

Figure 5. Influence of the arc current values *I* on the lamella thickness *t* in the structure of coatings obtained from three cored wires (indicated in the table 1 as *A*, *B*, *C*) with a diameter of 1.6 (light columns) and 2.4 (dark columns) mm: (*a*) CW 50CrNi2B5MnSi, (*b*) CW 250Cr21WVMnSi and (*c*) CW 50Cr6Mn2MoSi.

Since the coating process was carried out using an air jet, it turned out that in many cases the lamellae of the coatings are separated from each other by thin layers of oxide phases (figure 6). It was shown that with increasing the spraying distance from 80 to 120 mm (which is equivalent to an increase in the residence time of the droplets in the air stream), the content of the oxide phase in the coatings doubled. In the coatings formed with CW with a diameter of 2.4 mm, 40…100 % more oxide phase was found than in the coatings formed with CW with a diameter of 1.6 mm. This is due to the larger volume of the charge in the CW of larger diameter, the cavities between the powders of which are filled with air, which intensively oxidizes the melt droplets inside the CW and at their ends during spraying.

Phase analysis showed that the matrix phase of the coating structure with cored wire designated as *C* (50Cr6Mn2MoSi) was martensite with a small amount of iron oxide Fe_3O_4 (table 2). Whereas the matrix phase of the coating with cored wire designated as *B* (250Cr21WVMnSi) was residual austenite with a small amount of martensite and oxides of chromium and iron (Cr_2O_3 + Fe_3O_4). The rapid crystallization of droplets on the steel-bases contributed to the formation of a large amount of residual austenite in the coating structure. The matrix phase of the coating with cored wire designated as *A* 50CrNi2B5MnSi CW was martensite, which revealed nanoscale inclusions of complex-doped boride

FeCrB and a small amount of iron oxide Fe_3O_4. In this case, the rapid crystallization of the droplets led to the formation of a large number of crystallization centers of borides, which due to lack of time managed to grow only to nanoscale.

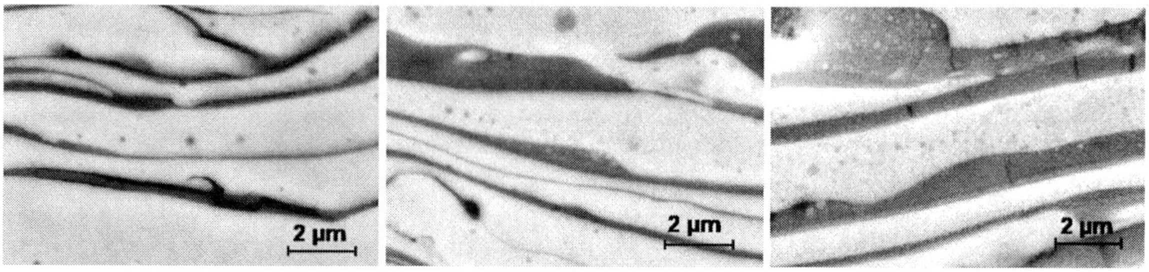

a *b* *c*

Figure 6. Oxide layers between the lamellae in the structure of the coatings sprayed of the *A* flux-cored wire from a distance of 80 (*a*), 100 (*b*), 120 (*c*) mm.

Table 2. Phase composition of structural elements of coatings from the investigated CW.

CW type (CW designation)	α-Fe (wt. %)	γ-Fe (wt. %)	oxide (wt. %)	FeCrB (wt. %)
50CrNi2B5MnSi (*A*)	47 martensite	-	Fe_3O_4	40
250Cr21WVMnSi (*B*)	28 martensite	60 residual austenite	Cr_2O_3+Fe_3O_4	-
50Cr6Mn2MoSi (*C*)	90 martensite	-	Fe_3O_4	-

3.2. Influence of lamella thickness in the structure of coatings on their physical and mechanical properties

The porosity of the coatings of all alloying systems increased with increasing thickness of the lamellae in the structure of the sprayed coatings and, accordingly, the diameter of the CW (figure 7). This pattern is due to the increase in the diameter of the droplets that formed the coating. With an increase in the diameter of the droplets they bounce off the substrate surface more intensively (figure 8), which led to the appearance of microcavities between the lamellae in the coating structure.

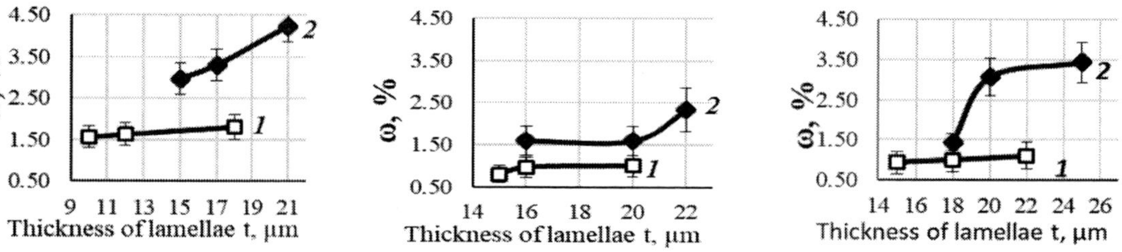

a *b* *c*

Figure 7. Effect of lamella thickness *t* on the porosity ω in the structure of coatings sprayed with cored wire 1.6 (*1*) and 2.4 (*2*) mm in diameter, designated as *A* (CW 50CrNi2B5MnSi, *a*), *B* (CW 250Cr21WVMnSi, *b*) and *C* (CW 50Cr6Mn2MoSi, *c*).

It was found that the microhardness of coatings sprayed with CW designated by *A* (CW 50CrNi2B5MnSi) and *C* (CW 50Cr6Mn2MoSi) reused with increasing thickness of lamellae in the structure of coatings, which was especially evident when used for spraying CW of larger diameter (figure 9*a*, *c*). This was explained by the evaporation and burnout of alloying elements from droplets, especially carbon during their spraying. As the diameter of the droplets increased, the amount of carbon burned out of the droplets decreased. And the higher carbon content in the martensitic matrix phase of the coatings was responsible for increasing their microhardness.

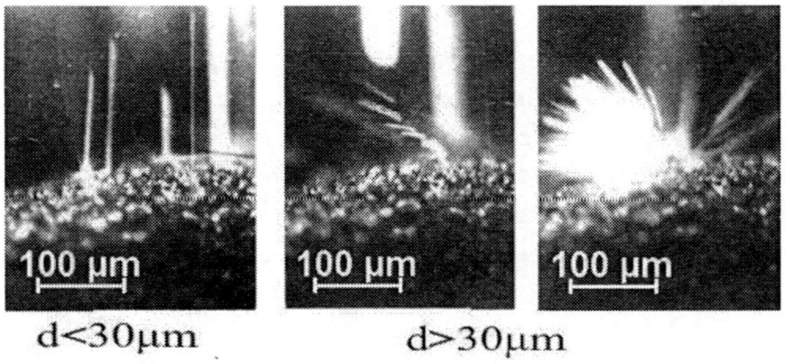

Figure 8. A typical picture of the bouncing of melt droplets of different diameters from the sprayed surface.

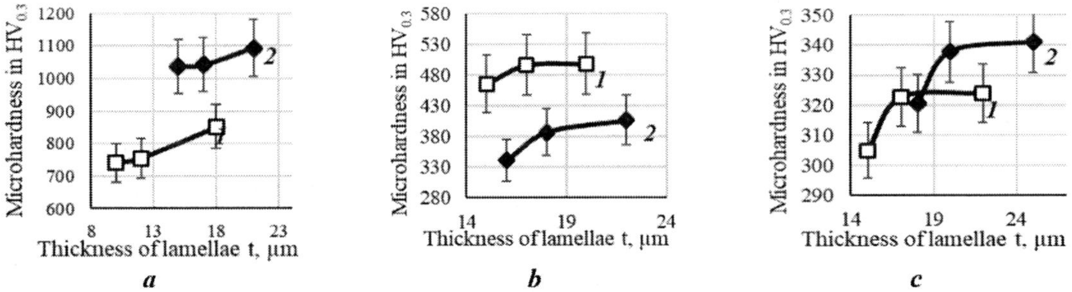

Figure 9. Influence of lamella thickness t on the microhardness $HV_{0.3}$ of coatings sprayed with cored wire 1.6 (*1*) and 2.4 (*2*) mm in diameter, designated as *A* (CW 50CrNi2B5MnSi, *a*), *B* (CW 250Cr21WVMnSi, *b*) and *C* (CW 50Cr6Mn2MoSi, *c*).

However, in coatings with 250Cr21WVMnS CW found the opposite trend of the influence of the thickness of the lamellae on the microhardness using a wire of larger diameter (figure 9*b*). This is due to the phase composition of these coatings, namely the significant amount of residual austenite in them, which has a lower microhardness than high-carbon martensite. Of the small droplets formed during the spraying of CW of smaller diameter, much more carbon burned out (figure 10), which means that the content of austenite in the lamellae of the coating was lower, and martensite higher.

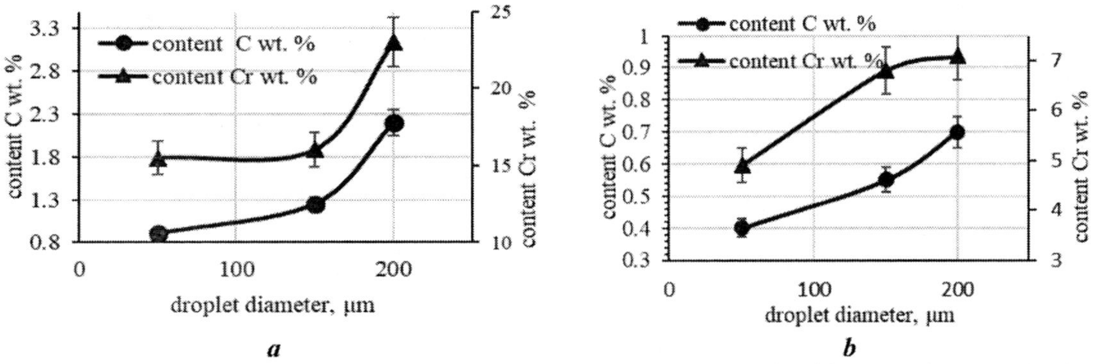

Figure 10. Influence of the diameter of molten metal droplets obtained by spraying cored wires *B* (CW 250Cr21WVMnSi, *a*) and *C* (CW 50Cr6Mn2MoSi, *b*) on the content of carbon and chromium in them.

The adhesion of coatings with 50CrNi2B5MnSi CW and 250Cr21WVMnSi CW (CWs designated *A* and *B*) slightly decreased with increasing lamella thickness in the structure of coatings, caused especially by the use of CW of larger diameter (figure 11*a*, *b*). This is due to the occurrence of tensile

IOP Conf. Series: Materials Science and Engineering 1147 (2021) 012033 doi:10.1088/1757-899X/1147/1/012033

stresses in the sprayed coatings. As a result of their influence, microcracks appeared in the coatings or even their network was formed, which significantly reduced their adhesion to the steel base

For the coating with 50Cr6Mn2MoSi CW (CW designated *C*), the inverse regularity was found (figure 11*c*). As the thickness of the lamellae in the structure of the coating and the diameter of the CW increased, its adhesion to the steel base increased. This was explained by the higher thermal energy of the droplets, the lower combustion of carbon from them and the formation of coatings with the matrix phase of high-carbon martensite. The more carbon in martensite, the lower the coefficient of thermal expansion of the coating, which causes a lower level of residual tensile stresses in it.

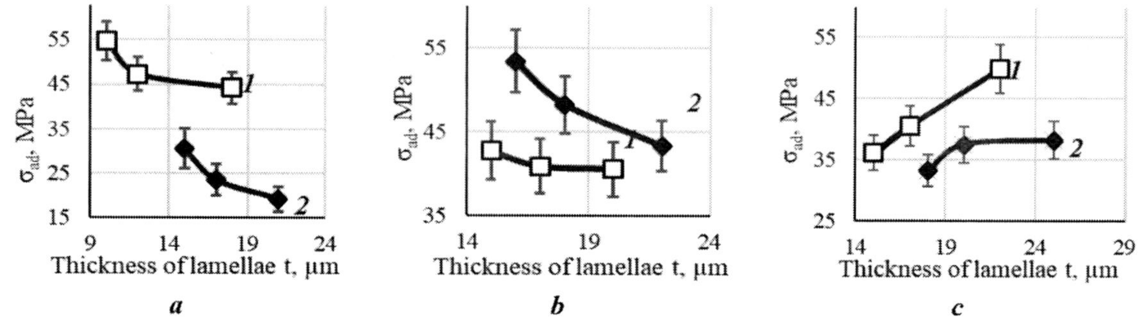

Figure 11. Relationship of the lamella thickness t in the structure of coatings sprayed with cored wires *A* (CW 50CrNi2B5MnSi, *a*), *B* (CW 250Cr21WVMnSi, *b*) and *C* (CW 50Cr6Mn2MoSi, *c*) with a diameter of 1.6 (*1*) and 2.4 (*2*) mm and of the adhesion values σ_{ad} of the coatings to the steel substrate.

3.3. Influence of lamella thickness in the structure of coatings on their wear resistance

With increasing thickness of the lamellae in the structure of coatings sprayed 250Cr21WVMnSi CW (especially larger diameter), their abrasive wear resistance decreases (figure 12). This result was explained by the greater amount of residual austenite in the coating (table 2), which has a lower microhardness than martensite. Smaller droplets burn out more carbon than larger ones. Therefore, in the coating sprayed with CW of a diameter of 1.6 mm less carbon than in the coating with CW of a diameter of 2.4 mm, and hence less residual austenite

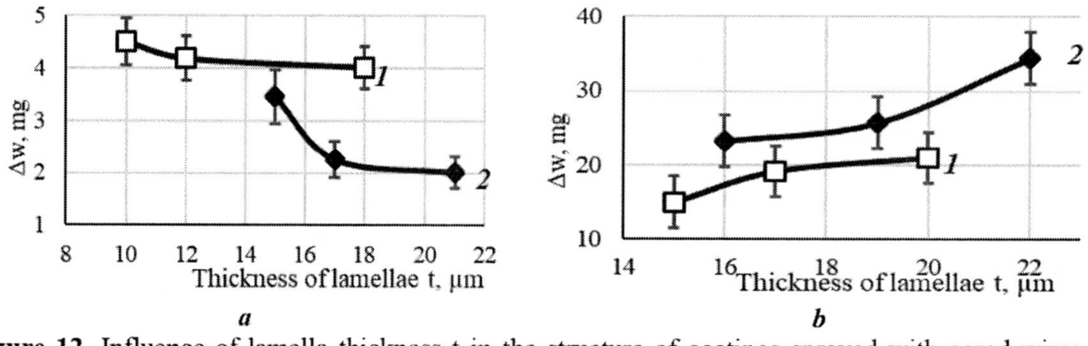

Figure 12. Influence of lamella thickness t in the structure of coatings sprayed with cored wires *B* (CW 250Cr21WVMnSi, *a*) and *C* (CW 50Cr6Mn2MoSi, *b*) with a diameter of 1.6 (*1*) and 2.4 (*2*) mm on their abrasive wear resistance by weight loss of samples Δw.

With increasing thickness of the lamellae in the coatings sprayed with CW 50CrNi2B5MnSi (CW designated *A*) especially larger diameter), their abrasive wear resistance increased. This is due to the increase in the hardness of such coatings and the lower level of residual tensile stresses (figure 13). In such coatings, tensile stresses are relaxed due to the formation of microcracks, which, however, do not reduce abrasive resistance.

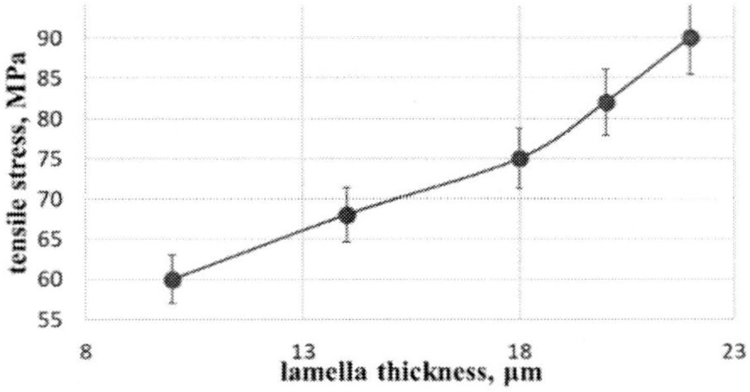

Figure 13. Influence of lamella thickness on the level of residual tensile stresses in the coating with CW 50CrNi2B5MnSi (CW designated *A*).

The nature of the change in gas-abrasive wear resistance of coatings from these CW is similar to that obtained for abrasive wear resistance. Gas-abrasive wear resistance of coatings with 250Cr21WVMnSi CW (CW designated *B*) decreased with increasing thickness of lamellae and diameter of CW (fgure 14*a*), while for coatings with CW 50CrNi2B5MnSi (CW designated *A*) – it increased and for coatings with CW of larger diameter was higher (figure 14*b*).

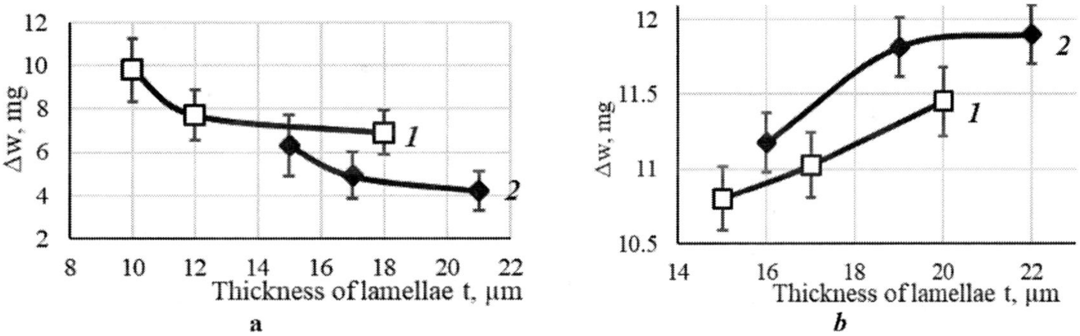

Figure 14. Influence of lamella thickness t in the structure of coatings with cored wires CW 50CrNi2B5MnSi (*a*) and CW 250Cr21WVMnSi (*b*) marked as *A* and *B* respectively with a diameter of 1.6 (*1*) and 2.4 (*2*) mm on their gas-abrasive wear resistance by loss of sample mass Δw.

3.4. Corrosion resistance of coatings sprayed with flux-cored wire 250Cr21WVMnSi

To ensure high corrosion resistance of the coatings, it is important that their microheterogeneity in terms of chromium content is low, and that its content in each lamella of the coating exceeds 12 wt. %. (first threshold of corrosion resistance) To fulfill this condition, it is necessary to ensure a sufficient number of chromium-containing components in the composition of the CW charge, which have achieved by an increase in the diameter of the CW, and complete fusion of the chromium-containing charge and steel shell. To meet the requirement for complete fusion of CW components, a eutectic mixture of powders (FeSi + FeMn) was added to the composition of its charge.

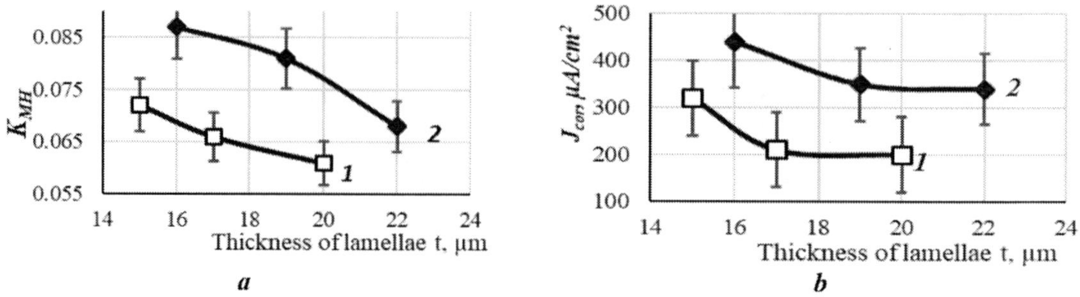

Figure 15. Influence of lamella thickness *t* in the structure of coatings with CW 250Cr21WVMnSi (CW designated *A*) with diameter 1.6 (*1*) and 2.4 (*2*) mm on the coefficient of microheterogeneity of coatings K_{MH} (*a*) and their corrosion resistance in 3% aqueous NaCl solution by corrosion current density j_{cor} (*b*).

On the example of coatings with 250Cr21WVMnSi CW of both diameters it was found that the microheterogeneity of coatings decreased and corrosion resistance increased with increasing thickness of the lamellae in their structure, due to a more complete fusion of the charge and shell (figure 15*a*). CW of larger diameter contains a larger amount of charge, which causes insufficient fusion of its chromium-containing charge and shell, so coatings with CW of larger diameter are more heterogeneous and less corrosion resistant (figure 15*b*).

4. Conclusions

It was found that the thickness of the lamellae in the coatings of all analyzed alloying systems increased with increasing volume of molten metal at the ends of the CW with the corresponding formation of droplets of larger diameter during air dispersion, which was realized through the use of CW of larger diameter with a correspondingly high filling factor of its shell with a charge and increased arc current

It is shown that the amount of oxide phase in the structure of coatings doubles with increasing spray distance CW from 80 to 120 mm. More oxide phase (40… 100%) was found in coatings formed with CW with a diameter of 2.4 mm with a higher filling factor of its shell charge compared with coatings formed with CW with a diameter of 1.6 mm with lower filling. This is due to the larger volume of the charge in the CW of larger diameter, the cavities between the powders in which are filled with air, which intensively oxidizes the melt droplets inside the CW and at their ends during spraying.

It is established that the porosity and microhardness of coatings from the developed CW increases with increasing thickness of the lamellae in their structure and, accordingly, the diameter of the CW. The increase in microhardness is due to less evaporation and burnout of alloying elements from the droplets forming the coating. Increase in porosity is caused by of the bouncing of melt droplets of different diameters from the sprayed surface, which contributes to microcavities between the lamellae of the coating.

It was found that with increasing the thickness of the lamellae in the structure of coatings with 250Cr21WVMnSi CW and 50CrNi2B5MnSi CW their adhesion to the steel base decreased slightly due to the occurrence of tensile stresses in coatings, which cause microcracks or cracks in their structure. At the same time, the adhesion of coatings with 50X6MГ2C CW to steel increased, which was associated with higher carbon content in large lamellae and favorable conditions for the formation of high-carbon martensite, which has the lowest coefficient of thermal expansion and causes the lowest residual tensile stresses in coatings.

It is established that the abrasive and gas-abrasive wear resistance of coatings with 250Cr21WVMnSi CW decreased, and with 50CrNi2B5MnSi CW increased due to the increase in the thickness of the lamellae in their structure with the use of CW of larger diameter.

It was found that the corrosion resistance of coatings from the developed CW correlates with the coefficient of microheterogeneity and chromium content in the lamellae of coatings: the higher the coefficient of microheterogeneity and chromium content in the lamellae, the higher the corrosion resistance of coatings.

References

[1] Dallaire S and Levert H 1997 Development of cored wires for improving the abrasion wear resistance of austenitic stainless steel. *J. Thermal Spray Technology* 6(4) 456–461 https://doi.org/10.1007/s11666-997-0031-7
[2] Liu S-G, Wu J-M, Zhang S-C, Rong S-J and Li Z-Z 2007 High temperature erosion properties of arc-sprayed coatings using various cored wires containing Ti-Al intermetallics *Wear* 262 (5-6) 555-561 https://doi.org/10.1016/j.wear.2006.06.021
[3] Arizmendi-Morquecho A, Campa-Castilla A, Leyva-Porras C, Martinez J A A, Gutiérrez G V, Bello K J M and López L 2014 Microstructural characterization and wear properties of Fe-based amorphous-crystalline coating deposited by twin wire arc spraying *Advances in Materials Science and Engineering* Article ID 836739 http://dx.doi.org/10.1155/2014/836739
[4] Korobov Yu, Filippov M, Makarov A, Malygina I, Soboleva N, Fantozzi D, Andrea M, Koivuluoto H and Vuoristo P 2018 Arc-sprayed Fe-based coatings from cored wires for wear and corrosion protection in power engineering *Coatings* 8 (2) 71 https://doi.org/10.3390/coatings8020071.
[5] Pokhmurskii V, Student M, Gvozdeckii V, Stypnutskyy T, Student O, Wielage B and Pokhmurska H 2013 Arc-sprayed iron-based coatings for erosion-corrosion protection of boiler tubes at elevated temperatures *Journal of Thermal Spray Technology* 22, 5 808-819
[6] Wielage B, Pokhmurska H, Student M, Gvozdeckii V, Stupnyckyj T and Pokhmurskii V 2013 Iron-based coatings arc-sprayed with cored wires for applications at elevated temperatures *Surface and Coatings Technology* 220 27–35
[7] Student M, Pokhmurs'ka H, Hvozdets'kyi V, Holovchuk M and Romaniv M 2009 Effect of high-temperature corrosion on the gas-abrasive resistance of electric-arc coatings *Materials Science* 45 481-489
[8] Student M, Pokhmurs'ka H and Zadorozhna K 2018 Structure and wear resistance of VC–FeCr and VC– FeCrCo coatings obtained by supersonic flame spraying *Materials Science* 54(4) 535–541
[9] Posuvailo V M, Klapkiv M D, Student M M, Pokhmurska H V and Sirak Y Y 2017 Gibbs energy calculation of electrolytic plasma channel with inclusions of copper and copper oxide with Al-base *IOP: Materials Science and Engineering* 181 012045 157–168
[10] Stupnyts'kyi T, Student M, Pokhmurs'ka H and Hvozdets'kyi V 2016 Optimization of the chromium content of powder wires of the Fe–Cr–C and Fe–Cr–B systems according to the corrosion resistance of electric-arc coatings *Materials Science* 52 (2) 165-172
[11] Hvozdets'kyi V M, Sirak Ya Ya, Zadorozhna Kh R and Dem'yanchuk Ya M 2018 Influence of the size of drops and the velocity of flow on the structure and properties of electric-arc coatings *Materials Science* 53(5) 702–708
[12] Pokhmurs'kyi V, Student M, Formanek B, Serivka V, Dz'oba Yu, Dovhunyk V and Sydorak I 2003 Heat resistance of electric arc coatings made of Fe-Cr-B-Al powder wire *Materials Science* 39(6) 829-834
[13] Pokhmurs'ka H V, Holovchuk M Y, Dz'oba Yu V, Hvozdets'kyi V M and Dzyubyk L V 2018 Influence of the composition of charge of powder wires on the structure and properties of electric-arc coatings *Materials Science* 53(6) 868-874
[14] Khoma M S, Sysyn H M, Radkevych O I and Holovchuk M Ya 2012 Electrochemical properties of welded joints of 17G1SU steel *Materials Science* 47(5) 696-702

Advanced corrosion resistant cylinder-bore coatings

B Schilder[1,*], A Garling[1], U Mayer[1], R Grabs[1] and T Lampke[2]

[1] Process Development, Mercedes-Benz AG, 70546 Stuttgart, Germany
[2] Institute of Materials Science and Engineering, Chemnitz University of Technology, 09125 Chemnitz, Germany

[*]e-mail: bernd.schilder@daimler.com

Abstract. This paper presents a novel approach to enhance the corrosion resistance of cylinder-bore coatings for internal combustion engines by changing surface characteristics and microstructures of the coating and utilizing alternative materials. The approach of this study refers to the knowledge of the complex gas-flow conditions during the internal-diameter coating of cylinder-bores using a twin-wire arc-spray process. Selective influencing of the gas flow enables the preparation of sample coatings with inhomogeneous and more homogeneous microstructures and surfaces. The various coatings were exposed to corrosive environments using the Kesternich testing method and an alternating immersion test. Significant improvements in corrosion resistance have been demonstrated by up to 50 % through altering the microstructure of the coating and almost 100 % by additional changing the spraying material. The study reveals an increased protection of cylinder-bore coatings against corrosive attacks through a homogeneous microstructure prepared in oxygen-poor coating conditions.

1. Introduction

In 2019, the European Union tightened the regulation for average CO_2 emissions for new car fleets by 15 % in 2025 and by 37.5 % in 2030 based on the average fleet target of 95 g CO_2/km for 2021 [1]. New driving concepts such as electric drives are currently developed to achieve the emission standards. These alternative drive concepts include battery electric vehicles (BEV), fuel-cell technology and hybrid technology. One major aspect for hybrid technology applications - and also for still produced conventional driving systems - is the continuous improvement to enhance the efficiency of internal combustion engines and consequently reduce emissions to achieve the aims. A method to improve internal combustion engines is the substitution of heavy gray-cast iron liners by thin, smooth functional cylinder-bore coatings with a coating thickness less than 100 μm. The specific coatings reduce frictional losses significantly compared to gray-cast iron liners with their typical honing structures at the surface [2–4]. Cylinder-bore coatings can be produced by several procedures. These include e.g. plasma-transferred wire arc (PTWA) [5], rotating single wire (RSW) [6], twin-wire arc-spray (TWAS) [2, 3], cold spray (CS) [7], high-velocity oxygen fuel (HVOF) [8] and atmospheric plasma spray (APS) [9, 10].

Additional techniques to reduce emissions for example nitrogen oxides (NO_x) emissions include an increase of exhaust-gas recirculation (EGR) rates [11, 12]. Other options to decrease emissions are water injection [13, 14] and diesel exhaust fluid (DEF) injection [15]. All efforts to lower emissions result in new conditions within the combustion chamber of the engine. These conditions can lead to a higher corrosion rate at the cylinder-bore coating. Figure 1a shows possible causes of corrosion on thermally sprayed cylinder-bore coatings.

Besides the emission-reducing methods, the wide variety of fuel qualities also has an influence on corrosion phenomena within the cylinder-bore. High sulfur contents in fuels can lead to the formation of sulfuric acid (H_2SO_4) inside the cylinders of combustion engines, resulting in a high corrosion potential [16]. An overview of the maximum sulfur content in diesel fuels worldwide is displayed in figure 1b. In a lot of countries especially in South America, Africa, the Middle East and South East Asia, high sulfur contents are still permitted [17]. In addition to the poor diesel fuel quality in numerous regions, high sulfur gasolines are also present in many countries around the world [18]. Moreover, biogenic fuels are increasingly added to fossil fuels to reduce emissions. Due to their ethanol content, biofuels exhibit hygroscopic behavior that can lead to corrosive reactions [19].

Furthermore, the use of hybrid technology can lead to situations where the internal combustion engine is rarely used. These can occur for particular driving profiles that exclusively use the electric powertrain for short distance drives for example in urban traffic. This circumstance cause the pistons to remain in a certain position for a longer period than in a conventional combustion engine with the same driving profile, which can lead to corrosion between the piston rings and the cylinder-bore coating. To meet all the increasing technical requirements for modern internal combustion engines and to further advance the efficiency and performance, Mercedes-Benz AG advances the NANOSLIDE® technology.

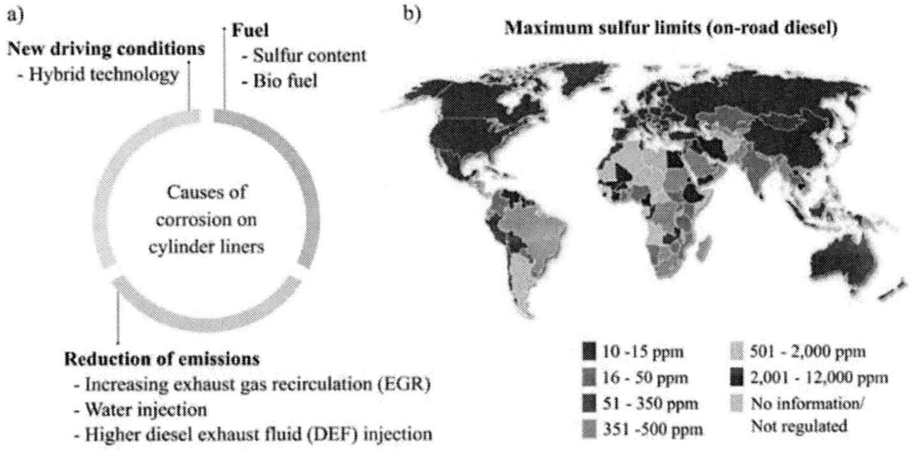

Figure 1. a) Sources for corrosion on cylinder liner coatings and b) worldwide maximum sulfur limits for diesel fuel (2020) reproduced with permission from Stratas Advisors [17].

1.1. NANOSLIDE® technology - coating of cylinder liners

The NANOSLIDE® technology contains all process operations to produce thin functional cylinder-bore coatings in modern aluminum crankcases for diesel and gasoline engines. The start of the process chain is a mechanical roughening process of the cylinder liner surface. The process, also called Nissan mechanical roughening process (NMRP), produces a characteristic cylinder surface with micro- and macrostructures to create ideal conditions for adhesion between the coating and the substrate by mechanical interlocking and metallurgical bonding [3]. Subsequently, the roughened surfaces are coated with a twin-wire arc-spray process (Germ. LDS). The as-sprayed coatings are machined by a fine boring process followed by several honing steps to create the final mirror-like surface with finely distributed pores, serving as oil-retention volumes.

This study presents the latest research and development results to enhance the corrosion resistance of thermally sprayed cylinder-bore coatings for internal combustion engines. The paper proposes an approach to improve the corrosion behavior for low and high alloy spraying materials by altering the microstructure and the functional surface of the coating based on computational fluid dynamic (CFD) simulations.

2. Experimental and materials

The spray materials used in this study were 1.0479 (13Mn6), 1.4021 (X20Cr13) [20] and 1.4122 (X39CrMo17-1) [21]. 13Mn6 represents the state of the art that is utilized for coating of cylinder-bores. Table 1 displays the three coating materials with their individual chemical composition.

Table 1. Chemical composition of the spraying materials 1.0479 (13Mn6) , 1.4021 (X20Cr13) and 1.4122 (X39CrMo17-1) [22, 23].

	C	Mn	Si	P	Cr	Ni	Cu	S	Mo
13Mn6 (% by mass)	0.08 - 0.14	1.35 - 1.65	0.30 - 0.45	≤ 0.025	≤ 0.12	≤ 0.12	≤ 0.17	≤ 0.25	-
X20Cr13 (% by mass)	0.16 - 0.25	≤ 1.50	≤ 1.00	≤ 0.04	12.0 - 14.0	-	-	≤ 0.03	-
X39CrMo17-1 (% by mass)	0.33 - 0.45	≤ 1.50	≤ 1.00	≤ 0.04	15.5 - 17.5	≤ 1.0	-	≤ 0.03	0.08 - 1.30

2.1. Coating experiments

All coating experiments in this study were conducted on aluminum crankcases (AlSi8Cu3) for diesel engines with a cylinder diameter of 82 mm. The twin-wire arc-spray coating system was a CBC200 module from Gebr. Heller Maschinenfabrik GmbH, Germany. The primary and secondary atomizing gas for the coating experiments was nitrogen with a ratio of 3:1. Special masks were placed at the cylinder head sealing surface and at the lower end of the cylinder liner near the crankshaft space to protect these surfaces of the crankcase from overspray and coating material. A vane anemometer (Model Testo 445, Testo AG, Germany) was used to measure the extraction velocity at the cylinder head sealing surface for one cylinder bore while the remaining cylinders were covered to reproduce the conditions equal to the coating process. The extraction velocity was adjusted to 2.0 m/s ± 0.5 m/s. These configurations result in oxygen-rich coating conditions in the top and bottom region and oxygen-poor conditions in the middle of the cylinder bore during the coating process [3]. Figure 2a shows a part of an image of a honed cylinder bore surface with different porous sections.

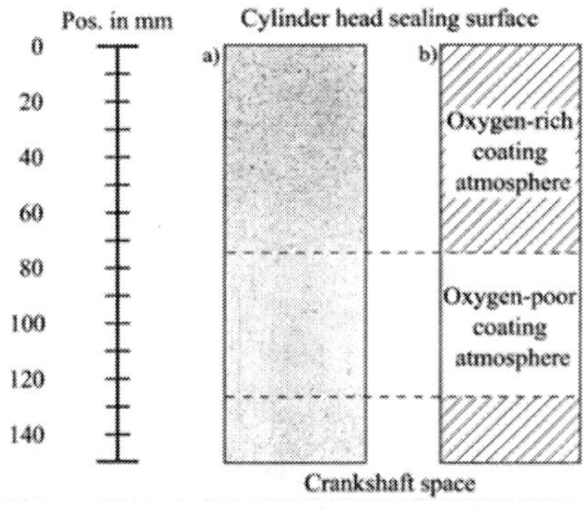

Figure 2. a) Partial image of a coated cylinder-bore surface with porous (oxygen rich) and less porous (oxygen poor) sections, b) schematic overview of the coating atmosphere for an extraction velocity of 2 m/s.

The expansion of the porous and less porous surface is rotationally symmetric. Samples cut out from the top of the cylinder bore (oxygen rich) exhibit a layer structure with many inhomogeneities like pores, lamellar structures and splat boundaries at the surface and throughout the layer structure. Instead, specimens taken from the middle of the cylinder bore (oxygen poor) show a more homogeneous layer structure with a considerably reduced level of lamellar structures, pores and boundaries between the individual splats. The experimental setup and the coating system are equivalent to the current large-scale production process at Mercedes-Benz AG.

2.2. Characterization of the corrosion behavior

For a quantitative analysis of the corrosion resistance of different coating materials, samples were examined in an alternating immersion test. The specimens were exposed to a synthetic exhaust-gas condensate (K1.2). The condensate was used according to VDA 230-214 for applications with high sulfuric fuels. It consists of nitric acid (HNO_3), sulfuric acid (H_2SO_4), sodium chloride (NaCl) and demineralized water [24]. The alternating immersion test is based on an internal Mercedes-Benz test guideline. Cylinder liners were cut out of the crankcases and prepared to an outer diameter of 95 mm. The samples were taken from the top of the cylinder bore (oxygen-rich coating atmosphere) and from the middle of the cylinder-bore (position 80 mm – oxygen-poor coating atmosphere). The height of each sample was 25 mm. All samples were cleaned in a bath with petroleum gasoline for 24 h to create comparable test conditions. The mass of each sample was measured with an analytical balance (Sartorius RC250S, Sartorius AG, Germany) before and after the test to determine the mass loss. The aluminum substrate was covered with a corrosion-resistant polyester masking tape BT-220 (Christian Zang GmbH, Germany) to ensure that the corrosive agent exclusively attacks at the cylinder-bore coating. All samples were alternately immersed in the synthetic condensate for a period of 60 s and placed above the immersion bath for 180 s to drip off. This procedure was repeated for a total time of 8 h. The temperature of the bath was adjusted to 60 °C. After the alternating cycles, a 16 h ageing phase took place above the heated exhaust-gas condensate. The entire cycle was repeated over a duration of five days. Then, the samples were removed from the test facility after the immersion loop and dried for another 48 h at ambient temperature. After the test, the weakly adherent corrosion products were removed by mechanical cleaning under water. The still remaining corrosion products were removed by chemical cleaning methods [25]. The stripping solution for the coating material 13Mn6 contained hexamethylenetetramine ($C_6H_{12}N_4$), hydrochloric acid (HCl) and distilled water and for X20Cr13 and X39CrMo17-1 nitric acid and distilled water [26]. The cylinder segments were weighted again to determine the mass loss of each sample. The alternating immersion test was performed for each coating material and microstructural condition in a double determination with three differential measurements of each sample mass.

The qualitative analysis of corrosion phenomena on cylinder-bore coatings was accomplished with the Kesternich test method. Samples were exposed to a saturated atmosphere in the presence of sulfur dioxide (SO_2). One cycle of the test was divided into two sections. In the first section, the specimens were heated to a temperature of 40 °C and exposed to the sulfur dioxide atmosphere for a period of 8 h. Then, the test chamber was ventilated for 16 h during the second phase. The temperature was set to ambient temperature. The samples were cut out of the cylinder bore at a position of 6 mm (oxygen rich) and 91 mm (oxygen poor) below the cylinder head sealing surface in order to get liner segments with different layer structures. The specimens′ dimension was 11.95 mm x 14.30 mm. All surfaces of the tested samples were cleaned with ethanol prior the corrosion test and subjected to a two cyclic test.

2.3. Characterization of the functional properties of the coating

Contact-angle measurements were done with the OCA 200 system (DataPhysics Instruments GmbH, Germany) to analyze the surface wetting behavior of the various cylinder-bore coatings. The static contact-angle measurements were carried out with the test liquids ethylene glycol, diiodomethane and thiodiglycol with a drop volume of 2 µl. In addition, dynamic contact-angle analyses were done to determine the roll-off behavior of liquid droplets from the surface of the coatings. Therefore, the samples were tilted in a range between 0° and 87.5° with a tilting base unit TBU 100 (DataPhysics Instruments GmbH, Germany) in combination with the OCA 200 system. The water drop volume was set to 22 µl.

Specimens were taken from the top of the cylinder bore (oxygen rich) and from position 80 mm (oxygen poor). The size was 75 mm x 50 mm.

The functional description of the surface structure was determined by the characteristic $V_{Oil\text{-}pores}$-value. This parameter represents a measure for the oil-retention volume of the pores at the surface of the cylinder-bore coating and is normalized to an area of 1 mm². The tactile measurements of the surface characteristic were done with the roughness measuring system Hommel-Etamic T8000 (Jenoptik Industrial Metrology Germany GmbH, Germany). A structural separation of the measurement profile is used in order to distinguish the machining profile of the surface and the pore profile [27]. Each measurement consists of three measuring paths with a respective length of 15 mm.

Supplementary analyses of the cylinder liner were done with the Visionline B100 system (Jenoptik Industrial Metrology Germany GmbH, Germany). The 360° lens system creates high-resolution images from the surface of the entire length of the cylinder-bore coatings. The images were used to map and quantify the expansion of the porous surface structures (e.g. figure 2a).

The microstructural analyses of the coatings were performed on cross section specimens cut out of the cylinder bores, embedded in resin (Allylharz Glasfiber, Presi GmbH, Germany) and polished (Swing-Plus, Microdiamant AG, Switzerland). An Axio Observer microscope (Carl Zeiss AG, Germany) was used to take light microscope (LM) images from the microstructures. Furthermore, investigations were carried out with a field emission scanning electron microscope (SEM) (Mira3 XMU, Tescan Orsay Holding, Czech Republic) (15 kV, *WD* = 15 mm) in combination with a Quantax XFlash®5030 detector (Bruker Corporation, USA) for energy dispersive X-ray spectroscopy (EDS) to analyze the microstructure and the element distribution.

3. Results and discussion

The investigations in this study focus on the characterization and improving of the corrosion resistance properties for cylinder-bore coatings. Quantitative and qualitative analysis of the microstructure and surface of the coatings were carried out in order to gain more insights in the various layer systems.

3.1. *Quantitative analysis of the corrosion behavior*

Since the cylindrical samples for the alternating immersion tests consist of the aluminum substrate and the coating, the measured loss of mass was related to a theoretical coating thickness of 100 µm. The theoretical coating mass m_t is estimated from the volume V and density ρ_c (without pores) of the coating material as shown in equation (1). The coating mass loss m_{loss} is determined by the difference between the mass of the sample before (m_1) and after (m_2) the alternating immersion test in relation to the theoretical coating mass (equation (2)).

$$m_t = V \cdot \rho_c = \pi \cdot (R^2 - r^2) \cdot h \cdot \rho_c \tag{1}$$

$$m_{loss} = (m_1 - m_2) \cdot (m_t)^{-1} \tag{2}$$

R corresponds to the cylinder radius without coating (41.1 mm), r represents the final measure of the cylinder bore after the honing process (41 mm) and h stands for the sample height of 25 mm. Figure 3 displays the percentage mass loss of the three analyzed materials and the two coating structures (oxygen rich and oxygen poor) based on a theoretically layer thickness of 100 µm (minimum distance between coating surface and the top of the roughening structure). Two alternating immersion tests were performed for each material and layer structure. The largest mass losses (> 40 %) occurred with the material 13Mn6 of the coatings separated from the top of the cylinder which represents the oxygen-rich atmosphere during the coating process. The minimal mass loss was found for the material X39CrMo17-1 coated in oxygen-poor conditions. The mass loss decreased from 13Mn6 to X20Cr13 to X39CrMo17-1. X20Cr13 and X39CrMo17-1 are capable to create a protective passive layer against corrosion in contrast to the low-alloy material 13Mn6 [21]. The analysis in figure 4 presents the mass loss of the coatings normalized to the state of the art material and microstructure. By altering the wire material, improvements of the corrosion resistance up to 66 % for X20Cr13 and 97 % for X39CrMo17-1 were achieved without changing the coating atmosphere. An additional change in the coating atmosphere lead to a reduction in mass loss by 81 % for the coating material X20Cr13 and 98 % for X39CrMo17-1.

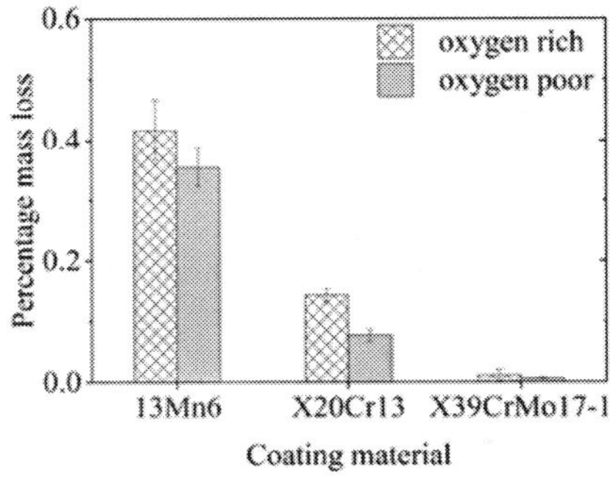

Figure 3. Absolute percentage mass loss of coatings related to a theoretical coating thickness of 100 μm.

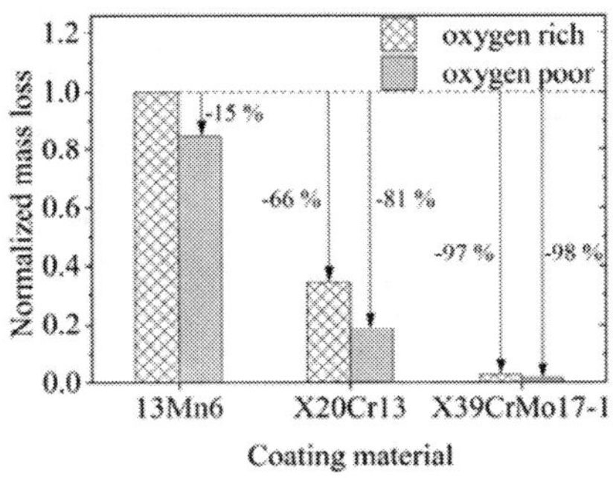

Figure 4. Normalized mass loss based on the sample with the low-alloy coating material (13Mn6) and sprayed in an oxygen-rich atmosphere.

The results show that the specimen from the middle of the cylinder bore coated in an oxygen-poor atmosphere generally exhibit lower mass losses in the alternating immersion test regardless of the material. Figure 5 indicates the percentage of improvement for the individual materials determined by equation (3). By changing the coating atmosphere, the microstructure of the coatings was influenced in such a way that the corrosion resistance properties were increased by 15 % for the material 13Mn6, 45 % for X20Cr13 and 46 % for X39CrMo17-1.

$$m_{\text{rel}} = (\Delta m_{O_2-\text{poor}} - \Delta m_{O_2-\text{rich}}) \cdot (\Delta m_{O_2-\text{rich}})^{-1} \tag{3}$$

The negative gradient of the two linear regressions lines in figure 6 demonstrate that the mass loss decreases depending on the chromium content of the coating. The higher the chromium content in the alloy materials, the lower the mass loss regardless of the coating conditions. These results reveal that the stability of the protective passive layer depends on the chromium content in the alloy. Happek (2015) determined similar relations for twin-wire arc sprayed metallic coatings, Zhang et al. (2016) indicated an increased stability of the passive layer for porous coatings with a higher chromium content [21, 28].

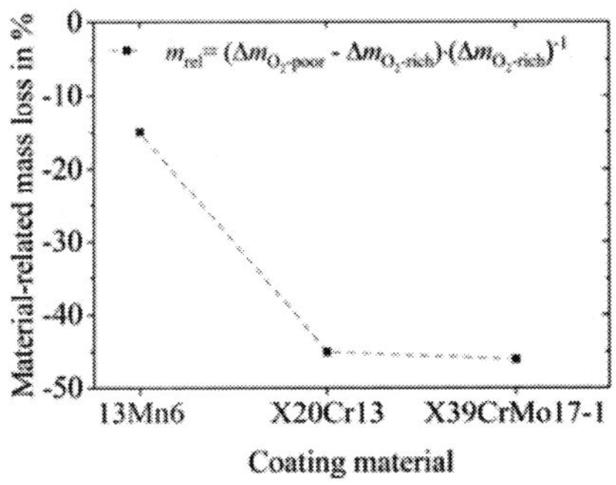

Figure 5. Decrease of mass loss due to spraying in an oxygen-poor atmosphere relative to spraying in an oxygen-rich atmosphere in relation to the spray material.

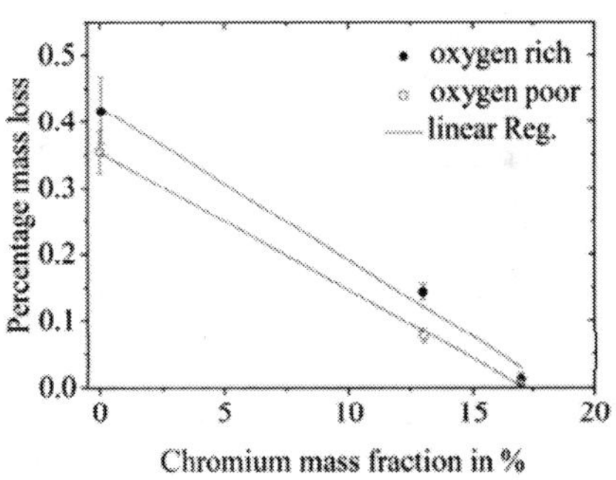

Figure 6. Mass loss of the coating related to a theoretical layer thickness of 100 μm in relation to the chromium content of the spray material.

3.2. Analysis of the functional surface

Optical contact-angle measurements were carried out to analyze the wetting properties of cylinder-bore coating surfaces in order to infer the corrosion behavior. Therefore static contact-angles (CA) were investigated to determine the surface energy of the samples using the Owens, Wendt, Rabel and Kaelble (OWKR) model. Figure 7 shows that the surface energy for the different coating materials is almost the same for surfaces with $V_{Oil\text{-}pores}$-values less than 100 µm³/mm² and tends to decrease for higher values. The analyses of the individual coating materials reveal that the samples produced in an oxygen-poor atmosphere exhibit a slightly higher surface energy. Pores at the surface act as energy barriers, which influence the spreading behavior of the fluid drops [29]. The high surface porosity inhibit the propagation and thus the flattening of the drops more than smooth surfaces with less inhomogeneity. This effect leads to larger contact-angles that were interpreted to smaller surface energies in the OWKR-model. Nota bene: all analyzed samples show a hydrophilic characteristic (CA < 90°) independent of the material and surface properties of the coatings [30, 31].

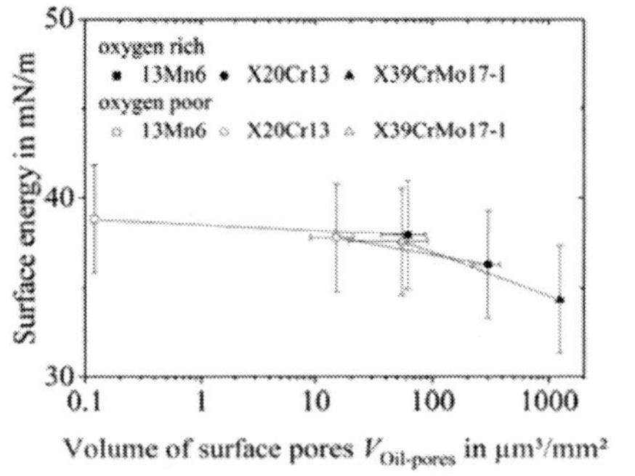

Figure 7. Surface energy of the various materials and surface textures in relation to the specific volume of the surface pores based on an area of one square millimeter.

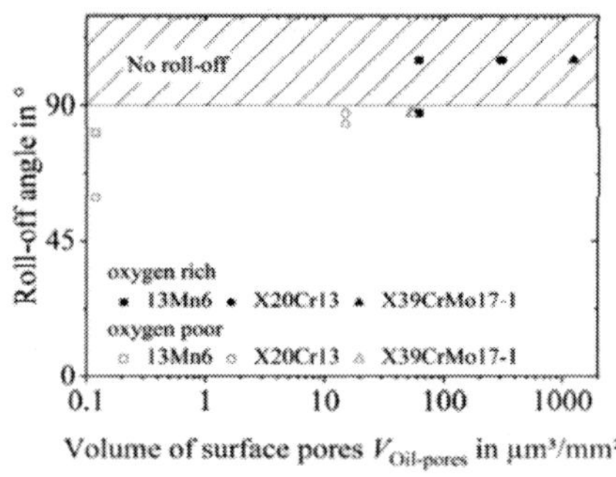

Figure 8. Roll-off angle of water droplets for the three coating materials and surface textures in relation to the specific volume of the surface pores $V_{Oil\text{-}pores}$.

In addition to the static contact-angle measurements, dynamic measurements were conducted to characterize the roll-off behavior of droplets from the various cylinder-bore coatings. The roll-off angle α was determined by the first occurrence of a contraction from the base diameter of the drop. The base diameter represents the lateral expansion of the drop between the advancing contact-angle (θ_{adv}) and receding contact-angle (θ_{rec}). The results of the dynamic contact-angle measurements indicate that the rolling behavior of the drops depends on the structure of the coating surface (figure 8). The structure influences the adhesion properties and thus the hydrophobicity of the surface [32]. The water drop adhered at the surface for all samples coated in an oxygen-rich atmosphere, except one specimen of 13Mn6. However, all coatings sprayed in an oxygen-poor atmosphere showed a sliding of the drop after the tilting process. The fewer the amount of pores and inhomogeneity at the surface, the faster the roll-off behavior. The roll-off behavior thus influences the contact time between the corrosive agent and the coating. A longer contact time increases the period to cause corrosion reactions [31].

The analysis of the dynamic contact-angle measurement demonstrates that the roll-off behavior can be strongly influenced by the atmosphere in which the coating process occurs. An oxygen-poor spraying process positively effects the corrosion resistance properties by reducing the contact time between the less porous surface and a corrosive medium.

3.3. Microstructure-corrosion-relationship

The Kesternich test was utilized to systematically corrode the various cylinder-bore coating materials under comparable conditions. Afterwards, the microstructure of cross sections and the surface of the coatings were examined. Figure 9 displays the surface of the samples after two cycles of Kesternich test. The coating material 13Mn6 shows in both samples from the top (oxygen-rich coating condition, figure 9a) and from the middle of the cylinder bore (oxygen-poor coating condition, figure 9b) loosely adherent corrosion products on the entire surface. The coating material X20Cr13 also reveals distinct corrosion reactions on the sample in figure 9c, which are reduced on the less porous surface (figure 9d). The fewest visible corrosion products were found on the surface of the material X39CrMo17-1 with the best result on the less porous surface (figure 9f). The corrosion attack depends on the coating material and the surface structure. An increased surface porosity ($V_{Oil\text{-}pores}$) results in a greater active surface area that is exposed to the corrosion agent. Thus, the growth of the solid-liquid interface will lead to enhanced corrosion reactions on the coating surfaces. This phenomena was also identified for increased surface roughness of coatings [30].

These results coincide with the alternating immersion test and the contact-angle measurements. The corrosion resistance increases due to a higher chromium content in the coating material and denser surfaces with fewer pores.

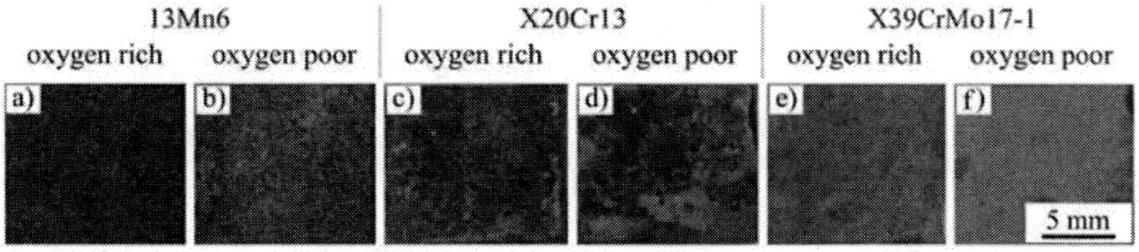

Figure 9. LM images of cylinder-bore coating surfaces after a two-cycle Kesternich test procedure.

After the Kesternich test method, cross sections of the coatings were made in order to further investigate the microstructure. The cross sections were analyzed by light microscope and scanning electron microscope. Figure 10a presents a LM image of a cross section from the coating material 13Mn6. The specimen cut out of the cylinder bore at position 6 mm below the cylinder head sealing surface (oxygen rich). The cross section shows a morphology with porous and lamellar structures that is characteristic for twin-wire arc-sprayed metallic coatings [3, 33–35]. This coating structure is also evident for the materials X20Cr13 (figure 10c, d) and X39CrMo17-1 (figure 10e). On the other hand, the samples that were created from the middle of the cylinder bore (position 91 mm – oxygen-poor coating conditions) show a more homogeneous layer structure with fewer lamellar structures and oxidized splat boundaries. Figure 10f displays an example of this microstructure for the coating material X39CrMo17-1. A detailed description for the effects that lead to the creation of the various microstructures can be found in [3]. Prior to the Kesternich test, a portion of the samples were covered with an anti-corrosion pad to provide a non-corroded reference section. These reference areas displays the typical smooth honed surface. The part of the 13Mn6 sample that was not covered with the anti-corrosion pad and exposed to the corrosive environment exhibits surface corrosion across the entire specimen as shown in figure 9a. The depth of the shallow corrosion attack is on average between 5 µm and 10 µm, also for the specimen coated in oxygen-poor spraying conditions. The SEM image in figure 10b presents a detailed view of the corroded surface of the material 13Mn6 with the resulting corrosion products on top of the coating.

An EDS study of the microstructure shows the composition of corroded and non-corroded regions. The EDS spectrum of a non-corroded splat inside the coating (figure 10b-1) reveals the chemical composition of the coating material 13Mn6 (table 1). In contrast, oxygen and sulfur could be detected in the darker area at the top of the coating (figure 10b-2). Sulfur originates from the corrosion agent sulfur dioxide (SO_2) used for the Kesternich test and oxygen from the oxidation reaction between the coating and the electrolyte. In contrary to 13Mn6, no surface corrosion with a quantifiable level of depth could be identified for the coating material X20Cr13 (figure 10c, d) and X39CrMo17-1 (figure 10e, f).

All high-alloyed coatings show the smooth honed surface. However, if defects occur on the surface for instance in terms of pores or lamellae structures, the corrosion medium can penetrate inside the layer and cause corrosion independent of the coating material composition and the general constitution of the microstructure (figure 10c-f) [28, 36].

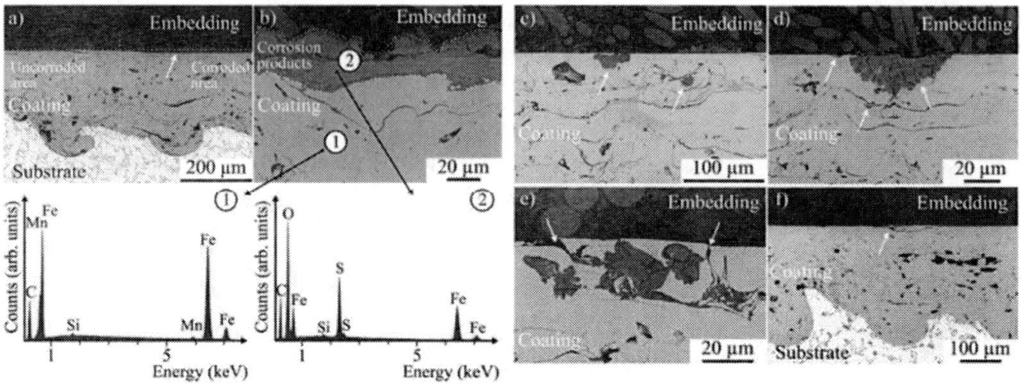

Figure 10. Microstructural analysis of cross sections from twin-wire arc-sprayed coatings. a) LM image, 13Mn6, b) SEM, 13Mn6, corrosive attack at the surface coating, EDS spectra of a homogeneous splat (1) and of corrosion products (2), c) and d) SEM, X20Cr13, e) SEM, X39CrMo17-1, f) LM, X39CrMo17-1. Cross sections a) - e) show microstructures coated in oxygen-rich atmospheres, f) microstructure coated in oxygen-poor conditions.

Inside the layer, corrosion occurs preferentially at lamellar structures, splat boundaries and pores (e.g. figure 10c) [37]. These regions often exhibit a serrated and frayed contour. Defect structures, pores and oxidized splat boundaries that form the elongated lamellar structures are susceptible to chromium depletion and therefore vulnerable to corrosion [28, 38]. Furthermore, oxide layers at splat boundaries, pores and structural defects can lead to localized corrosion elements [36]. Coatings sprayed in an oxygen-rich atmosphere exhibit a higher oxide content within the microstructure and thus an enhanced susceptibility to corrosion. Figure 10c-e represent examples for corroded lamellar structures and pores even within the coating. These irregularities build potential paths for the corrosive agent. For specimen with an inhomogeneous microstructure, corrosion could often be detected even in deeper areas near the substrate, whereas homogeneous layers showed only corrosion attacks close to the surface. In general, a homogeneous layer structure reduces the probability that a connected network of lamellar structures and pores will be formed where corrosive media can penetrate inside the layer and in the worst case reach the interface between the substrate and the coating [28, 36].

Changing the mask geometry can significantly increase the oxygen-poor region during the coating process [3]. The development and implementation of an advanced coating process for the entire cylinder length and a large-scale application are currently in realization.

4. Conclusions

This paper presents a novel method to enhance the corrosion resistance of cylinder-bore coatings for internal combustion engines by changing the surface characteristics and microstructures of the coating and utilizing alternative materials. The results from the alternating immersion test demonstrate a significant improvement of the corrosion resistance up to 50 % through altering the microstructure of the coating and almost 100 % by additional changing the spraying material. The corrosion resistance property of the low-alloy material 13Mn6, that is currently applied for cylinder-bore coatings, could also been enhanced by 15 % through changing the microstructure to a more homogeneous layer formation. Contact-angle measurements indicate that the wetting behavior strongly depends on the oil-retention volume of the coating. Less porous surface produced in an oxygen-poor coating atmosphere reveal shorter contact times and a decreased surface area between the coating and a corrosive agent and therefore have a positive influence on the corrosion resistance of the coatings. The results of the Kesternich test in combination with the analysis of the coating microstructures show that inhomogeneous areas (e.g. pores, lamellar structures, splat boundaries) at the surface and inside the layer are susceptible to corrosion in both low and high-alloy coating materials. The more homogeneous the microstructure at the surface and inside the coating, the lower is the probability of corrosive media penetrating into the layer system and causing corrosion.

The corrosion resistance of cylinder-bore coatings can be enhanced in general by specifically influencing the gas-flow conditions during the coating process to ensure a low-oxygen coating atmosphere. In addition to the presented model-based test methods further validation should be done by engine-test runs and vehicle tests to gain more knowledge about the corrosion behavior in true engine operation conditions.

Acknowledgments

This study was carried out as part of a research and development project at Mercedes-Benz AG. The authors acknowledge the Institute of Materials Science and Engineering, Chemnitz University of Technology and the Mercedes-Benz AG, Stuttgart, for supporting the research. We would also like to thank all colleagues from process development and materials technology department at Mercedes-Benz AG for the intensive and interdisciplinary cooperation.

References

[1] European Parliament and the Council of the European Union 2019 *Regulation (EU) 2019/ 631 of the European Parliament and of the Council - of 17 April 2019 - setting CO2 emission performance standards for new passenger cars and for new light commercial vehicles, and repealing Regulations (EC) No 443 / 2009 and (EU) No 510 / 2011*

[2] Schmidt M, Spieth H, Haubach C and Kühne C 2019 *100 Pioneers in Efficient Resource Management: Best practice cases from producing companies* (Berlin, Heidelberg: Springer Berlin Heidelberg)

[3] Schilder B, Garling A, Reimer F, Hamann M, Joos R, Hüger J, Pöhlmann M and Lampke T 2020 CFD Enhanced Thermal Spray Process for Coating of Cylinder Bores of Car Engines *J. Therm. Spray Technol.* **29** 546–59

[4] Schommers J, Scheib H, Hartweg M and Bosler A 2013 Minimising Friction in Combustion Engines *MTZ Worldwide* 28–35

[5] Bobzin K, Öte M and Königstein T 2018 Effect of Different Atomization Gases on the Properties of Cylinder Bore Coatings *Adv. Eng. Mater.* **21** 1800853

[6] Hanke W 2019 Friction Reduction in Power Cylinder Systems of Commercial Vehicle Engines *MTZ Worldwide* **80** 18–23

[7] Aubanel L, Lefeivre L, Delloro F, Jeandin M and Sura E 2019 Cold Spray Coatings for Automotive Cylinder Block Application *ITSC 2019: International Thermal Spray Conference and Exposition : 2019 New Waves of Thermal Spray Technology for Sustainable Growth ITSC 2019 (Yokohama, 26.05.2019-29.05.2019)* ed F Azarmi (Materials Park, OH: ASM International) 433-440

[8] Hahn M and Fischer A 2010 Characterization of Thermal Spray Coatings for Cylinder Running Surfaces of Diesel Engines *J. Therm. Spray Technol.* **19** 866–72

[9] Barbezat G 2005 Advanced thermal spray technology and coating for lightweight engine blocks for the automotive industry *Surf. Coat. Technol.* **200** 1990–3

[10] Grund T, Paczkowski G, Lampke T, Liborius H, Nestler A and Schubert A 2020 Finish Turning of FeCr17Ni2C0.2 Iron-based Sprayed Coatings: Influences of Substrate Preparation, Cutting Speed and Feed on the Coating and Surface Properties *J. Therm. Spray Technol.* **29** 308–18

[11] Praveena V and Martin M L J 2018 A review on various after treatment techniques to reduce NOx emissions in a CI engine *J. Energy Inst.* **91** 704–20
[12] Nun J, Wieler A, Andres C and Betz T 2019 The First Mercedes-Benz Four-cylinder Diesel Engine Complying with RDE Step 2 *MTZ Worldwide* **80** 46–53
[13] Mingrui W, Thanh Sa N, Turkson R F, Jinping L and Guanlun G 2017 Water injection for higher engine performance and lower emissions *J. Energy Inst.* **90** 285–99
[14] Fan Y, Wu T, Di Xiao, Xu H, Li X and Xu M 2021 Effect of port water injection on the characteristics of combustion and emissions in a spark ignition direct injection engine *Fuel* **283** 119271
[15] Nova I and Tronconi E 2014 *Urea-SCR Technology for deNOx After Treatment of Diesel Exhausts* (New York, NY: Springer New York)
[16] Distelrath-Lübeck A, Scherer V and Aumiller M 2017 Corrosion of Cylinder Bore Materials *MTZ Worldwide* **78** 52–7
[17] Stratas Advisors 2020 *Five Countries Move Up in Top 100 Ranking on Diesel Sulfur Limits* (Stratas Advisors)
[18] Stratas Advisors 2019 *Stratas Advisors Ranks Top 100 Countries with Lowest Sulfur in Gasoline* (Stratas Advisors)
[19] Käfer S, Melz T, Engler T and Oechsner M 2019 Impact of Biogenic Fuels on the Fatigue Behavior of Steels *MTZ Worldwide* **80** 100–5
[20] Özdeniz E A 2016 Entwicklung korrosions- und verschleißbeständiger thermisch gespritzter Zylinderlaufbahnen für Verbrennungsmotoren *Dissertation* Fakultät für Maschinenbau, Technische Universität Chemnitz
[21] Happek B 2015 Tribologische und elektrochemische Untersuchungen an lichtbogendrahtgespritzten Beschichtungen aus nichtrostendem Stahl für die Zylinderlaufbahnen von PKW-Dieselmotoren *Dissertation* Maschinenbau und Verfahrenstechnik, Universität Duisburg-Essen
[22] DIN Deutsches Institut für Normung e. V. 2014 Stainless steels: Part 3: Technical delivery conditions for semi-finished products, bars, rods, wire, sections and bright products of corrosion resisting steels for general purposes (DIN EN 10088-3) (Berlin: Beuth Verlag GmbH)
[23] DIN Deutsches Institut für Normung e. V. 1980 Round wire rod for welding filler metals: Technical terms of delivery (DIN 17145) (Berlin: Beuth Verlag GmbH)
[24] Verband der Automobilindustrie e. V. 2018 Resistance of metallic materials to condensate corrosion in exhaustgas-carrying components (VDA 230-214) (Berlin)
[25] DIN Deutsches Institut für Normung e. V. 2009 Corrosion of metals - Corrosion testing: Part1: General guidance (DIN 50905-1) (Berlin: Beuth Verlag GmbH)
[26] DIN Deutsches Institut für Normung e. V. 2020 Corrosion of metals and alloys: Removal of corrosion products from corrosion test specimens (DIN EN ISO 8407) (Berlin: Beuth Verlag GmbH)
[27] Boehm J, Hartweg M, Hercke T, Izquierdo P, Michel M, Schweickert S and Rau G 2012 Thermisch beschichtetes Bauteil mit einer reibungsoptimierten Laufbahnoberfläche und Verfahren zur Bauteil-Beschichtungsoberflächensimulation eines thermisch beschichteten Bauteils *DE102012002766 (A1)*
[28] Zhang S D, Wu J, Qi W B and Wang J Q 2016 Effect of porosity defects on the long-term corrosion behaviour of Fe-based amorphous alloy coated mild steel *Corros. Sci.* **110** 57–70
[29] Duursma G R, Sefiane K and David S 2010 Advancing and receding contact lines on patterned structured surfaces *Chem. Eng. Res. Des.* **88** 737–43
[30] Ramachandran R and Nosonovsky M 2015 Coupling of surface energy with electric potential makes superhydrophobic surfaces corrosion-resistant *Phys. Chem. Chem. Phys.* **17** 24988–97

[31] Tejero-Martin D, Rezvani Rad M, McDonald A and Hussain T 2019 Beyond Traditional Coatings: A Review on Thermal-Sprayed Functional and Smart Coatings *J. Therm. Spray Technol.* **28** 598–644
[32] Mohamed A M A, Abdullah A M and Younan N A 2015 Corrosion behavior of superhydrophobic surfaces: A review *Arabian J. Chem.* **8** 749–65
[33] Jandin G, Liao H, Feng Z Q and Coddet C 2003 Correlations between operating conditions, microstructure and mechanical properties of twin wire arc sprayed steel coatings *Materials Science and Engineering: A* **349** 298–305
[34] Newbery A P and Grant P S 2006 Oxidation during electric arc spray forming of steel *J. Mater. Process. Technol.* **178** 259–69
[35] Gedzevicius I and Valiulis A V 2006 Analysis of wire arc spraying process variables on coatings properties *J. Mater. Process. Technol.* **175** 206–11
[36] Wu J, Zhang S D, Sun W H and Wang J Q 2018 Influence of oxidation related structural defects on localized corrosion in HVAF-sprayed Fe-based metallic coatings *Surf. Coat. Technol.* **335** 205–18
[37] Sadeghi E, Markocsan N and Joshi S 2019 Advances in Corrosion-Resistant Thermal Spray Coatings for Renewable Energy Power Plants. Part I: Effect of Composition and Microstructure *J. Therm. Spray Technol.* **28** 1749–88
[38] Zeng Z, Sakoda N and Tajiri T 2006 Corrosion Behavior of Wire-Arc-Sprayed Stainless Steel Coating on Mild Steel *J. Therm. Spray Technol.* **15** 431–7

Corrigendum: Formation of corundum-rich alumina coatings on low-carbon steel by plasma electrolytic oxidation (2021 *IOP Conf. Ser.: Mater Sci Eng.* 1147 *012007*)

F Simchen, N Masoud-Nia, T Mehner, and T Lampke
Materials and Surface Engineering Group, Institute of Materials Science and Engineering, Chemnitz University of Technology, Gemany

The surface area of the test sample depicted in figure 2 was calculated wrong (without the contribution of the lateral area). That's why the current densities in Table 3, Figure 4 and Figure 5 were wrong also. We adjusted the mentioned article elements as well as the corresponding statement within the abstract:

Abstract: As a result of this, it was possible to obtain alumina layers of 80 micrometers in thickness, with a high corundum content of approximately 50 to 90%, after 37 minutes of treatment time, at a current density below 10 (instead of 25) A/dm² on C8C-steel.

Table 3. Parameters of the symmetric current-controlled alternating rectangular pulse for PEO.

I_1 / A	i_1/A·dm^{-2}	t_1/ms	U_1/V	I_2 / A	i_2/A·dm^{-2}	t_2/ms	U_2/V
1	10.85	5	500	-0.5	-5.42	5	-500

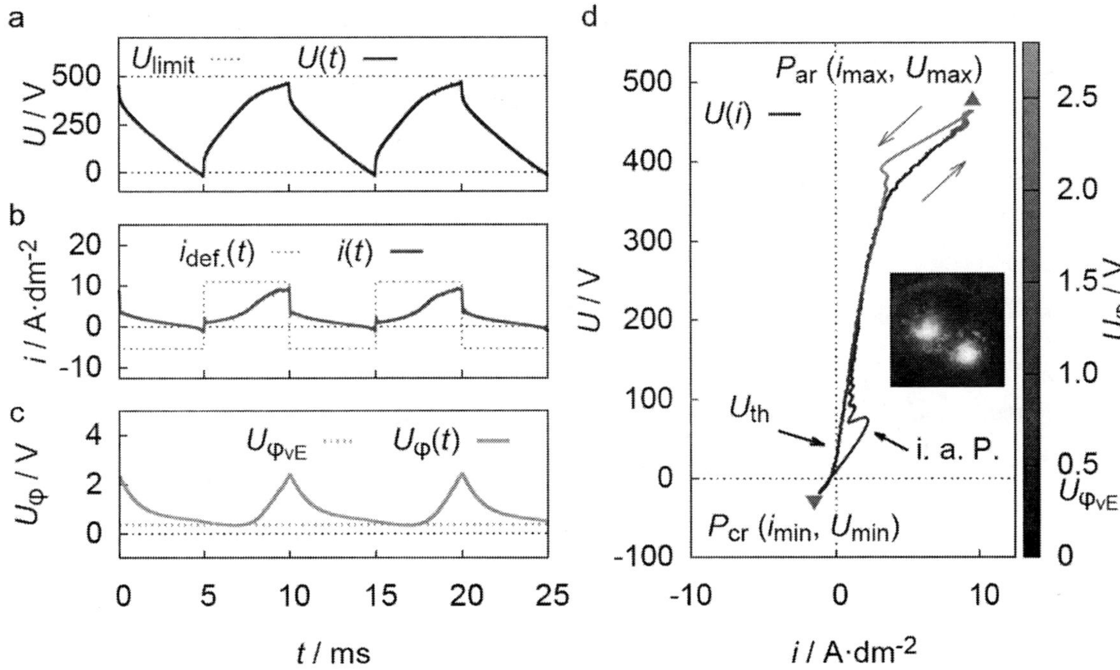

Figure 4. Graphical representation of the electrical pulse regime at 15 minutes of PEO treatment after initial voltage staircase: process-voltage (a), current-density (b), and photo-voltage (c) along with the associated limits, definitions, and pre-discharge values. These (a-c) are then summarised into a voltage current density-characteristic (d), with photo-voltage indicated by colour, combined with an image of the process.

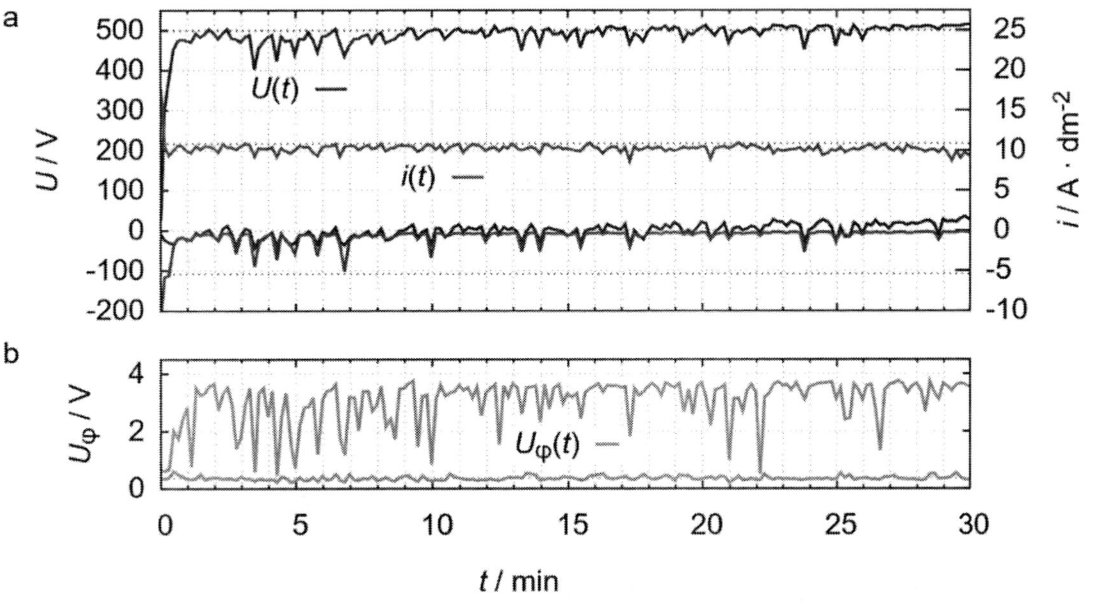

Figure 5. Course of the process voltage and current-density extremes (a) as well as the maximum and minimum values of the photo-voltage (b) as dependent on the process time.

IOP Conf. Series: Materials Science and Engineering 1147 (2021) 012036 doi:10.1088/1757-899X/1147/1/012036

Corrigendum: Influence of metal matrix powder size on the tensile strength of a SiC_p/AlSi7Mg0,6 composite produced by field assisted sintering technique (2021 *IOP Conf. Ser.: Mater Sci Eng.* 1147 *012020*)

R Pippig, S J Hirsch, T Grund and T Lampke
Chemnitz University of Technology, Chair of Materials and Surface Engineering

In the section Results and Discussion of our published paper the units for the heating energy and the toughness contain a mistake.

Page 4:
In the Results and Discussion section, the following unit appear:

"J"

This should read:

"kJ"

Figure 2(a):
In the right vertical axis, the following unit appear:

"J"

This should read:

"kJ"

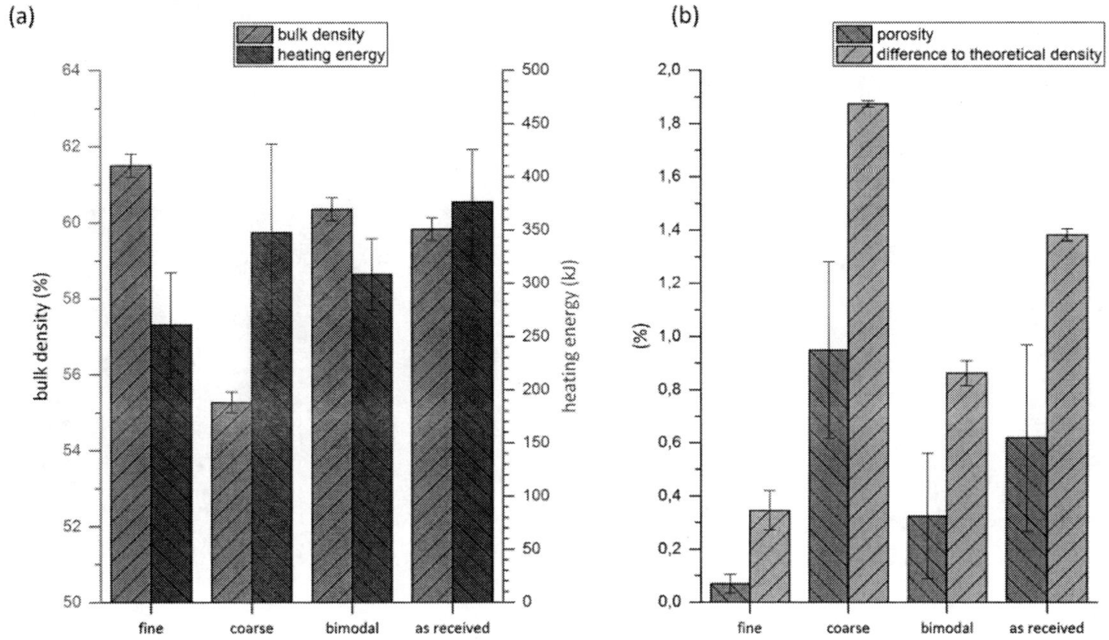

Page 6:
In the Results and Discussion section, the following unit appear:

"J cm^{-3}"

This should read:

"cJ cm^{-3}"

Figure 4(b):
In the right vertical axis, the following unit appear:

"J/cm^3"

This should read:

"10^{-2} J/cm^3"

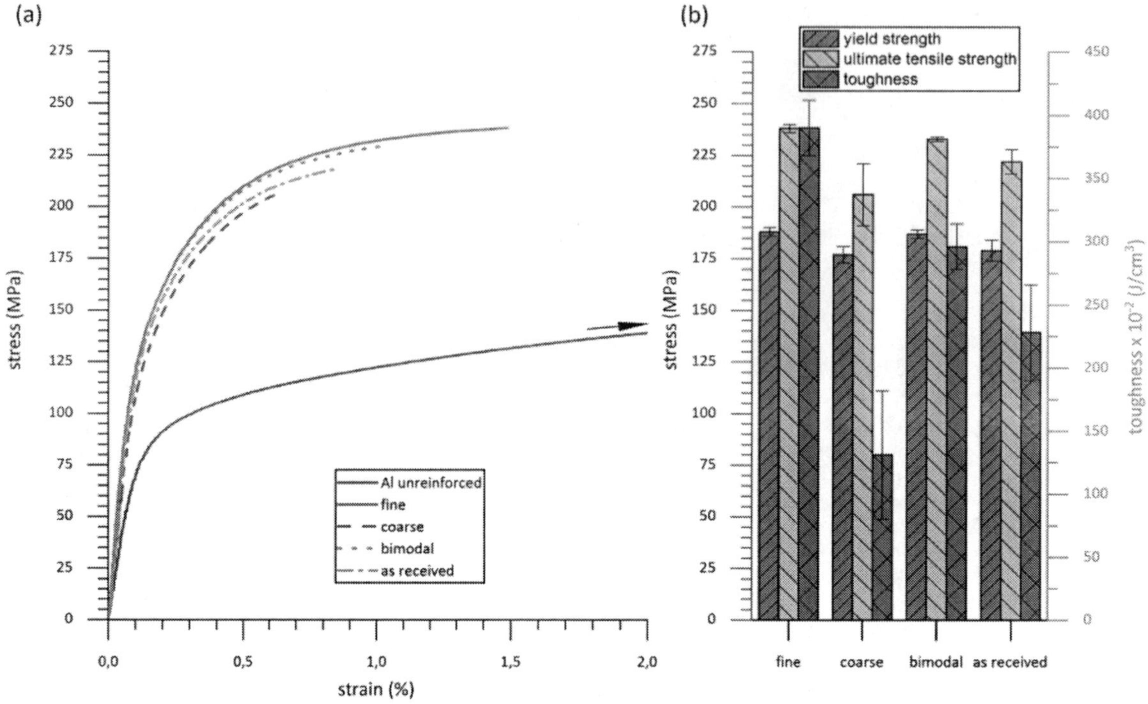
(a)
stress (MPa)
strain (%)
Al unreinforced
fine
coarse
bimodal
as received
(b)
yield strength
ultimate tensile strength
toughness
stress (MPa)
toughness x 10^-2 (J/cm^3)
fine
coarse
bimodal
as received

AUTHOR INDEX